GREEN AVIATION: REDUCTION OF ENVIRONMENTAL IMPACT THROUGH AIRCRAFT TECHNOLOGY AND ALTERNATIVE FUELS

# Sustainable Energy Developments

*Series Editor*

Jochen Bundschuh
*University of Southern Queensland (USQ), Toowoomba, Australia*
*Royal Institute of Technology (KTH), Stockholm, Sweden*

ISSN: 2164-0645

# Volume 14

# Green Aviation: Reduction of Environmental Impact Through Aircraft Technology and Alternative Fuels

*Editors*

Emily S. Nelson & Dhanireddy R. Reddy

*NASA Glenn Research Center, Cleveland, OH, USA*

CRC Press
Taylor & Francis Group
Boca Raton   London   New York   Leiden

CRC Press is an imprint of the
Taylor & Francis Group, an **informa** business

A BALKEMA BOOK

Published by:
CRC Press/Balkema
Schipholweg 107C, 2316 XC Leiden, The Netherlands

First issued in paperback 2020

© 2017 Taylor & Francis Group, London, UK
*CRC Press/Balkema is an imprint of the Taylor & Francis Group, an informa business*

No claim to original U.S. Government works

ISBN-13: 978-0-367-57304-1 (pbk)
ISBN-13: 978-0-415-62098-7 (hbk)

**Visit the Taylor & Francis Web site at**
**http://www.taylorandfrancis.com**

**and the CRC Press Web site at**
**http://www.crcpress.com**

Typeset by MPS Limited, Chennai, India

Although all care is taken to ensure integrity and the quality of this publication and the information herein, no responsibility is assumed by the publishers nor the author for any damage to the property or persons as a result of operation or use of this publication and/or the information contained herein.

*Library of Congress Cataloging-in-Publication Data*

# About the book series

Renewable energy sources and sustainable policies, including the promotion of energy efficiency and energy conservation, offer substantial long-term benefits to industrialized, developing, and transitional countries. They provide access to clean and domestically available energy and lead to a decreased dependence on fossil fuel imports and a reduction in greenhouse gas emissions.

Replacing fossil fuels with renewable resources affords a solution to the increased scarcity and price of fossil fuels. Additionally, it helps to reduce anthropogenic emission of greenhouse gases and their impacts on climate. In the energy sector, fossil fuels can be replaced by renewable energy sources. In the chemistry sector, petroleum chemistry can be replaced by sustainable or green chemistry. In agriculture, sustainable methods can be used to enable soils to act as carbon dioxide sinks. In the construction sector, sustainable building practice and green construction can be used, replacing, for example, steel-enforced concrete by textile-reinforced concrete. Research and development and capital investments in all these sectors will not only contribute to climate protection but will also stimulate economic growth and create millions of new jobs.

This book series will serve as a multidisciplinary resource. It links the use of renewable energy and renewable raw materials, such as sustainably grown plants, with the needs of human society. The series addresses the rapidly growing worldwide interest in sustainable solutions. These solutions foster development and economic growth while providing a secure supply of energy. They make society less dependent on petroleum by substituting alternative compounds for fossil-fuel-based goods. All these contribute to minimize our impacts on climate. The series covers all fields of renewable energy sources and materials. It addresses possible applications not only from a technical point of view, but also from economic, financial, social, and political viewpoints. Legislative and regulatory aspects, key issues for implementing sustainable measures, are of particular interest.

This book series aims to become a state-of-the-art resource for a broad group of readers including a diversity of stakeholders and professionals. Readers will include members of governmental and non-governmental organizations, international funding agencies, universities, public energy institutions, the renewable industry sector, the green chemistry sector, organic farmers and farming industry, public health and other relevant institutions, and the broader public. It is designed to increase awareness and understanding of renewable energy sources and the use of sustainable materials. It also aims to accelerate their development and deployment worldwide, bringing their use into the mainstream over the next few decades while systematically replacing fossil and nuclear fuels.

The objective of this book series is to focus on practical solutions in the implementation of sustainable energy and climate protection projects. Not moving forward with these efforts could have serious social and economic impacts. This book series will help to consolidate international findings on sustainable solutions. It includes books authored and edited by world-renowned scientists and engineers and by leading authorities in economics and politics. It will provide a valuable reference work to help surmount our existing global challenges.

Jochen Bundschuh
(Series Editor)

# Editorial board

This book is dedicated to our families, especially
Dr. L. Grace and Dr. F. Burton Nelson
Geeta Dhanireddy, and our children, Dr. Soni, Dr. Swati, and Srikant

# Table of contents

About the book series                                                          vii

Editorial board                                                                ix

List of contributors                                                           xxix

Foreword by Lourdes Maurice                                                    xxxi

Foreword by Thomas B. Irvine                                                   xxxiii

Editors' preface                                                               xxxv


Part I. Environmental impacts of aviation                                      1

1.  Noise emissions from commercial aircraft                                   3
    *Edmane Envia*
    1.1   Introduction                                                         3
    1.2   Sources of aircraft noise                                            10
          1.2.1   Engine noise sources                                         10
                  1.2.1.1   Fan noise                                          13
                  1.2.1.2   Propeller noise                                    13
                  1.2.1.3   Core noise                                         14
                  1.2.1.4   Jet noise                                          14
          1.2.2   Airframe noise sources                                       15
                  1.2.2.1   Trailing edge noise                                15
                  1.2.2.2   Landing gear noise                                 16
                  1.2.2.3   Slat noise                                         16
                  1.2.2.4   Flap (side-edge) noise                             17
                  1.2.2.5   Other noise sources                                18
    1.3   Aircraft component noise levels (example)                            19
    1.4   Summary                                                              22

2.  Aircraft emissions: gaseous and particulate                                25
    *Changlie Wey and Chi-Ming Lee*
    2.1   Introduction                                                         25
    2.2   Gaseous emissions                                                    25
          2.2.1   Carbon dioxide                                               27
          2.2.2   Oxides of nitrogen                                           28
          2.2.3   Carbon monoxide                                              29
          2.2.4   Unburned hydrocarbons                                        29
          2.2.5   Water vapor                                                  33
          2.2.6   Sulfur oxides                                                35
          2.2.7   Methane and nitrous oxide                                    37
    2.3   Particle emissions                                                   37
    2.4   Alternative fuels                                                    39
    2.5   Summary                                                              42

3.  Improvement of aeropropulsion fuel efficiency through engine design                49
    *Kenneth L. Suder and James D. Heidmann*
    3.1   Introduction                                                                  49
    3.2   Early history of NASA Glenn Research Center aeropropulsion
          fuel efficiency efforts, 1943 to 1958                                        54
    3.3   Introduction of turbofan engines and Improved propulsive efficiency          56
    3.4   Energy crisis of 1970s and NASA Aeronautics Response                         56
    3.5   NASA's role in component test cases and computational fluid
          dynamics development                                                         59
    3.6   Current NASA efforts at reduced fuel consumption                             63
          3.6.1   Open-rotor propulsors                                                65
          3.6.2   Ultra-high-bypass engine cycle research                              68
          3.6.3   Boundary-layer-ingesting engines                                     69
          3.6.4   Engine core research                                                 71
                  3.6.4.1   NASA ERA core compressor technology development efforts    73
                  3.6.4.2   NASA ERA core hot-section technology development efforts   74
    3.7   Summary                                                                      77

Part II. Technologies to mitigate environmental impacts                                81

4.  Noise mitigation strategies                                                        83
    *Dennis L. Huff*
    4.1   Introduction                                                                 83
    4.2   Noise reduction methods                                                      83
          4.2.1   Engine components                                                    84
                  4.2.1.1   Fan noise                                                  84
                  4.2.1.2   Propeller noise                                            88
                  4.2.1.3   Core noise                                                 88
                  4.2.1.4   Jet noise                                                  91
          4.2.2   Airframe components                                                  92
                  4.2.2.1   Trailing edge noise                                        93
                  4.2.2.2   Landing gear noise                                         94
                  4.2.2.3   Slat noise                                                 94
                  4.2.2.4   Flap (side-edge) noise                                     96
                  4.2.2.5   Other noise                                                97
    4.3   Future Noise-Reduction Technologies                                          97
          4.3.1   Engine noise reduction                                               98
                  4.3.1.1   Engine mid-term technologies                               98
                  4.3.1.2   Engine long-term technologies                              99
          4.3.2   Airframe noise reduction                                            100
                  4.3.2.1   Airframe mid-term technologies                            100
                  4.3.2.2   Airframe long-term technologies                           100
          4.3.3   Aircraft noise-reduction goals                                      101
                  4.3.3.1   Mid-term goals                                            101
                  4.3.3.2   Long-term goals                                           101
    4.4   Summary                                                                     102

5.  Advanced materials for green aviation                                             105
    *Ajay Misra*
    5.1   Introduction                                                                105
    5.2   Lightweight materials                                                       106
          5.2.1   Polymer matrix composites                                           106
          5.2.2   Carbon nanotubes for structural applications                        109

|  |  |  |  |
|---|---|---|---|
|  | 5.2.3 | Multifunctional materials | 112 |
|  | 5.2.4 | Cellular materials for sandwich structures | 114 |
| 5.3 | | Smart materials | 116 |
|  | 5.3.1 | Shape memory alloys | 116 |
|  | 5.3.2 | Piezoelectric materials | 119 |
| 5.4 | | High-temperature materials | 123 |
|  | 5.4.1 | High-temperature Ni-base superalloys | 123 |
|  | 5.4.2 | Ceramic matrix composites | 125 |
|  | 5.4.3 | Environmental degradation challenges with increase in turbine operating temperatures | 128 |
| 5.5 | | Materials for electric aircraft | 129 |
|  | 5.5.1 | Advanced magnetic materials | 130 |
|  | 5.5.2 | Materials with high electrical conductivity | 130 |
|  | 5.5.3 | Advanced insulation materials | 131 |
|  | 5.5.4 | Advanced capacitors for power electronics | 131 |
| 5.6 | | Summary | 131 |

6. Clean combustion and emission control — 137
*Changlie Wey and Chi-Ming Lee*

|  |  |  |  |
|---|---|---|---|
| 6.1 | | Introduction | 137 |
| 6.2 | | Products of combustion | 137 |
|  | 6.2.1 | Fundamentals of $NO_x$ formation | 139 |
|  | 6.2.2 | $NO_x$ emissions standards | 141 |
| 6.3 | | Emissions control | 142 |
| 6.4 | | Engine $NO_x$ control strategies | 145 |
|  | 6.4.1 | Rich-burn, quick-quench, lean-burn combustion | 147 |
|  | 6.4.2 | Lean-burn, premixed, prevaporized combustors | 149 |
|  | 6.4.3 | Lean-burn staged combustors | 150 |
|  |  | 6.4.3.1 Rolls-Royce lean-burn combustor | 152 |
|  |  | 6.4.3.2 GE Aviation's dual-annular combustor | 153 |
|  |  | 6.4.3.3 GE Aviation's twin-annular premixed system combustor | 154 |
|  | 6.4.4 | N+2 advanced low-$NO_x$ combustors | 155 |
|  |  | 6.4.4.1 GE Aviation technology | 155 |
|  |  | 6.4.4.2 Pratt & Whitney | 158 |
| 6.5 | | Tradeoffs involved in reducing $NO_x$ emissions | 159 |
| 6.6 | | Summary | 161 |

7. Airspace systems technologies — 165
*Banavar Sridhar*

|  |  |  |  |
|---|---|---|---|
| 7.1 | | Introduction | 165 |
| 7.2 | | Current airspace operations | 166 |
|  | 7.2.1 | Separation assurance | 167 |
|  | 7.2.2 | Traffic flow management | 167 |
|  |  | 7.2.2.1 National traffic flow management | 167 |
|  |  | 7.2.2.2 Regional traffic flow management | 169 |
|  | 7.2.3 | Terminal area operations | 169 |
|  | 7.2.4 | Surface traffic operations | 171 |
|  | 7.2.5 | Environmental operations | 171 |
| 7.3 | | Advanced airspace operations concepts | 171 |
|  | 7.3.1 | Separation assurance | 173 |
|  | 7.3.2 | Traffic flow management | 173 |

|  |  |  |  |
|---|---|---|---|
|  | 7.3.3 | Terminal area | 173 |
|  | 7.3.4 | Surface operations | 174 |
|  | 7.3.5 | Environmentally friendly operations | 174 |
| 7.4 | Next generation air transportation system technologies | | 176 |
|  | 7.4.1 | Automatic dependent surveillance—broadcast | 176 |
|  | 7.4.2 | Performance-based navigation | 176 |
|  | 7.4.3 | Weather integration | 177 |
|  | 7.4.4 | Data communication | 178 |
|  | 7.4.5 | Operations | 178 |
|  |  | 7.4.5.1 Optimal descent trajectories | 178 |
|  |  | 7.4.5.2 Wind-optimal and user-preferred routes | 178 |
|  |  | 7.4.5.3 Surface | 179 |
|  | 7.4.6 | Integrated technologies | 179 |
|  | 7.4.7 | Global harmonization | 179 |
| 7.5 | Conclusions | | 180 |

8. Alternative fuels and green aviation — 183
*Emily S. Nelson*
| | 8.1 | Introduction | 183 |
|---|---|---|---|
| | 8.2 | Aviation fuel requirements | 186 |
| | | 8.2.1 Jet fuel specifications | 186 |
| | | 8.2.2 Alternative jet fuel specifications | 194 |
| | 8.3 | Fuel properties | 197 |
| | | 8.3.1 Effect of composition on fuel properties | 197 |
| | | 8.3.2 Emissions | 208 |
| | 8.4 | Biofuel feedstocks for aviation fuels | 209 |
| | | 8.4.1 Crop production for oil from seeds | 209 |
| | | 8.4.2 Crop production for oil from algae | 212 |
| | 8.5 | Manufacturing stages | 216 |
| | | 8.5.1 Dewatering, crude oil extraction, and preprocessing | 217 |
| | | 8.5.2 Biorefining processes | 219 |
| | | 8.5.2.1 Transesterification | 219 |
| | | 8.5.2.2 Hydroprocessing | 221 |
| | | 8.5.2.3 Other strategies | 222 |
| | | 8.5.3 Coproducts | 222 |
| | 8.6 | Life cycle assessment | 223 |
| | 8.7 | Conclusions | 227 |
| | | Appendix. Basic terminology and concepts in hydrocarbon chemistry | 236 |

9. Overview of alternative fuel drivers, technology options, and demand fulfillment — 239
*Kirsten Van Fossen, Kristin C. Lewis, Robert Malina, Hakan Olcay and James I. Hileman*
| | 9.1 | Introduction | 239 |
|---|---|---|---|
| | 9.2 | Alternative fuel drivers | 239 |
| | | 9.2.1 Environment and human health | 239 |
| | | 9.2.2 Economy | 240 |
| | | 9.2.3 Energy security | 242 |
| | 9.3 | Technology options | 242 |
| | 9.4 | Meeting demand for alternative jet fuel | 243 |
| | 9.5 | Conclusions | 244 |

10. Biofuel feedstocks and supply chains: how ecological models can assist with
    design and scaleup                                                                      247
    *Kristin C. Lewis, Dan F.B. Flynn and Jeffrey J. Steiner*
    10.1  Introduction                                                                       247
    10.2  Challenges of developing an agriculturally based advanced biofuel industry         248
    10.3  Potential benefits of scaled-up biofuel feedstock production                       251
    10.4  Regionalized biomass production and linkage to conversion technology               253
    10.5  Applying ecological models to biofuel production                                   256
          10.5.1  Description of on ecological model for biofuel production                  260
          10.5.2  Scenarios to test biofuel production                                       261
    10.6  Summary                                                                            265

11. Microalgae feedstocks for aviation fuels                                                 269
    *Mark S. Wigmosta, Andre M. Coleman, Erik R. Venteris and Richard L. Skaggs*
    11.1  Introduction                                                                       269
    11.2  Algae growth characteristics                                                       271
          11.2.1  Biophysics                                                                 271
          11.2.2  Climate variability                                                        273
          11.2.3  Productivity                                                               274
    11.3  Large-scale production potential and resource constraints                          275
          11.3.1  Land                                                                       279
          11.3.2  Water                                                                      280
          11.3.3  Nutrients                                                                  283
    11.4  Two-billion gallon per year case study                                            284
          11.4.1  Results                                                                    285
          11.4.2  Discussion and conclusions                                                 287
    11.5  Summary and conclusions                                                            289

12. Certification of alternative fuels                                                       295
    *Mark Rumizen and Tim Edwards*
    12.1  Introduction                                                                       295
    12.2  Background                                                                         295
          12.2.1  Aviation fuel specifications                                               295
          12.2.2  Foundational elements of the approval process                             296
    12.3  ASTM certification process                                                         297
          12.3.1  Basis of the approval process                                             297
          12.3.2  Guidebook for the approval process (ASTM D4054)                           297
          12.3.3  Drop-in fuel specification (ASTM D7566)                                   300
          12.3.4  Fischer-Tropsch–synthesized paraffinic kerosene, 2006 to 2009            301
          12.3.5  Hydroprocessed esters and fatty acids                                      301
          12.3.6  Synthesized isoparaffin                                                    302
    12.4  U.S. Federal Aviation Administration certification                                 302
    12.5  Future pathways                                                                    303

13. Environmental performance of alternative jet fuels                                       307
    *Hakan Olcay, Robert Malina, Kristin Lewis, Jennifer Papazian, Kirsten van Fossen,*
    *Warren Gillette, Mark Staples, Steven R.H. Barrett, Russell W. Stratton and*
    *James I. Hileman*
    13.1  Introduction                                                                       307
    13.2  Evaluating greenhouse gas emissions and impacts of alternative fuels on
          global climate change                                                              309
          13.2.1  Greenhouse gas life cycle analysis background                             309

|  | 13.2.2 | Overview of greenhouse gas life cycle analysis results for drop-in jet fuels | 312 |
|  |  | 13.2.2.1 Conventional jet fuel from crude oil | 313 |
|  |  | 13.2.2.2 Fischer-Tropsch jet fuel | 315 |
|  |  | 13.2.2.3 Hydroprocessed esters and fatty acids jet fuel | 316 |
|  |  | 13.2.2.4 Renewable jet fuel from sugars | 317 |
|  | 13.2.3 | Land-use change | 319 |
|  | 13.2.4 | Discussion of greenhouse gas life-cycle analysis results | 321 |
|  | 13.2.5 | Addition of non-$CO_2$ combustion emissions | 321 |
| 13.3 | Water |  | 322 |
|  | 13.3.1 | Water use and consumption | 322 |
|  | 13.3.2 | Water quality | 325 |
| 13.4 | Biodiversity |  | 326 |
| 13.5 | Conclusions |  | 329 |

14. Perspectives on the future of green aviation — 335

*Jay E. Dryer*

| 14.1 | Introduction |  | 335 |
| 14.2 | Key factors affecting the future of green aviation |  | 335 |
|  | 14.2.1 | Technology trends | 335 |
|  | 14.2.2 | Economic trends | 336 |
|  | 14.2.3 | Policy and regulatory trends | 338 |
|  | 14.2.4 | Social trends | 339 |
| 14.3 | Required technology for aircraft development and design |  | 339 |
| 14.4 | Required technology for greater alternative fuel utilization |  | 341 |
| 14.5 | Possible disruptive technologies |  | 342 |
| 14.6 | Forecast |  | 343 |
|  | 14.6.1 | The steady-state case (or business as usual) | 343 |
|  | 14.6.2 | The pessimistic case | 344 |
|  | 14.6.3 | The optimistic case | 344 |
| 14.7 | Summary |  | 345 |

| Acronym list | 347 |
| Subject index | 353 |
| Book series page | 357 |

# List of contributors

| | |
|---|---|
| Steven R. H. Barrett | Massachusetts Institute of Technology, Cambridge, Massachusetts, USA |
| Andre M. Coleman | Pacific Northwest National Laboratory, Richland, Washington, USA |
| Jay E. Dryer | National Aeronautics and Space Administration, Headquarters, Washington, DC, USA |
| Tim Edwards | U.S. Air Force Research Laboratory, Wright-Patterson Air Force Base, Ohio, USA |
| Edmane Envia | National Aeronautics and Space Administration, Glenn Research Center, Cleveland, Ohio, USA |
| Dan F. B. Flynn | Environmental Biologist, CORA/iBiz at Volpe National Transportation Systems Center, Cambridge, Massachusetts, USA |
| Warren Gillette | Environmental Protection Specialist, U.S. Federal Aviation Administration, Washington, USA |
| James D. Heidmann | National Aeronautics and Space Administration, Glenn Research Center, Cleveland, Ohio, USA |
| James I. Hileman | Chief Scientist, Office of Environment and Energy, U.S. Federal Aviation Administration, Washington, DC, USA |
| Dennis Huff | National Aeronautics and Space Administration, Glenn Research Center, Cleveland, Ohio, USA |
| Chi-Ming Lee | National Aeronautics and Space Administration, Glenn Research Center, Cleveland, Ohio, USA (retired) |
| Kristin C. Lewis | Environmental Biologist, Volpe National Transportation Systems Center, Cambridge, Massachusetts, USA |
| Robert Malina | University of Hasselt, Hasselt, Belgium |
| Ajay Misra | National Aeronautics and Space Administration, Glenn Research Center, Cleveland, Ohio, USA |
| Emily S. Nelson | National Aeronautics and Space Administration, Glenn Research Center, Cleveland, Ohio, USA |
| Hakan Olcay | Massachusetts Institute of Technology, Cambridge, Massachusetts, USA |
| Jennifer Papazian | Environmental Protection Specialist, Volpe National Transportation Systems Center, Cambridge, Massachusetts, USA |
| Dhanireddy R. Reddy | National Aeronautics and Space Administration, Glenn Research Center, Cleveland, Ohio, USA |
| Mark Rumizen | Senior Technical Specialist, Aviation Fuels, Aircraft Certification Service, U.S. Federal Aviation Administration, Burlington, Massachusetts, USA |
| Richard L. Skaggs | Pacific Northwest National Laboratory, Richland, Washington, USA |
| Mark Staples | Massachusetts Institute of Technology, Cambridge, Massachusetts, USA |
| Russell W. Stratton | Massachusetts Institute of Technology, Cambridge, Massachusetts, USA |
| Banavar Sridhar | National Aeronautics and Space Administration, Ames Research Center, Moffett Field, California, USA |

Jeffrey J. Steiner       Colorado State University, Fort Collins, Colorado, USA
Kenneth L. Suder        National Aeronautics and Space Administration, Glenn Research Center, Cleveland, Ohio, USA
Kirsten Van Fossen      National Transportation Systems Center, Cambridge, Massachusetts, USA
Erik R. Venteris        Pacific Northwest National Laboratory, Richland, Washington, USA
Changlie Wey            Vantage Partners, Brook Park, Ohio USA (retired)
Mark Wigmosta           Pacific Northwest National Laboratory, Richland, Washington, USA

# Foreword by Lourdes Maurice

Aviation environmental issues are not new. Noise complaints arose hand in hand with the new air machines of the early 1900s, and concerns about sonic boom facets plagued the Concorde. We have had concerns about the air quality and climate effects of aviation activities for decades. What is perhaps new is the intensity of concerns about aviation environmental issues in the last few years. Many factors have come together, including the quick communications enabled by social media, increased aviation activity and growing expectations and awareness about our environment.

If we look at aviation's environmental performance, the record of improvement is impressive. Through a balanced approach including new technology and international standards, operational improvements and judicious land use, exposure of citizens to significant noise has been reduced by about 90% since the 1970s, despite growth. Aviation emissions have decreased as we have tightened international standards, and aircraft fuel efficiency has increased almost 20% over the last 7 years alone. Impressive? Yes. Is it enough? No, there is more to do—and we have the tools to do it.

The benefits of aviation are many, including economic and social aspects. Aviation fuels the global economy, and environmental awareness accompanies improved economic conditions. The aviation enterprise, including governments, industry, and civil society share a common goal of improving aviation's environmental performance.

2016 was an exciting year. In February the International Civil Aviation Organization (ICAO) agreed on the first-ever fuel efficiency standard. ICAO is poised to adopt an agreement on a global-market-based measure (GMBM) at its 39th General Assembly and is also tackling the next frontier of aviation air quality, with a standard on particulate matter expected in 2019.

Sustainable alternative fuels, a theoretical concept just a decade ago, are now being used in commercial airliners. As is often the case, aviation is a pioneering leader in biofuels.

Much remains to be done, however. And the aviation industry is well poised to do it. Aviation environmental professionals as well as novices need the latest information. This volume provides such information, written by the leaders in the industry.

The future is most often what we imagine it to be. Aviation's future is very green.

Lourdes Maurice
*Executive Director*
*Office of Environment and Energy*
*U.S. Federal Aviation Administration*
*Washington, DC*

# Foreword by Thomas B. Irvine

When I was in high school in the 1970s, Civics was a required course in my school's curriculum. One assignment in that class was to read articles from either *Time* or *Newsweek* magazine and do an oral report to the class. A recurring cover story in both magazines from that era was, "What Cost Energy?" In 1973, the Arab oil embargo had caused a dramatic spike in the cost of fuel. As a typical high schooler who had just acquired a driver's license, and as one whose parents expected me to pay for my share of the cost of an automobile, I was keenly aware of the cost of a gallon of gasoline. What I wasn't aware of at that time was the effect that the oil embargo and increasing fuel costs were having on airline operators. Witnesses at Congressional hearings held during that time on the state of the airline industry painted a bleak picture, with various testimony using terms such as "immediate crisis condition," "long-range trouble," and "serious danger."

As policy, the U.S. Government responded to this crisis by, among other things, initiating the Aircraft Energy Efficiency (ACEE) Program within the aeronautics research part of the National Aeronautics and Space Administration (NASA). This program set an audacious goal of reducing aircraft fuel consumption by an astonishing 50%. Divided into six projects that focused on propulsion system efficiency, aerodynamic efficiencies, and aircraft structural weight reduction technologies, the program would ultimately prove successful in several fundamental ways. First, many of the research results provided for technology and even product development paths to achieving the 50% fuel burn reduction goals. Second, for the first time since the founding of the National Advisory Committee for Aeronautics (NASA's predecessor) in the 1910s, U.S. Government engineers turned from aircraft performance (faster and higher) and found satisfaction and fulfillment working on aircraft efficiency. Third, and perhaps most importantly, the dire economic situation motivated U.S. industry and Government to work together to successfully transfer technologies out of the Government labs and into the engines and airframes being designed and built by U.S. industry. According to Joseph Chambers, a NASA Langley research engineer during that era, the ACEE Program owed its success to a perfect mixture of "funding, world economics, and technology readiness." While at the 2003 International Society of Air Breathing Engines meeting in Cleveland, Ohio, no less an expert than Dr. Mike Benzakein, then the GE Aeronautical Engines Advanced Product Development Manager, said that without the Energy Efficient Engine, or E3, project (one of the six projects under the ACEE Program), there would not have been a GE90 engine. Today, the GE90 engine and its variants and derivatives power much of the world's Boeing 777 and 787 fleets.

Since the 1970s, many other government programs or projects have sought to make aviation more efficient and thus reduce aviation's impact on the environment. Industry innovations, many coming from manufacturers' own in-house research and development (R&D) efforts, some in collaboration with government research programs, and some through successful transition of technologies from government R&D to successful product design and development, have resulted in the significantly lower fuel consumption that we see in today's airplanes. Depending on how one measures aircraft fuel efficiency, modern aircraft such as the Boeing 787 and the Airbus A350 are between 50% and 70% more efficient than the Boeing 707, an airplane that entered into service over 50 years ago. Reductions in aircraft weight, including the growing widespread use of high-strength, low-weight composites in both primary and secondary structures, improvements in vehicle aerodynamics, and improved jet engine performance have all been contributors. The use in aviation of alternative or biofuels, while still evolving, is a path to future green aviation that is being worked by the aviation industry and by adjacent industries such as energy and agriculture.

This is an area that requires serious and sober R&D policy and investment as we go about determining the viability of replacing fossil fuels with alternative sources.

So globally, in what direction is society heading, and how will that direction prescribe that aeronautical engineers design and manufacture efficient and environmentally friendly aircraft? The urbanization of the planet's population is well documented, as is the burgeoning middle class, especially in China and elsewhere in Asia. The resultant affluence associated with these phenomena is driving an increasing demand for air travel. Some projections indicate a doubling, or even tripling, of air traffic in the next 20 to 30 years. But it all starts with the economics of flight. Consumers continue to show preferences for cost-effective—read that as low-cost—air travel. Operators respond to this consumer demand by in turn demanding aircraft with ever-increasing fuel efficiency. And regulators, from international bodies, such as the International Civil Aviation Organization (ICAO), to localities, squeeze the airplane manufacturers from the regulatory side on both emissions and noise. Such regulatory controls have been in place for some time at all levels of government, but now for the first time, the ICAO has passed an aircraft $CO_2$ emission standard—in essence a fuel-efficiency standard. Although the air transportation system's contribution to pollutants that affect the environment, relative to contributions from industrial sources and other transportation, are estimated to be only approximately 3%, no one questions that sustainable aviation will be necessary to meet the projected increasing air travel demand in the coming decades.

The challenges ahead are not trivial. This book describes many of the technologies that are in the R&D phase today but that will need to find their way into next generations of aircraft. Each and every player in this endeavor has a key role to play—from the flying public, to the regulators, to the financiers, to the R&D engineers toiling in government and industry labs and research centers. It is imperative that the R&D results are such that designers can engineer the results and corresponding knowledge out of R&D efforts into useful and salable products—products that increasingly environmentally conscious consumers will ultimately want and support.

Our college and university engineering school deans and administrators report that many of today's engineering students are motivated to make useful and lasting contributions to society. For such students who also have a passion for aviation, this an exciting time to be entering the aerospace profession. One can anticipate that today's students will make many contributions to efficient, environmentally friendly, and sustainable air travel, as will the many talented and dedicated professionals currently working in the aeronautics and aviation industries. I highly recommend this book to anyone who wants, or needs, to know more on the myriad topics and technologies that contribute to green aviation. We all have a stake in a future where air travel is a viable means of transportation. Contributions to green aviation, as described and discussed in this book, will ensure that future.

Thomas B. Irvine
*Managing Director, Content Development*
*American Institute of Aeronautics and Astronautics*
*Former Deputy Associate Administrator, National Aeronautics and Space Administration,*
*Aeronautics Research Mission Directorate*

# Editors' preface

Emily S. Nelson and Dhanireddy R. Reddy

Efficient, reliable, and safe air transport is considered vital to global economic stability, growth, and security. Although humankind has made enormous advances in aviation technology, there are still many challenges and opportunities that stem from the availability of energy, engine efficiency, pollutant emissions, noise, reliability, and safety. It has been a well-recognized challenge for technology advancement community as well as policy makers to ensure that the technology advancements in air transportation would not lead to adverse impacts on the environment. The United States, being the world leader in the development of aerospace technologies, has long recognized the crucial role aviation plays in the modern world, and as a result, the National Aeronautics Research and Development Policy was established by an Executive Order for the first time in December 2006 to help guide the conduct of U.S. aeronautics research and development programs through 2020 (Marburger, 2006). The plan includes coordination with other government agencies, academia, and industry through collaborative partnerships and ensures the availability of a world-class resource (personnel, facilities, knowledge, and expertise) ready to be drawn upon by all the partners. Energy efficiency and environmental compatibility have been highlighted in the plan as the key factors to be considered in the development of technology roadmaps for future air transport in both civil and military applications.

Aircraft noise and emissions can have significant adverse impact on the environment. The aviation industry's greenhouse gas production and stratospheric ozone depletion, both of which are implicated in climate change, can have long-lasting repercussions. There are also more immediate impacts on the health of humans, animals, and ecosystems in the vicinity of airports.

In light of the global implications of aviation emissions, the United Nations (UN) in 1947 established an agency, the International Civil Aviation Organization (ICAO), to provide guidance to its member nations on the regulation of aircraft noise and emissions. Currently, there are 191 ICAO members, consisting of 190 of the 193 UN members (all but Dominica, Liechtenstein, and Tuvalu), plus the Cook Islands. Liechtenstein has delegated Switzerland to implement the treaty to make it applicable in the territory of Liechtenstein. The Committee on Aviation Environmental Protection (CAEP) of the ICAO has specified a series of increasingly stringent emission standards for oxides of nitrogen ($NO_x$) over the years. These standards have served as the basis for regulation of aviation emissions below 3,000 feet (about 914 m) in altitude. The standards govern emissions that are generated during the landing and take-off cycle, which includes takeoff, climb, descent, and taxiing/ground idle phases of the engine operation. CAEP is also responsible for recommending international noise standards. In recommending noise and emission standards, CAEP relies on technology forecasts as well as the economic costs of adopting new standards (ICAO, 2010; and ICAO, 2014). Aircraft noise is specified as a function of the maximum takeoff weight of the aircraft, the number of the engines, and the operating conditions. These operating conditions, referred to as certification points, are called the approach reference, the flyover reference, and the lateral (or sideline) reference. All commercial aircraft must be certified as having met the regulation noise levels for each before they can enter service.

Recent concerns regarding long-term damage to the global environment have led the ICAO to consider the regulation of gaseous and particulate emissions during cruise in addition to the already established $NO_x$ and noise regulation in and around the airports. Water vapor in the form of contrails may also have an adverse impact on the environment. At this time, it is difficult to precisely estimate the impact of these factors on the environment because of uncertainty in the accuracy of prediction methods (Singh, 2015). However, there are potentially grave consequences if the impact of these emissions is ignored. With the current trend of commercial air traffic growth world-wide, it is estimated that emissions will roughly triple by 2030. Over the next 35 years, if no new abatement technology is employed and the current growth rate continues unabated, the impact of emissions on the environment may swell to about 6 times that of the present—even with the best current state-of-the-art air transportation technology (Singh, 2015). Hence, technology advancement is considered crucial in combating environmental impact. Leading government, industrial, and academic research organizations across the world that are involved in aviation technology advancement have set long-term quantifiable goals for energy efficiency, as well as noise and emissions reductions. For example, the U.S. Government Agency, the National Aeronautics and Space Administration (NASA), the world leader in the aviation technology advancement, has set long-term goals of

1. Energy-efficiency improvement of 60% compared to the Boeing 737-800 with CFM56-7B engines
2. Noise reduction to 52 dB, cumulative margin relative to stage 4 noise limit
3. $NO_x$ emissions reduction of 80% compared to CAEP6 limits

As part of their Flight Path 2050 initiative, the European Union Advisory Council for Aviation Research in Europe has also identified targets of 75% reduction in $CO_2$ per passenger kilometer, 90% reduction in $NO_x$ emissions, and 65% reduction in noise by 2050 compared to the datum of 2000.

To meet the aggressive objectives in reducing the environmental impact of aviation as outlined above, a number of technologies covering a wide range of disciplines and scope must be developed and matured for application to future vehicle architectures, which would drastically depart from the current architectures. Even though the targets for energy efficiency, noise, and emissions are set for the long term, the technology roadmaps to achieve those targets are phased in three different timeframes: near-, mid-, and long-term plans with intermediate target values identified for the near- and mid-term time frames. The time frames for NASA's research plans are expressed in terms of proximity to the current state of the art, defined as N: N+1 denotes the near term (i.e., next 10 years; N+2, the mid term (i.e., next 20 years); and N+3, the long term (i.e., next 30 years).

Technologies to improve fuel and energy efficiencies are aimed at improving the overall efficiency of the propulsion system. Advancements in airframe architecture are designed to reduce both the aerodynamic drag and net weight of the aircraft. The propulsion system efficiency consists of two main components: thermal efficiency, which is a measure of how effectively the chemical energy in the fuel is converted into mechanical energy, and propulsive efficiency, which indicates how effectively the mechanical energy is converted to the motive or propulsive energy used to propel the aircraft. Thermal efficiency increases as the overall pressure ratio (OPR) of the engine cycle, resulting in higher turbine inlet temperature. In practical terms, the OPR is limited by the sensitivity of combustor and turbine materials to the high temperature in the hot section of the turbine engine. Propulsive efficiency is maximum at exhaust velocities that are very close to the forward velocity of motion: when the thrust is produced by moving a large quantity of air through the engine with little velocity increment. Hence, the propulsive efficiency of a typical turbofan engine increases with the bypass ratio, defined as the ratio of the mass flow rate through the fan annulus passage to that through the core.

The noise generated by air traffic can pose a nuisance to people and wildlife, particularly in the regions surrounding airports. In an effort to minimize the disturbance, noise reduction technologies aim to identify the primary sources of noise and to reduce the noise generated by various components of the propulsion system and the airframe through engine and airframe design.

Whereas fuel efficiency and noise reduction contributions can come from both airframe and propulsion systems technologies, emissions reduction can only be achieved through propulsion technology advancement which is aimed at the reduction of $NO_x$ produced per unit weight of the fuel burnt. The $NO_x$ emissions level is the time integral of its formation rate, which is an exponential function of the air temperature. The $NO_x$ emissions level correlates very well to the fuel injector's ability to prepare the fuel-air mixture. Mixing the fuel with air as quickly and uniformly as possible before burning starts is a crucial factor in reducing hot spots (near-stoichiometric pockets of fuel-air mixtures), which will produce $NO_x$ very quickly. The NASA Glenn Research Center, with air-breathing propulsion as one of its core competencies, has played a key role in advancing low-emissions combustion technology.

For the near-term (N+1) time frame, it is not economically feasible to drastically alter the conventional architecture for the aircraft; that is, the tube and wing configuration with podded engine installation. Instead, the focus is placed on improved propulsive efficiency through higher bypass ratio. To meet the noise, emissions, and performance goals, very high bypass engine cycles are areas of active research. Some of the concepts include (1) the geared turbofan engine, which has the benefit of low noise generation due to lower fan tip speeds, and (2) the low-pressure turbine, which can rotate at a higher speed. The concept of the variable area nozzle for the fan bypass stream is also under consideration, which can enable the engine to produce adequate thrust for takeoff but with enough margin for stable operation. The geared turbofan concept has been championed by the aeroengine company Pratt & Whitney, who has advanced the technology in collaboration with NASA Glenn. Another concept to realize a very high bypass ratio is the open-rotor concept, sponsored by General Electric based on their earlier technology development effort of the Advanced Turboprop (ATP) (i.e., unducted fan, UDF) program in collaboration with NASA. However, the biggest challenge for the open-rotor concept is noise generation, which currently exceeds noise regulations. Better blade design obtained through advanced computational fluid dynamics methods and the considerable progress made in the noise reduction technologies are expected to permit the open-rotor concept to overcome the noise challenge and meet the targets set for fuel burn, emissions, and noise reduction.

For the mid-term (N+2) time frame, consensus among industry and government laboratories is that the hybrid wing-body (or blended wing-body) airframe with an embedded distributed propulsion system is needed to meet noise, emissions, and fuel burn reduction goals. In addition to the benefits of the large bypass ratio that is characteristic of distributed propulsion and of the higher propulsion efficiency due to boundary-layer ingestion, a number of technical challenges remain. Major concerns include reduced aerodynamic efficiency due to flowfield distortion caused by the embedded nature of the engine inlets, as well as the detrimental impact on engine performance and lifetime due to boundary-layer ingestion. In addition, high-pressure core engine technology (OPR of 60 to 70), which is attractive due to its higher thermal efficiency, is complicated by technical challenges such as small blade heights and tight clearance requirements in the aft stages of compressors. Another challenge is undesirable $NO_x$ production, due to high temperatures in combustion chambers.

For the long-term (N+3) time frame, a number of radically new propulsion concepts are being considered because of the long lead time available and the potential maturation of some competitive technologies such as high-temperature superconductors. These alternative propulsion concepts include electric drive and hybrid propulsion devices. Even in the conventional gas turbine area, advances in materials technology are anticipated to result in new classes of lightweight materials that can accommodate aggressive heat exchange such as recuperative cycles involving inter-stage cooling of the compressor, leading to very high overall OPRs of 100 and above. Other concepts that may reach fruition over the long term are constant-volume combustion to increase thermal efficiency, integrated energy optimization for propulsion and other power-consuming devices such as the auxiliary power unit, and the recycling of energy dissipated during landing. Alternative fuels with low carbon content, combined with optimal operational options such as flying at lower cruise speeds and formational flying, can be candidates in the broader effort of reducing the overall carbon footprint. Alternative power-plant concepts such as fuel cells are also being studied. They need to have power densities (power per unit weight) comparable to those of modern gas turbine engines to be considered for aircraft propulsion, however, and even though these exciting new technologies are being actively pursued for terrestrial applications (e.g., automotive and electric power), the power densities of fuel cells are not likely to be comparable to those of gas turbines, even in the N+3 time frame, based on the current rate of technology advancement.

Another avenue for reducing the environmental impact of aviation is the consideration of alternative fuels. Since aircraft have no feasible alternatives to the internal combustion engine over the near term, a practical strategy is to use "drop-in" alternative fuels, which can be used in existing aircraft engines, but reduce the net carbon footprint of aviation by employing a greener replacement. For example, if jet fuel is derived from vegetative sources, or a blend of petroleum and plant-derived fuel, the net impact of aircraft emissions is partially offset by the environmental carbon dioxide taken in by the vegetation during growth. Inflight testing of such fuels has been successful, and specifications on fuel composition and properties have been comprehensively developed for drop-in compatibility.

The net impact of air travel can be formally quantified using life cycle assessment (LCA), which is a holistic accounting of the environmental or economic costs and benefits at every stage of the process: from growth and harvest or mining of raw material; to refinement of the crude into usable jet fuel; to transportation of feedstock, waste products, and refined fuel; and finally to end use during combustion in flight. It can incorporate the positive effect of creatively using the left-over processing waste to produce new products, to employ as an energy source, or to sequester pollutants from the environment. LCAs can be used to compare and contrast different production pathways, and they provide credible evidence for promoting rational regional solutions for alternative fuel production.

Some plant-derived fuels, such as ethanol from corn or sugar cane, have been successfully scaled up from the laboratory to economically viable agricultural production. The environmental impact of such fuel, however, is significant, rendering its value as an alternative to petroleum-derived fuel less attractive. A more enlightened approach is to develop farming techniques for plant-based feedstock that do not compete with resources better used for food production. Vegetation that does not require arable land, fresh water, or intensive fertilization and growth management practices are far more appealing. Crops such as *Jatropha*, *Camelina*, and the many strains of algae are promising as potential feedstock, although they require more applied research to advance them to commercial scale.

In the chapters that follow, the book presents these concepts, discussing them in detail and concluding with an analysis of the future of green aviation.

## ACKNOWLEDGEMENTS

The editors are grateful to NASA Glenn Research Center for providing support in developing this book. We would like to acknowledge our chapter authors for their generosity in sharing their wisdom and experience. We also thank our support staff for their tireless efforts, in particular: Caroline Rist, for her management capabilities; Laura Becker and Nancy O'Bryan, for their careful editing; Jaime Scibelli and Robin Pertz, for their thoroughness in checking citations and obtaining copyright permissions; Nancy Mieczkowski and Lorraine Feher for their expertise in manuscript preparation; Wendy Berndt for her efficient clerical support; and Lorie Passe, for her skillfulness in developing layouts and just about everything else.

## REFERENCES

International Civil Aviation Organization. 2010. Report to CAEP by the CAEP Noise Technology Independent Expert Panel: aircraft noise technology review and medium and long term noise reduction goals. Doc 9943, ISBN 978-92-9231-594-8.

International Civil Aviation Organization. 2014. Report to CAEP by the CAEP Noise Technology Independent Expert Panel: novel aircraft-noise technology review and medium- and long-term noise reduction goals. ICAO Doc 10017, ISBN 978-92-9249-401-8.

Marburger, J. 2006. *National Aeronautics Research and Development Policy.* Washington, DC: Executive Office of the President, National Science and Technology Council. *https://www.hq.nasa.gov/office/aero/releases/national_aeronautics_rd_policy_dec_2006.pdf* (accessed March 27, 2012).

Singh, R. 2015. The future journey of civil aerospace weather, environment and climate. Presented at the 22nd International Symposium of Air Breathing Engines (ISABE) Conference, Phoenix, AZ. *http://isabe2015.org/InvitedLectures.aspx* (accessed March 27, 2017).

*Part I*
*Environmental impacts of aviation*

# CHAPTER 1

# Noise emissions from commercial aircraft

Edmane Envia

## 1.1 INTRODUCTION

Aircraft noise[1] is an aerodynamically generated phenomenon, a byproduct of the interaction of the airstream with the various engine and airframe components as well the flow processes within an aircraft's engine exhaust plume. These interactions and processes generate pressure fluctuations in the ambient atmosphere that radiate away from the aircraft as noise. The portion of the aircraft noise that reaches the ground affects the populated communities and is the topic of this chapter.[2] Depending on the particular airframe or engine component, and the nature of its interaction with the airstream, tone and broadband noise is generated over a wide range of frequencies. Aircraft noise is the amalgam of these various component noise contributions whose relative intensities are dictated by the engine architecture and power level as well as the airframe design and whether or not the landing gear and/or the high-lift devices (i.e., the flaps and slats) are deployed. Figure 1.1 lists some of the important sources of aircraft noise during the landing approach for large commercial jet aircraft. Depending on the type of aircraft, not all of these noise sources may be present or significant. The various aircraft noise sources and their relative importance will be discussed later in this chapter.

For a typical jet-powered commercial aircraft, the total acoustic power radiated from all of its noise sources is a minute fraction (often much less than 0.01%) of the aerodynamically generated power produced by its engines. Yet, aircraft noise is quite discernible and is often a cause for complaints by people living near airports. This is especially true within an airport's "noise impact zone," which is the area surrounding an airport where the noise level from commercial aircraft operations exceeds 65 DNL. Usually expressed in dBA[3] units, DNL is the day-night average sound level. This metric, which sometimes is denoted by $L_{dn}$, was developed by the U.S. Environmental Protection Agency for the purpose of quantifying the impact of multiple noise events that vary in level and duration. As far as the impact of the aviation noise on the community is concerned, the DNL metric is used to quantify the cumulative noise level from all takeoffs and landings at an airport during a day. Specifically, DNL is the sum of annually averaged noise levels over a 24-hour period divided by the number of seconds in a day.

The annual average denotes the 365-day average of (i) the noise levels for each aircraft type and (ii) the time of day at which the operation occurred. The time-of-day information is important in order to account for the increased sensitivity to the external sounds during the night (i.e., sleeping hours) when the ambient noise level is low. In calculating DNL, each nighttime event is considered the equivalent of 10 daytime events, which translates to adding a 10-dB penalty to account for the intrusiveness of each nighttime event occurring between the hours of 10 p.m. and

---

[1] Noise is defined as unpleasant or objectionable sound.
[2] Aerodynamic noise (especially that generated by the airflow over the fuselage) could also have an adverse effect on the ambient noise level in the aircraft cabin.
[3] dBA, the A-weighted decibel level, expresses the relative loudness of sound as perceived by the human ear. It corresponds to people's natural auditory response by de-emphasizing very low and very high frequencies. Given the human perception of sound, a 10-dB increase in the sound level is perceived, subjectively, by most people as the doubling of the loudness of the sound.

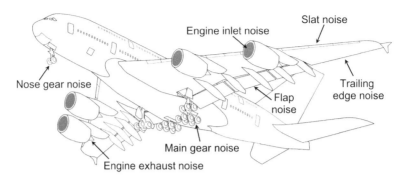

Figure 1.1.    Principal sources of aircraft noise at landing approach. (Adapted from Airbus, 2008. Airbus A380 during landing approach. France: Toulouse.)

7 a.m. The Federal Aviation Administration (FAA) considers 65 DNL to be a significant level of aircraft noise and incompatible with residential areas. In contrast, 55 DNL is considered a noise exposure level that does not compromise public health and welfare, and it is considered a good long-term goal for aircraft noise reduction.

An example of a noise impact zone is depicted in Figure 1.2, which shows the DNL contours around the Cleveland Hopkins International Airport overlaid on the municipal map of the local area. Note that the 65-DNL contour extends beyond the airport boundary into the surrounding communities. Clearly, in addition to the neighborhoods situated along the immediate perimeter of the airport, other nearby neighborhoods are also exposed to aircraft noise levels that are 65 DNL or higher. This is often the case for urban airports that have commercial jet aircraft traffic of any significant volume. It should be noted that the shape and extent of the noise impact zone depend, among other factors, on the type of aircraft, number of runways, main arrival and departure routes, and the prevailing wind direction(s). As such, they vary from airport to airport.

For the purpose of making policies that address the impact of noise (including aviation noise) on the environment, objective measures such as DNL must ultimately be associated with the reaction of the public to noise. In fact, the very choice of the 65 DNL as a significant threshold noise level has its root in such subjective measures of noise.

The most widely accepted subjective measure is the high level of annoyance caused by noise. In 1992 the U.S. Federal Interagency Committee on Noise (FICON)[4] selected "the percentage of the area population characterized as "highly annoyed" (%HA) by long-term exposure to noise of a specified level (expressed in DNL)" as its preferred indicator of people's adverse reaction to noise (Hegland et al., 1992). A widely used relationship between the high level of annoyance experienced by people and the noise exposure dose (Finegold et al., 1994) is shown in Figure 1.3. This relationship, which is based on curve fit of the data from surveys of community reaction to transportation noise conducted over several studies, was endorsed by FICON and has since formed the basis of aviation noise policies adopted by the Federal government. As the graph shows there is a nonlinear relationship between the noise exposure level and the predicted occurrence of high levels of annoyance in the community. At low noise exposure levels (say, <55 DNL), less than 3% of population is predicted to express high levels of annoyance, but as

---

[4] The Federal Interagency Committee on Aviation Noise (FICAN) was formed in 1993 in response to the recommendations of the FICON 1992 report that "a standing Federal interagency committee should be established to assist agencies in providing adequate forums for discussion of public and private sector proposals, identifying needed research, and in encouraging the conduct of research and development in these areas". FICAN member agencies include the Department of Defense, Department of Interior, Department of Transportation, Environmental Protection Agency, National Aeronautics and Space Administration (NASA), and Department of Housing and Urban Development.

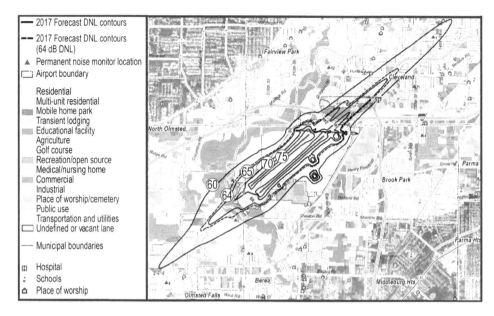

Figure 1.2.  Noise impact zone for the Cleveland Hopkins International Airport. Day-night average sound level (DNL) contours are shown in the range 60 to 75 dBA. (From Cleveland Hopkins International Airport. 2016. Noise compatibility report. *http://www.clevelandairport.com/company/environment/noise-compatibility-reports* (accessed March 8, 2017). With permission.)

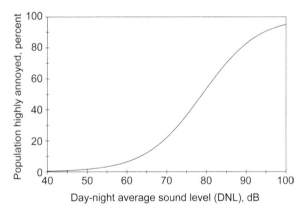

Figure 1.3.  Relationship between percent population highly annoyed (%HA) by noise exposure and the DNL noise metric. The curve is given by the expression %HA = $100(1 + e^{11.13-0.0141\,DNL})$. (From Hegland, J., W. Dickerson, and R. Danforth. 1992. Federal agency review of selected airport noise analysis issues. Washington, DC: Federal Interagency Committee on Noise. Work of the U.S. Government.)

the noise level increases there is a rapid rise in the predicted occurrence of high annoyance level (e.g., at 70 DNL over a quarter of the population would indicate high level of annoyance with noise).

However, the choice of 65 DNL as a significant noise exposure threshold is not immediately obvious from this graph since it does not seem to represent a special point, either quantitatively (the value of the function) or qualitatively (the slope of the function). The choice seems to be driven by the tacit correlation that exists between the annoyance level experienced by the public as a result of long-term cumulative exposure to noise and the rate at which the public complains about it or takes legal action as a result of it. Although the data do not always indicate a straightforward, causal relationship between annoyance level and rate of complaints by the public,[5] it is nonetheless true that noise complaints and legal action have played an important role in the decision-making process about airport design and operation (Fidell, 2003). Viewed in that way, the choice of 65 DNL as a significant noise level is traceable to the U.S. Department of Transportation's Aviation Noise Abatement Policy of 1976 (FAA, 1976), which identified 65 DNL as the level at which individual complaints or even group action from the public are possible.[6] The choice of the DNL 65 guideline was also driven in part by "a compromise between what was environmentally desirable and what was economically and technologically feasible at the time" (Eagan and Gardner, 2009).

The FAA has identified short- and long-term targets to alleviate the impact of commercial aviation noise on the community (FAA, 2012; Pickard, 2008). In the short term, the FAA's goal is to maintain the 4% compound annual reduction rate in the number of people exposed to 65 DNL through 2018 using the 2005 numbers as the baseline exposure count.[7] In the long term (beyond 2025), the FAA aims to contain the 65 DNL, and ultimately the 55 DNL, contours primarily within the airport boundary. The 65 DNL "containment" goal effectively implies a 10-dB reduction in the DNL across the board which is a challenging target even for a moderately sized airport like Cleveland Hopkins. The 55-DNL containment target (i.e., a 20-dB reduction) is obviously even more challenging and would likely require a confluence of technological, operational, and land-use planning solutions.

To reduce the impact of commercial aviation noise on the environment, aircraft noise is regulated and must not exceed predefined levels as a function of the maximum takeoff weight (MTOW), number of the engines, and the operating conditions of the aircraft. These operating conditions, referred to as the certification points, are the approach reference, the flyover reference, and the lateral (or sideline) reference as depicted in Figure 1.4. All commercial aircraft must be certified as having met the regulation noise levels before they can enter service.[8]

In the United States, the regulation noise levels for commercial aircraft and the rules governing the certification process are set by the FAA. The regulation, now called Federal Aviation Regulation (FAA, 2014), was first promulgated in Title 14, Code of Federal Regulations (14 CFR) Part 36, "Noise Standards: Aircraft Type Certification" in 1969. The current U.S. noise rule, 14 CFR Part 36, Stage 4, is essentially identical to the internationally adopted standard called Annex 16, Chapter 4 (ICAO, 2008).[9] The entity responsible for recommending the

---

[5] There is evidence that the number of individual noise events can influence the perception of sound by the noise-sensitive residents. For example, data suggest that people often find noise from a large number of aircraft operations with a low DNL rating more objectionable than fewer aircraft operations with a high DNL rating (Albee *et al.*, 2006). This seems to be the case if the DNL exceeds the threshold of noise annoyance for an individual even though his/her noise exposure is technically less at the lower DNL.

[6] In the 1976 Aviation Noise Abatement Policy document the noise level was expressed in terms of noise exposure forecast (NEF) in dBA. The report called out NEF 30 as the threshold level that would precipitate many individual or group complaints from the public. The NEF and DNL metrics are highly correlated, and their equivalency relationship is often taken to be DNL = NEF + 35. One NEF unit reduction in the level of noise is equivalent to 2% reduction in the number of people who are highly annoyed by the noise exposure. One NEF unit reduction is also equivalent to 14% reduction in the area of land that is exposed to a given level of noise; see (FAA, 1976). It is also worth noting that FAA uses 65 DNL as the threshold level for sound insulation upgrades for residential homes affected by high noise exposure levels around airports.

[7] FAA estimated that 500,000 people in the United States were impacted by 65 DNL in the 2000 to 2004 time frame.

[8] Each aircraft type (i.e., airframe and engine combination, such as Boeing 737-800 with CFM-56-7B27, for example), is noise certified—not each individual aircraft in service.

[9] This wasn't always the case. The first U.S. aircraft noise level regulation, which became effective on December 1, 1969, was more stringent by as much as 5 EPNdB compared with the ICAO regulation of that time, namely, Annex 16, Chapter 2. Today, however, the gap between the two has closed and now CFR

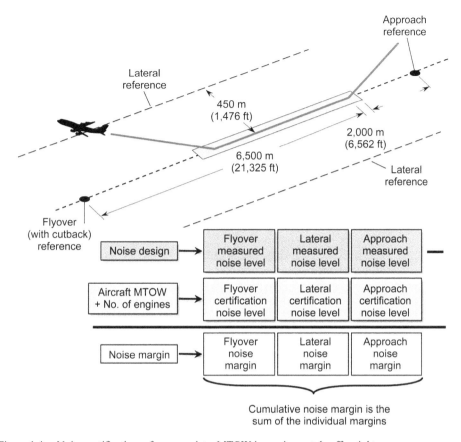

Figure 1.4.   Noise certification reference points. MTOW is maximum takeoff weight.

international noise standards is the Committee on Aviation Environmental Protection (CAEP), which was established by the International Civil Aviation Organization (ICAO) and of which United States is a member. In recommending the noise standards, CAEP relies on technology forecasts as well as the economic costs of adopting new standards (ICAO, 2014).

A common way to express the performance of an aircraft relative to the noise standard is via the so-called cumulative noise margin. As depicted in Figure 1.4, the cumulative noise margin is the sum of the noise margins at the three certification points, with the noise margin at each certification point being the difference between the noise standard and the measured noise of the aircraft. A negative noise margin means that the aircraft is quieter than the noise standard. Because of the peculiar spectral, persistence, and annoyance characteristics of aircraft noise, a specially designed metric called the effective perceived noise level (EPNL), expressed in EPNdB units, is used to quantify aircraft noise levels for the certification purposes. The metric takes into account human response to such aspects of aircraft noise as spectral shape, tonal content, intensity, and duration. As such, EPNL cannot be directly measured and is instead calculated according to a procedure described in the Part 36 or Annex 16 regulation. An example of EPNdB levels for an aircraft will be discussed later in this chapter. It should be noted that EPNL is a measure of a single noise event; that is, the noise level of a single aircraft under a specific operating condition.

---

Part 36 and the ICAO rules are identical. This is a consequence of the fact that since 1999 United States has participated in the rule making process for establishing the international standards for aviation noise.

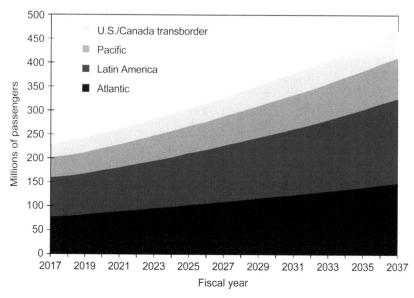

Figure 1.5.　Growth forecast of passenger traffic to and from the United States from various regions via U.S. and foreign flagged commercial air carriers between years 2017 and 2037. (Data from Federal Aviation Administration. 2017. FAA aerospace forecast, FY 2013–2033. *https://www.faa.gov/data_research/ aviation/aerospace_forecasts/media/FY2017-37_FAA_Aerospace_Forecast.pdf* (accessed March 8, 2017). Work of the U.S. Government.)

DNL, however, is a measure of average daily noise exposure from many aircraft operating in and out of a given airport. As such, there is no explicit mathematical relationship between the two metrics, although if the cumulative noise margins on all aircraft in the fleet were increased (they were made quieter) there should be an overall reduction in DNL for the same number of aircraft operations (departures and arrivals).

Since the advent of the first jetliners in the 1950s, commercial aviation has enjoyed almost unabated growth worldwide. Over the last 60 years, the number of commercial jet aircraft in service worldwide has steadily increased. Now the number is around 20,000 aircraft, but that number is expected to more than double in the next two decades (Boeing, 2012) to keep up with the anticipated increase in demand for air travel for business and leisure. Figure 1.5 shows growth forecasts by the FAA for the number of passengers traveling to and from the United States via both domestic and international commercial air carriers over a 20-year time period between the years 2017 and 2037 (FAA, 2017). If the predicted trend holds, the passenger load will increase by more than a factor of 2 in 20 years. Clearly, if no action is taken to offset the impact of the increased commercial air traffic, the community noise exposure will inevitably increase from its current levels.

In order to limit projected increases in noise exposure, more stringent limits on aircraft noise are periodically enacted. In fact, in 2013 CAEP recommended a more stringent noise standard, called Chapter 14, to begin to address the anticipated increase in noise exposure in the next two decades. The new standard is to take effect starting in 2018 for new commercial aircraft entering service. Under this standard, new aircraft will have to be 7 EPNdB quieter on a cumulative basis than the Annex 16, Chapter 4 (ICAO, 2008), noise limits. Starting in 2021, the new standard will also apply to aircraft with MTOW less than 120,000 lb (54 t) to encourage the development of quieter aircraft in smaller class sizes (ICAO, 2013).

Figure 1.6 shows the historical evolution of the commercial aircraft noise level over time in response to the increasingly more stringent community noise regulations. The aircraft noise

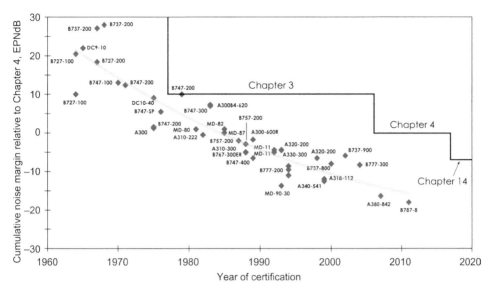

Figure 1.6. The historical trend in aircraft noise levels from 1960 onwards. Aircraft noise levels are expressed in terms of cumulative noise margins relative to ICAO Annex 16, Chapter 4, noise regulation. The evolution of the noise regulation with time is also shown. The data point labels indicate aircraft model.

data shown are by no means all inclusive, but serve to express the general trend in aircraft noise reduction that has been achieved over the last few decades.

Improvements in the design of the engine and airframe have been responsible for this trend, and much of the reduction is due to the evolution of the aircraft engine architecture. The focus on engines stems from the fact that they are the dominant source of aircraft noise. The principal breakthrough in this area came as the very noisy turbojet engines of the early jet age were replaced with much quieter turbofan engines.[10] The reason for the changeover was twofold. Turbofans have higher propulsive efficiency than turbojets and, hence, burn less fuel to generate the same level of thrust. They are also quieter by virtue of having relatively lower engine exhaust velocity compared with the turbojets. It is a feature of aerodynamic noise generation process that lower velocities generally result in lower noise levels.

Together with the improvements in the engine design over the years, a great deal of research effort has also been invested in the development of noise reduction technologies for mitigating both the engine and airframe noise sources. Significant progress has been made in this area leading to a number of improved design practices for low-noise engine components. For example, it has been shown that a judicious choice of blade counts for the fan, compressor, and turbine components has a beneficial effect by reducing tone noise generation through a mechanism called cut-off. Other design improvements include incorporating chevrons on engine nozzle trailing edge to enhance turbulent mixing of the exhaust flow to reduce jet noise (see Fig. 1.7(a)), and the use of swept outlet guide vanes to reduce fan noise (see Fig. 1.7(b)). These and other noise reduction technologies will be discussed in more detail in Chapter 4 (Part II) of this book.

It should be noted that in addition to the technical solutions such as changes to the engine architecture and the technology developments aimed at mitigating the aircraft noise levels, a great deal of research has also been conducted in the development of noise abatement procedures

---

[10] The reduction in noise is achieved by directing a portion of the ingested airflow so that it bypasses the engine core and is instead discharged through the bypass nozzle at much lower velocity than the core flow. The reduced velocity increases the propulsive efficiency and reduces noise. The turbojets have no bypass flow, so their exhaust velocity is significantly higher than turbofans of equivalent thrust.

Figure 1.7.   Two aircraft engine noise reduction technologies. (a) Chevrons on General Electric GEnx engine nozzles. (b) Swept bypass guide vanes on the 22-in. NASA-Allison fan stage.

(ICAO, 2007) that seek to reduce the impact of aviation noise on the environment independent of the certification rules. Examples include noise-abatement flight procedures like the continuous descent approach, spatial management strategies like the noise-preferred arrival and departure routes, and ground management practices like the aircraft taxi and queue management.

## 1.2   SOURCES OF AIRCRAFT NOISE

Practically speaking, it is not easy or always possible to directly measure, isolate, and character-ize the noise associated with an installed aircraft component (in flight or on the ground), given the overlapping nature of the various aircraft noise sources. However, aided by a large body of theoretical research as well as extensive component testing in rigs and wind tunnels carried out over the last 60 years, much is known about the various sources of aircraft noise and their under-lying generation mechanisms. It has been a common practice to divide aircraft noise sources into two major categories: propulsion (engine) noise sources and airframe noise sources. However, there are occasions where unintended noise sources are introduced as a result of the integration of the engines with the airframe. Examples include the interaction of the engine jet exhaust flow with the flaps and flow distortions produced by the airframe that are ingested by the engine inlet. Aircraft designers strive to optimize engine-airframe integration in order to eliminate these potential extraneous noise sources. The study of this problem is called propulsion airframe aero-acoustics (PAA). The importance of PAA is likely to increase for unconventional aircraft design of the future where the engines may be more closely integrated with the airframe than they are in the "tube and wing" designs of today. Figure 1.8 depicts two future aircraft concepts for which PAA would play a crucial role.

### 1.2.1   *Engine noise sources*

Engine noise is the aggregate contribution of many constituent sources internal and external to the engine nacelle. For a modern turbofan engine (see Fig. 1.9(a)) these components include fan noise and core noise, which are produced inside the engine, and jet exhaust noise (from both the core and bypass nozzles), which is produced external to the engine. Core noise is the collective name used for the noise generated by the compressor, combustor, and turbine components. A pictorial description of the sources of a turbofan engine is shown in Figure 1.10. As was men-tioned earlier, the relative noise levels from the various engine components depend on the engine architecture and power setting. For a modern turbofan engine operating at high power condition

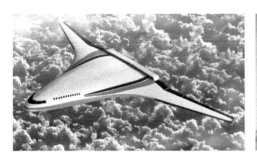

Figure 1.8.   Concept aircraft for which propulsion airframe aeroacoustics (PAA) is particularly important because of the high level of engine-airframe integration.

Figure 1.9.   Three types of aircraft engines. (a) High-bypass-ratio (HBR) turbofan. (b) Turboprop. (c) Counter-rotating open rotor (CROR). ((a) From Wikimedia Commons. 2006. Close up view of Rolls-Royce Trent 900 on Airbus A380-ILA 2006. Photographed by Kolossos. *https://commons.wikimedia.org/ wiki/File:A380-trent900.JPG*. Creative Commons CC BY-SA 3.0 license, *https://creativecommons.org/li-censes/by-sa/3.0/legalcode* (accessed April 27, 2017). (b) From Wikimedia Commons. 2004. Hercules pro-pellers. Photographed by Adrian Pingstone. *https://commons.wikimedia.org/wiki/File:Hercules.propeller. arp.jpg*. Public Domain. (accessed April 27, 2017). (c) From Domke, B. 2007. The GE36 Unducted Fan (UDF) used two contra-rotating propellers with eight blades each. *http://www.b-domke.de/AviationImages/ Rarebird/0809.html* (accessed April 28, 2017). Copyright 2007. With permission.)

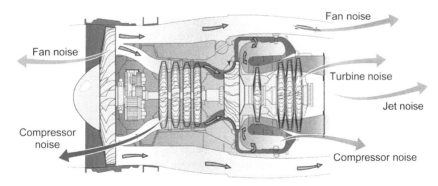

Figure 1.10.   Sources of turbofan engine noise.

(e.g., takeoff), the fan and jet noise sources are significantly louder than the other engine noise sources. At low engine power condition (e.g., approach), the fan is typically the most significant engine noise source since jet noise is considerably reduced because of the lower jet exhaust velocities at approach throttle setting. The contribution of the core noise is less than that from the fan and jet sources for current-generation turbofan engines. The forward-radiated engine noise (inlet noise) is mainly that produced by the fan, although some level of compressor noise

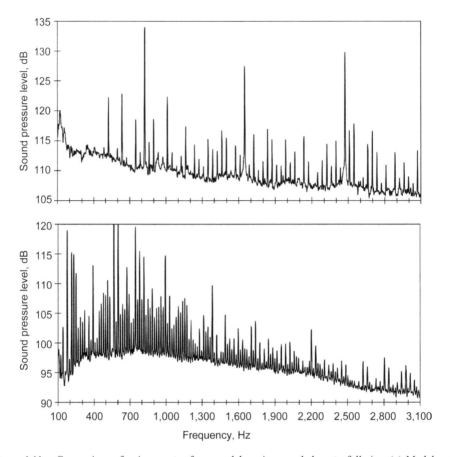

Figure 1.11.   Comparison of noise spectra from model engines, scaled up to full size. (a) Model-scale fan. (b) Counter-rotating open rotor. (Based on data from Woodward, R. P., C. E. Hughes, R. J. Jeracki, *et al.* 2002. Fan noise source diagnostic test—far-field acoustic results. AIAA 2002-2427; Elliott, D. M. 2011. Initial investigation of the acoustics of a counter rotating open rotor model with historical baseline blades in a low-speed wind tunnel. AIAA 2011-2760. Work of the U.S. Government.)

may also be discernible. The aft-radiated noise (exhaust noise) principally comprises fan and jet noise though there may also be significant combustor and turbine noise present for some engines at some engine power conditions.

Turboprop and open-rotor engines produce propeller noise and core noise. The term "propeller noise" here is used as a generic label for the noise produced by the propulsor in these engines, whether the engine has a single-rotation rotor (Fig. 1.9(b)) or two counter-rotating rotors (Fig. 1.9(c)). One qualitative difference between the fan and propeller noise is related to their tone content. Whereas the fan noise spectrum typically contains a relatively small number of strong tones (i.e., the harmonics of the blade-passing frequency and some multiple pure tone (MPT) content if the fan tip speed is supersonic), the propeller noise spectrum exhibits significantly higher tone content. In fact, the noise spectrum of a counter-rotating open rotor may contain several dozen tones generated by the aerodynamic interaction of the two rotors. This is especially true if the blade counts of the two rotors are different and/or if they are run at different rotational speeds. Figure 1.11 shows a qualitative comparison of the noise spectra from a fan and an open rotor.

The noise spectra shown are from model-scale rig tests scaled up to full scale.[11] The open-rotor spectrum has a much higher preponderance of tones even though the fan spectrum has MPT content in this case.

### 1.2.1.1   *Fan noise*

The fan is the largest component of a modern high-bypass-ratio (HBR) turbofan engine and the primary source of engine thrust. It is also a major source of engine noise produced mainly as a result of the interaction of fan (i.e., rotor) wakes with the fan exit guide vanes (i.e., stator). This so-called rotor-stator interaction noise has both tone and broadband content with the tone content comprising the harmonics of the blade-passing frequency.

Another source of fan noise is that caused by the interaction of inflow distortions with the rotor blades. This source is usually not significant for conventional inlets, since for performance reasons the inlets are designed to minimize the inflow distortion as much as possible.[12] Fan noise can also be self-generated as in the case of the noise produced by the local separation of the flow on the airfoils, or the scattering of blade boundary layer turbulence and vortex shedding at the fan blade trailing edges. Other fan noise sources are associated with the flow within the fan stage. These include the interaction of the fan with the steady potential field of the engine struts and the multiple pure tones of the fan that are produced as a result of the nonuniformities in the rotor-locked pressure field. The fan-strut interaction produces discrete tones at the harmonics of the blade-passing frequency, whereas MPT noise is produced at the harmonics of the fan shaft rotational frequency. The turbulent flow within the fan stage is potentially an important source of fan noise at high frequencies and fan aerodynamic loading conditions. Finally, it should be noted that the presence of the nacelle has an influence on the noise produced by fans. The nacelle acts as a filter in that it effectively cuts off a portion of the acoustic field. The nacelle also provides a location for adding acoustic treatment (liners) to absorb a portion of the sound generated inside the engine. Liners are commonly used in turbofan engine inlets to mitigate fan noise.

The strengths of the various fan noise sources depend on the fan operating condition. As an example, the MPT noise is only significant when the fan tip speed is supersonic. At subsonic tip speeds, the rotor-locked pressure field is evanescent and contributes very little to the noise signature of the fan. However, this particular source may be important at subsonic speeds for short inlets where the field may not have decayed sufficiently before it is released at the inlet. As the fan tip speed and aerodynamic loading vary across the operating regime of the engine, the proportion of the contributions from various fan noise sources changes. Measurements of fan noise from scale-model rigs and full-scale engines have provided indications of the relative importance of some of the fan noise sources as a function of the fan tip speed, but because of the overlapping nature of the sources, it is not always possible to clearly differentiate between them.

### 1.2.1.2   *Propeller noise*

Turboprop and open-rotor engines do not benefit from the shielding effect of a nacelle. As a result, their noise spectra have higher tone content (see Fig. 1.11(b)). There are three principal mechanisms that contribute to the noise of a propeller. These are commonly referred to as the thickness noise, loading noise, and quadrupole noise. Thickness noise is generated by the volume displacement effect as the blade cuts through the air. It is strictly tonal in nature and is a weak noise source for the thin blades typically used in modern turboprops and open rotors. Loading noise is produced by the unsteady aerodynamic forces on the blades and exhibits both tonal and

---

[11] It should be noted that although these model spectra have been scaled up to the same geometric size, the thrust levels produced by the fan and open rotor are different in this case. As such, these levels cannot not be compared on an equal footing. This comparison is only meant to illustrate the difference in the spectral characteristics of the fan and open rotor noise.

[12] With the advent of ultra-high-bypass-ratio (UHBR) turbofans, which have a larger fan diameter than a high-bypass-ratio turbofan, there is need to reduce the length of the nacelle to offset the weight increase of a larger diameter nacelle. As such, inlet distortion noise may not be a negligible source for UHBR fans.

broadband content. It is a strong tone source comprising the harmonics of the blade-passing frequency of the rotor(s) and, in the case of counter-rotating open rotors, also the interaction tones associated with the sum and difference of the harmonics of the blade-passing frequencies of the front and aft rotors. The quadrupole noise source is related to the fluctuating shear stresses in the volume of fluid surrounding the rotor. It is usually of secondary importance compared to the loading noise at low flight speeds typical of takeoff and landing, but it is an important source at speeds typical of the cruise condition.

### 1.2.1.3   *Core noise*

Similar to fan noise, compressor noise and turbine noise are principally produced by the rotor-stator interaction mechanism. Both sources exhibit tonal and broadband characteristics, although typically compressor and turbine tone frequencies are much higher than the fan tones because of their higher rotational speeds and significantly higher blade counts. Unlike fan tone noise, the compressor noise and turbine tone noise are generated not only at the blade-passing frequencies of the respective stages, but also at the combination frequencies associated with the aerodynamic and acoustic interaction of adjacent stages. However, in the engine far-field noise spectra, typically only the blade-passing tones from individual compressor stages are discernible. In contrast, the blade-passing tones from multiple stages of the turbine as well as the associated combination tones are detectable in the far-field engine noise spectra. Another difference between compressor noise and turbine noise is that the turbine tones are refracted when they pass through the exhaust shear layer and exhibit haystacking (frequency broadening) in the far field. Also, because of the acoustic blockage effect of the adjacent stages (called acoustic transmission loss), the engine noise spectra typically do not exhibit tone noise from all compressor or turbine stages. Usually, only blade-passing tones (and their first harmonics) from the low-pressure compressor and turbine stages as well as some difference tones from the low-pressure turbine stages are observed in the far field noise spectra. The compressor and turbine broadband noise content are also present in the engine spectra, but they do not significantly contribute to the overall engine noise levels.

Combustion noise is the byproduct of the burning of the air-fuel mixture in the engine combustor. It is produced by two distinct mechanisms generally referred to as 'direct' and 'indirect' combustion noise. Direct combustion noise is the name given to the pressure fluctuations that arise within the combustor as a result of the fluctuating heat release within the flame zone. Indirect combustion noise is produced as a result of the passage of hot streaks (i.e., temperature nonuniformities inside the combustor) through the flow gradients present at the entrance to the first high-pressure turbine stage. The resulting interaction converts some of the energy in the hot streaks into sound. Since both combustion noise sources are produced around the same nominal fundamental frequency (typically around 400 Hz irrespective of the engine size), it has proven difficult to ascertain their relative contribution to the total combustion noise level or clearly distinguish their contribution from jet noise that can occur at that frequency range. Core noise is generally broadband in character, although for lean-burning combustors there is the possibility that combustion instabilities may occur at low power condition, leading to tone noise generation.

### 1.2.1.4   *Jet noise*

Unlike fan and core noise, jet noise is principally produced external to the engine by flow processes within the engine exhaust plume.[13] The main source of jet noise produced by modern commercial aircraft engines is the turbulent mixing between the high-speed exhaust stream and the ambient atmosphere. Mixing noise is broadband in nature, and its characteristics depend on the jet velocity. When the velocity is lower than the local speed of sound, the fine-scale turbulence produced as a result of the mixing process is the source of jet noise. The resulting noise mainly radiates towards the broadside and upstream angles. When the jet velocity exceeds the local speed of sound, large-scale turbulent structures produced by the jet also become efficient

---

[13] For engines with internal forced mixers, some high-frequency jet noise is generated inside the engine.

Figure 1.12.   Segmented wing flap system on an aircraft wing. (From Wikimedia Commons. 2010. The escape rope bracket (yellow) on an Easyjet Airbus A319-100 (G-EZAV) during a flight to Bristol, England from Palma Airport, Majorca, Spain. Photographed by Adrian Pingstone. *https://commons.wikimedia.org/ wiki/File:Easyjet_a319_wing_g-ezav_arp.jpg.* Public Domain. (accessed April 27, 2017).)

radiators of sound. The mechanism by which the large-scale structures radiate sound is called the Mach wave emission. Large-scale mixing noise is more directional than the fine-scale noise and is mainly radiated towards the downstream angles. The presence of any solid surfaces near the jet plume could significantly increase jet noise, so the integration of the engine with the airframe and wing is crucial for avoiding such potential extraneous noise problems.[14]

### 1.2.2   *Airframe noise sources*

In the case of engine noise, power generation processes inside the engine, namely, compression, combustion and expansion, are mainly responsible for generation of pressure fluctuations in the airstream. In contrast, airframe noise is nonpropulsive and "self-generated" through the creation of unsteady flow and turbulence when the airstream interacts with the landing gear and the high-lift devices (i.e., trailing edge flaps and leading edge slats). Even when the landing gear and high-lift devices are retracted (the so-called clean configuration), the airflow over the fuselage and wing generates noise. However, airframe noise level increases substantially (by as much as 10 dB or more) when the landing gear and/or high-lift devices are deployed. The noise produced by the airframe components is mostly broadband, although some tones could also be generated by vortex shedding from sharp edges or from flow over cavities like fuel and anti-ice vents on the airframe. The relative importance of various airframe noise sources depends on the type of the aircraft.

#### 1.2.2.1   *Trailing edge noise*

The broadband noise generated by the interaction of unsteady flow with the wing trailing edge, or more precisely, with the wing flap trailing edge (see Fig. 1.12), represents the minimum noise

---

[14]When the airplane is at cruise, another mechanism of jet noise is present: the so-called shock-associated noise. Shock noise is caused by the interaction of jet plume turbulence with the shock system that is created in the nonideally expanded jet plume. Shock-associated noise is broadband in nature, though at certain conditions tones (called screech) could also be generated under rare circumstances. Shock-associated noise for commercial jets is a concern for cabin noise and does not contribute to the community noise.

Figure 1.13.    A large aircraft's main landing gear. (From Wikimedia Commons. 2008. Boeing-777-300 chassis. Photographed by Dmitry A. Mottl. *https://commons.wikimedia.org/wiki/File:Boeing-777-300_ chassis_.jpg*. Creative Commons CC BY-SA 3.0 license, *https://creativecommons.org/licenses/by-sa/3.0/ legalcode.* (accessed April 27, 2017).

for an aircraft in flight. Like the fan blade trailing edge noise, flap trailing edge noise is produced as result of the scattering of some of the energy of the wing turbulent boundary layer into sound. However, unlike the rigid rotor blade, the relatively flexible surface of a wing flap could potentially also contribute a flow-structure interaction component to the noise-generation process. The trailing edge noise (i.e., airframe in clean configuration) has a relatively uniform directivity.

### 1.2.2.2    *Landing gear noise*

A commercial aircraft landing gear mechanism typically has many constituent components few of which are streamlined (see Fig. 1.13). Consequently, when the landing gear is extended into the airstream, a significant amount of turbulence is created by the interaction of the airflow with these components. The resulting aerodynamic interaction generates mostly broadband noise, though exposed cavities in the landing gear mechanism may also result in the generation of some tones. Some of the broadband noise is generated as a result of turbulent flow separation from the large components like the main strut (oleo) or the wheel trucks. Noise is also generated as a result of the interaction of the turbulent wakes from upstream components like the oleo with the downstream components like the retraction elements. Landing gear noise is more or less omni-directional, radiating uniformly at all angles. For small aircraft (i.e., single aisles and regionals) landing gear noise level is on par with the noise level generated by the high-lift devices, but for large aircraft (i.e., wide bodies), landing gear is the most significant source of airframe noise.

### 1.2.2.3    *Slat noise*

Slats are high-lift devices installed on the wing leading edge to enhance the wing's lift by allowing it to operate at higher angles of attack at low flight speeds. Slats are deployed during landing to allow the aircraft to operate close to its stall speed. As shown in Figure 1.14, when slats are deployed a gap is generated between the slats and the main wing, which allows a highly unsteady flow to be established inside the slat cavity called the slat cove. A main feature of that flow is the cove vortex whose impingement on the leeward face of the cove is thought to be a major source of slat noise. Added to this source is the vortex shedding that occurs at the slat trailing edge and,

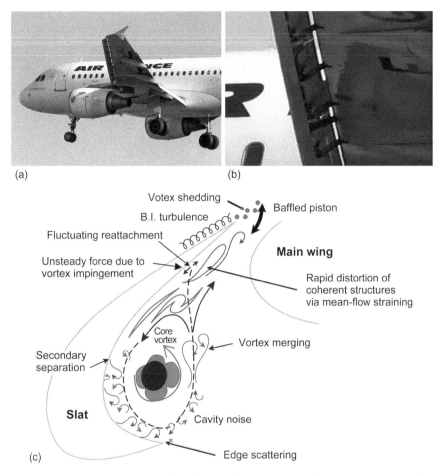

Figure 1.14.   Slat noise sources. (a) Leading edge slats. (b) Slat close up. (c) Noise generation mechanisms. ((a) and (b) From Wikimedia Commons. 2006. Air France Airbus A318-100 (F-GUGJ) lands at London Heathrow Airport, England. Photographed by Adrian Pingstone. Public Domain. *https://commons. wikimedia.org/wiki/File:Airfrance.a318-100.f-gugj.arp.jpg.* (accessed April 27, 2017). (c) From Choudhari, M. M., and M. R. Khorrami. 2006. Slat cove unsteadiness: effect of 3D flow structures. AIAA 2006−0211. Work of the U.S. Government.)

potentially, the unsteady flow interactions with the main wing's leading edge. Slat noise is broad-band but more directional than the landing gear noise with its dominant radiation in a quadrant directly below and just to the aft of the aircraft.

### 1.2.2.4   *Flap (side-edge) noise*

Like slats, flaps are high-lift devices that are installed on the trailing edge of the wing to improve its lift characteristics at the reduced speeds typical of takeoff and landing. When deployed, flaps also increase the drag of the aircraft, allowing it to slow down more rapidly during landing. In addition to the ever-present noise produced by the flap trailing edge, there is also another source of noise generation associated with the flaps when they are deployed. This source is called the flap side-edge noise, or more commonly, flap noise. As can be seen in Figure 1.15, the deployed flap presents a sharp side edge to the flow that is absent when the flap is retracted. When the flap side edge is exposed to the flow, the pressure difference between the suction and pressure sides of the flap results in a crossflow normal to the flap surface. This sets up a complex vortex flow (Streett, 1998) whose interaction with the flap's upper surface is the cause of the flap side-edge

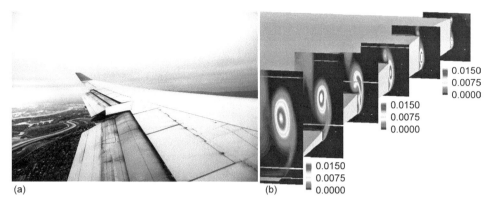

Figure 1.15.   Sources of flap noise. (a) Deployed trailing edge flaps and flap side-edge vortex. (b) Streamwise vorticity contours. ((a) From Wikimedia Commons. 2013a. Angelo DeSantis. United 747-400 leaving a cloudy NRT bound for SFO. Photographed by Angelo DeSantis. *https://commons.wikimedia.org/wiki/File:Massive_747-400_wing_with_flaps_deployed_%287747239584%29_%282%29.jpg.* Creative Commons CC BY 2.0 license, *https://creativecommons.org/licenses/by/2.0/legalcode* (accessed April 27, 2017). (b) From Streett, C.L. 1998. Numerical simulation of fluctuations leading to noise in a flap-edge flowfield. AIAA–98–0628. Work of the U.S. Government.)

Figure 1.16. Wing spoilers (airbrakes) shown deployed on landing. (From Wikimedia Commons. 2013b. A Qantas Boeing 737-800 enroute from AKL-MEL deploys it's spoilers to slow down and prepare for descent into Melbourne. Photographed by Jg4817. *https://commons.wikimedia.org/wiki/File:Qantas_Boeing_737-800_spoiler_deployed_for_descent.jpg.* Creative Commons CC BY-SA 3.0 license, *https://creativecommons.org/licenses/by-sa/3.0/* (accessed April 27, 2017).)

noise. It has also been suggested that the turbulence in the flap side-edge vortex may also be another source of flap noise. Flap noise is generally broadband in nature with a complicated directivity pattern (Brooks and Humphreys, 2000).

### 1.2.2.5   *Other noise sources*

Airbrakes or wing spoilers (shown in Fig. 1.16) can also be a source of airframe noise. The source mechanisms are similar to the flap noise, though for typical landing approach operation they are only sporadically deployed and hence do not constitute a major airframe noise source. However, in circumstances for which they would be used more often on approach, (e.g., a continuous descent approach, they could become a significant source. As for other sources, depending on the type of the aircraft, there is also the potential for additional noise generated as a result of the aerodynamic interaction between the turbulent wake of the main landing gear and the flaps or even the engine exhaust plume with the flaps. These may be of secondary importance, but they cannot always be ignored.

Figure 1.17 shows visual representations of the various airframe noise sources for a Gulfstream G550 regional jet.[15] Each photograph in the figure shows the aircraft with one or more of the airframe components deployed. The color contours below each photograph show the associated acoustic source localization map for that component. The maps were created using a microphone phased-array system deployed on the ground over which the aircraft flew a number of times in the course of a joint NASA-Gulfstream test in 2006 (Khorrami *et al.*, 2008). The maps show the locations and the spatial extents of the various airframe noise sources. Note that the magnitude of one noise source relative to another source cannot be inferred from these maps, since the noise levels in each map are referenced to the peak noise level in that particular map. Even so, the maps provide a clear and useful visual description of the airframe noise sources.

## 1.3   AIRCRAFT COMPONENT NOISE LEVELS (EXAMPLE)

In order to provide a more quantitative comparison of the relative importance of the various aircraft noise sources, the predicted component noise levels for the Boeing 737-800 aircraft equipped with two CFM56-7B27 turbofan engines are plotted in Figure 1.18. These predictions, given in terms of the component effective perceived noise level, were computed using empirically based aircraft component noise models.[16] As was discussed earlier, for modern high-by-pass-ratio turbofan engines like the CFM56, fan (i.e., fan inlet and fan exhaust) and jet are generally the dominant sources of engine noise at all power conditions (see Fig. 1.18(a)). For this engine, core noise is between 5 to 14 EPNdB below either the fan or jet level except at the approach condition where it is nearly on par with the level of jet noise, since jet noise is significantly diminished at this relatively low-engine-power condition. As the engine architecture evolves, the relative balance between the engine sources will likely change. For example, for the ultra-high-bypass-ratio turbofan engines, which potentially offer additional fuel burn and noise reduction advantage over HBR engines, the hierarchy of the source levels is predicted to be the fan, followed by the core, followed by the far less significant jet (Berton *et al.*, 2009). As for the airframe noise sources, they are only significant at the approach operating condition where the total airframe noise is about 6 EPNdB below the total engine noise level (see the bottom portion of Fig. 1.18). At the lateral and flyover conditions, when the airframe is in the clean (or nearly clean) configuration, engines are by far the dominant sources of aircraft noise, exceeding the airframe total noise levels by 14 to 20 EPNdB.

A comparison of the predicted noise levels and those certified by the Federal Aviation Administration for the Boeing 737-800 aircraft with CFM56-7B27 engines are shown in Table 1.1. The total noise level is the aggregate of all engine and airframe component noise source contributions and is the level that is calculated, according to the (Title 14, Code of Federal Regulations) Part 36 procedures, from the predicted or measured data. The first line in Table 1.1 shows the predicted aircraft noise level at each certification point together with the cumulative

---

[15] Like many regional and single-aisle aircraft, the Gulfstream 550 does not have leading edge slats.

[16] Predicted levels were computed using the NASA Aircraft Noise Prediction Program (ANOPP). The 737-800 aircraft was assumed to have a MTOW of 174,200 lb and the CFM56-7B27 engines to have a thrust rating of 27,300 lb each. As was noted earlier, direct measurement of the level and spectral content of noise generated by an installed individual aircraft component is quite difficult in practice, so component-based noise estimates are usually calculated from empirical models for the purpose of source separation and/or for estimating the relative source contributions to the aircraft EPNL.

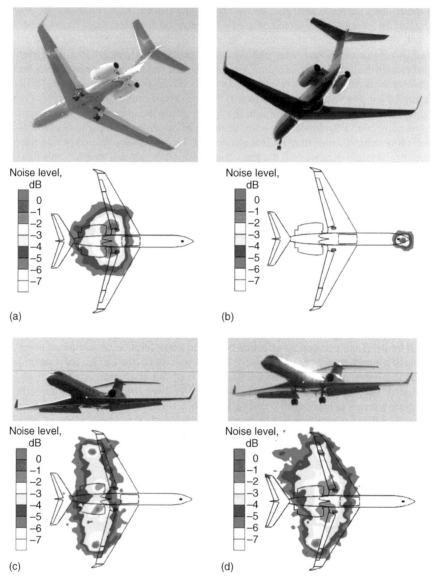

Figure 1.17.   Phased-array source localization of the important airframe noise sources for a Gulfstream G550 aircraft at a frequency of ~1.4 kHz. Noise levels are normalized by the peak level in each map. (a) Main gear noise. (b) Nose gear noise. (c) Flap noise. (d) Main gear plus flap noise. (From Khorrami, M. R., D. P. Lockard, W. M. Humphreys, Jr., *et al*. 2008. Preliminary analysis of acoustic measurements from the NASA-Gulfstream airframe noise flight test. AIAA 2008–2814. Work of the U.S. Government.)

noise level. The uncertainty bands in that row indicate the level of variation in the prediction based on a reasonable statistical scatter in input parameters used in the noise predictions (Berton and Burley, 2014). The second line of the table shows the certification data for that aircraft (FAA, 2001). The data and the corresponding uncertainty bands are based on the average of eight published certification noise levels for this airframe/engine combination that exactly match the maximum takeoff weight and engine thrust levels assumed in the Berton and Burley study. When

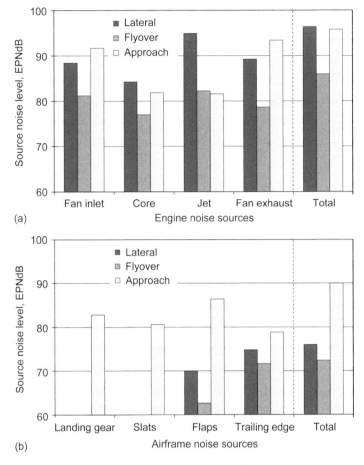

Figure 1.18.   Predicted component noise levels at the three certification points for Boeing 737-800 with CFM56-7B27 turbofan engines. (a) Engine component noise levels. (b) Airframe component noise levels. Note that the landing gear and high-lift devices are retracted/stowed at the flyover and lateral conditions and hence do not contribute to airframe noise. (From Berton, J. J., and C. L. Burley. 2014. System noise prediction of a 737 using NASA tools. Unpublished results.)

the uncertainties in both the predictions and measurements are taken into account, the absolute cumulative level comparisons between the predicted and measured noise levels show an excellent agreement. The third line in the table shows the Stage 3 noise limits for this aircraft. There are no corresponding lateral, flyover or approach noise limits defined under Stage 4 regulations. Instead, Stage 4 is defined as being 10 EPNdB lower than Stage 3 on the cumulative basis. So, the cumulative noise margin with respect to the Stage 4 limit can only be calculated from the Stage 3 limits as follows. First the noise margin at each certification point is computed by subtracting the Stage 3 limit from the corresponding predicted level and the three resulting margins are summed to calculate the cumulative margin relative to Stage 3 noise limits (i.e., the fourth line). Then, 10 EPNdB is subtracted from that number to provide the cumulative noise margin relative to Stage 4 (the fifth line). The result is −0.8±2.2 EPNdB. The same procedure is carried out for the FAA certification data as shown on the sixth and seventh lines in the table resulting in a cumulative noise margin of −2.4±2.2. With the uncertainties included, there is good agreement between the predicted and certified cumulative margins. This and other comparable examples provide confidence in the analytical procedures that are used to compute the component and

Table 1.1.   Predicted and certified noise margins for Boeing 737-800 with CFM56-7B27 engines.

|  | Lateral | Flyover | Approach | Cumulative |
|---|---|---|---|---|
| Predictions (absolute levels) | 96.4±1.1 | 86.1±1.0 | 96.3±0.8 | 278.8±2.2 |
| Federal Aviation Administration (FAA) Certification data (absolute levels) | 94.6±0.6 | 86.0±0.7 | 96.6±0.2 | 277.1±1.0 |
| Stage 3 noise limits | 97.0 | 91.9 | 100.7 | 289.6 |
| Stage 3 margins (predicted) | −0.6±1.1 | −5.8±1.0 | −4.4±0.8 | −10.8±2.2 |
| Stage 4 margins (predicted) | – | – | – | −0.8±2.2 |
| Stage 3 margins (certified) | −2.4±1.1 | −5.9±1.0 | −4.1±0.8 | −12.4±2.2 |
| Stage 4 margins (certified) | – | – | – | −2.4±2.2 |

*Source for predicted noise levels*: Berton, J. J., and C. L. Burley. 2014. System noise prediction of a 737 using NASA tools. Unpublished results.
*Source for noise certification data*: Federal Aviation Administration. 2001. Noise levels for U.S. certified and foreign aircraft. Advisory Circular 361H. Work of the U.S. Government.


system noise levels estimates. It should be noted that empirical models used in this analysis are tailored to conventional "tube and wing" aircraft that are in service today. The reliability of these tools in predicting component- or system-level noise from unconventional aircraft configurations such as those shown in Figure 1.8 is questionable.


1.4   SUMMARY

The commercial aviation enterprise has inevitable consequences for the environment. The most immediately obvious of these consequences is the noise experienced by the communities living around the airports. Data suggest that aircraft noise continues to be the number one cause of the adverse community reaction to the expansion of airports and the associated increases in the air traffic. To limit the growth of the noise exposure problem, by international agreements, limits are placed on the level of noise that aircraft can generate when operating in and out of airports. To keep up with the growth of the commercial fleet, these limits are periodically revised downward by the International Civil Aviation Organization (ICAO) based on technology forecasts and economic feasibility. Once approved by the ICAO member countries, these new limits are adopted for implementation by the respective national regulatory agencies of the member countries. In the United States, the Federal Aviation Administration (FAA) is the responsible government agency for enforcing the aircraft noise limits. In addition, through other measures, like land-use management and changes to the operational procedures, the FAA seeks to reduce, over time, the number of people in the United States who are exposed to objectionable aircraft noise levels.

The sources of aircraft noise are many, and almost all are associated with the pressure disturbances that are generated as a result of the interaction of the airflow with the various engine and airframe components during flight. Much of the progress made in the last six decades in reducing the aviation noise has been the direct result of technological breakthroughs in designing quieter aircraft for which the intensity of the dominant aircraft noise sources has been mitigated. Nowhere this is more evident than the progress made in designing quieter engines. However, as the aircraft noise limits become even more stringent with the passage of time, it is likely that the future breakthroughs in aircraft noise reduction will also have to come from radically different airframe designs for which much of the engine noise is shielded or contained by the airframe.

Ultimately, the solution to the aviation noise problem is to contain all objectionable noise emissions from commercial aircraft within the boundary of the airport so that the communities around the airport are no longer exposed to high levels of aircraft noise. This is a challenging goal and would likely require a combination of technological, operational, and land-use planning solutions.

If additional information is needed about the subject of aircraft noise, there is a wealth of information on the subject that can be consulted online, in books, and in numerous papers and reports. For further reading about the sources of aircraft noise and their underlying generation mechanisms, the following three references on the subject are recommended as the starting point:

1. The book by Smith, M. J. T. (1989) on aircraft noise is an excellent technical resource on all aspects of aircraft noise.
2. Volume 1 of a two-volume NASA Reference Publication edited by H. Hubbard (1991) on the sources of aircraft noise provides a more in-depth discussion of the underlying mechanism of noise generation mechanisms and processes.
3. The extensive review article by W. Dobrzynski (2010) is an informative and up-to-date resource on the airframe noise sources and mechanisms.

All three references have extensive reference lists for additional material.

More detailed information about aircraft noise limits, the rulemaking process, and the commercial aircraft noise certification levels may be found on the ICAO, FAA, and the European Aviation Safety Agency Web sites:

http://www.icao.int/environmental-protection/Pages/noise.aspx,
http://www.faa.gov/regulations_policies/advisory_circulars/index.cfm/go/document.information/
    documentID/22942,
www.easa.europa.eu/certification/type-certificates/noise.php.

# REFERENCES

Airbus. 2008. Airbus A380 during landing approach. France: Toulouse.
Albee, W., T. Connor, R. Bassarab, *et al.* 2006. What's in your DNL? El Segundo, CA: Wyle Laboratories.
Berton, J. J., and C. L. Burley. 2014. System noise prediction of a 737 using NASA tools. Unpublished results.
Berton, J. J., E. Envia, and C. L. Burley. 2009. An analytical assessment of NASA's N+1 subsonic fixed wing project noise goal. AIAA 2009−3144.
Boeing. 2012. Current market outlook 2013–2032. *http://www.boeing.com/boeing/commercial/cmo/* (accessed March 8, 2017).
Brooks, T. F., and W. M. Humphreys, Jr. 2000. Flap edge aeroacoustics measurements and predictions. AIAA 2000−1975.
Choudhari, M. M., and M. R. Khorrami. 2006. Slat cove unsteadiness: effect of 3D flow structures. AIAA 2006−0211.
Cleveland Hopkins International Airport. 2016. Noise compatibility report. *http://www.clevelandairport. com/company/environment/noise-compatibility-reports* (accessed March 8, 2017).
Dobrzynski, W. 2010. Almost 40 years of airframe noise research: what did we achieve? *AIAA Journal of Aircraft* 47:353−367.
Domke, B. 2007. The GE36 Unducted Fan (UDF) used two contra-rotating propellers with eight blades each. *http://www.b-domke.de/AviationImages/Rarebird/0809.html* (accessed April 28, 2017). Copyright 2007.
Eagan, M. E., and R. Gardner. 2009. Compilation of noise programs in areas outside DNL 65—a synthesis of airport practice. In *Airport Cooperative Research Program (ACRP) synthesis 16*, Washington, DC: Transportation Research Board. *http://onlinepubs.trb.org/onlinepubs/acrp/acrp_syn_016.pdf* (accessed March 8, 2017).
Elliott, D. M. 2011. Initial investigation of the acoustics of a counter rotating open rotor model with historical baseline blades in a low-speed wind tunnel. AIAA 2011–2760.
Federal Aviation Administration. 1976. Federal Aviation Administration aviation noise abatement policy. *http://airportnoiselaw.org/faanap-1.html* (accessed March 8, 2017).
Federal Aviation Administration. 2001. Noise levels for U.S. certified and foreign aircraft. Advisory Circular 361H.
Federal Aviation Administration. 2012. Destination 2025 Next Level of Safety. In *Portfolio of goals—FAA performance metrics FY 2012*. Washington, DC: Federal Aviation Administration. *http://www.faa.gov/ about/plans_reports/media/FY12_POG.pdf* (accessed March 8, 2017).

Federal Aviation Administration. 2014. Part 36—noise standards: aircraft type and airworthiness certification. In *Electronic Code of Federal Regulations*. Washington, DC: Federal Aviation Administration. *http://www.ecfr.gov/cgi-bin/text-idx?c=ecfr&rgn=div5&view=text&node=14:1.0.1.3.19&idno=14* (accessed March 8, 2017).

Federal Aviation Administration. 2017. FAA aerospace forecast, FY 2013–2033. *https://www.faa.gov/data_research/aviation/aerospace_forecasts/media/FY2017-37_FAA_Aerospace_Forecast.pdf* (accessed March 8, 2017).

Fidell, S. 2003. The Schultz curve 25 years later: a research perspective. *J. Acoust. Soc. Am.* 114:3007.

Finegold, L. S., C. S. Harris, and H. E. von Gierke. 1994. Community annoyance and sleep disturbance: updated criteria for assessing the impacts of general transportation noise on people. *Noise Control Eng. J.* 42:25–30.

Hegland, J., W. Dickerson, and R. Danforth. 1992. Federal agency review of selected airport noise analysis issues. Washington, DC: Federal Interagency Committee on Noise.

Hubbard, H. H. 1991. Aeroacoustics of flight vehicles: theory and practice. NASA Reference Publication 1258, Vol. 1, WRDC Technical Report 90–3052.

International Civil Aviation Organization. 2007. Review of noise abatement procedure research & development and implementation results—discussion of survey results. *http://www.icao.int/environmental-protection/Documents/ReviewNADRD.pdf* (accessed March 8, 2017).

International Civil Aviation Organization. 2008. Annex 16 to the convention on international civil aviation. In *Environmental protection*, Vol. II, Third ed., Quebec: International Civil Aviation Organization. *https://law.resource.org/pub/us/cfr/ibr/004/icao.annex.16.v2.2008.pdf* (accessed March 8, 2017).

International Civil Aviation Organization. 2013. Environment—2013 environmental report. Chapter 2, 58–60. *http://cfapp.icao.int/Environmental-Report-2013/* (accessed March 8, 2017).

International Civil Aviation Organization Council. 2014. Committee on Aviation Environmental Protection (CAEP). *http://www.icao.int/environmental-protection/pages/CAEP.aspx* (accessed March 8, 2017).

Khorrami, M. R., D. P. Lockard, W. M. Humphreys, Jr., *et al.* 2008. Preliminary analysis of acoustic measurements from the NASA-Gulfstream airframe noise flight test. AIAA 2008–2814.

Pickard, L. 2008. NextGen environmental goals and targets. Washington, DC: Federal Aviation Administration ACI–NA Environmental Committee. *https://aci-na.org/static/entransit/sunday_faa_nextgen_pickard.pdf* (accessed March 8, 2017).

Smith, M. J. T. 1989. *Aircraft Noise*. Cambridge: Cambridge University Press.

Streett, C. L. 1998. Numerical simulation of fluctuations leading to noise in a flap-edge flowfield. AIAA–98–0628.

Wikimedia Commons. 2004. Hercules propellers. *https://commons.wikimedia.org/wiki/File:Hercules.propeller.arp.jpg* (accessed April 27, 2017).

Wikimedia Commons. 2006. Air France Airbus A318-100 (F-GUGJ) lands at London Heathrow Airport, England. *https://commons.wikimedia.org/wiki/File:Airfrance.a318-100.f-gugj.arp.jpg* (accessed April 27, 2017).

Wikimedia Commons. 2006. Rolls-Royce Trent 900 on Airbus A380-ILA 2006. *https://commons.wikimedia.org/wiki/File:A380-trent900.JPG* (accessed April 27, 2017).

Wikimedia Commons. 2008. Boeing-777-300 chassis. *https://commons.wikimedia.org/wiki/File:Boeing-777-300_chassis_.jpg* (accessed April 27, 2017).

Wikimedia Commons. 2010. The escape rope bracket (yellow) on an Easyjet Airbus A319-100 (G-EZAV) during a flight to Bristol, England from Palma Airport, Majorca, Spain. *https://commons.wikimedia.org/wiki/File:Easyjet_a319_wing_g-ezav_arp.jpg* (accessed April 27, 2017).

Wikimedia Commons. 2013a. United 747-400 leaving a cloudy NRT bound for SFO. *https://commons.wikimedia.org/wiki/File:Massive_747-400_wing_with_flaps_deployed_%287747239584%29_%282%29.jpg* (accessed April 27, 2017).

Wikimedia Commons. 2013b. A Qantas Boeing 737-800 enroute from AKL-MEL deploys it's spoilers to slow down and prepare for descent into Melbourne. *https://commons.wikimedia.org/wiki/File:Qantas_Boeing_737-800_spoiler_deployed_for_descent.jpg.* (accessed April 27, 2017).

Woodward, R. P., C. E. Hughes, R. J. Jeracki, *et al.* 2002. Fan noise source diagnostic test—far-field acoustic results. AIAA 2002–2427.

# CHAPTER 2

## Aircraft emissions: gaseous and particulate

Changlie Wey and Chi-Ming Lee

### 2.1 INTRODUCTION

Aircraft emit gases and particles directly into the atmosphere, and these become unique pollutant sources at cruise altitude, the upper troposphere, and the lower stratosphere. The gaseous emissions of aircraft engines contain mostly greenhouse gases, such as carbon dioxide ($CO_2$), and water ($H_2O$) vapor. They also contain species that occur in smaller amounts but have particular health and environmental concerns. These species include nitric oxide (NO) and nitrogen dioxide ($NO_2$) (together termed nitrogen oxides, $NO_x$), sulfur oxides ($SO_x$), carbon monoxide (CO), and unburned hydrocarbons (UHCs) of many species lumped together. Aircraft also emit soot particles and some gaseous species that may act as precursors of fine particles. Many of these emissions affect human health when they are inhaled at the ground level. They also impact the air quality near airports; for example, $NO_x$ contributes to photochemical smog.

Commercial aircraft cruise at altitudes between 8 and 13 km (26,250 and 42,650 ft) (upper troposphere and lower stratosphere), with a tendency over time for higher average cruise altitudes. These gaseous and particulate emissions at high altitude change the concentration of atmospheric greenhouse gases, including $CO_2$, ozone ($O_3$), and methane ($CH_4$). They also trigger the formation of contrails and may increase cirrus cloudiness, which contributes to climate change. One critical challenge in comparing the impacts of these emissions on climate is that these pollutants have very different residence or reaction times. For example, emitted $CO_2$ has a residence time of many decades; hence, it makes little difference where the $CO_2$ was initially emitted. In contrast, $NO_x$ is relatively reactive and short-lived; therefore, the altitude at which it is emitted is more relevant to its radiative forcing effect on the formation of ozone, considering the sensitivity of the atmosphere at these flight altitudes. The impact of aircraft $NO_x$ emissions on global climate change is therefore considered to be potentially more significant than a simple proportion of all anthropogenic combustion sources would suggest (Committee on Aviation Environmental Protection (CAEP), 2006). Figure 2.1 illustrates the atmospheric impacts of aircraft exhaust emissions.

Currently, gaseous emissions are regulated in a prorated fashion below an altitude of 0.914 km (3000 ft), which covers the takeoff, climb, descent, taxiing, and ground idle phases of engine operation—the so-called landing and takeoff cycle. We anticipate that gaseous emissions at cruise conditions will be regulated in the near future. Originally, particulate emissions were defined simply as "smoke," and restrictions focused primarily on larger particles. Recent scientific findings indicate that fine particles may also have a significant adverse impact on the global climate and an even more adverse effect on human health.

The study of aircraft particulate emissions began about a decade ago. We anticipate that more refined regulations on particulate emissions will be established in the foreseeable future. Technology trends are periodically reviewed so that reasonable recommendations on achievable emission limits can be provided to regulators.

### 2.2 GASEOUS EMISSIONS

Aeroengines burn kerosene-type HC fuels. These middle distillates contain about a thousand species, but most are some form of normal and branched alkanes, alkenes, cycloparaffins, and aromatics.

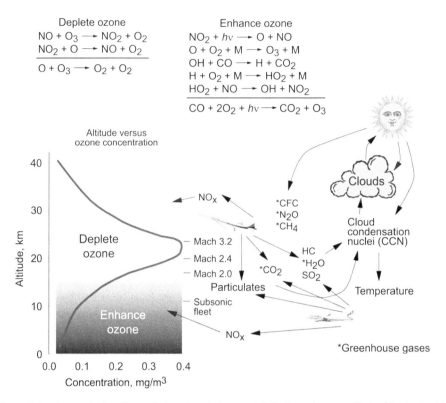

Figure 2.1. Atmospheric effects of aircraft emissions and their dependency on flight altitude. $hv$, Planck constant times frequency of electromagnetic wave. CFC, chlorinated fluorocarbon. HC, hydrocarbon. $HO_2$, hydroperoxyl. $N_2O$, nitrous oxide.

Under ideal conditions, the combustion of pure kerosene-type fuels produces only $CO_2$ and water vapor ($H_2O$) in proportions that depend on the specific fuel-carbon-to-hydrogen ratio. In general, the average chemical form of kerosene-based jet fuel can be expressed as $C_{12}H_{23}$. The perfect reaction with air, in which all reactants are used up, can then be expressed as

$$C_{12}H_{23} + 17.75(O_2 + 3.76N_2) \rightarrow 12CO_2 + 5.75H_2O + 66.74N_2 \qquad (2.1)$$

This perfect fuel-air mixture condition is referred to as the "stoichiometric" condition. At this condition, the mass ratio of fuel to air is about 0.068. The ratio of any fuel-air mixture to this perfect condition is referred to as the equivalence ratio. If the equivalence ratio value is greater than 1, there is excess fuel; this is referred to as burning fuel rich, or simply "rich." If the equivalence ratio is less than 1, oxygen is left—a mixture referred to as "lean."

In reality, however, aircraft emissions are more complicated for a number of reasons, including the following: (1) jet fuel typically contains impurities such as sulfur, which produces $SO_x$; (2) incomplete combustion results in residues such as CO, unburned hydrocarbons, and carbons (soot); and (3) high temperatures oxidize the nitrogen in the air to create $NO_x$. Figure 2.2 shows the ideal and real combustion processes. It also illustrates the scale of the combustion products by showing that, at cruise conditions, they constitute only about 8.5% of total mass flow emerging from the engine. Of these residual combustion products, only a very small volume (about 0.4 v/v%) arise from nonideal combustion processes (soot, UHC, and CO) and the oxidation of nitrogen ($NO_x$).

Aircraft are a unique anthropogenic source of pollution because they emit at both ground level and high altitude. Ground-level emissions affect the air quality of areas surrounding airports, whereas emissions at altitude impact the global climate by increasing cloud cover (through the

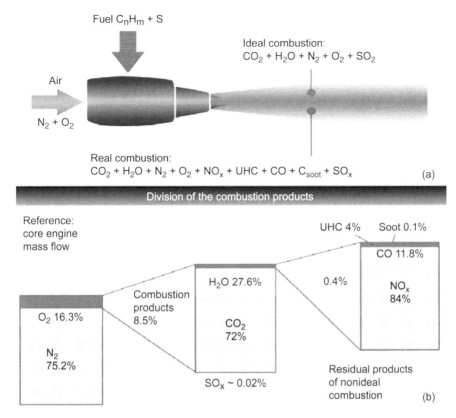

Figure 2.2.   Aircraft engine combustion products. UHC, unburned hydrocarbon. (a) Comparison of ideal and real combustion products. (b) Scale of combustion products at cruise condition. (From Vedantham, A. 1999. Aviation and the global atmosphere: a special report of IPCC Working Groups I and III. *http://repository.upenn.edu/library_papers/61* (accessed August 12, 2016). With permission.)

formation of contrails and aircraft-induced cloudiness) and higher background levels of black carbon (estimated to be the second greatest contributor to current global warming) (Ramanathan and Carmichael, 2008).

Table 2.1 clearly illustrates that the emissions indices for $CO_2$, $H_2O$, and $SO_x$ are approximately constant throughout the flight cycle. These emissions are directly related to the fuel consumption of the engine in its various flight phases. In contrast, $NO_x$, CO, UHC, and soot emissions are strongly influenced by a wide range of variables—particularly the engine power setting and the ambient engine inlet conditions. CO and UHC are products of incomplete combustion. They are highest at low power settings when the temperature of the air is relatively low and the fuel atomization and mixing processes are least efficient. This problem area is proving responsive to improvements linked to detailed studies of basic fuel-air mixing processes. In contrast, $NO_x$ and soot (not shown in Table 2.1) are highest at high power settings.

## 2.2.1   Carbon dioxide

$CO_2$ is the principal product of the combustion of fossil fuels. $CO_2$ emissions, which are the most significant and best understood element of aviation's total contribution to climate change, have been estimated at approximately 2% of all such anthropogenic emissions. Because $CO_2$ has a very long lifetime in the atmosphere, the level and effects of $CO_2$ emissions are currently believed to be broadly the same regardless of altitude; that is, inflight and ground-based emissions have the same atmospheric effects.

Table 2.1.   Typical emissions indices (EI) levels for engine operating regimes, g/kg.

| Species | Idle | Takeoff | Cruise |
|---|---|---|---|
| $CO_2$ | 3,160 | 3,160 | 3,160 |
| $H_2O$ | 1,230 | 1,230 | 1,230 |
| CO | 25 (10 to 65) | 1 | 1 to 3.5 |
| HC[a] (as $CH_4$) | 4 (0 to 12) | 0.5 | 0.2 to 1.3 |
| $NO_x$ | | | |
| Short haul | 4.5 (3 to 6) | 32 (20 to 65) | 7.9 to 11.9 |
| Long haul | 4.5 (3 to 6) | 27 (10 to 53) | 11 to 15 |
| $SO_x$ (as $SO_2$) | 1.0 | 1.0 | 1.0 |

[a]Hydrocarbons
*Source:* Vedantham, A. 1999. Aviation and the global atmosphere: a special report of IPCC Working Groups I and III. *http://repository.upenn.edu/library_papers/61* (accessed August 12, 2016).

Aircraft engine emissions are directly related to the fuel burn. The key to minimizing their environmental impacts is to use fuel more efficiently. For $CO_2$, the environmental benefit is a constant value: each kilogram of fuel saved reduces $CO_2$ emissions by 3.16 kg.

### 2.2.2   *Oxides of nitrogen*

Oxides of nitrogen consist of NO, $NO_2$, nitrous acid (HONO), and nitric acid ($HNO_3$), as well as aerosol nitrate ($NO_3$), peroxyacetyl nitrate (PAN), and other organic nitrates—dinitrogen pentoxide ($N_2O_5$) and peroxynitric acid ($HNO_4$). Some of these compounds may only become important long after they are emitted into the atmosphere. NO and $NO_2$ emissions are typically grouped together as $NO_x$, and $NO_y$ represents the sum of all reactive gaseous nitrogen oxides.

$NO_x$ emissions are generated in the highest temperature regions of the combustor, usually in the primary combustion zone, before the products are diluted. The fundamental processes of $NO_x$ formation are well known and documented (Bowman, 1992) and are best expressed as a function of local combustion temperature, pressure, and time. Combustion zone temperature depends on combustor inlet air temperatures and pressure as well as the fuel-air mass ratio. Figure 2.3 shows the dependence of $NO_x$ on fuel-air ratio. As illustrated, peak $NO_x$ formation coincides with peak temperature, which occurs close to the stoichiometric fuel-air ratio (i.e., equivalence ratio = 1). In current gas turbine engine combustors, there are always some regions of the flame that burn stoichiometrically. Therefore $NO_x$ formation is very strongly linked to combustor inlet temperature.

The gaseous and particulate emissions of a CFM International aircraft engine, CFM56–2C1, were measured during the National Aeronautics and Space Administration (NASA)-led Aircraft Engine Particle eXperiment (APEX). The $NO_x$ emissions indices ($EINO_x$) shown in Figure 2.4 (Wey *et al.*, 2006) are consistent with the International Civil Aviation Organization (ICAO) certification measurements for the same engine model. As expected, $EINO_x$ increased (gradually) with increasing engine thrust, from about 3.0 g/kg fuel at ground idle (4% of maximum rated thrust) to about 18 g/kg fuel at maximum power (100% of maximum rated thrust).

At higher power, aircraft $NO_x$ is dominated by NO, up to 93% at the highest thrust level. At low power, $NO_2$ can contribute more than 80% of the total $NO_x$. In this case, NO is, in part, formed in the turbine through a reaction with a hydroxyl radical (OH) (Fig. 2.5).

HONO also can be significant (up to 7% of $NO_x$ in the observations in Fig. 2.6), although HONO is a relatively small fraction of the total $NO_y$ emission at any engine power setting. HONO does have considerable importance as a diagnostic of the reactive trace species evolution in aircraft engine exhaust (Brundish *et al.*, 2007; Wormhoudt *et al.*, 2007).

HONO results from NO oxidation via the hydroxyl radical, indicating the importance of OH-driven oxidation through the engine. These results indicate that the chemical and

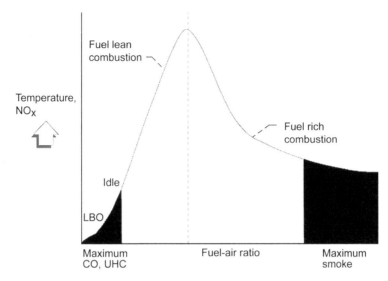

Figure 2.3.   NO$_x$ and temperature dependency on fuel-air ratio. Highest temperature and NO$_x$ at stoichiometric ratio; design approaches for NO$_x$ reduction take advantage of rich or lean combustion. LBO, lean boilout; UHC, unburned hydrocarbon. (From Vedantham, A. 1999. Aviation and the global atmosphere: a special report of IPCC Working Groups I and III. *http://repository.upenn.edu/library_papers/61* (accessed August 12, 2016). With permission.)

physical processes occurring in the turbine are important in determining aircraft engine emissions. Underlying this statement are two key assumptions: (1) observed HONO concentrations are representative of the true exhaust composition, and (2) differences in engine chemistry are indeed manifested as differences in HONO emissions.

The EI for HONO (EIHONO) increases approximately sixfold from idle to takeoff and then plateaus at engine power settings between 60% and 100%, whereas EINO$_x$ increases continuously. EIHONO did not exhibit a dependence on fuel type (traditional, synthetic, or synthetic blend), ambient temperature, relative humidity, or presence of sunlight, whereas EINO$_x$ showed a positive dependence on the ambient temperature at the maximum rated engine thrust.

### 2.2.3   Carbon monoxide

CO is formed when the carbon in fuel is not completely burned. Figure 2.7 (Wey *et al.*, 2006) shows that emissions index carbon monoxide (EICO) decreased sharply at high power settings, from up to 65 g/kg fuel at ground idle to about 4 g/kg fuel at 30% maximum rated thrust, to less than 1 g/kg fuel above 60% of maximum rated thrust.

### 2.2.4   Unburned hydrocarbons

UHCs, like CO, are associated with combustion inefficiency. The UHC total consists of many HC species, including $CH_4$, formaldehyde (HCHO), ethylene ($C_2H_4$), acetaldehyde ($CH_3CHO$), acrolein ($C_3H_4O$), 1,3-butadiene ($C_4H_6$), benzene ($C_6H_6$), toluene ($C_7H_8$), and higher aromatic species.

EIUHC follows the same trend as CO: highest at the lowest power and decreasing sharply at higher power settings. Although all aircraft engine emissions decrease when less fuel is burned, total UHC emissions are highest at idle (when fuel consumption is lowest) instead of at takeoff conditions (when fuel consumption is highest) (Fig. 2.8). Individual HC species exhibited similar engine power dependencies.

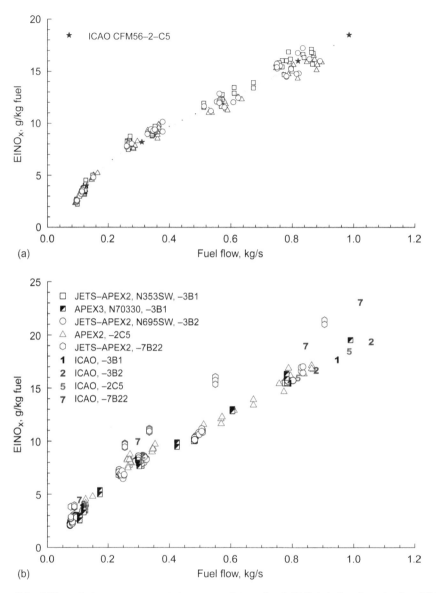

Figure 2.4.   $NO_x$ emissions measurement data versus International Civil Aviation Organization (ICAO) certification data. EI, emissions index. JETS, Jet Emissions Testing for Speciation. APEX, Aircraft Particle Emissions eXperiment. (a) APEX CFM56–2C1 (b) APEX CFM56–2C1, JETS–APEX2 CFM56–3B1 and CFM56–3B2, and APEX3 CFM56–3B1.

Depending on the nature of the chemical composition of the UHC emission products, this may be of concern because operating commercial aircraft spend the bulk of their time at airports under low engine power conditions such as taxiing or idle (Figs. 2.9 and 2.10). Only a limited number of studies have addressed the chemical speciation of the UHC emissions at relevant airport operating conditions (Anderson *et al.*, 2006; Spicer *et al.*, 1994). One of these studies has raised concern because it suggests that at engine idle, the airborne toxins HCHO, $CH_3CHO$, $C_3H_4O$, $C_4H_6$, and $C_6H_6$ compose a significant fraction ($\sim$26%) of the total nonmethane HC emissions (Spicer *et al.*, 1994).

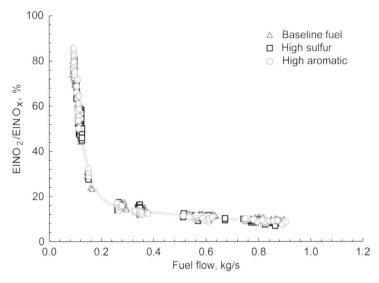

Figure 2.5.    APEX data: $NO_2/NO_x$ ratio versus fuel flow. EI, emissions index. At low powers, $NO_2$ can contribute to more than 80% of the total $NO_x$, but the total $NO_x$ is dominated by NO at high power. From Wey, C. C., B. E. Anderson, C. Hudgins, *et al.,* 2006. Aircraft Particle Emissions eXperiment (APEX). NASA/ TM—2006-214382 (ARL–TR–3903). Work of the U.S. Government.)

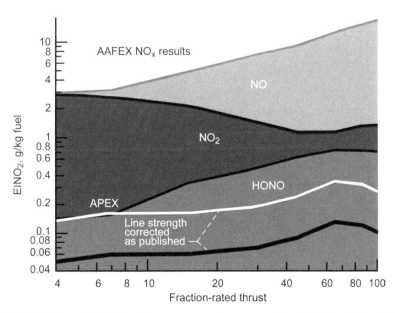

Figure 2.6.    NO, $NO_2$, and HONO emissions versus engine power setting. HONO emissions indices (EIs) from Aircraft Particle Emissions eXperiment 3 (APEX–3), corrected for line strength, are overlaid on observations from Alternative Aviation Fuel Experiment (AAFEX). Emissions indices for $NO_2$ and NO from APEX–3 are overlaid on those for HONO. (Miake-Lye, R. C. 2010. AAFEX gaseous emissions: criteria, greenhouse gases, and hazardous air pollutants. Paper presented at the AAFEX Workshop AIAA Meeting, Orlando, FL. With permission.)

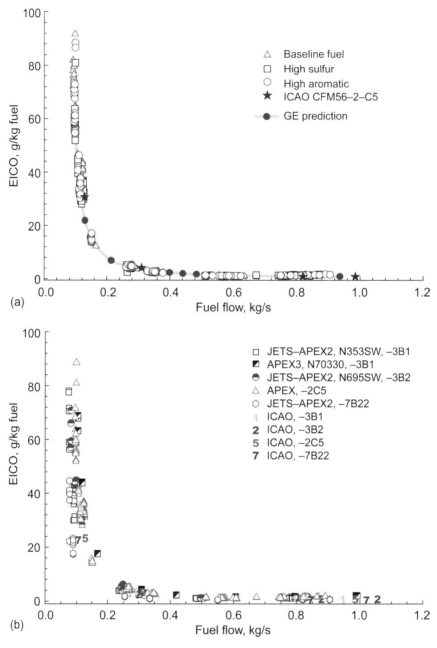

Figure 2.7.  CO emissions versus fuel flow data. CO, as the product of incomplete combustion, drops sharply with increasing power. ICAO, International Civil Aviation Organization; JETS, Jet Emissions Testing for Speciation. (a) APEX CFM56–2C1 with three different fuels. (b)APEX CFM56–2C1, JETS–APEX2 CFM56–3B1 and CFM56–3B2, APEX3 CFM56–3B1. ((a) From Wey, C. C., B. A. Anderson, C. Wey, *et al.*, 2007. Overview on the Aircraft Particle Emissions eXperiment (APEX). *J. Propul. P.,* 23:898–905. Work of the U.S. Government. (b) From Wey, C., and C. C. Wey. 2007. Gas emissions acquired during the Aircraft Particle Emission eXperiment (APEX) series. ISABE–2007–1277. Work of the U.S. Government.)

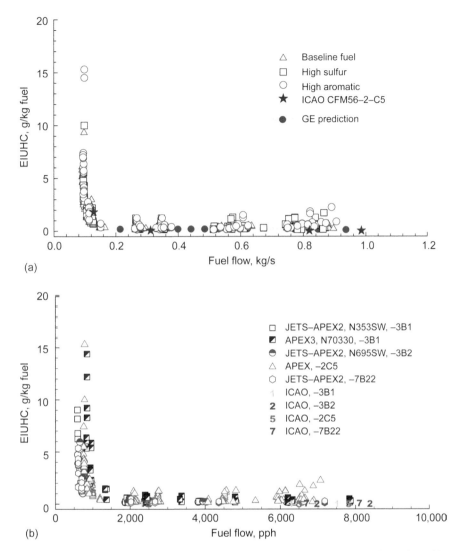

Figure 2.8.    Unburned hydrocarbon (UHC) emissions versus fuel flow data. UHC, as the product of incomplete combustion, drops sharply with increasing power. ICAO, International Civil Aviation Organization. JETS, Jet Emissions Testing for Speciation. (a) APEX CFM56–2C1 with three different fuels. (b) APEX CFM56–2C1, JETS–APEX2 CFM56–3B1 and CFM56–3B2; and APEX3 CFM56–3B1. ((a) From Wey, C. C., B. A. Anderson, C. Wey, *et al.*, 2007. Overview on the aircraft particle emissions experiment (APEX). *J. Propul. P.*, 23:898–905. Work of the U.S. Government. (b) From Wey, C., and C. C. Wey. 2007. Gas emissions acquired during the Aircraft Particle Emission eXperiment (APEX) series. ISABE–2007–1277. Work of the U.S. Government.)

## 2.2.5    *Water vapor*

Contrails, or condensation trails, are clouds produced by aircraft engine exhaust. Of all aircraft emissions at altitude, those of interest for their effect on contrail formation are water vapor, sulfur gases, and fine particles. Water vapor increases humidity, and sulfur species contribute to the formation of small particles that can serve as sites for water droplet growth. When enough

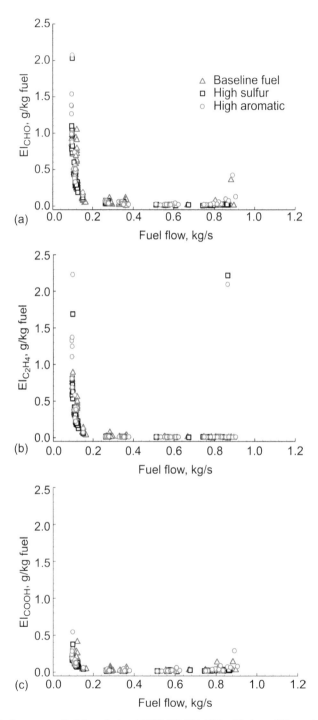

Figure 2.9.   Emissions versus fuel flow during APEX CFM56–2C1 with three different hydrocarbon fuels. All show the same trend as unburned hydrocarbons (UHCs)—decreasing sharply with increasing power setting power. (a) Emissions index (EI) of HCHO ($EI_{CHO}$). (b) $EI_{C_2H_4}$. (c) $EIH_{COOH}$. (From Wey, C. C., B. E. Anderson, C. Hudgins, *et al.*, 2006. Aircraft Particle Emissions eXperiment (APEX). NASA/TM—2006-214382 (ARL–TR–3903). Work of the U.S. Government.)

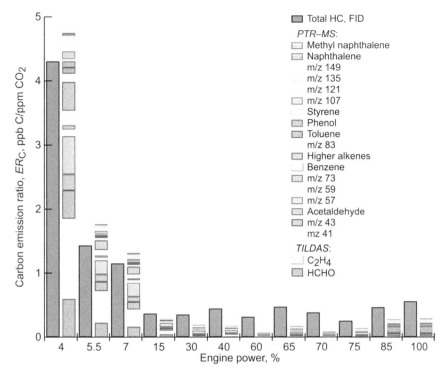

Figure 2.10. Comparison of sum of speciated hydrocarbons from Aerodyne tunable infrared differential absorption spectroscopy (TILDAS) and proton-transfer reaction mass spectrometry (PTR–MS) with total hydrocarbon (HC) measurement from NASA continuous flame ionization detector (FID) during APEX measurements. m/z, mass-to-charge ratio. (From Wey, C. C., B. E. Anderson, C. Hudgins, *et al.*, 2006. Aircraft Particle Emissions eXperiment (APEX). NASA/TM—2006-214382 (ARL–TR–3903). Work of the U.S. Government.)

humidity is present in the engine exhaust plume, water condenses on small particulate matter, forming liquid droplets that quickly freeze as the exhaust air mixes with the colder ambient air. The resulting ice particles create a visible contrail. The humidity level depends on the external air temperature, the amount of water already present in the surrounding air, and the amount of water and heat emitted in the engine exhaust (EPA, 2000).

At low humidity, a contrail is short-lived, extending just a short distance behind the aircraft and quickly evaporating (Fig. 2.11). High humidity leads to persistent contrails—line-shaped clouds that continue to grow as their ice particles take water from the surrounding atmosphere. Persistent contrails can extend behind aircraft for large distances (Fig. 2.12), lasting for hours and growing to several kilometers in width and 200 to 400 m in height. Because contrails add to cloudiness in the Earth's atmosphere, they could potentially play a role in the Earth's temperature and climate (EPA, 2000).

In persistent contrails, only a small percentage of the water that forms ice comes from jet engine exhaust; most of the water is already present in the surrounding atmosphere. If temperature and humidity conditions are known, contrail formation for a specific aircraft during flight can be accurately predicted (EPA, 2000).

### 2.2.6 *Sulfur oxides*

Jet fuels commonly contain impurities such as sulfur and trace metals. All sulfur emissions in the engine exhaust are caused by the combustion of sulfur introduced into the fuel. Like other

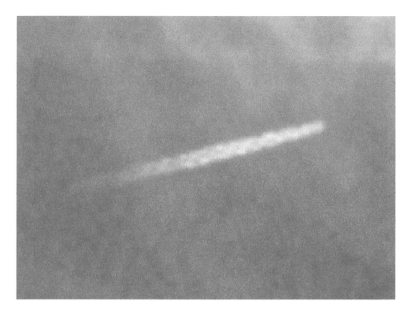

Figure 2.11.    Photograph of a short-lived contrail that, because of low atmospheric humidity, evaporated soon after being formed (Photo: J. Holecek, National Oceanic and Atmospheric Administration (NOAA) Aeronomy Laboratory, Boulder, CO. From Environmental Protection Agency (EPA). 2000. Aircraft contrails factsheet. EPA430–F–00–005. Work of the U.S. Government.)

Figure 2.12.    Multiple persistent contrails formed in high humidity; some are spreading into cirrus clouds. (Photo: L. Chang, Office of Atmospheric Programs, U.S. EPA. (From Environmental Protection Agency (EPA). 2000. Aircraft contrails factsheet. EPA430–F–00–005. Work of the U.S. Government.)

gases in the $SO_x$ family, sulfur dioxide ($SO_2$) dissolves easily in water. When $SO_2$ dissolves in water vapor, it forms acid and interacts with other gases and particles in the air to form sulfates and other products that can be harmful to people and the environment.

Due to its high reactivity, the sulfur in fuel is fully oxidized. However, these $SO_x$ emissions are in the form of both $SO_2$ and sulfur trioxide ($SO_3$), and measurements show that the ratio of $SO_3$ to $SO_2$ varies. APEX reports that 95% of the sulfur in the engine is converted to $SO_2$ in the exhaust. The remaining sulfur is converted into $SO_3$, which combines with water vapor in the exhaust to form sulfuric acid ($H_2SO_4$).

### 2.2.7 *Methane and nitrous oxide*

$CH_4$ and $N_2O$ have significant global warming potentials. $CH_4$ has a larger potential than $CO_2$ does, and it plays an important role in determining tropospheric OH concentrations. $N_2O$ has a potential even higher than that of $CH_4$. $N_2O$ has a very long atmospheric lifetime and is one of the primary sources of stratospheric NO.

In 2002, the NASA Glenn Research Center acquired a limited set of $CH_4$ and $N_2O$ emissions data via its combustion facilities, flame tubes, and sector test rigs. The analyzer used was based on a Fourier transform infrared spectrometer that measured the infrared absorption of the exhaust gas to identify and quantify a variety of compounds.

The nominal value for $CH_4$ concentration in the atmosphere is typically 1.8 parts per million (ppm). All measurements were made with a low-emission-concept fuel injector that assumed a background concentration of 2 to 3 ppm. The measured $CH_4$ emissions were higher than the background values only at lower temperatures, close to the equivalent "idle" condition.

Because 0.3 ppm is typically referenced as the nominal value for $N_2O$ in the atmosphere, all tests were performed with a low-emission-concept fuel injector that assumed a background concentration of 0.25 to 0.35 ppm. Data showed no significant dependency on the temperature of the inlet, exit, or the flame.

### 2.3 PARTICLE EMISSIONS

Aircraft particulate emissions were first noted as the black exhaust that came out of aircraft. Consequently, the smoke number (SN) measurement described in SAE International Aerospace Recommended Practice (ARP) 1179 (1970) was developed to ensure compliance with an aircraft engine exhaust visibility criterion. The SN regulation has significantly reduced the black smoke—that is, the larger particles from the aircraft emissions. However, over the last decade, it has become evident that smaller particles (with diameters less than 1 mm) are of particular concern due to their detrimental effects on human health, local air and water quality, and global climate change (scattering or absorbing solar radiation and promoting the formation of contrails and clouds). Thus, the SN does not address the current need to quantitatively measure the mass and number of smaller particles emitted from the engine.

Gaseous emissions measurement procedures were developed by SAE's E–31 Aircraft Exhaust Emissions Measurement Committee and for many decades have been adopted by the U.S. Environmental Protection Agency (EPA) and the International Civil Aviation Organization as standard practice for measuring aircraft exhaust gaseous species. The E–31 particulate matter (PM) subcommittee was established about a decade ago with the same final goal: the establishment of an international standard for measuring particulate emissions from aircraft turbine engines.

Two challenges were obvious and fundamental: (1) where to measure and (2) what to measure. However, it took several years of discussions to reach agreements on these topics.

The final agreement dictated that the measurement was to be taken at the same location as for gaseous emissions—the engine exit plane. However, the measurement environment behind a gas turbine engine places significant challenges on the sampling system for particulate emissions (Figs. 2.13 and 2.14). A sophisticated sampling system that controls the temperature, dilution, and residence time to minimize perturbations to the PM in the exhaust sample prior to measurement has yet to be thoroughly defined and standardized to minimize variability between test facilities and operators.

Because of the high temperatures at the typical aircraft engine exit plane, only very small nonvolatile particles and precursor gases can form volatile particles. These tiny particles do not

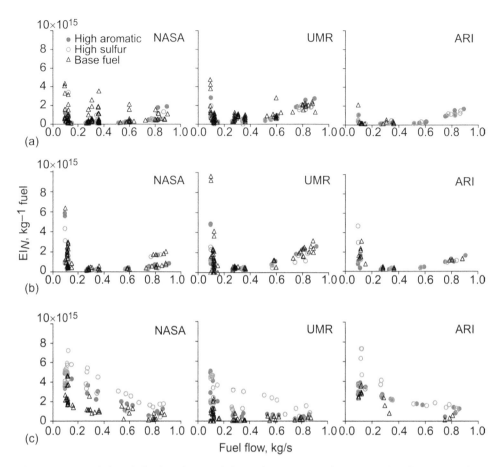

Figure 2.13.   Emissions index based on particle number, EI$_N$, as a function of fuel flow rate at three sampling locations using three types of fuel during the APEX) Sampling locations: NASA, University of Missouri-Rolla (UMR), and Aerodyne Research, Inc. (ARI) (a) Sampled at 1 m. (b) Sampled at 10 m. (c) Sampled at 30 m. (From Wey, C. C., B. A. Anderson, C. Wey, *et al.*, 2007. Overview on the aircraft Particle Emissions experiment (APEX). *J. Propul. P.*, 23:898–905. Work of the U.S. Government.)

coincide with the EPA's existing definition of PM or with regulations for local air quality. This led to the other challenge—what to measure.

Many discussions with various groups had concluded that both the nonvolatile and volatile particles are of critical importance, and after many years of discussions, the complexity of measuring PM was finally successfully communicated and accepted by all relevant parties. To consider both nonvolatile and volatile particles, the E–31 PM subcommittee took the approach of "divide and conquer," first tackling the measurement of nonvolatile particles. In April 2010, the subcommittee formed four teams (sampling, number, mass, and calculations) focused on establishing an ARP for measuring aircraft engine nonvolatile particulate matter (nvPM) mass and particle number emissions. Particle transport loss through the sampling system needs to be well understood for nvPM to be measured.

Based on techniques pioneered by the particle research community (AIR5892, 2004; AIR6037, 2010) to measure the mass and number of the emitted particles, the PM subcommittee leveraged the existing ICAO Annex 16 Volume II engineering principles (International Civil Aviation Organization, 2008) for gaseous emissions measurements (the same as ARP1533B,

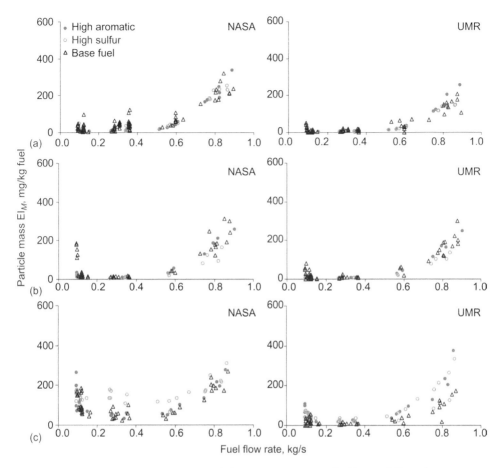

Figure 2.14. Emissions index based on particle mass, $EI_M$, measured with two different instruments—NASA's scanning mobility particle sizer and University of Missouri-Rolla (UMR)'s DMS500—as a function of fuel flow rate using three different fuels. Values were calculated using a particle mass density of 1 g = cm³ during the APEX measurement. (a) Sampled at 1 m. (b) Sampled at 10 m. (c) Sampled at 30 m. (From Wey, C. C., B. A. Anderson, C. Wey, *et al.* 2007. Overview on the aircraft particle emissions experiment (APEX). *J. Propul. P.,* 23:898–905. Work of the U.S. Government.)

2013) and developed a methodology. A well-defined nvPM sampling system that accommodates both mass and number measurement instrumentation was published by SAE International in an Aerospace Information Report (AIR6241, 2013) in November 2013.

## 2.4  ALTERNATIVE FUELS

Alternative fuels are being considered as an alternative method for reducing greenhouse gas emissions and are being studied as replacements for current conventional-petroleum-derived aircraft engine fuels (Carter *et al.*, 2011; Hileman *et al.*, 2010). Fuels used in commercial aircraft must meet the very stringent standards specified by ASTM International D1655 (2014). A number of research activities have been and continue to be conducted to evaluate and certify new fuels for use in commercial aviation. ASTM D7566 (2014) recently adopted synthetic paraffinic kerosene (SPK) fuels, allowing up to 50 v/v% Fischer-Tropsch (FT), hydroprocessed esters and

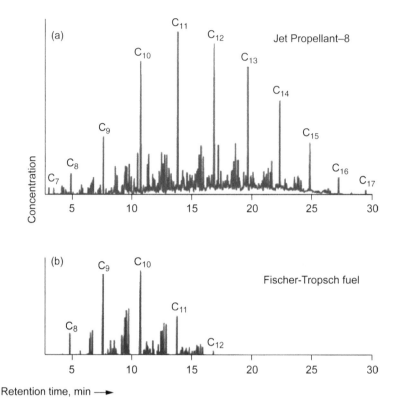

Figure 2.15.   Gas chromatography and mass spectometry chromatograms of JP–8 and FT fuel, showing that FT fuel does not contain larger hydrocarbons. (a) JP–8. (b) FT.

fatty acids (HEFA), or hydrotreated renewable jet (HRJ) fuels to be mixed with conventional jet fuel. Figure 2.15 shows the fundamental differences between conventional jet fuel and alternative FT fuel.

The current focus is on using "drop-in" fuels to replace current conventional jet fuels. Such fuels are designed to function in current engines and aircraft without any hardware modification. Engine tests using FT fuels are reported by Bester and Yates (2009), Corporan *et al.* (2007), and Bulzan *et al.* (2010). Their data indicate that alternative fuels had small or no effects on engine performance and only minor differences from conventional fuels in gaseous emissions, but there were fairly large reductions in combustion-generated particulate emissions with the various SPK fuels. Engine performance of a microturbine operated on FT and HRJ fuels is reported in Quintero *et al.* (2012).

Combustion tests, fuel properties, and the effects of FT fuels are reported in Rizk *et al.* (2008), DeWitt *et al.* (2008), and Moses and Roets (2008). Combustor flame tube studies using FT fuel are reported in Thomas *et al.* (2012). Tests were performed on a sector rig with a section from a CFM56 combustor using FT and JP–8 fuels. Significant differences were reported in the particulates and liner temperatures for the two fuels. However, very limited gaseous emissions data were reported.

To examine the effects of two bio-based jet fuels on gaseous emissions, NASA used a nine-point, lean-direct-injection, low-emissions combustion concept that had been studied for fuel injector performance (Wey and Bulzan, 2013). The two fuels used were HRJ fuel made from tallow and Amyris fuel made from the fermentation of sugar. A standard JP–8 fuel was used as the baseline and the blending component. A low-power or pilot-only mode was simulated by fueling only the center injector in the nine-point array. These fuels had large differences in

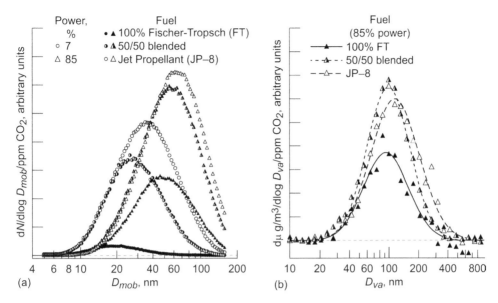

Figure 2.16.   Particle size distribution data for combustion of JP–8, 100% FT fuel, and blended fuel during the Pratt & Whitney PW308 engine tests. (a) Scanning mobility particle sizer size-resolved number density at 7% (idle) and 85% (takeoff) power settings. (b) Alpha magnetic spectrometer size-resolved volatile particulate matter (PM) mass at 85% power. (Reprinted (adapted) with permission from Timko, M. T., Z. Yu, T. B. Onasch, *et al.*, 2010. Particulate emissions of gas turbine engine combustion of a Fischer-Tropsch synthetic fuel. *Energy & Fuels* 24:5883–5896. Copyright 2010 American Chemical Society.)

aromatic, sulfur, and paraffinic contents but smaller differences in other properties. The Amyris fuel differed markedly from conventional hydrocarbon fuels. However, no significant differences in gaseous emissions were observed. In summary, in all of the tests performed thus far by various research groups, there have been no significant differences in gaseous emissions, but there have been some notable effects on particulate emissions.

Additional fundamental combustion studies of SPK fuels have been performed by a number of investigators to focus more on particulate emissions (Allen *et al.*, 2011; Mondragon *et al.*, 2012). Effects on drop size, spray angle, autoignition delay times, and jet penetration have been reported for a number of alternative fuels, including HRJ fuels. Some of the differences found in drop size, spray angle and breakup, jet penetration, autoignition times, and vaporization rate may affect combustion performance and emissions. A test was performed on a Pratt & Whitney Canada Corporation's PW308® engine (Longueuil, Quebec, Canada) to study the impacts of alternative fuels on particulate emissions. In general, the use of alternative fuel reduced particle size and total particulate emissions.

To further study the effects of FT fuel usage on aircraft gaseous and particulate emissions (Fig. 2.16), NASA sponsored the Alternative Aviation Fuel Experiment (AAFEX). Measurements were made behind a commercial aircraft that was burning standard JP–8, FT, and JP–8 FT blended fuels. In addition to measurements at the engine exit plane, downwind plumes were sampled to determine the "aging" effects on aerosol concentrations and composition as well as the role of ambient temperature. Gaseous emissions are presented in Figure 2.17, and particulate emissions are presented in Figure 2.18 (Anderson *et al.*, 2011) and Figures 2.19 and 2.20 (Beyersdorf *et al.*, 2014). Emissions from the aircraft's auxiliary power unit are discussed in Kinsey *et al.* (2012).

The particulate emission reductions associated with FT fuel usage may have significant impacts on air quality and climate change (via changing the radiation budgets). Reduced ground-level particulate emissions are beneficial for local air quality, and the reductions in soot size and $SO_x$ emissions will lower the potential for particles to act as cloud condensation nuclei (CCN).

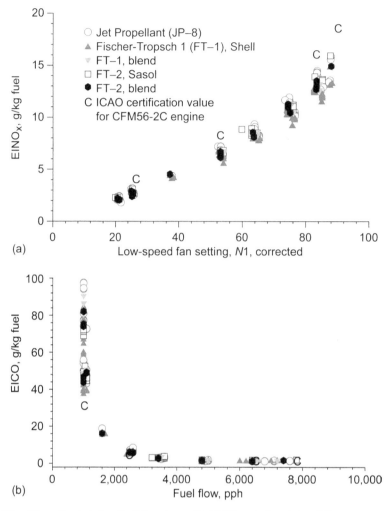

Figure 2.17.   Alternative Aviation Fuel Experiment (AAFEX) emissions index (EI) versus corrected low-speed fan ($N1$) settings and fuel flow rate. (a) EINO$_x$ versus corrected $N1$. (b) Emissions index carbon monoxide (EICO) versus corrected fuel flow rate. The figure shows that FT fuels reduced notable NO$_x$, CO, and unburned hydrocarbon (UHC) emissions. ICAO, International Civil Aviation Organization.

Along with the reduction in total soot emissions, fewer CCN should decrease contrail formation. However, the smaller particle size and reduced CCN activation will lengthen the lifetime of soot in the atmosphere. Both effects—lower CCN potential and longer lifetime—must be taken into account in modeling the impact of aircraft emissions on the radiation budget.

In addition, some benefits from the reductions of particle emissions may be offset by increased CO$_2$ emissions during FT fuel production. Emissions during both production and use must therefore be analyzed further to determine the true benefits of alternative aviation fuels.

## 2.5   SUMMARY

Gaseous and particulate emissions from aircraft directly impact the quality of the environment not only around the airport but also at cruise altitude, the upper troposphere, and the lower

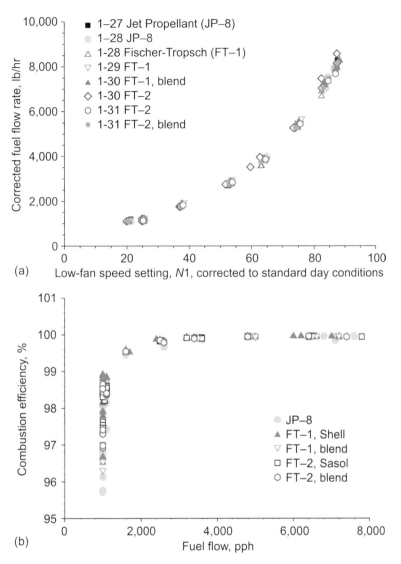

Figure 2.18. AAFEX fuel flow rates versus low-speed fan ($N1$) settings and combustion efficiencies. Data show insignificant differences between the levels required to produce the $N1$ settings with JP–8 and FT fuels, suggesting that the alternative fuels offer neither advantage nor penalty in terms of fuel economy. FT fuels did provide higher combustion efficiencies at low fuel flow rates, indicating that they are less polluting in general. (a) $N1$ settings. (b) Combustion efficiencies. (From Anderson, B. E., A. J. Beyersdorf, C. H. Hudgins, *et al.,* 2011. Alternative Aviation Fuel Experiment (AAFEX). NASA/TM—2011-217059. Work of the Government)

stratosphere. Oxides of nitrogen ($NO_x$) emissions from aircraft are currently regulated during takeoff and landing to meet local air quality standards. Particulate emissions are expected to be regulated in the very near future due to increased concerns about adverse health effects from the ingestion of fine particles. Standards for the certification of $NO_x$ levels are set by the International Civil Aviation Organization, a United Nations specialized agency established by member nations to reach consensus on international civil aviation standards. Levels are revised periodically based on technology advancements and the ability of airlines to adapt the technologies in their flight

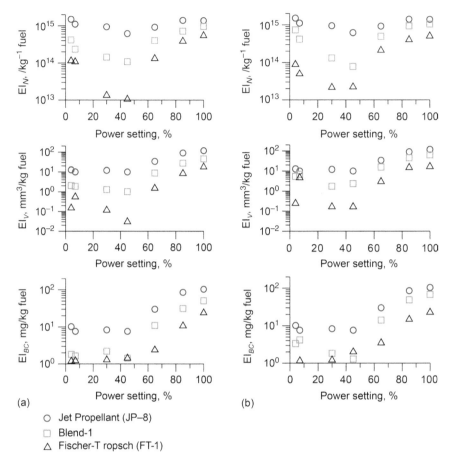

Figure 2.19.   Alternative Aviation Fuel Experiment (AAFEX) aerosol emissions indices (EIs) at 1 m behind engine 3 and fueled with JP–8, blended fuels, and pure FT fuels. FT fuels in general reduced total particulate emissions. (a) JP–8, Blend 1, and FT–1. (b) JP–8, Blend 2, and FT–2. (From Beyersdorf, A. J., M. T. Timko, L. D. Ziemba, *et al.,* 2014. Reductions in aircraft particulate emissions due to the use of Fischer-Tropsch fuels. *Atmos. Chem. Phys.* 14:11–23. *http://www.atmos-chem-phys.net/14/11/2014/.* Creative Commons CC BY 3.0 license, *https://creativecommons.org/licenses/by/3.0/legalcode* (accessed February 28, 2017)).

vehicles. As a result, technology development plays a key role in minimizing the impact of aviation on the environment. Over the last three decades, the certification level for $NO_X$ emissions has been revised downward three different times; the current level is about 55% less than that of 1986. Technology advancement efforts currently underway in the United States and Europe target a further reduction of 70% to 75% from the current $NO_X$ level for aircraft entering into service in the 2025 timeframe. Government-funded agencies such as NASA and the European Space Agency play a critical role in technology advancement, working in collaboration with academia and industry as well as other government agencies and effectively transferring the technology to industry to enable flight demonstration.

Low-emission combustor research and technology development efforts continue to influence advanced combustor designs to meet aggressive long-term goals regarding gaseous and particulate emissions from aircraft. Although gaseous emissions have been studied for the last five decades, the study of particulate emissions began only a decade ago, when scientific findings

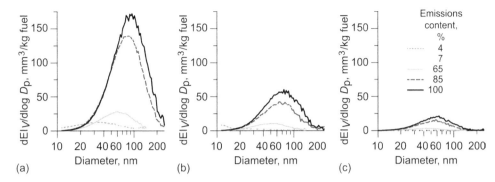

Figure 2.20.   Alternative Aviation Fuel Experiment (AAFEX) volume-weighted size distributions at 1 m behind the exhaust plane. Fischer-Tropsch (FT) fuels in general produce smaller particles. (a) JP–8. (b) Blend-1. (c) FT–1 fuels (size distributions for Blend-2 and FT–2 are similar). (From Beyersdorf, A. J., M. T. Timko, L. D. Ziemba *et al.,* 2014. Reductions in aircraft particulate emissions due to the use of Fischer-Tropsch fuels. *Atmos. Chem. Phys.* 14:11–23. *http://www.atmos-chem-phys.net/14/11/2014/.* Creative Commons CC BY 3.0 license, *https://creativecommons.org/licenses/by/3.0/legalcode* (accessed February 28, 2017)).

indicated that fine particles emitted from aircraft engine combustion might have a significant detrimental effect on human health in addition to their adverse impacts on the global climate. Research and development efforts to enhance the understanding of particulate emissions and to advance the state of measurement technology are currently underway in anticipation of a new standard for regulating particulate emissions. In an effort to reduce carbon footprint and greenhouse gas emissions, alternative fuels with lower life-cycle carbon are being considered as replacements for current conventional-petroleum-derived aircraft engine fuels. The current focus is on optimizing the use of drop-in fuels for 50% to 100% alternative fuel blends in current engines and aircraft without any hardware modification. Future studies will consider the feasibility and potential of non-drop-in fuels and will develop combustion and combustor concepts for these fuels with a long-term strategy of moving toward alternative propulsion system concepts utilizing nonconventional fuels.

Advancements in technology and dramatic reductions of emissions have been achieved through successful partnership of government agencies such as NASA with industry, and these partnerships are expected to continue in the future. Industry participation at the appropriate phases of technology progression, from fundamental understanding through the various Technology Readiness Levels, has been a vital factor in the successful advancement of key technologies to flight. Effective partnerships among government agencies, industry, and academia play a crucial role in advancing technology to meet performance requirements as well as the increasingly stringent environmental compatibility requirements of future aviation propulsion systems.

## REFERENCES

AIR5892. 2004. Nonvolatile exhaust particle measurement techniques. Aerospace Information Report, SAE International, Warrendale, PA.

AIR6037. 2010. Aircraft exhaust nonvolatile particle matter measurement method development. Aerospace Information Report, SAE International, Warrendale, PA.

AIR6241. 2013. Procedure for the continuous sampling and measurement of non-volatile particle emissions from aircraft turbine engines. Aerospace Information Report, SAE International, Warrendale, PA.

Allen, C. M., E. Toulson, and T. Lee. 2011. An experimental investigation of the autoignition characteristics of camelina-based hydroprocessed renewable jet fuel. AIAA 2011–604.

Anderson, B. E., G. Chen, and D. R. Blake. 2006. Hydrocarbon emissions from a modern commercial airliner. *Atmos. Environ.* 40:3601–3612.

Anderson, B. E., A. J. Beyersdorf, C. H. Hudgins, *et al.* 2011. Alternative Aviation Fuel Experiment (AAFEX). NASA/TM—2011-217059.

ARP1179. 1970. Aircraft gas turbine engine exhaust smoke measurement. Aerospace Recommended Practice, SAE International, Warrendale, PA.

ARP1533B. 2013. Procedure for the analysis and evaluation of gaseous emissions. Aerospace Recommended Practice, SAE International, Warrendale, PA.

ASTM D1655. 2014. Standard specification for aviation turbine fuels. ASTM International, West Conshohocken, PA.

ASTM D7566. 2014. Standard specification for aviation turbine fuel containing synthesized hydrocarbons. ASTM International, West Conshohocken, PA.

Bester, N., and A. Yates. 2009. Assessment of the operational performance of Fischer-Tropsch synthetic-paraffinic kerosene in a T63 gas turbine compared to conventional jet A–1 fuel. ASME GT2009–60333.

Beyersdorf, A. J., M. T. Timko, L. D. Ziemba, *et al.* 2014. Reductions in aircraft particulate emissions due to the use of Fischer-Tropsch fuels. *Atmos. Chem. Phys.* 14:11–23. *http://www.atmos-chem-phys. net/14/11/2014/* (accessed February 28, 2017).

Bowman, C. T. 1992. Control of combustion-generated nitrogen oxide emissions: technology driven by regulation. In *Twenty-Fourth Symposium (International) on Combustion.* Pittsburgh: The Combustion Institute, 859–878.

Brundish, K. D., A. R. Clague, C. W. Wilson, *et al.* 2007. Evolution of carbonaceous aerosol and aerosol precursor emissions through a jet engine. *J. Propul. P.* 23:959–970.

Bulzan, D., B. Anderson, C. Wey, *et al.* 2010. Gaseous and particulate emissions results of the NASA Alternative Aviation Fuel Experiment (AAFEX). ASME GT2010–23524.

CAEP. 2006. Report of the independent experts to the long term technology task group on the 2006 LTTG NOx review and the establishment of medium and long term technology goals for NOx. CAEP/7 Working Paper, CAEP/7-WP/11. *http://www.faa.gov/about/office_org/headquarters_offices/apl/research/science_ integrated_modeling/media/Independent%20Experts%20Report.pdf* (accessed February 28, 2017).

Carter, N. A., R. W. Stratton, M. K. Bredehoeft, *et al.* 2011. Energy and environmental viability of select alternative jet fuel pathways. AIAA 2011–5968.

Corporan, E., M. J. DeWitt, V. Belovich, *et al.* 2007. Emissions characteristics of a turbine engine and research combustor burning a Fischer-Tropsch jet fuel. *Energy Fuels* 21:2615–2626.

DeWitt, M. J., E. Corporan, J. Graham, *et al.* 2008. Effects of aromatic type and concentration in Fischer-Tropsch fuel on emissions production and material compatibility, *Energy Fuels* 22:2411–2418.

Environmental Protection Agency. 2000. Aircraft contrails factsheet. EPA430–F–00–005.

Hileman, J. I., R. W. Stratton, and P. E. Donohoo. 2010. Energy content and alternative jet fuel viability. *J. Propul. P.* 26:1184–1196.

International Civil Aviation Organization. 2008. Annex 16 to the Convention on International Civil Aviation: Environmental Protection, Volume II—Aircraft Engine Emissions. Third edition, ICAO, Montréal, Quebec. *https://archive.org/details/gov.law.icao.annex.16.v2.2008* (accessed February 28, 2017).

Kinsey, J. S., M. T. Timko, S. C. Herndon, *et al.* 2012. Determination of the emissions from an aircraft auxiliary power unit (APU) during the Alternative Aviation Fuel Experiment (AAFEX). *J. Air Waste Manage. Assoc.* 62:420– 430.

Miake-Lye, R. C. 2010. AAFEX gaseous emissions: criteria, greenhouse gases, and hazardous air pollutants. Paper presented at the AAFEX Workshop AIAA Meeting, Orlando, FL.

Mondragon, U. M., C. T. Brown, and V. G. McDonell. 2012. Evaluation of spray and combustion behavior of alternative fuels for JP–8. AIAA 2012–348.

Moses, C. A., and P. N. J. Roets. 2008. Properties, characteristics, and combustion performance of Sasol fully synthetic jet fuel. ASME GT2008–50545.

Quintero, S. A., M. Ricklick, and J. Kapat. 2012. Synthetic jet fuels and their impact in aircraft performance and elastomer materials. AIAA 2012–3965.

Ramanathan, V., and G. Carmichael. 2008. Global and regional climate changes due to black carbon. *Nat. Geosci.* 1:221–227.

Rizk, N., R. Moritz, and A. Blackwell. 2008. Combustion and material compatibility tests of syntroleum Fischer-Tropsch iso-paraffinic kerosene. Paper presented at the Alternative Energy Now Conference, Lake Buena Vista, FL.

Spicer, C. W., M. W. Holdren, R. M. Riggin, *et al.* 1994. Chemical composition and photochemical reactivity of exhaust from aircraft turbine engines. *Ann. Geophys.* 12:944–955.

Thomas, A. E., N. T. Saxena, D. T. Shouse, *et al.* 2012. Heating and efficiency comparison of a Fischer-Tropsch (FT) fuel, JP-8+100, and blends in a three-cup combustor sector. ASME GT2012–70008.

Timko, M. T., Z. Yu, T. B. Onasch, *et al.* 2010. Particulate emissions of gas turbine engine combustion of a Fischer-Tropsch synthetic fuel. *Energy Fuels* 24:5883–5896.

Vedantham, A. 1999. Aviation and the global atmosphere: a special report of IPCC Working Groups I and III. *http://repository.upenn.edu/library_papers/61* (accessed August 12, 2016).

Wey, C., and C. C. Wey. 2007. Gas emissions acquired during the Aircraft Particle Emissions eXperiment (APEX) series. ISABE–2007–1277.

Wey, C., and D. Bulzan. 2013. Effects of bio-derived fuels on emissions and performance using a 9-point lean direct injection low emissions concept. ASME GT2013–94888.

Wey, C. C., B. E. Anderson, C. Hudgins, *et al.* 2006. Aircraft Particle Emissions eXperiment (APEX). NASA/TM—2006-214382 (ARL–TR–3903).

Wey, C. C., B. A. Anderson, C. Wey, *et al.* 2007. Overview on the Aircraft Particle Emissions eXperiment (APEX). *J. Propul. P.* 23:898–905.

Wormhoudt, J., S. C. Herndon, P. E. Yelvington, *et al.* 2007. Nitrogen oxide (NO/NO$_2$/HONO) emissions measurements in aircraft exhausts. *J. Propul. P.* 23:906–911.

# CHAPTER 3

## Improvement of aeropropulsion fuel efficiency through engine design

Kenneth L. Suder and James D. Heidmann

### 3.1 INTRODUCTION

National Aeronautics and Space Administration (NASA) has had a long and successful history of contributions toward the Nation's goal of improved aircraft fuel efficiency. Specifically, the NASA Glenn Research Center (GRC) has played a direct role in commercial aircraft propulsion system improvements through concept development, component testing, analysis, and model development for aircraft engine inlets, fans, compressors, turbines, and nozzles. This chapter will focus primarily on aerothermodynamic improvements to aircraft engine systems that were enabled through NASA research efforts.

There are multiple motivations for improving aircraft fuel efficiency and thereby reducing fuel consumption. There is an economic motivation in that fuel consumption is a large factor in the cost of operation of aircraft, which directly impacts the profitability of airlines, aircraft and engine manufacturers, and associated industries. The public benefits from improved fuel efficiency through more affordable travel. Reduced fuel consumption is also related to increased U.S. energy security and lower reliance on foreign oil. Finally, in recent years environmental concerns over global warming and air quality have increased the motivation to reduce fuel burn and the associated carbon dioxide and other emissions. As the Nation's civil aeronautics research agency, NASA has a large stake in ensuring improvements in fuel efficiency of the aviation sector.

Aircraft fuel burn per seat-mile has decreased dramatically over the last 50+ years (Fig. 3.1) (Rutherford, 2012). This improvement can be traced to many aircraft improvements including those in aircraft aerodynamics, vehicle weight, and aircraft engine efficiency. A large fraction of this improvement can be traced to aircraft engine fuel efficiency improvements enabled by increases in engine bypass ratio (BPR), cycle pressure ratio, turbine inlet temperature, and component efficiencies over the past 70 years. Many of these improvements were enabled by research efforts at GRC working in collaboration with NASA Langley Research Center, NASA Ames Research Center, universities, aircraft suppliers, and aircraft engine industry partners.

Before discussing details of the history of GRC contributions to improved fuel efficiency for aircraft engines, it is important to understand the underlying physics motivating these research and development efforts. The following discussion will walk through the mathematical equations describing aeropropulsion fuel efficiency before the remainder of the paper delves into specific contributions of GRC in this area.

The high-level starting point for any discussion of aircraft fuel efficiency is the well-known Breguet range Equation (3.1). In this equation, increased aircraft range can be viewed as a surrogate for reduced aircraft fuel burn for a fixed mission. The Breguet range equation shows that propulsion system contributions to improved aircraft range (and reduced fuel burn) come primarily through reduction in the propulsion system thrust-specific fuel consumption ($TSFC$, sometimes denoted as SFC), as well as secondarily from reductions in the engine weight:

$$\text{Aircraft range} = \frac{\text{Velocity}}{TSFC}\left(\frac{\text{Lift}}{\text{Drag}}\right)\ln\left(1 + \frac{W_{\text{fuel}}}{W_{PL} + W_0}\right) \qquad (3.1)$$

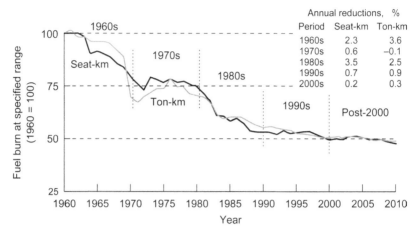

Figure 3.1.   Average fuel burn for new jet aircraft, 1960 to 2010. (From Rutherford, D. 2012. Overturning conventional wisdom on aircraft efficiency trends. The International Council on Clean Transportation, *http://www.theicct.org/blogs/staff/overturning-conventional-wisdom-aircraft-efficiency-trends*. Creative Commons CC BY-SA 3.0 license, *https://creativecommons.org/licenses/by-sa/3.0/legalcode* (accessed August 31, 2016).)

where *TSFC* represents engine thrust-specific fuel consumption, $W_{\text{fuel}}$ is fuel weight, $W_{PL}$ is payload weight, and $W_0$ is aircraft empty weight. The Velocity is that of the aircraft, and Lift and Drag represent the aerodynamic quantities of the aircraft performance. As will be explained later in Section 3.6, engine architectural changes can have a dramatic impact on the engine *TSFC*. Figure 3.2 plots the *TSFC* benefits resulting from some of these architectural changes such as the major trend from turbojet to low- and high-bypass turbofan engines that began in the 1960s and has continued to the present day.

TSFC can be further decomposed as shown in Equation (3.2). For a given aircraft flight velocity and fuel energy per unit mass it can be seen that *TSFC* is inversely proportional to overall efficiency $\eta_o$:

$$TSFC = \frac{\text{Velocity}}{\eta_o(\text{Fuel energy per unit mass})} \qquad (3.2)$$

Overall efficiency $\eta_o$ is primarily a function of propulsive $\eta_{pr}$ and thermal $\eta_{th}$ efficiencies (Eq. (3.3)), as transmission efficiency $\eta_{tr}$ is generally close to 1.0:

$$\eta_o = (\eta_{th})(\eta_{pr})(\eta_{tr}) \qquad (3.3)$$

Figure 3.3 shows a plot of thermal and propulsive efficiency trends spanning the history of jet engine development and highlights the mutual contributions of thermal and propulsive efficiency to the overall efficiency. It is also worth noting that even modern aircraft/engine combinations such as the Boeing 777/GE90 and more recently developed fuel-efficient aircraft including the Boeing 787 still leave room for future gains in both thermal and propulsive efficiency. It is also clear from the figure that the gains in overall efficiency are increased most by simultaneous improvements in the core engine and the propulsor. Section 3.6 will discuss some of the future concepts toward capturing those potential gains.

The gas turbine engine Brayton cycle ideal thermal efficiency $\eta_B$ is set by the pressure ratio of the cycle (*PR*, or also known as the overall pressure ratio (OPR)):

$$\eta_B = 1 - \frac{1}{TR} = 1 - \frac{1}{PR^{(\gamma-1)/\gamma}} \qquad (3.4)$$

where *TR* is the temperature ratio and $\gamma$ is the ratio of specific heats. This ideal thermal efficiency, however, assumes that the compression system components have no aerodynamic loss.

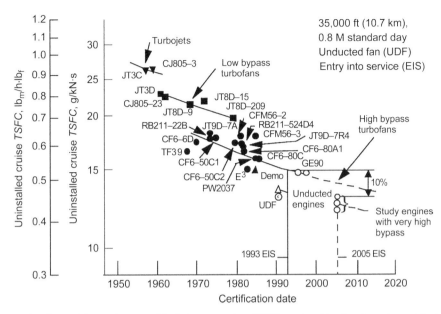

Figure 3.2.   State-of-the-art thrust-specific fuel consumption (*TSFC*) trends with subsonic engine architecture.

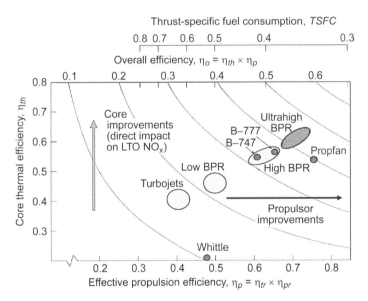

Figure 3.3.   Comparison of historical engine thermal ($\eta_{th}$) and effective propulsion ($\eta_p$) efficiency improvements. BPR, bypass ratio; LTO, landing and takeoff; $\eta_{tr}$, transmission efficiency. (From Epstein, A. H. 2014. Aeropropulsion for commercial aviation in the twenty-first century and research directions needed. *AIAA J.* 52:901–911. Reproduced by permission of United Technologies Corporation, Pratt & Whitney.)

The actual thermal efficiency of the cycle will depend on the component efficiencies in both the compressor and turbine. As will be seen later in Section 3.6.4 of this chapter, much work at GRC has supported improvements in these component efficiencies. In addition, this ideal thermal efficiency requires increased turbine inlet temperatures to fully realize the thermal efficiency potential. This is illustrated in Figure 3.4, which shows trends of ideal thermal efficiency and specific power for various cycle pressure ratios and turbine inlet temperatures. This figure

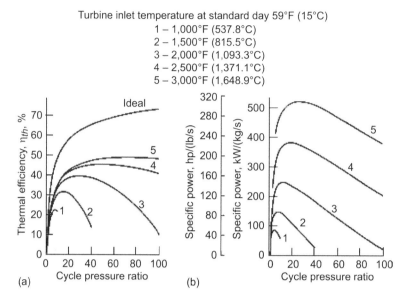

Figure 3.4.   Brayton cycle thermal efficiency and specific power trends.

illustrates that there is a synergistic relationship between cycle pressure ratio and turbine inlet temperature. Higher pressure ratios, and the accompanying advantages in thermal efficiency, must be coupled with complementary increases in turbine inlet temperature, or the pressure ratio advantage is lost. Additionally, Figure 3.4(b) demonstrates that an increased turbine inlet temperature results in a higher power density, higher thrust-to-weight engine regardless of pressure ratio. This explains the especially strong emphasis in military engines on increased turbine inlet temperature. OPRs have continued to rise for both aircraft and power turbine applications, reflective of the direct impact of this parameter on fuel burn reduction.

Figure 3.5 shows a historical progression of increased turbine inlet temperatures enabled by advanced cooling strategies as well as advanced materials. Figure 3.6 demonstrates these materials improvements, and includes the more recent application of thermal barrier coatings to further increase turbine inlet temperature capability. It should be noted that thermal barrier coatings are a technology that work synergistically with turbine internal cooling to reduce the underlying turbine metal temperature. Beginning with uncooled metals before 1960, turbine inlet temperatures have progressively increased due to the introduction of increasingly more sophisticated cooling designs and advanced materials, including thermal barrier coatings. It can be seen from Figures 3.5 and 3.6 that approximately two-thirds of the historical increase in turbine inlet temperature has been enabled by improved turbine cooling schemes and about one-third by improved turbine materials. As discussed in Chapter 5, the introduction of ceramic-based turbine base materials will offer a step-change in turbine inlet temperatures in the future.

Figure 3.7 highlights the strong benefit of increased turbine inlet temperature enabled by this cooling and materials development in the increased core specific power and resultant thrust-to-weight of the engine. In addition to enabling the benefits of higher engine cycle OPR and thermal efficiency, raising turbine inlet temperature ($T_4$) increases the thrust-to-weight of the engine which has dramatic benefits at the aircraft system level, particularly for military and high-speed flight. Consider also that as engine thermal efficiency and overall pressure ratio increase, the compressor exit temperature (combustor inlet temperature), $T_3$, increases due to increased compressive heating. For a fixed $T_4$, the amount of allowable energy addition in the combustor decreases and the thrust of the engine must decrease for a given engine core flow rate. Therefore, increasing allowable $T_4$ enables engines having acceptable thrust-to-weight and core power density. This also becomes important as the overall size of the engine is limited by airframe mounting considerations—for

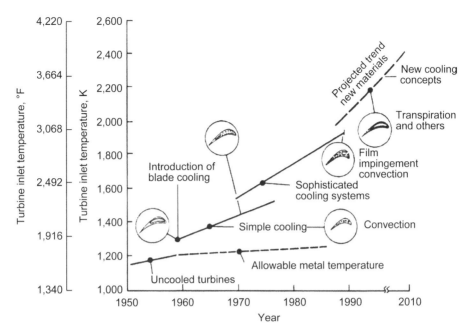

Figure 3.5. Turbine inlet temperature trends with technology improvements. (From Ballal, D., and J. Zelina. 2003. Progress in aero-engine technology, 1939–2003. AIAA 2003–4412.2.)

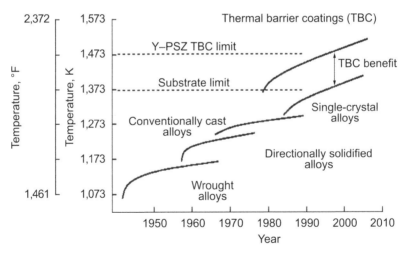

Figure 3.6. Turbine component material temperature capability improvements, showing increase in operational temperature of turbine components. Y-PSZ, yttrium partially stabilized zirconia. (Adapted from Schultze, U., C. Leyens, K. Fritscher, *et al.,* 2003. Some recent trends in research and technology of advanced thermal barrier coatings. *Aerosp. Sci. Technol.* 7:73–80. Copyright 2003, published by Elsevier Masson SAS. All rights reserved.)

a given thrust, higher $T_4$ can improve integration of the engine with the airframe. This becomes particularly important with the rise of very high-BPR engines and the resulting large fans used to provide thrust. Similar to the OPR trend discussed earlier, the trend toward higher turbine inlet temperatures for both aviation and industrial ground power applications of gas turbines has continued to the present day because of the dramatic fuel burn benefits.

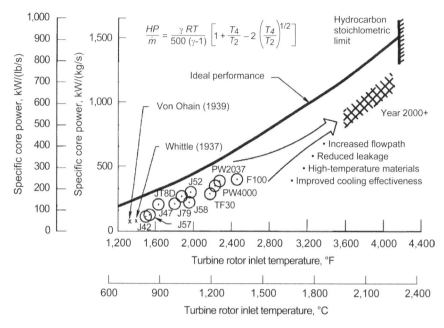

Figure 3.7.   Engine-specific power and thrust increase with turbine rotor inlet temperature, $T_4$. $HP$ is power, $\dot{m}$ is mass flow rate, $\gamma$ is ratio of specific heats, $R$ is ideal gas constant, $T$ is gas temperature, and $T_2$ is compressor inlet temperature. (From Koff, B. L. 1991. Spanning the globe with jet propulsion. AIAA 91–2187.)

Another technology which can potentially reduce aircraft fuel burn is the idea of engine boundary layer ingestion (BLI). It can be shown by control volume analysis that the ingestion of airframe boundary layer fluid into the engines can result in a net aircraft fuel efficiency benefit if the detrimental effects of the resulting non-uniform velocity profile entering the engine can be mitigated. For aircraft architectures which are proposed to benefit from BLI, a smaller, higher power density core can enable a greater percentage of BLI along with a higher BPR, both of which can further reduce fuel burn.

The propulsive efficiency of any thrust-producing device is given by

$$\eta_{pr} = \frac{2}{1 + \dfrac{c}{v}} \tag{3.5}$$

where $v$ is the flight speed of the vehicle and $c$ is the velocity of the air ejected by the thrust-producing device. This equation demonstrates that it is more propulsively efficient to eject a large quantity of low velocity air rather than a smaller quantity of high velocity air for a given thrust requirement. Engine propulsive efficiency is strongly dependent on engine architecture and flight speed with high-bypass turbofan engines being the architecture of choice for cruise speeds typical of the large commercial aviation market.

## 3.2   EARLY HISTORY OF NASA GLENN RESEARCH CENTER AEROPROPULSION FUEL EFFICIENCY EFFORTS, 1943 TO 1958

Improving aircraft fuel efficiency has been an economic consideration since the earliest days of aviation, beginning with the development of improved engines based on the reciprocating engine and propeller combination. Great strides were made in the early development of these propulsion systems, but it was the impetus given by World War II that accelerated the need for higher performance aircraft and initiated the formation of NASA Glenn Research Center. From its genesis in 1943 until its absorption into the new NASA Agency in 1958, GRC was instrumental in the development of turbojet and early turbofan engine technology. When Frank Whittle and Hans von Ohain independently began to develop the gas turbine jet engine or turbojet engine for flight application in the late 1930s, it sparked a revolutionary new means of aircraft propulsion, offering advantages for higher-speed flight over the conventional reciprocating engine and propeller combination. Both men realized that the jet engine was uniquely suited to providing power for flight because it was compatible with the flow of air through the engine, unlike a reciprocating engine (Conner, 2001).

In the early stages of turbojet engine development, the military was keenly aware of the obvious advantages of its high-speed flight capability and heavily supported development in Germany, Great Britain, and then in the United States. Because of fears of a German invasion, the British entered into an agreement to send plans for the Whittle engine to the United States in 1941. General Electric (GE) was chosen to develop the engine for production, owing to their expertise in superchargers. In March 1942, the General Electric I-A Whittle-derived engine ran on the test stand, and it flew in a two-engine arrangement on the Bell Airacomet XP-59A aircraft in October 1942. Although GRC had initially focused solely on air- and liquid-cooled reciprocating engines, the dawn of the jet engine age became a reality at the center with the delivery of the GE I-A for testing in the new Static Test Laboratory in 1943 (Dawson, 1991).

The GRC Altitude Wind Tunnel was completed in 1944 and was the first wind tunnel designed to test aircraft engines at simulated altitude conditions. The facility was large enough for both propeller and engine mount and was initially conceived for piston engine research, but was quickly converted to test turbojet and turboprop engines upon their introduction. Between 1944 and its conversion to a vacuum facility for rockets under the new NASA space directive in 1958, a great number of engine performance tests were conducted in the facility and led to dramatic improvements in turbojet and turboprop engine fuel efficiency. Among the success stories attributable to work in this important facility were the solution of cooling problems for the R-3350 engine in 1944, GE TG-180 and TG-190 (also known as the J-47) engine and afterburner performance tests from 1945 to 1950, Westinghouse 24C-7 and 24C-8 engine and afterburner performance and cooling tests from 1950 to 1952, and Allison J-71 and T-38 engine tests from 1952 to 1955 (Dawson, 1991).

One of the primary design choices in the development of early turbojet engines was between axial and centrifugal compressors. The centrifugal compressor of the Whittle and von Ohain concepts was simpler and more reliable, but the multistage axial compressor offered potential advantages in efficiency and pressure ratio if the complex aerodynamics and mechanical design issues could be mastered. The axial compressor quickly became the compression system of choice in the United States, but not without initial troubles refining early designs. These early challenges with multistage axial compressors highlighted the need for component research to improve efficiencies and enable the higher pressure ratios promised by the architecture. GRC took a leading role in component development in its Compressor and Turbine Division during the 1940s and 1950s. Many single-stage and multistage compressor tests were conducted at the Center in the Engine Research Building (ERB), providing essential data to the industry to validate both industry and NASA-developed models. The ERB was completed in 1942, again predominantly for piston engine component research, but was upgraded in 1944 to enable testing of compressors and turbines for jet engines. This unique facility is still in use today for testing of turbomachinery, combustion, heat transfer and other engine components. Among the turbomachinery component tests conducted in ERB in the late 1940s and 1950s were numerous experiments related to the Wright J-65 and GE J-47 engine series. In fact, testing in the Altitude Wind Tunnel and ERB enabled the success of the GE J-47 turbojet (General Electric company designation TG-190) as the first axial-flow turbojet approved for civil use in the United States in 1949. It was used in many types of military aircraft, and more than 30,000 were manufactured before production ceased in 1956. It saw continued service in the U.S. military until 1978. A culmination of this early period of compressor testing at GRC was published as a series of classified reports in 1956 and eventually declassified and republished in 1965 as "Aerodynamic Design of Axial-Flow Compressors," NASA SP-36 (Bullock and Johnsen, 1965). This NASA publication has provided great value to the axial compressor design community for many years and is still considered the authoritative publication on multistage axial compressor design theory and practice.

A similar research trajectory was playing out in the area of fundamental turbine heat transfer and cooling research during the period 1943 to 1957. A key figure in this effort was Ernst Eckert. Eckert had joined von Ohain and other German scientists in the United States after World War II and worked initially at the U.S. Air Force's Wright Field in Dayton, Ohio. In 1949 he came to GRC and led the Center's efforts at improving turbine cooling methods. Referring to Figure 3.5, the work conducted at GRC in this time period was instrumental in the industry acceptance of increasingly complex turbine cooling methods beginning with internally cooled hollow turbine blades and continuing toward more exotic cooling schemes such as film cooling and transpiration cooling.

In 1957, the Compressor and Turbine Division was disbanded as GRC moved toward nuclear and space research in response to the Soviet launch of Sputnik. Although turbojet and early turbofan engine development continued in the aviation industry during this period, it was not until 1966 that the Center turned its attention back to aeronautics research. By 1966, the commercial aviation industry had grown to the point that issues of capacity, congestion, noise, and pollution associated with airports had become a major issue. GRC was called upon to reinitiate aeronautics research and to help solve some of these growing issues.

## 3.3    INTRODUCTION OF TURBOFAN ENGINES AND IMPROVED PROPULSIVE EFFICIENCY

In the early 1960s, turbofan engines began to emerge—an engine architecture that would dramatically reduce fuel burn over the next several decades. As compressor pressure ratios were increased to enable improved thermal efficiency for turbojet engines, designers began to incorporate dual-spool compressor concepts to allow for better efficiency through optimal design speed of each spool. Initially, this design change was not intended to create a low pressure spool capable of providing an appreciable bypass flow and thrust, but slowly the additional benefits of increasing low spool bypass flow and the incorporation of a fan stage for thrust were introduced into production engines. Figure 3.8 shows cross-sectional schematics of low-bypass and the high-bypass turbofan engine architecture that has become the predominant large commercial aircraft propulsion architecture today.

When NASA reinitiated aeronautics research in 1966, turbofan engine development became a large part of the research focus. The Quiet Engine Program looked at engine noise benefits that would be enabled along with the fuel burn reduction benefits of higher bypass turbofans. Meanwhile, high bypass turbofan engines were being developed by the industry. General Electric was developing the CF6 engine, often considered the first high bypass commercial turbofan engine. The GE CF6 was developed out of their military TF-39 engine, and both the high-bypass fan used in the CF6 engine and the installation technology for high-bypass turbofan engines in general were based on developments made by NASA and military programs (U.S. Government, 1991). Figure 3.2 demonstrates that a step change in engine thrust-specific fuel consumption was enabled by the introduction of the JT9D engine in 1970 and the CF6 engine in 1971 as well as subsequent high-bypass-ratio turbofan engines.

## 3.4    ENERGY CRISIS OF 1970S AND NASA AERONAUTICS RESPONSE

Interest in aircraft fuel efficiency increased dramatically in the 1970s because of the sharp rise in jet fuel prices and their effect on the airline industry. Oil (and jet fuel) prices remained relatively

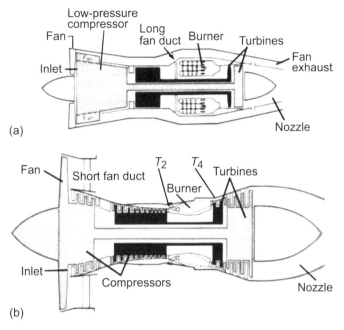

Figure 3.8. Turbofan engine architectures. (a) Low-bypass turbofan. (b) High-bypass turbofan. $T_2$ is compressor inlet temperature and $T_4$ is turbine inlet temperature.

stable for a long period from 1945 through the early 1970s, followed by spikes that increased fuel prices by a factor of 4 in the mid-1970s and a factor of 8 by 1980 (Shetty and Hansman, 2012). This dramatic increase in fuel prices put the airlines under deep financial pressure, driving Pan American World Airlines to the brink of bankruptcy in 1974 and ultimately to bankruptcy filing in 1991 (Bowles, 2010).

In response to this rapid increase in fuel prices, NASA established the Aircraft Energy Efficiency (ACEE) Program in 1975. The goal of this program was to accelerate the development of various aeronautical technologies that would make future transport aircraft up to 50% more fuel efficient. The baseline engines used for this goal were the Pratt & Whitney JT9D-7A and General Electric CF6-50C. The ACEE program was composed of six projects, three of which related to engine technology and led by NASA Glenn Research Center (Bowles, 2010).

The first of these projects was the Engine Component Improvement (ECI) Project. The goal of this project was to increase aircraft engine fuel efficiency by 5% through redesign of specific engine components. The second project was the Energy Efficient Engine (E3) Project. The E3 Project had more aggressive goals than the ECI Project in that the goal was to design a new engine rather than simply improve existing components. A goal of 12% fuel reduction (installed thrust-specific fuel consumption compared to the GE CF6-50C was established along with improvements in direct operating cost, noise, emissions, and performance retention. An often-overlooked goal of the E3 Project was a 50% reduction in the rate of performance deterioration compared to the CF6-50C. Since aircraft engines have a long lifespan in the commercial aviation fleet, the rate of performance deterioration can have a dramatic impact on overall fleet fuel consumption. This aspect is missed if one only compares performance of new engines. The third GRC project under the ACEE Program was the Advanced Turboprop Project. This project proposed to take the dramatic step of incorporating large, unducted propellers as the main propulsor for high-subsonic (Mach number $M$ approximately 0.8) commercial aircraft. It is well known that at lower flight speeds propellers offer lower SFC because of their low pressure ratio

Composite-bladed fan       Dual-annular combustor

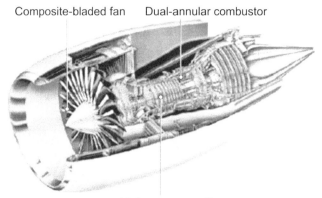

High-pressure-ratio compressor

Figure 3.9.   General Electric GE90 engine cross section.

and high effective bypass ratio. However, at $M = 0.8$ these benefits typically diminish dramatically because of high relative Mach numbers, and acoustic issues become problematic. But with fuel prices spiking, the promise of reduced fuel burn engine concepts such as the advanced turboprop was very enticing to the aviation industry.

The E3 Project from 1975 to 1984 developed many engine core technologies that were introduced into engine products into the 1990s and beyond. Specifically, GE's large GE90 engine (Fig. 3.9), which powers the Boeing 777 aircraft, benefited greatly from the E3 Project efforts. To summarize, the E3 Project goals were to (1) reduce SFC by 12%, (2) reduce SFC performance deterioration by 50%, (3) reduce direct operating costs by 5%, (4) meet Federal Aviation Administration noise regulations, and (5) meet EPA then-proposed emissions standards (Ciepluch *et al.*, 1987). The E3 Project achieved higher propulsive efficiency by using a low-pressure-ratio fan and higher thermal efficiency by using higher overall pressure ratio, higher turbine inlet temperatures, and improved component efficiencies. These are common themes in the effort to reduce SFC, and continue to be the main drivers for such efforts even today under NASA projects such as Environmentally Responsible Aviation (ERA) and the Subsonic Fixed Wing (SFW) Project.

Some of the features of the GE E3 effort included a 10-stage, 23:1 pressure ratio compressor (note that the compressor pressure ratio is only a part of the cycle OPR—one must include the fan, low-spool pressure ratio to arrive at the OPR), a highly efficient two-stage high-pressure turbine (HPT) and five-stage low-pressure turbine (LPT), and component efficiencies above the previous state of the art. Along with increased cycle temperatures, reduced turbine cooling flows were achieved through a combination of materials development and cooling concept improvement (Davis and Stearns, 1985).

The Advanced Turboprop Project under ACEE had a vision for a 20% to 30% fuel consumption reduction relative to then-current engines. Major challenges existed in making such an architecture viable for large civilian aircraft. Like propellers, turboprops (or "propfans") were more efficient at lower flight speeds because of the high relative tip Mach numbers associated with such large-diameter propulsors. The challenge was to enable highly efficient turboprop operation at Mach 0.8 flight speeds and higher altitude flight as well as to mitigate the noise issues inherent in unducted configurations, having no nacelle to shield and absorb radiated noise. The technical solution to both the noise and high-speed efficiency problems was to use swept blades more representative of fan blades than typical propeller blade shapes—hence the commonly used term "propfan." The swept blade geometry would result in a lower tip Mach number for a given flight speed and would potentially be able to offset the noise disadvantage of propfans (Bowles, 2010).

Figure 3.10. General Electric unducted fan engine. (From Domke, B. 2007. The GE36 Unducted Fan (UDF) used two contra-rotating propellers with eight blades each. *http://www.b-domke.de/AviationImages/ Rarebird/0809.html* (accessed April 28, 2017). Copyright 2007. With permission.)

Although much progress was made on the development of a viable propfan through both the NASA/Allison/Pratt & Whitney/Hamilton Standard single rotation concept and the later counterrotating GE "unducted fan" (UDF) concept (Fig. 3.10), various factors kept these concepts from coming to fruition in the market. First, potential negative public perception of propellor-like engine architectures made the airframers reluctant to deviate from their established commitment to turbofan engines, despite the large benefits in fuel burn reduction. Perhaps more importantly, fuel prices by 1986 had retreated back to nearly pre-1970 values in inflation-adjusted terms. This greatly reduced the urgency for the airline industry to adopt a radical change in engine architecture and ended heavy NASA investment in unducted configurations by the late 1980s. The idea would however return in the mid-2000s with the spike in fuel prices.

## 3.5 NASA'S ROLE IN COMPONENT TEST CASES AND COMPUTATIONAL FLUID DYNAMICS DEVELOPMENT

Throughout NASA Glenn Research Center's history, the use of the Center's unique experimental capabilities for compressor and turbine testing and the emphasis on providing return to the Nation on its taxpayer-funded research has resulted in the production of open experimental datasets. In the 1970s and 1980s GRC produced a number of compressor datasets that have been used by the turbomachinery community as a basis for the validation and development of turbomachinery analysis tools, including the growing field of computational fluid dynamics (CFD) codes. Laser Doppler velocimetry (LDV) was customized to measure the axial and tangential velocity inside the rotating passages of transonic compressors. The transonic fan NASA Rotor 67 was the first major dataset acquired with a single-channel LDV, which captured the shock and wake structure in an isolated transonic fan (Hathaway *et al.*, 1986; Strazisar, 1985, 1989; Wood *et al.*, 1987). Subsequently, NASA Stage 67 (Rotor 67 + Stator 67) was the first dataset that captured the unsteady fan rotor/stator blade row interactions with the same single-channel LDV system (Hathaway *et al.*, 1987; Suder *et al.*, 1987). A two-channel laser anemometer system was later developed and utilized to measure both axial and tangential velocity components simultaneously in NASA Rotor 37 (Reid and Moore, 1978). NASA Rotor 37 is perhaps the most widely referenced compressor geometry for such datasets, having been the basis for the American Society of Mechanical Engineering's (ASME's) International Gas Turbine Institute CFD blind test case. NASA Rotor 37 has an extensive set of LDV data across the rotor operating range from maximum flow to near-stall conditions at 70% speed (fully subsonic), 80%

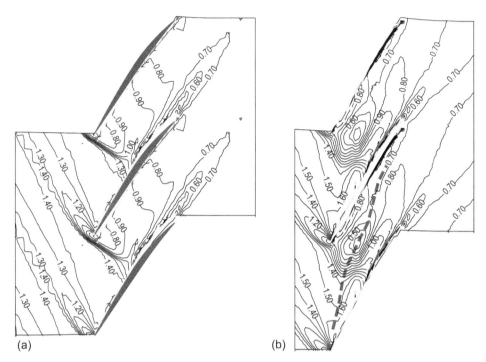

(a)                                                    (b)

Figure 3.11.    NASA Rotor 37 Laser Doppler velocimetry data. (From Suder, K. L. 1996. *Experimental investigation of the flow field in a transonic, axial flow compressor with respect to the development of blockage and loss*. Ph.D. Thesis (NASA TM–107310), Case Western Reserve University, Cleveland, OH.)

and 90% speed (transonic), and 100% design rotor speed (fully supersonic in the rotor frame of reference). The data are best summarized in Suder (1996), and an example of the measurement detail is provided in Figure 3.11, which shows the shock boundary layer interaction at 70% span and shock/tip leakage vortex interaction at 95% span for a 0.5% span rotor tip clearance. The ASME blind test case results, shown in Figure 3.12, compare the NASA Rotor 37 experimental and CFD results of overall performance at 100% design speed as well as the radial distribution of pressure ratio, temperature ratio, and efficiency. The eight CFD codes in Figure 3.12 represented the state-of-the-art (SOA) prediction tools from around the globe in 1994. Note the discrepancies not only in the level of the performance parameter but also the shape of the radial distribution, which indicated the codes were not accurately predicting the flow physics of this compressor rotor in isolation. The Advisory Group for Aerospace Research and Development also used the NASA Rotor 37 benchmark data set to compare results from a large number of Navier-Stokes CFD codes (Dunham, 1998). These test case activities highlighted the large range of results produced by the various codes, some of which is attributable to how the codes were employed in addition to the underlying code algorithms and methods. These discrepancies between the CFD and experimental results have led to significant improvements in CFD mesh generation, turbulence model implementation, and tip clearance modeling.

Additional experimental test cases produced by GRC include the NASA Stage 35 (Van Zante *et al.*, 2002) which incorporates a full compressor stage versus the rotor-only approach of the Rotor 37 test case. In addition, NASA built a 5-foot diameter (5 ft = 1.524 m) centrifugal compressor to make detailed measurements for code validation and the results are summarized in Hathaway *et al.* (1993). Centrifugal compressor scaling studies (Skoch and Moore, 1987) and code validation datasets (Skoch *et al.*, 1997) were used to improve centrifugal compressor CFD codes and the resulting designs. In the turbine area, an example of one of the widely employed

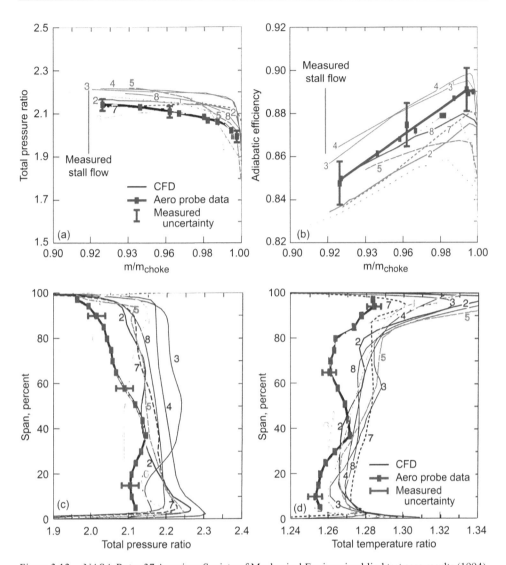

Figure 3.12.   NASA Rotor 37 American Society of Mechanical Engineering blind test case results (1994). CFD is computational fluid dynamics, $m$ is the mass flow rate and $m_{choke}$ is the choking mass flow rate. (From Suder, K. L. 1996. *Experimental investigation of the flow field in a transonic, axial flow compressor with respect to the development of blockage and loss.* Ph.D. Thesis (NASA TM–107310), Case Western Reserve University, Cleveland, OH.)

test cases is the NASA Transonic Cascade Heat Transfer dataset (Giel *et al.*, 1999), which has been used to validate turbine heat transfer tools across the community (Fig. 3.13). For example, these endwall heat transfer data were instrumental in the development and assessment of the v2-f and Spalart-Allmaras turbulence models (Durbin and Reif, 2001).

NASA has also directly contributed to CFD analysis improvement through development of NASA in-house turbomachinery codes that have contributed to the body of knowledge in the field. A prime example of this contribution is the APNASA code (Adamczyk, 1984). This Navier-Stokes code offers the ability to accurately model the deterministic impact of blade rows throughout a multistage turbomachine without the massive time and expense that would be required to resolve the unsteady full-wheel flowfield for all stages. This is particularly important

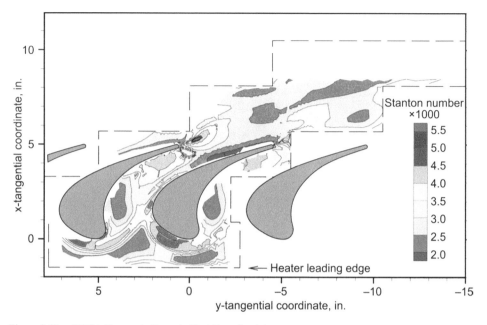

Figure 3.13.    NASA Transonic Cascade Heat Transfer data.

for multistage compressors, where such an unsteady calculation would be prohibitive, even with today's computers. The APNASA code has been distributed to the U.S. aircraft and industrial gas turbine industry and is in common use today. Other NASA-sponsored Navier-Stokes CFD codes that have made a substantial impact on the turbomachinery analysis field include Glenn-HT, TURBO, H3D, ADPAC, and SWIFT. The Glenn-HT code development has focused on turbine cooling and heat transfer applications. It has incorporated the ability to resolve the complicated turbine cooling passages and film cooling holes that were discussed earlier in this chapter as methods to increase turbine inlet temperatures (Fig. 3.14). Several first-of-their-kind demonstrations of turbine heat transfer analyses have been carried out using the Glenn-HT code including internal passage heat transfer, film-cooled external heat transfer, and turbine tip clearance heat transfer. The TURBO code was developed under GRC funding and enables full unsteady Navier-Stokes simulations of multistage compressors and turbines. This kind of unsteady analysis capability has found excellent application in studying the impact of distorted inlet flows on downstream fan aerodynamic performance.

APNASA, TURBO, Glenn-HT, H3D, and SWIFT were all recently validated against NASA rotor 37 and NASA stage 35 test cases as part of a NASA turbomachinery code assessment activity. The results were reported at the 2009 AIAA Aerospace Sciences Meeting (Ameri, 2009; Celestina and Mulac, 2009; Chima, 2009; Hah, 2009, Herrick *et al.*, 2009). The results indicated strong agreement among the codes for compressor speedline and stall, with some of the codes highlighting detailed flow phenomena such as leakage flows, resulting in better exit flow profile prediction using advanced modeling techniques such as unsteady Reynolds-averaged Navier-Stokes equations, large eddy simulation, and detailed spatial resolution of small geometric and flow features. NASA CFD developments and applications to turbomachinery problems have contributed significantly to turbomachinery flow physics insight from synergistic computational and experimental investigations. Such turbomachinery flow physics features as shock structure, tip leakage flows, turbine cooling flows, blade row interaction, stall inception and flow control have been studied and better understood through GRC efforts.

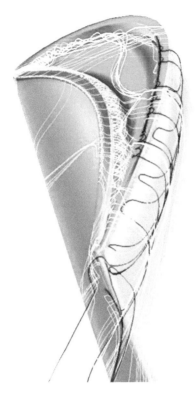

Figure 3.14.   Turbine tip flow structures predicted with modern computational fluid dynamics (CFD).

## 3.6   CURRENT NASA EFFORTS AT REDUCED FUEL CONSUMPTION

Recent NASA system studies conducted under the Subsonic Fixed Wing and Environmentally Responsible Aviation Projects indicate that the propulsion system plays a large role in the predicted improvement in aircraft fuel burn for the N+2 timeframe (engine technology readiness level (TRL) 6 by 2020 with potential entry into service (EIS) by 2025) (Fig. 3.15), as well as for the N+3 timeframe (about 5 years beyond N+2). In Figure 3.15, advanced engine technologies of all kinds, including both core and propulsor improvements, are included in the large bar representing engines. A smaller bar of 3.3% represents the potential benefit of boundary layer ingestion for the "accelerated technology development" configuration. Airframe technologies represented include large contributions from hybrid laminar flow control (a way to reduce airframe drag by reducing turbulent boundary layer shear), Pultruded Rod Stitched Efficient Unitized Structure—an advanced composite structure that may enable the hybrid wing-body (HWB) concept, and the large effect of the HWB concept itself as a fuel burn reduction technology owing to its improved lift-to-drag ratio relative to the traditional tube-and-wing configuration. Note that the reduction in aircraft size, drag, and weight due to engine fuel burn reduction is categorized under aircraft improvements for bookkeeping purposes. Therefore, the engine technology plays a more significant role than Figure 3.15 portrays to provide more fuel-efficient aircraft. This fact is evidenced by the latest trend to re-engine commercial aircraft such as the Airbus 320 and Boeing 737 as opposed to developing a new aircraft.

The SFW Project's N+3 studies have focused primarily on advanced aircraft configurations, which can serve as "collectors" for technologies that may apply to multiple long-term aircraft concepts. Among the key propulsion technologies identified by these N+3 studies are more

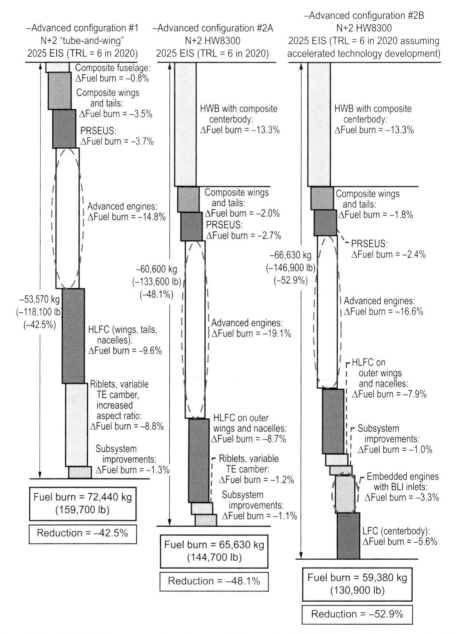

Figure 3.15.   NASA fuel burn reduction estimates for future aircraft. BLI, boundary layer ingestion; EIS, entry into service; HLFC, hybrid laminar flow control; HWB, hybrid wing-body; LFC, laminar flow control; PRSEUS, Pultruded Rod Stitched Efficient Unitized Structure; TE, trailing edge; TRL, technology readiness level.

compact, high-efficiency gas generators, higher bypass ratios enabled by various methods of distributed propulsion, boundary-layer-ingesting engines, and hybrid turbo-electric engines using either battery or fuel cell energy sources that have potential for significant reduction in emissions, fuel burn, and noise.

Much of the "leveling-off" of aircraft fuel burn reductions seen in Figure 3.1 from 1990 onward are attributable to the relatively stable price of jet fuel from 1985 to the early 2000s. However, increased energy prices tends to place a renewed emphasis on both alternative aircraft and engine architectures as well as more aggressive engine core technology development leading to higher overall pressure ratio and turbine inlet temperature ($T_4$) cycles. In addition, global warming fears have risen during this past decade, a factor that places additional emphasis on reducing aircraft fuel burn and resultant carbon dioxide emissions. The following sections will describe recent NASA and industry efforts at meeting this need for the aviation sector.

### 3.6.1 *Open-rotor propulsors*

Propulsion systems incorporating open rotors have the potential for game-changing reductions in fuel burn because of their low fan pressure ratio and thus increased propulsive efficiency. To reduce aircraft fuel burn, open rotors (or propfans or unducted fans, as they were known then) were studied in the late 1980s under the NASA Advanced Turboprop Project as a result of the aforementioned oil price spikes of the previous decade. The UDF, or GE36 engine, was one example of this development effort. The UDF was installed on the MD-80 aircraft as a flight demonstration of the technology. Because of limitations of the design and modeling methodology, it was necessary to compromise the GE36 aerodynamics so the engine could meet noise goals. When oil prices dropped in the 1990s, technology development in the area of high-speed propellers ended. Recent uncertainty in oil prices in combination with climate change concerns and the desire for reduced emissions has resulted in a renewed interest in open-rotor systems.

NASA has been collaborating with General Electric Aviation and the Federal Aviation Administration to explore the design space for lower noise while maintaining the high propulsive efficiency from a counter-rotating open-rotor system. Candidate technologies for lower noise were investigated as well as installation effects such as pylon and fuselage integration. Advances in computational fluid dynamics over the last 20 years enable three-dimensional (3D) tailoring of blade shapes to minimize noise while still maintaining efficiency. These modeling advances increase the possibility of meeting both noise and efficiency goals simultaneously for the new generation of open-rotor designs. Figure 3.16 shows an open-rotor model tested at NASA Glenn Research Center recently.

During the test campaign six different blade sets or unique combinations of fore and aft blades were evaluated for their aerodynamic performance and acoustic characteristics. One of the blade sets, the Historical Baseline blade set, is representative of 1990s blade design. Aerodynamic and acoustic measurements of the Historical Baseline blade set were used as a benchmark dataset to improve modeling and simulation capabilities for open rotors. The other five blade sets represent modern designs that incorporate various 3D design features and other strategies to reduce the acoustic signature but maintain performance. The open-rotor test campaign is documented in Van Zante (2013) and Van Zante *et al.* (2014), and the following paragraphs provide a brief synopsis of the activity.

The open-rotor test program consists of three phases: (1) takeoff and approach aerodynamics and acoustics, (2) diagnostics, and (3) cruise performance. For phases 1 and 2 the Open Rotor Propulsion Rig (ORPR) is installed in the 9- by 15-Foot Low-Speed Wind Tunnel (9×15 LSWT) at GRC. ORPR was completely refurbished for the current test entry and also underwent significant upgrades such as a new digital telemetry system for rotor force and strain gage monitoring. For the third phase of testing the rig was installed in the 8- by 6-Foot Supersonic Wind Tunnel (8×6 SWT) for cruise performance testing.

NASA acquired a substantial amount of aerodynamic and acoustic data on a variety of blade geometries for an isolated configuration during the phase 1 testing. Figure 3.17 (Suder *et al.*, 2013) compares the fuel burn and noise levels of the GE36 (1980s open rotor) and turbofan engine to a modern open-rotor design. It is clear from Figure 3.17 that the modern open-rotor designs provide significant improvements in both fuel burn and noise relative to the 1980s GE36 UDF design; thereby making them a viable propulsor concept for the next generation of fuel-efficient aircraft.

Figure 3.16.   NASA-General Electric open-rotor testing configuration.

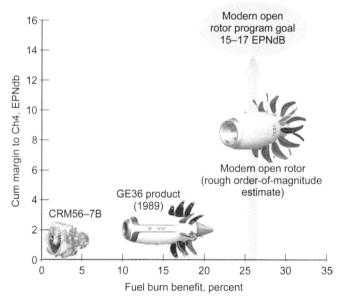

Figure 3.17.   Modern open rotor designs provide greater than 25% reduction in fuel burn and about 15 EPNdB noise margin to International Civil Aviation Organization Chapter 4 standard (ICAO, 2008). (From Suder, K. L., J. Delaat, C. Hughes, D. Arend, *et al.*, 2013. NASA environmentally responsible aviation project's propulsion technology phase I overview and highlights of accomplishments. AIAA 2013–0414. Work of the U.S. Government.)

The diagnostics program acquired a comprehensive, detailed data set, which is not only useful for modeling these systems but also for understanding how future progress is possible. Measurements were acquired in an isolated configuration as well as with the generic pylon installed upstream of the rotors. The pylon installed data will be needed to assess the aerodynamic and acoustic penalties associated with an aircraft installation. Four different measurement techniques were applied during the diagnostics testing, each with a specific objective. The acoustic phased-array technique identified noise source locations on the blades as well as on the trailing edge of the pylon. Farfield acoustic data were acquired with the pylon installed to determine the acoustic "adder" that must be applied to account for a realistic installation on the aircraft. Pressure-sensitive paint was used to quantify the magnitude and infer information about the time history of static pressure fluctuations on the forward and aft rotor airfoils as well as the trailing edge of the generic pylon. Stereo particle image velocimetry is the fourth measurement technique and was used to quantify the velocity characteristics and trajectory of the forward rotor wakes and tip vortex in support of tone noise predictions. In addition, second-order statistics (turbulence intensities) were determined from the measurements in support of broadband noise predictions.

Phase 3 of the test campaign determined the rotor aerodynamic performance at cruise Mach number of approximately 0.78 in the GRC 8×6 SWT. In addition, unsteady pressure field measurements were acquired near the rotor tips from a linear array of pressure transducers mounted in a translating plate. This type of data is useful in analyzing the rotor pressure field interaction with the aircraft fuselage.

The data gathered and understanding obtained from the testing will be instrumental in solving some of the challenges in making open-rotor systems viable. The future design intent is to use improved aero and acoustic tools to mitigate the installation effects. In order to perform a direct comparison of an open-rotor system to a high-BPR ducted propulsor, NASA designed a common aircraft platform to compare the tradeoff between fuel burn and noise reduction (see Hendricks *et al.*, 2013). The NASA notional aircraft design was a modern 162-passenger airplane with rear fuselage-mounted engines and having a cruising Mach number of 0.78 at 35,000 ft and a mission range of 3250 nautical miles. A comparison of the fuel burn and noise for the open-rotor and ducted high-bypass propulsors are shown in Figure 3.18. The aircraft with the open-rotor

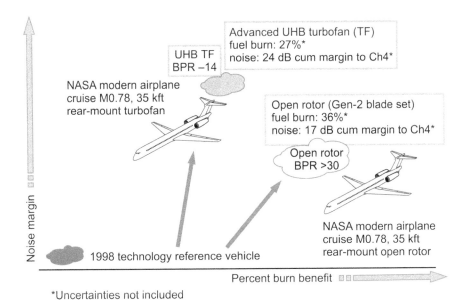

Figure 3.18.   Comparison of advanced turbofan and open rotor on common aircraft platform. BPR, bypass ratio; TF, turbofan; UHB, ultrahigh-bypass. (From Hendricks, E. S., J. J. Berton, W. J. Haller, *et al.*, 2013. Updated assessments of an open-rotor airplane using advanced blade designs. AIAA 2013–3628. Work of the U.S. Government.)

propulsor provided an additional 9% reduction in fuel burn despite the increased weight of the engine and at the expense of an increase of 7 dB cum in noise relative to the ducted propulsor for this notional aircraft size and mission.

In summary, the modern open-rotor designs provide significant margin in Stage 4 noise requirements and offer substantial reductions in fuel burn. In addition, more research on installation effects and certifications must be addressed before open-rotor propulsion systems are installed on commercial aircraft. Also, it is unlikely that open-rotor systems will be able to match the acoustic margin of ducted systems because open-rotor systems by definition have no duct (and acoustic liner), and as a result, have greater flow and acoustic interactions with the airframe. The next section discusses the development of the ultrahigh-bypass (UHB) ducted propulsor, where the question to consider is, "Will the modern geared-turbofan engine, once optimized, provide comparable fuel burn reductions as an open-rotor system?"

### 3.6.2    *Ultra-high-bypass engine cycle research*

NASA's aggressive noise and fuel burn reduction goals are driving aircraft engine designs to higher bypass ratios and larger fan diameters. Aircraft engine noise and fuel burn reduction are directly correlated to fan size, fan pressure ratio and fan bypass ratio. As the fan size increases, there is a corresponding drop in fan pressure ratio and an increase in fan BPR. At some point, as the fan size continues to increase, a minimum is reached between fan size and weight and drag. The larger, heavier nacelle produces more drag during flight, and overcomes the advantages of a larger fan. Hence, a technology paradigm shift is needed to reduce the minimum point, which is produced by introducing advanced fan and core technology. A shift of this type was produced by Pratt & Whitney (P&W) with their geared-turbofan (GTF) UHB engine design. UHB engines are defined as engines with a fan BPR equal to or greater than 12. NASA in cooperation with P&W has been investigating UHB technology over the last 20 years, but the GTF is the first generation of UHB engines that will see EIS with an aircraft manufacturer. The paradigm shift produced by the GTF is achieved by operating the fan and core in such a way as to optimize the performance of both. Direct-drive turbofans necessarily operate the fan and low-pressure turbine at the same speed. At low fan speeds, the LPT is operating at faroff-design conditions, and its efficiency goes down, increasing fuel burn. P&W introduced a gearbox into their GTF engine design that allows the fan and LPT to operate at different speeds—thus more optimum, higher efficiency conditions—and so reduced fuel burn. As BPR increases, the mean radius ratio of the fan and LPT increases. Consequently, if the fan is to rotate at its optimum blade speed, the LPT will spin slowly so that additional LPT stages will be required to extract sufficient energy to drive the fan. Introducing a planetary reduction gearbox with a suitable gear ratio between the low-pressure shaft and the fan enables both the fan and LPT to operate at their optimum speeds. A geared turbofan uses a larger fan that moves more air at a lower speed, allowing the same thrust as its nongeared counterpart, but with less energy expended.

Fan propulsive efficiency increases with decreasing fan pressure ratio, but direct-drive turbofans are limited in their ability to operate at very low fan pressure ratios. The GTF architecture can enable further reductions in fan pressure ratio compared with direct-drive turbofans, thereby increasing propulsive efficiency and reducing fuel burn. The fan pressure ratio curve for the first generation GTF is between 1.2 and 1.5, but as the fan BPR increases the fan pressure ratio decreases. So the next-generation GTF will be required to operate at the lower end of the fan pressure ratio curve, and at a significant increase in fan BPR, to achieve the second paradigm shift necessary to reduce the fuel burn minimum point even further.

The UHB engine technology associated with the first-generation P&W GTF was close to reaching NASA's N+1 noise and fuel burn reduction goals, but additional technologies are needed to achieve the N+2 goals. Figure 3.19 illustrates the technology roadmap NASA is following and the additional technologies that will be needed for not only the propulsion system but for the aircraft system as well. Whereas the first-generation GTF operated at a BPR of around 9 to 12, the second-generation GTF will necessarily need to operate at a BPR from 15 to 18,

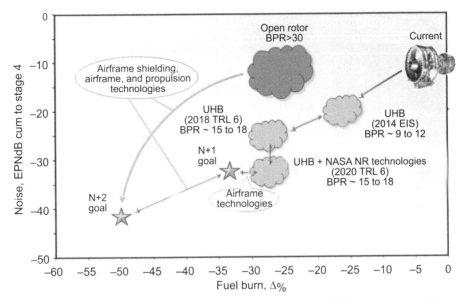

Figure 3.19.   Ultrahigh bypass (UHB) propulsion technology roadmap. BPR, bypass ratio; EIS, entry into service; TRL, technology readiness level. (From Suder, K. L., J. Delaat, C. Hughes, *et al.*, 2013. NASA environmentally responsible aviation project's propulsion technology phase I overview and highlights of accomplishments. AIAA 2013–0414. Work of the U.S. Government.)

possibly as high as 20, with correspondingly lower fan pressure ratios between 1.2 and 1.4. As a result, NASA and P&W have again teamed to develop propulsion and noise reduction technology for the next-generation GTF.

NASA and P&W have been collaboratively designing a scale model of the GTF Gen 2, with a 22-in. (56-cm) fan diameter, for testing in the GRC 9- by 15-Foot Low-Speed Wind Tunnel. This test will investigate new three-dimensional fan geometries and advanced inlet designs to increase propulsive efficiency and lower nacelle weight. At the same time, new variable-area nozzle (VAN) technologies are being investigated. Because of the wide range of flight conditions the UHB propulsion cycle must operate over, the fan nozzle area is required to vary as much as 50% to achieve the proper fan operating conditions. However, traditional VAN designs are heavy, and so NASA is investigating advanced, lighter weight designs using shape memory alloy technology. Investigating advanced noise-reduction technologies are also in the NASA plans to meet the aggressive noise goals. The next generation of over-the-rotor acoustic treatment (OTR) and acoustically treated soft vanes (SVs) is focusing on achieving 3 to 4 dB of noise reduction with a minimal impact on aerodynamic performance, optimally less than 0.5% in fan efficiency, including testing of the advanced OTR/SV designs using existing 22-in.-scale-model turbofan test hardware.

### 3.6.3   *Boundary-layer-ingesting engines*

Embedded engines with boundary layer ingestion offer an additional fuel burn benefit of up to 5% to 10% because of their reacceleration of fluid slowed by the viscous drag of the vehicle. This technology benefits the propulsive efficiency of the vehicle as described in Equation (3.5) by reducing the jetting velocity ($c$) compared to a podded engine and reducing the vehicle wake deficit (see Fig. 3.20). The potential benefit depends upon the percentage of the boundary layer from the vehicle ingested into the engines, so some concepts attempt to capture a larger percentage of this boundary layer by using distributed propulsors across the upper surface of the vehicle. Blended-wing-body vehicles offer an attractive method to leverage boundary-layer-ingesting

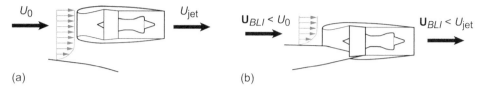

(a)                                        (b)

Figure 3.20.   Propulsion benefits of boundary layer ingestion (BLI), in terms of blade tip speed $U$ relative to station 0, upstream of engine. (a) Conventional (jet) propulsion. (b) BLI propulsion.

Figure 3.21.   NASA-Boeing blended wing-body concept.

Figure 3.22.   NASA, Massachusetts Institute of Technology, and Pratt & Whitney double bubble aircraft concept.

engines because of their larger surface area, which results in a larger boundary layer and more flexibility in engine mounting on the upper surface of the lifting body. Figure 3.21 shows a concept from ingesting the boundary layer on the NASA-Boeing blended wing-body aircraft. Figures 3.22 and 3.23 show additional BLI-related concepts, including the NASA-Massachusetts Institute of Technology-Pratt & Whitney "double-bubble" configuration and the NASA in-house Turboelectric Distributed Propulsion concept, respectively.

One of the challenges for BLI engines, however, is the potential loss in fan efficiency and degradation of life due to the periodic distortion experienced by the rotating fan. NASA and United Technologies Research Center (UTRC) are jointly investigating fan designs that can mitigate this problem and flow control technologies that can make the fan inflow more uniform. The goal is to demonstrate an embedded integrated inlet and distortion-tolerant fan system that provides the identified aircraft benefits by achieving less than a 2% loss in fan efficiency while maintaining ample stability margin. The study used an existing NASA Research Announcement (NRA)

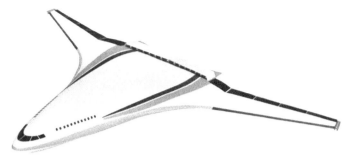

Figure 3.23.    NASA turboelectric distributed propulsion concept.

sponsored blended-wing-body design such as is depicted in Figure 3.21, to define the design constraints for the inlet boundary layer and the requirements for a relevant embedded engine configuration. NASA partnered with UTRC, Pratt & Whitney Aircraft Engines and Virginia Polytechnic and State University (Virginia Tech) through the NRA to exploit the optimal design space and to design and build an integrated inlet and fan embedded system. A sampling of the relevant publications supporting this activity inclusive of simulated aircraft boundary layer, the embedded inlet and distortion tolerant fan design, and the aeromechanics analysis is found in Arend *et al.*, 2012; Florea *et al.*, 2012; Bakhle *et al.*, 2012; Bakhle *et al.*, 2014; Tilman *et al.*, 2011; Ferrar *et al.*, 2009; Florea *et al.*, 2009.

NASA is testing a distortion-tolerant fan with a relevant boundary layer inflow field in the 8- by 6-Foot Supersonic Wind Tunnel at GRC. The arrangement of this embedded propulsor experiment is shown in Figure 3.24 where a false floor was inserted in the tunnel to mount the inlet/fan hardware. Note the rods located far upstream of the embedded fan inlet to provide a thick inlet boundary layer. Downstream of the rods and upstream of the inlet, the false floor contains a porous section to provide bleed control to adjust the incoming fan/inlet boundary layer to simulate that of a HWB vehicle such as the one shown in Figure 3.21. The main objective of the test is to assess the ability of the fan to sustain high performance with minimal loss and to maintain a sufficient stability margin. The test is in progress as of this writing and is expected to finish before summer 2017. Through this effort, distortion-tolerant fan technology and system-level benefits will be validated along with the design and analysis tools required to model the relevant physics.

### 3.6.4    *Engine core research*

The previous sections on open-rotor propulsors, ultrahigh-bypass engines, and boundary-layer ingesting engines addressed improvements in propulsive efficiency. Returning to Figure 3.3, recall that it is imperative to make improvements in both propulsive efficiency and thermal efficiency in order to make the biggest impact on overall engine efficiency and resulting fuel burn reductions. In this section the areas of NASA research and development to improve thermal efficiency are presented.

In the core turbomachinery area, the emphasis is on increasing the overall pressure ratio of the compression system while maintaining or improving aerodynamic efficiency and increasing the turbine inlet temperature ($T_4$) while reducing nitrogen oxide ($NO_x$) emissions from the combustor. These are challenging goals, because in both cases these are competing constraints. Another challenge is related to the need for more compact, high-OPR, high-bypass-ratio engines. These competing demands require ever-smaller rear compressor stage blade heights along with increased combustor inlet temperature $T_3$ values. One potential solution to these demands is an axi-centrifugal compressor, whereby the rear axial stages of the multistage compressor are

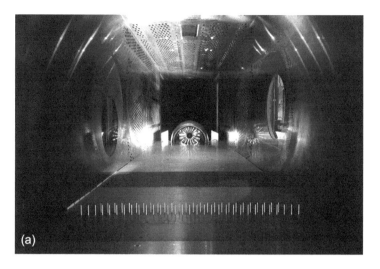

Figure 3.24.    Boundary layer ingestion (BLI) fan test rig installed in NASA Glenn 8- by 6-Foot Supersonic Wind Tunnel (8×6 SWT). (a) Bars upstream of fan are used to thicken boundary layer, and downstream bleed plates are used to customize boundary layer upstream of fan inlet. (b) close-up of the integrated inlet and fan installation in the 8×6 SWT.

replaced by a centrifugal rear stage that would be able to operate at a higher efficiency for the small corrected mass-flow values required of such cycles. Higher temperature materials and/or innovative cooling schemes would potentially be required to enable this concept. NASA is currently studying this and other potential solutions to this challenging problem. NASA has also recently funded a set of NASA Research Announcement awards focusing on better understanding and mitigating turbine and compressor tip clearance flows, which can enable reduced aerodynamic loss and increased pressure ratio cycle engines. The awards are also producing experimental data for use in computational fluid dynamics validation efforts across the turbomachinery community, in the ongoing spirit of NASA-led development of turbomachinery experimental databases. Refer to Reid and Key (2015), Volino (2017), and Katz (2017).

Increasing compressor OPR either drives the design toward more stages or higher stage loading, in which case the overall efficiency of the compression system tends to suffer because of

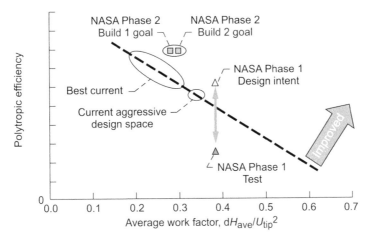

Figure 3.25. Compressor design space for Environmentally Responsible Aviation Phase 1 and Phase 2 relative to the state-of-the-art best current practices as indicated by the dashed line, representing the change in enthalpy $dH_{ave}$ divided by the square of the rotor tip rotational speed $U_{tip}^2$. (From Van Zante, D. E., and K. L. Suder. 2015. Environmentally responsible aviation: propulsion research to enable fuel burn, noise, and emissions reduction. ISABE 2015–20209. Work of the U.S. Government.)

either increased wetted area and drag losses or increased boundary layer separation and mixing loss, respectively. Overall engine size and weight constraints, engine operability, and rotor dynamics issues can also limit the use of additional compressor stages, so often the solution to increased OPR is higher stage loading. The emphasis within the industry and in the NASA research programs is to push the component efficiency-loading curve higher such that either a higher efficiency can be attained at a given loading or a higher loading can be achieved for a given component efficiency.

### 3.6.4.1 *NASA ERA core compressor technology development efforts*

The NASA Environmentally Responsible Aviation Project focused on the compressor technologies to enable high-efficiency and high Overall Pressure Ratio core engines. Specifically, the goal of the ERA highly loaded compressor activity was to increase efficiency and to increase pressure rise by 30% relative to the ERA baseline engine (GE90 engine on the 777-200) to achieve a 2.5% reduction in engine specific fuel consumption. Refer to Suder *et al.* (2013) and Van Zante and Suder (2015) for background on the NASA ERA Propulsion activities. Two test and analysis campaigns explored the design space to improve the compressor OPR (blade loading) and efficiency without negatively impacting weight, length, diameter, and operability. The first test campaign (NASA ERA Phase 1) investigated the front two stages of a legacy high-pressure ratio six-stage core compressor to determine what limits blade loading. The second test campaign (NASA ERA Phase 2) focused on two builds of the front stages of a new compressor design. A pictorial view of the design space explored is found in Figure 3.25. The dashed line represents state of the art for blade loading (represented as the change in enthalpy divided by the square of the rotor tip rotational speed) and efficiency. As shown, the higher that the blade loading is the more difficult it is to achieve high efficiency. Any compressor with a design point above the dashed line would represent a design that was better than the SOA.

In ERA Phase 1, a legacy high-OPR compressor design that fell short of the efficiency design goals was investigated. This design pushed the SOA design space to higher blade-loading levels (pressure rise per stage) with increased efficiency relative to the best current designs. Unfortunately, the efficiency goals were not obtained at this high blade loading (refer to Fig. 3.25).

The high losses were attributed to the front two stages of this highly loaded six-stage compressor design. The front two stages are transonic across the span, and therefore their performance is very sensitive to variations in the effective flow area, which can affect the location and strength of the passage shocks and further impact flow separations and/or low momentum and loss regions due to the shock and/or blade row interactions. Therefore, the goals in ERA Phase 1 were to isolate, analyze, and test the first two stages of a transonic SOA high-pressure compressor in order to (1) understand the flow physics that resulted in high losses, (2) characterize the blade row inter-actions and their impact on loss, and (3) validate the design methodology and capability of the prediction tools by comparisons with the experimental results.

NASA tested the first two stages using SOA research instrumentation to investigate the loss mechanisms and interaction effects of embedded transonic highly loaded compressor stages. The high-speed multistage compressor test facility, W7 in the Engine Research Building at NASA Glenn Research Center, was used to run this test. The inlet to the core compressor modeled the inlet conditions to a high-pressure compressor (HPC) of an engine, inclusive of fan frame struts and a transition duct from the low-pressure compressor (LPC) to the HPC compressor. The test plan focused on making steady and unsteady measurements for the single stage and then again after adding the second stage to enable evaluation of the performance and losses in each stage. This approach enabled the ability to sort out the loss contributions from each stage and provided detailed data to define the inlet boundary conditions to the compressor.

For both 1- and 2-stage configurations, detailed data were taken at 97% design speed, acquiring data from leading-edge (LE) instrumentation, wall statics, over the rotor Kulites (a piezo-electric device to measure instantaneous pressure), and traversing probes. The results indicated that stage 2 was choking at a mass flow rate that prevented stage 1 from reaching its peak efficiency point, leading to a stage mismatch issue. The mismatch is thought to be due to a loss in the first stage that was not predicted by design tools. Assessment of Stator 1 LE measurements in both test con-figurations revealed that the level of performance at this location is unaffected by the presence of the second stage. Therefore, the major source of unexplained loss resulted from the first stage of the compressor. For additional details and discussion of the CFD analysis and experimental test results refer to Celestina, *et al.* (2012), and Prahst, *et al.* (2015).

ERA Phase 2 utilized a completely new core compressor design strategy and leveraged lessons learned from the Phase 1 compressor design. The Phase 2 compressor was designed for increased efficiency and blade loading. Refer to Figure 3.25 and note that the Phase 2 compressor efficiency levels are higher than those of Phase 1 and that the blade-loading levels were increased relative to best current design but not to the higher levels of blade loading that were attempted in the Phase 1 design (discussed in the previous paragraphs). For ERA Phase 2, NASA tested the first three stages of a high-efficiency, high-OPR core compressor design in the same NASA facility as the Phase 1 testing. The Phase 2 compressor test campaign consisted of a Build 1 test and a Build 2 test where the primary difference is that Build 2 was designed to achieve higher compressor blade loading (pressure rise per stage) at the same efficiency levels of Build 2, as shown in Figure 3.25. The higher blade loading of Build 2 provides an overall system benefit because it allows for the compressor bleed locations to be moved further upstream, thereby reducing the compressor work required to provide the bleed flow. Extensive CFD simulations that have been conducted are not only in agreement with each other but are also in agreement with the design intent. Build 2 testing is complete, and initial results indicate the compressor has met its design intent.

### 3.6.4.2   *NASA ERA core hot-section technology development efforts*

The ERA goal is a 50% reduction in fuel burn below current technology aircraft, while achiev-ing a 75% reduction in landing and take-off (LTO) nitrogen oxides ($NO_x$) below Committee on Aviation Environment Protection CAEP-6 standard requirements (Suder *et al.*, 2013). Achieving this goal requires development of high-power-density, high-thermal-efficiency cores. High-power-density cores enable UHB systems by increasing the bypass ratio with minimal changes in engine diameter required to achieve the higher bypass. Not only does this enable the

UHB engines to be installed under the wing, but this also contributes to the reduced drag and weight associated with the larger diameter UHB engines. The technical challenges associated with high-power-density, highly efficient cores are that they result in (1) higher combustor inlet pressures and temperatures, which encourages $NO_x$ production and (2) higher engine exhaust temperatures and jet velocities, which increase noise and add weight.

The approach is to

1. Select NASA-unique capabilities to increase thermal efficiency (fuel burn), minimize core diameter, and increase power density to enable high-BPR engines
2. Maximize engine OPR and turbine inlet temperature, $T_4$, and reduce cooling flow and weight by
   a. Developing integrated ceramic-matrix composite (CMC) high-pressure turbine (HPT) vanes and blades that are integrated with environmental barrier coatings (EBCs),
   b. Maximizing HPC loading and performance, and
   c. Developing and demonstrating low-weight durable oxide nozzles
3. Interact with the U.S. Department of Defense to insure NASA-unique capabilities in compression systems and EBC coatings are exploited.

In the following sections the results will be divided into two elements: (1) one that addresses increase in engine OPR by working on the compressor technologies and (2) CMC material development to increase $T_4$ and reduce cooling flow. The benefits of these technologies to reduce fuel burn are illustrated by the system study results shown in Figure 3.26 (Tong, 2010).

One of the constraints on ever-increasing OPR and $T_4$ for reduced fuel burn is the increased emphasis on emissions. It was noted earlier that the Energy Efficient Engine Project anticipated revisions to the emissions regulations and included the meeting of these regulations in their goal set. Since then, emissions standards have become even more stringent because of local air quality concerns near airports. Specifically, $NO_x$ emissions are a major concern, and this presents a challenge for increasing OPR and $T_4$ as shown in Figure 3.27. For a given level of combustor technology, $NO_x$ emissions increase dramatically with increasing OPR and cycle temperature.

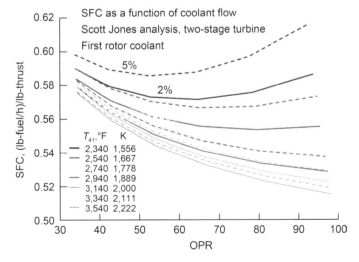

Figure 3.26. Specific fuel consumption (SFC) reduction due to increased overall pressure ratio (OPR) and increased turbine blade inlet temperature $T_{41}$ as a function of reduced coolant flow (From Tong, M. T. 2010. An assessment of the impact of emerging high-temperature materials on engine cycle performance. ASME Paper GT2010–22361. Work of the U.S. Government.)

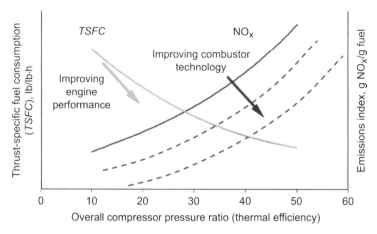

Figure 3.27.   Trade space between engine overall compressor pressure ratio and nitrogen oxide ($NO_x$) emissions. (From Suder, K. L., J. Delaat, C. Hughes, *et al.*, 2013. NASA environmentally responsible aviation project's propulsion technology phase I overview and highlights of accomplishments. AIAA 2013–0414. Work of the U.S. Government.)

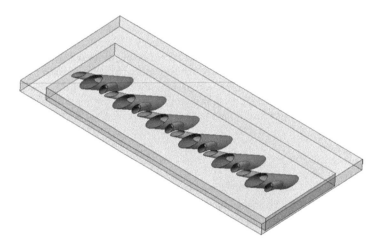

Figure 3.28.   NASA anti-vortex film cooling concept with bifurcated exits.

The strategy is to advance combustor mixing technology in concert with OPR and $T_4$ advances to maintain or reduce $NO_x$ along with thrust-specific fuel consumption.

NASA is addressing these challenges of higher OPR and higher $T_4$ through a combination of materials development, compressor testing, and computational analysis. In the materials area, high-temperature CMC combustor, turbine vane, and engine nozzle components are being developed to allow for higher engine temperatures and reduced cooling flow requirements. The reduction of cooling flow in the HPT vane additionally reduces $NO_x$ emissions by reducing the combustor exit temperature for a given turbine rotor inlet temperature and freeing coolant usage for combustor dilution jets. The plan is to advance the technology readiness (TRL) level of CMC components through design and fabrication of larger, more complex models than have been currently demonstrated and to test these models in a relevant environment in NASA and out-of-house laboratories.

The NASA Glenn Research Center is continuing to develop advanced turbine cooling concepts, including ideas such as an "antivortex" row of film cooling holes having bifurcated exits (Fig. 3.28), which can offer dramatic improvements in film-cooling effectiveness and reduced cooling flows. A recent area of research delves into the optimized cooling of ceramic-based turbine materials, which have unique constraints for cooling compared to metal parts. Because of their reduced thermal gradient capability, CMCs and other ceramic-based turbine components may need to de-emphasize internal cooling and rely more on external film cooling. This combined cooling/materials problem continues the historical trend of synergistic turbine cooling and materials improvements toward reduced fuel burn engines. Development continues on providing robust EBCs for CMC components to protect the ceramic material from the erosive effects of high-temperature water-laden gas.

The NASA Subsonic Fixed Wing Project has recently initiated a number of research awards focusing on enabling continued improvements to engine OPR and thermal efficiency through a better understanding and mitigation of turbomachinery tip clearance and endwall flow losses. As engine cores increase in OPR and reduce in size to enable further increases in turbofan BPR, the tip, endwall, and leakage flows become dominant sources of loss in the engine core. The advocacy of this research topic was strengthened through a series of turbomachinery white papers that were developed under the auspices of the NASA-led Turbomachinery Technical Working Group, which continues today as a forum for better collaboration between NASA, industry, university, and other U.S. Federal Government Agencies.

## 3.7  SUMMARY

The history of the NASA Glenn Research Center (1943 to present) coincides with an era of dramatic improvement in aircraft fuel efficiency and performance. The Center has contributed greatly to this improvement through full engine testing, engine component testing and development, analytical tool and model development, and research investigating fundamental flow physics insight and computational fluid dynamics validation in partnership with the aircraft engine industry. Beginning with the early reciprocating engines with propellers and progressively through the development of turbojet, turbofan, and potential unducted fan concepts, GRC has played a leading role in advocacy for new engine architectures in fundamental and applied research programs. Through this partnership with industry, the fuel burn per passenger-mile and engine specific fuel consumption has been reduced by more than 50%, with similar improvements envisioned in the next decades through component, engine, and aircraft concepts championed by GRC and its research staff. This chapter has summarized a high-level view of these contributions and how they have been achieved through a consistent focus on concept development and research development based on the underlying physics of jet propulsion and improved aircraft engine efficiency.

## REFERENCES

Adamczyk, J. J. 1984. Model equation for simulating flows in multistage turbomachinery. NASA TM–86869.

Ameri, Ali A. 2009. NASA ROTOR 37 CFD CODE validation: Glenn-HT code. AIAA 2009–1060.

Arend, D. J., G. Tillman, and W. F. O'Brien. 2012. Generation after next propulsor research: robust design for embedded engine systems. AIAA 2012–4041.

Bakhle, M. A., T. S. R. Reddy, G. P. Herrick, *et al.* 2012. Aeromechanics analysis of a boundary layer ingesting fan. AIAA 2012–3995.

Bakhle, M. A., T. S. R. Reddy, and R. M. Coroneos. 2014. Forced response of a fan with boundary layer inlet distortion. AIAA Paper 2014–3734.

Ballal, D., and J. Zelina. 2003. Progress in aero-engine technology, 1939–2003. AIAA 2003–4412.

Berdanier, R. A., and N. L. Key. 2015. An experimental investigation of the flow physics associated with end wall losses and large rotor tip clearances as found in the rear stages of a high pressure compressor. NASA/CR—2015-218868.

Bowles, M. D. 2010. The "Apollo" of aeronautics: NASA's aircraft energy efficiency program, 1973–1987. NASA e-book, *http://www.nasa.gov/connect/ebooks/aero_apollo_detail.html* (accessed August 31, 2016).

Bullock, R. O., and I. A. Johnsen. 1965. Aerodynamic design of axial flow compressors. NASA SP–36.

Celestina, M. L., and R. A. Mulac. 2009. Assessment of Stage 35 with APNASA. AIAA 2009–1057.

Celestina, M. L., J. C. Fabian, and S. Kulkarni. 2012. NASA environmentally responsible aviation high overall pressure ratio compressor research—pre-test CFD. AIAA 2012–4040.

Chima, R. V. 2009. SWIFT code assessment for two similar transonic compressors. AIAA 2009–1058.

Ciepluch, C. C., D. Y. Davis, and D. E. Gray. 1987. Results of NASA's energy efficient engine program. *J. Propul. Power* 3:560–568.

Conner, M. 2001. Hans Von Ohain: elegance in flight. Reston, VA: AIAA.

Davis, D. Y., and E. M. Stearns. 1985. Energy efficient engine: flight propulsion system final design and analysis. NASA CR–168219.

Dawson, V. P. 1991. Engines and innovation: Lewis Laboratory and American propulsion technology. NASA SP–4306.

Domke, B. 2007. The GE36 Unducted Fan (UDF) used two contra-rotating propellers with eight blades each. *http://www.b-domke.de/AviationImages/Rarebird/0809.html* (accessed April 28, 2017).

Dunham, J., ed. 1998. CFD validation for propulsion system components. AGARD–AR–355.

Dunlap, S. F. 1991. Technology transfer in U.S. aeronautics: U.S. Government support of the U.S. commercial aircraft industry. Darby, PA: Diane Publishing.

Durbin, P. A., and B. A. Pettersson Reif. 2001. *Statistical Theory and Modeling for Turbulent Flows*. New York, NY: John Wiley & Sons.

Epstein, A. H. 2014. Aeropropulsion for commercial aviation in the twenty-first century and research directions needed. *AIAA J.* 52:901–911.

Ferrar, A. M., W. F. O'Brien, W. F. Ng, R. V. Florea, and D. J. Arend. 2009. Active control of flow in serpentine inlets for blended wing-body aircraft. AIAA 2009–4901.

Florea, R. V., R. Reba, P. R. VanSlooten, *et al.* 2009. Preliminary design for embedded engine systems. AIAA 2009–1131.

Florea, R. V., C. Matalanis, L. W. Hardin, *et al.* 2012. Parametric analysis and design for embedded engine inlets. AIAA Paper 2012–3994.

Giel, P. W., G. J. Van Fossen, R. J. Boyle, *et al.* 1999. Blade heat transfer measurements and predictions in a transonic turbine cascade. NASA/TM—1999-209296.

Hah, C. 2009. Large eddy simulation of transonic flow field in NASA Rotor 37. AIAA 2009–1061.

Hathaway, M. D., J. B. Gertz, A. H. Epstein, *et al.* 1986. Rotor wake characteristics of a transonic axial flow fan. *AIAA J.* 24:1802–1810.

Hathaway, M. D., T. H. Okiishi, K. L. Suder, *et al.* 1987. Measurements of the unsteady flow field within the stator row of a transonic axial-flow fan: Part II— results and discussion. ASME Paper 87–GT–227.

Hathaway, M. D., R. M. Chriss, J. R. Wood, *et al.* 1993. Experimental and computational investigation of the NASA low-speed centrifugal compressor flow field. *J. Turbomach.*, 115:527–542.

Hendricks, E. S., J. J. Berton, W. J. Haller, M. T. Tong, and M. D. Guynn. 2013. Updated assessments of an open rotor airplane using advanced blade designs. AIAA 2013–3628.

Herrick, G. P., M. D. Hathaway, and J.-P. Chen. 2009. Unsteady full annulus simulations of a transonic axial compressor stage. AIAA 2009–1059.

Katz, J. 2016. High resolution measurements of the effects of tip geometry on flow structure and turbulence in the tip region of a rotor blade. Final Report for NASA Grant/ Cooperative Agreement Number NNX-11AI21A for the period of June 6, 2011–July 31, 2016.

Koff, B. L. 1991. Spanning the globe with jet propulsion. AIAA 91–2987, 1991.

Prahst, P. S., S. Kulkarni, and K. H. Sohn. 2015. Experimental results of the first two stages of an advanced transonic core compressor under isolated and multi-stage conditions. ASME Paper GT2015–42727.

Reid, L., and R. D. Moore. 1978. Design and overall performance of four highly loaded, high speed inlet stages for an advanced, high-pressure-ratio core compressor. NASA TP–1337.

Rutherford, D. 2012. Overturning conventional wisdom on aircraft efficiency trends. The International Council on Clean Transportation blog, *http://www.theicct.org/blogs/staff/overturning-conventional-wisdom-aircraft-efficiency-trends* (accessed August 31, 2016).

Schulz, U., C. Leyens, K. Fritscher, *et al.* 2003. Some recent trends in research and technology of advanced thermal barrier coatings. *Aerosp. Sci. Technol.* 7:73–80.

Shetty, K. I., and R. J. Hansman. 2012. Current and historical trends in general aviation in the United States. Report No. ICAT–2012–6. MIT International Center for Air Transportation (ICAT). *https://dspace.mit. edu/bitstream/handle/1721.1/72392/ICAT%20REPORT%20SHETTY.pdf* (accessed May 2, 2017).

Skoch, G. J., and R. D. Moore. 1987. Performance of two 10-lb/sec centrifugal compressors with different blade and shroud thicknesses operating over a range of Reynolds numbers. NASA TM–100115 (AVSCOM–TR–87–C–21 and AIAA–87–1745).

Skoch, G. J., P. S. Prahst, M. P. Wernet, *et al.* 1997. Laser anemometer measurements of the flow field in a 4:1 pressure ratio centrifugal impeller. NASA TM–107541 (ARL–TR–1448 and ASME Paper 97–GT–342).

Strazisar, A. J. 1985. Investigation of flow phenomena in a transonic fan using laser anemometry. *J. Eng. Power—T. ASME* 107:427–436.

Strazisar, A. J., J. R. Wood, M. D. Hathaway, *et al.* 1989. Laser anemometer measurements in a transonic axial-flow fan rotor. NASA TP–2879.

Suder, K. L. 1996. *Experimental investigation of the flow field in a transonic, axial flow compressor with respect to the development of blockage and loss.* Ph.D. Thesis (NASA TM–107310), Case Western Reserve University, Cleveland, OH.

Suder, K. L., J. Delaat, C. Hughes, *et al.* 2013. NASA environmentally responsible aviation project's propulsion technology phase I overview and highlights of accomplishments. AIAA 2013–0414.

Suder, K. L., T. H. Okiishi, M. D. Hathaway, *et al.* 1987. Measurements of the unsteady flow field within the stator row of a transonic axial-flow fan: Part I—measurement and analysis technique. ASME Paper 87–GT–226.

Tillman, G., L. W. Hardin, B. A. Moffitt, *et al.* 2011. System-level benefits of boundary layer ingesting propulsion. Invited Paper to the 49th AIAA Aerospace Sciences Meeting, Orlando, FL.

Tong, M. T. 2010. An assessment of the impact of emerging high-temperature materials on engine cycle performance. ASME Paper GT2010–22361.

U.S. Government. 1991. Technology transfer in U.S. aeronautics: U.S. Government support of the U.S. commercial aircraft industry. Darby, PA: Diane Publishing.

Van Zante, D. E. 2013. The NASA Environmentally Responsible Aviation Project/General Electric Open Rotor Test Campaign. AIAA 2013–0414.

Van Zante, D. E., and K. L. Suder. 2015. Environmentally responsible aviation: propulsion research to enable fuel burn, noise, and emissions reduction. ISABE 2015–20209.

Van Zante, D. E., J. J. Adamczyk, A. J. Strazisar, *et al.* 2002. Wake recovery performance benefit in a high-speed axial compressor. *J. Turbomach.* 124:275–284.

Van Zante, D. E., F. Collier, A. Orton, *et al.* 2014. Progress in open rotor propulsors: the FAA/GE/NASA open rotor test campaign. *Aeronaut. J.* 118:1181–1213.

Volino, R. J. 2016. Experimental and computational investigation of unsteady endwall and tip gap flows in gas turbine passages. Final Report for NASA Interagency Agreement NNC11IA11I.

Wood, J. R., A. J. Strazisar, and P. S. Simonyi. 1987. Shock structure measured in a transonic fan using laser anemometry. AGARD–CP–401.

*Part II*
*Technologies to mitigate environmental impacts*

# CHAPTER 4

## Noise mitigation strategies

Dennis L. Huff

### 4.1  INTRODUCTION

There has been tremendous progress reducing aircraft noise since the problem peaked with the introduction of the turbojet engine for commercial aviation in the 1950s. Significant growth in air travel during the 1960s caused public concern over noise and emissions. Government-funded research programs were initiated, aimed at reducing the environmental impact of air travel, followed by regulations. Regulations are based on what is technically feasible and economically viable. Aircraft noise levels have been reduced by over 20 dB since the 1960s.

Noise is one of many aircraft design parameters with others being safety, performance, emissions, cost, maintainability, etc. Of these, safety is always the most important and is never compromised. Most of the other parameters are traded throughout the design process, and historically, performance has been a key parameter because of the cost of fuel. But noise has been important with continued pressure to reduce community noise around airports. There have been recent cases where the engine fan diameters have been made larger to meet noise goals. Fortunately many of the technologies that reduce noise also improve performance such as the development of higher bypass ratio engines for higher propulsive efficiency. A goal of aircraft noise research is to find ways to reduce noise without significant impact on other design parameters.

Noise can be an issue for all air vehicles, and research programs have been conducted over many years to develop technologies for reducing noise for all sizes of fixed-wing aircraft and helicopters. This chapter will limit the scope to commercial fixed-wing aircraft powered by turbofans and turboprops. Noise mitigation strategies will be discussed, but this is not a complete review of noise reduction technologies. Brief overviews of noise reduction methods will be presented that have either been implemented or are being developed for future aircraft. A summary from recent noise assessments sponsored by the International Civil Aviation Organization (ICAO) will be presented that include noise projections for future aircraft.

### 4.2  NOISE REDUCTION METHODS

The focus of research programs has been on reducing the noise at the source or absorbing noise as it propagates, however there are other ways to mitigate noise impact on the community through the use of noise-optimized flight operations and by providing a larger buffer zone around airports. For example, throttling the engine to a lower speed just after takeoff (called "cutback") has been a common way to reduce engine noise. Flying aircraft over less populated areas is an example of an operational approach to reducing community noise. Flight control systems have enabled lower noise approaches into airports by reducing the changes in engine throttle and eliminating tiered altitude decents. This method is called "continuous descent approach". The remainder of this section will discuss source noise reduction. Mitigation methods will be presented for each component of the engine and airframe that contributes to the overall aircraft noise.

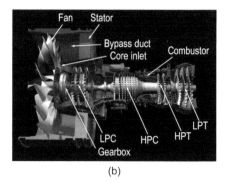

(a)                                                    (b)

Figure 4.1.   Pratt & Whitney 1000G Geared Turbofan™ (a) Photo. (b) Schematic. LPC and HPC are low- and high-pressure compressors, respectively, and HPT and LPT are high- and low-pressure turbines, respectively.

### 4.2.1  *Engine components*

Components of a turbofan engine that produce noise include the fan, jet, and core. The core typically consists of a low-pressure compressor (LPC), a high-pressure compressor (HPC), the combustor, a high-pressure turbine (HPT), a low-pressure turbine (LPT), and the exhaust. Some engines have an intermediate stage with a third spool (the term used to designate an assembly consisting of a shaft with a compressor and turbine). For turboprop engines, the fan is replaced with a propeller driven by the core through a gearbox. Newer high-bypass-ratio (BPR) turbofan engines also use a gearbox to reduce the speed of the fan and increase the speed of the turbine. The Pratt & Whitney Geared Turbofan™ (GTF™) (East Hartford, Connecticut) is an example of a high-BPR engine (Fig. 4.1(a)). A cutaway view of an early version of the GTF™ is shown in Figure 4.1(b).

#### 4.2.1.1  *Fan noise*

Fan noise contributes to the overall aircraft noise levels across all engine power settings. There is a distinct directivity pattern for fan noise consisting of noise radiated from the engine inlet and from the engine bypass duct exhaust. Acoustic source spectra contain a combination of broadband noise resulting from turbulent-flow interactions with surfaces as well as tones from fundamental and higher harmonic blade-passing frequencies (BPFs). For supersonic rotational tip speeds, multiple pure tones (MPTs) radiating from the fan inlet can be present in the spectra.

The most common way to reduce fan source noise is to reduce the rotational speed and pressure ratio. Older engines powering large commercial aircraft had takeoff fan rotational tip speeds exceeding 1,600 fps (488 m/s) and fan pressure ratios greater than 1.7. Modern engines for the same thrust class have takeoff fan tip speeds below 1,250 fps (381 m/s) and pressure ratios below 1.45. The fan diameters have increased to retain thrust because of the lower pressure ratio. The BPR, defined as the mass flow of air passing through the bypass duct divided by the mass flow of air passing through the engine core, increases as a result of the fan ducts being larger and the core sizes being smaller compared with the older engines.

Modern engines also have fewer fan blades. The number of fan blades is primarily determined by the aerodynamic design and desire to lower weight and cost. However, fan noise also contributes to the selection since the number of blades and operating speed determine the BPF. Fan noise can be reduced by shifting tones away from regions of the noise spectrum that contribute to the most annoying frequencies as determined by the perceived noise level (PNL) metric. PNL is used to compute the effective perceived noise level (EPNL) for the aircraft certification process described in Chapter 1. Figure 4.2 shows a picture of a fan from a CF6 engine compared with a modern engine design. The part-span shroud shown in Figure 4.2(a) is no longer used since modern wide-chord fans (Fig. 4.2(b)) have improved aeroelastic characteristics in addition to having higher efficiencies and lower noise.

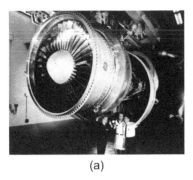
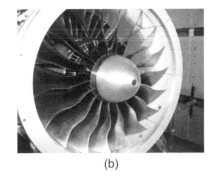

(a)                                                    (b)

Figure 4.2.   Representative aircraft engine fans from past and present. (a) CF6 engine. (b) Model fan for Pratt & Whitney Geared Turbofan™ engine.

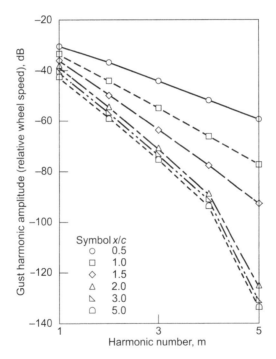

Figure 4.3.   Influence of fan/stator space on fan tone harmonics, spacing ($x/c$, the streamwise position $x$ normalized by the chord length $c$) defined in rotor chords. (From Majjigi, R. K., and P. R. Gliebe. 1994. Development of a rotor/wake-vortex model, Volume 1. NASA CR–174849. Work of the U.S. Government.)

Aerodynamic flow disturbances interacting with the fan, stators, and pylons are sources of fan noise. Engine inlets contain turbulent boundary layers with turbulent eddies that interact with the fan blades. Designers aim to minimize secondary flows ingested by the fan since they impact both the fan efficiency and noise. The wakes from the fan blades interact with the stators generating broadband and tone noise ("rotor-stator" interaction noise). This noise source is very important as the spacing between the fan and the stator is decreased. A common way to reduce interaction fan noise is to increase this spacing (defined as the distance between the rotor trailing edge (TE) and the stator leading edge) to values over two rotor chords as measured at the tip or ¾-span of the fan. Figure 4.3 shows a typical relationship between the fan-stator spacing and the resulting tone noise.

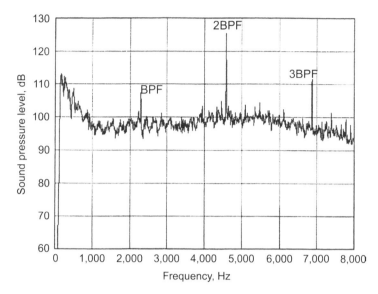

Figure 4.4.   Narrowband acoustic spectra for fan noise showing blade-passing frequency (BPF) cutoff. (From Dittmar, J. H., D. M. Elliott, and L. A. Bock. 1999. Some acoustic results from the Pratt & Whitney advanced ducted propulsor: fan 1. NASA/TM—1999-209049. Work of the U.S. Government.)

The fundamental blade-passing frequency tone resulting from the fan wake impinging on the stator can be controlled by selecting the blade/vane count ratio for tone cutoff. Cutoff is achieved if the number of stator vanes is sufficiently larger than the number of fan blades (typically greater than two times) such that the resulting acoustic waves are evanescent and decay exponentially (Tyler and Sofrin, 1962). This significantly reduces the strength of the BPF tone if the fan duct is sufficiently long. Most turbofan engines flying today use cutoff to reduce BPF fan tone level. Figure 4.4 illustrates how a cutoff fan design can reduce the BPF tone level. For this model fan there were 18 fan blades and 45 stator vanes. It should be noted, however, that cutoff is best achieved in an ideal condition such as for a wind tunnel model. Significant cutoff of the BPF tone in engine applications is more difficult due to other factors that generate a tone such as inlet instrumentation probes, inflow distortion, short inlets, pylons, and asymmetric nacelles.

Another strategy for reducing interaction fan noise is to select the blade/vane ratio such that the propagating acoustic modes can be more easily absorbed by acoustic treatment or blocked by the fan (transmission loss). This becomes a challenging optimization process when considering the number of acoustic modes for all dominant tones over a range of fan speeds. Acoustic waves that are directed towards the acoustic treatment installed in the fan nacelle exhibit higher absorption than those waves that graze the treatment. Each acoustic mode contributing to a tone has a direction and amplitude that can be tailored for acoustic treatment design. For fan designs with acoustic treatment between the fan and stator, the trapping modes that can be beneficial are those achieved by choosing blade/vane count ratios that favor spinning modes that are counterrotating relative to the direction of the fan rotation and are more easily reflected by the fan blades.

The purpose of the stator vanes is to remove the swirl induced by the fan to increase thrust. In several newer engines it also serves as a structural member to support the nacelle. During the 1960s the idea of sweeping and leaning the stator vanes was investigated to reduce noise, but experimental methods during that time failed to measure the acoustic benefits. Swept and leaned stators were investigated again in the 1990s in wind tunnel tests and were found to reduce both tone and broadband noise. Leaning the stators in the direction of the fan rotation was found to be best. Engine tests confirmed the noise reduction benefit (Fig. 4.5).

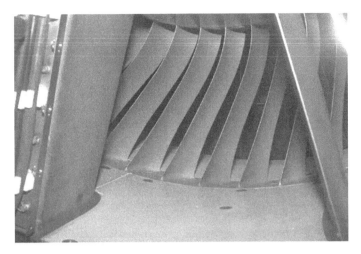

Figure 4.5.   Swept and leaned stators tested on CFM-56 engine.

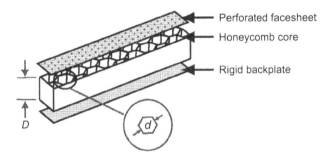

Figure 4.6.   Typical acoustic treatment used in aircraft engines. $D$ is depth of honeycomb core, and $d$ is diameter of cell. (From Jones, M. G., M. B. Tracy, W. R. Watson, *et al.*, 2002. Effects of liner geometry on acoustic impedance. AIAA 2002–2446. Work of the U.S. Government.)

In addition to interaction noise sources, the fan itself can produce noise especially as the rotational tip speed becomes supersonic. The boundary layers on the blade scattering off the rotor TE and vortex shedding from the TE contribute to fan noise similar to an aircraft wing or helicopter blade. But the noise levels tend to be lower than the interaction noise sources for a well-designed fan blade. MPT noise is associated with high-speed fans and is caused by shockwaves generated at fan blade leading edges. Slight blade-to-blade differences exaggerate differences between those shockwaves and distribute the sound power in shaft order tones (not just the blade-passing tones). A common way to reduce MPT noise is to design an inlet acoustic liner to reduce the tone levels. The fan speed can also be adjusted to minimize the tones for noise-sensitive conditions. Newer fans have incorporated forward sweep at the tip of the blades to increase the mass flow and decrease aerodynamic losses. Forward sweep can delay the onset of MPT noise. For smaller fans, a blisk design (i.e., manufacturing the blades and the hub as a single piece) is also known to reduce or eliminate MPT noise.

Acoustic treatment (also called "acoustic liner") is routinely used in turbofan engines to absorb fan noise as it propagates through the inlet and bypass duct. Liners are locally reactive resonators that are tuned to absorb one or more frequencies of sound. They consist of a face sheet that can be porous metal or wire mesh, a honeycomb structure, and a backing plate (Fig. 4.6). The depth of the honeycomb, $D$, and the cell diameter, $d$, are varied to control the tuning frequencies. The acoustic impedance of the face sheet, which is a measure of the resistance of the liner to the incident acoustic pressure, is selected to permit acoustic waves to penetrate into the liner core in the presence of grazing flow on the liner. Sometimes multiple layers are used to increase the number of absorption target frequencies and provide better broadband attenuation. Acoustic treatment is added to the inner surface of the nacelle upstream and downstream of the fan (Fig. 4.5). It is also added to the inner bypass duct if space permits. The construction of the liner can affect its performance. Seams resulting from splicing liners together in the circumferential direction can create noise due to impedance discontinuities. Newer liners minimize the extent of the seams and have been found to reduce the extraneous noise created by the seam.

### 4.2.1.2  *Propeller noise*

Unlike turbofans, propeller-driven aircraft do not have nacelles with acoustic treatment to reduce the rotor noise. Their noise spectra are dominated by tones (Fig. 1.11, Chapter 1) and their broadband noise is less important in determining their overall noise levels. For turboprops, the directivity peaks near the propeller plane of rotation. Propellers are typically installed on the wing and are exposed to cyclic loading variations once per revolution because of the angle of attack. There are fewer blades, and the tip rotational speed is less, for turboprops compared with turbofans. The BPF tone occurs at a lower frequency, and therefore the higher harmonics are important since they contribute to the community noise PNL. Additionally, lower frequency tones couple with the structural modes of the aircraft cabin, requiring treatment and vibration control to reduce interior noise.

The primary method for reducing propeller noise is to increase the number of blades and slow down their rotational speed. Increasing the number of blades reduces the aerodynamic loading per blade for a constant disk area and pressure ratio. Newer turboprops have between six and eight blades. Reducing the rotational speed also reduces the strength of the thickness noise (volumetric displacement of the air by the propeller blades) and the aerodynamic loading fluctuation. It also changes the distribution of tones in the spectra. Making the blades as thin as possible reduces the thickness noise, but this needs to be done without compromising the structural integrity of the blades. During the design process, the propeller's number of blades, rotational speed, and diameter are parametrically optimized for noise and aerodynamic performance.

Changing the planform of the blade can also help reduce propeller noise. For higher tip speed propellers, blade sweep is introduced to reduce the aerodynamic losses associated with transonic flows. This also helps reduce the noise. Figure 4.7 shows a modern propeller design for a turboprop aircraft.

### 4.2.1.3  *Core noise*

The term "core" refers to all components in the gas turbine engine that generate the power to drive a fan or propeller. These are typically the compressor, combustor, and turbine, and all of them generate noise. The compressor produces tones due to rotor-stator interaction across multiple blade rows that propagate forward from the engine inlet. The combustor produces low-frequency broadband noise that can occur at lower engine speeds and propagates aft. Turbine noise is similar to the aft-radiated fan noise and includes tones and broadband noise, both of which contribute to the overall PNL. For older engines, the fan and jet noise dominate over the core noise. Higher BPR engines have lower fan and jet noise levels, and thus no longer mask the core noise. In fact, newer low-emission combustors with lean direct injection can have measurably higher combustion noise levels.

Compressor noise is usually the result of blade row interaction within the LPC, particularly the booster stages closest to the inlet. Some turbofan architectures have the LPC booster stages located just downstream of the fan, which can increase interactions with the fan wakes. The interaction tones are easily identified in narrowband acoustic spectra, knowing the shaft rotational speed and the number of blades in each blade row. Figure 4.8 shows booster stage interaction tones from a model-scale wind tunnel test compared to the fan tones (Woodward and Hughes, 2012). In many engines the tone levels are low enough that they do not need to be reduced since

Figure 4.7. Dowty six-blade R408 swept propeller for Bombardier Q400 Dash 8. (From Dowty Propellers. 2016. R408 propeller system for the Bombardier Q400 Dash 8. *http://dowty.com/products/regional-airliners/* (accessed August 31, 2016). With permission.)

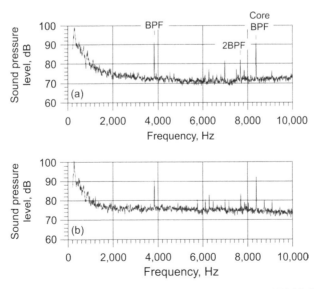

Figure 4.8. Compressor noise from low-pressure combustor booster stages. BPF is blade-passing frequency. (a) M4 metal rotor. (b) Modified aft-swept metal rotor (C-8). (From Woodward, R. P., and C. E. Hughes. 2012. Acoustic performance of the GEAE UPS Research Fan in the NASA Glenn 9- by 15-Foot Low-Speed Wind Tunnel. NASA/TM—2012-217450. Work of the U.S. Government.)

they do not contribute to the PNL and need not be mitigated. The tone levels can be controlled by careful selection of the blade/vane ratios and taking advantage of cutoff similar to fan noise reduction methods. Acoustic treatment has also been used to help mitigate the tone levels.

Combustion noise was a concern for gas turbine engines during the 1970s. Research was initiated to understand source mechanisms, develop noise prediction methods, and identify ways to reduce the noise. Combustion noise is proportional to the size of the combustor, fuel/air ratio (temperature rise), burner pressure, and burner flow parameters (Mathews *et al.*, 1977). Direct combustion noise is related to the fluctuating heat release of the flame front in each fuel injector. It generally peaks at an angle of 120° from the engine inlet axis with a broadband "hump" that can span the range of 300 to 1,000 Hz with a peak around 400 Hz, although there can also be multiple peaks in the combustion noise spectrum. The combustor noise is important at low to mid engine power until jet noise is prevalent at high powers. Indirect combustion noise occurs when the hot streaks downstream of fuel nozzles enter the accelerating flow region at the entrance to the first high-pressure turbine stage. The result is broadband noise over the same frequency range as the direct combustion noise. It has proved difficult to distinguish between these two source mechanisms, and research work continues. In addition to the combustion broadband noise, tones sometimes occur from resonances associated with combustor instabilities. They are usually mitigated during the engine development process by modifying the combustor geometry and fuel injection staging.

Modern low-emission combustors contain single- or dual-annular fuel injectors, which are quieter than older can-type combustors. Increasing the number of fuel nozzles has been a way to reduce combustion noise. The turbine and tailpipe nozzle reflect the combustion noise back toward the combustor and away from the community, which leads to adjusting the number of turbine blades and vanes, and the tailpipe length as another way to reduce combustion noise. The turbine and tailpipe provide a transmission loss mechanism for combustion noise similar to a muffler or acoustic filter. In some engines, acoustic treatment is added to the tailpipe plug to mitigate combustion noise. Single- and double-degree-of-freedom liners can be used to extend the liner effectiveness over a wide frequency range. The lower frequencies make this challenging since the liners use resonators, which require increased volume and deeper liner thickness. Tailcone resonators have been proposed to increase the volume and provide associated noise attenuation.

Turbine noise occurs across all engine power settings and radiates aft. The LPT is located closest to the core exit (Fig. 4.1(b)). The stages of the LPT near the exit are most important for noise and can be observed in the far-field acoustic spectra as tones with "haystacks." The tones are generated by the turbine rotor wakes interacting with the stator vanes and rotors from other stages. The wakes can also interact with support pylons at the engine exit. The mechanism is similar to fan noise, but the turbine tones propagate through the core nozzle shear layer causing "haystacks" in the acoustic spectra. The frequency range of the tones is high due to the spool speed and number of turbine blades. Fundamental BPF tones from various stages and their higher harmonics can contribute to the PNL. It is sometimes difficult to distinguish between fan broadband noise and turbine noise because of the frequency broadening from the nozzle shear layer requiring narrowband measurements to identify tones.

The most common noise reduction method for turbine noise is to carefully select blade and vane numbers for cutoff of acoustic modes. The objective is to optimize the vane/blade ratios to minimize the coupling with the unsteady aerodynamic response of the blade rows near the exit of the LPT. Since the turbine flow passage is a narrow annulus, only lower order modes need to be addressed. Further noise reduction can be achieved if all propagating modes are cut off (as described for fan noise reduction), but the vane/blade ratios need to be high. This makes cutoff impractical for many engines since the weight rises significantly with the number of blades and it becomes difficult to maintain high aerodynamic efficiencies. Newer engine designs use gearboxes to transmit power from the turbine to the fan. This allows an increase in the turbine rotational speed so the turbine tones are produced at higher frequencies that do not significantly impact the PNL. However, newer engines also have fewer turbine stages making cutoff difficult (and also consequently reducing the transmission loss for combustion noise). Increasing the spacing between blade rows helps to reduce interaction noise, but it also increases the engine length and weight.

Another noise source from the core is bleed valve noise. Bleed valves are used to dump high-pressure air from the compressor into the fan duct to maintain stability as the engine changes speed. Air is collected in a manifold and exits into the bypass duct through a nozzle with high flow velocity. These noise levels can impact the PNL during approach when the other engine noise sources are lower. Such noise spectra are similar to those of small jets with subsonic exit velocities. Mitigation methods include adding "teeth" or screens to the nozzles to break up the turbulence.

#### 4.2.1.4 *Jet noise*

Jet noise has been studied since the 1950s when turbojets were introduced for commercial aircraft. It is dominant for lower BPR turbofans during takeoff and is characterized as a broadband source with peak directivities aft of the engine. It is worth noting that the low-frequency rumble associated with jet noise can persist over long distances. The primary cause of jet noise is the turbulent mixing of the core and bypass flow streams exiting the engine. It is a distributed noise source external to the engine that persists for 15 to 20 nozzle diameters until the turbulence sufficiently decays. This makes jet noise reduction a challenge since modifications to the nozzles need to favorably alter turbulence well downstream of the engine. Higher frequency noise can sometimes be generated inside the nozzles by mixing devices and can impact the PNL.

There are two basic types of exhausts for turbofan engines: separate-flow nozzles (Fig. 4.9(a)) and mixed-flow nozzles (Fig. 4.9(b)). Separate-flow nozzles are common on higher thrust engines or smaller engines that require shorter lengths. For engines with longer ducts, lobed mixers are added to the core nozzle to promote mixing of the core and bypass flow streams before exiting the duct. With longer duct lengths, acoustic treatment can be added to absorb internal mixing noise.

Mixed-flow engines are quieter than engines with separate-flow nozzles. The current trend for larger engines is to use separate-flow nozzles since long-duct mixed-flow nozzles are heavier, increase cost, and if the BPR is high enough, mixing of the flow streams is not necessary. It is common for smaller engines with BPR < 6 to use mixers. There has been substantial research on lobed-mixer designs to optimize the geometry for lower noise and better aerodynamic performance. Figure 4.10 shows examples of lobed mixers that have been designed for model-scale acoustic tests.

The transition from turbojets to turbofans helped to not only increase the engine efficiency, but significantly reduced the aircraft noise since jet noise was the dominant source. Newer engines with higher BPR emit lower jet noise because of the lower jet exit velocities and temperatures. This has been the primary method for reducing jet noise. Over the past 50 years there has been a gradual trend toward increasing the BPR of turbofan engines and an associated decrease in jet velocities. The well-known correlation by Lighthill (1952) showing that jet noise is proportional to the jet velocity raised to the eighth power illustrates how even small changes in the jet exit velocity can significantly reduce noise.

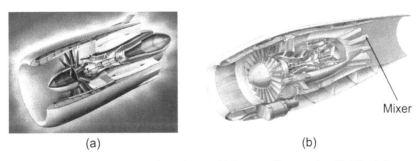

(a)                                        (b)

Figure 4.9.    Basic types of turbofan engine exhausts. (a) Separate-flow nozzles. (b) Mixed-flow nozzles.

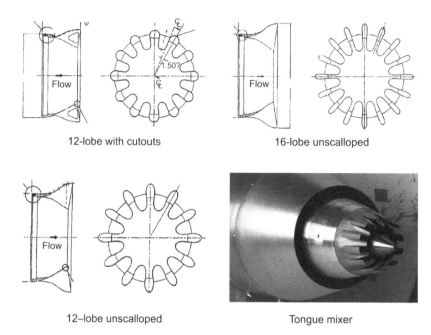

12-lobe with cutouts                    16-lobe unscalloped

12–lobe unscalloped                    Tongue mixer

Figure 4.10.   Mixers used in model-scale jet noise tests. (From Mengle, V. G., V. D. Baker, *et al.*, 2002. Lobed mixer design for noise suppression: plume, aerodynamic and acoustic data. NASA/CR—2002-210823/VOL2. Work of the U.S. Government.)

Modern engines with separate-flow nozzles have added chevrons to the core nozzle to reduce jet noise. Chevrons mix the core and bypass flow by using a distinctive sawtooth pattern at the exit of the core nozzle. A slight penetration of the chevron into the core flow reduces the low-frequency jet noise with little to no high-frequency penalty from the mixing process. This is important since there is no acoustic treatment surrounding separate flow nozzles to absorb high-frequency mixing noise. The thrust loss from chevron nozzles is low and they have been implemented on several engines. Figure 4.11 shows the acoustic spectra from a flight test using a chevron nozzle compared to a separate-flow baseline nozzle. Chevrons have also been used on the bypass duct to decrease noise inside the cabin at cruise. At higher altitudes, shocks can form at the nozzle exit due to high-flow Mach numbers creating another noise source that can impact interior noise.

The newest engines entering service today on medium-size aircraft have a high BPR (10 to 13) with jet velocities that are low enough to make jet noise a secondary source. Examples include the new Pratt & Whitney Geared Turbofan™ engine (Fig. 4.1).

### 4.2.2   *Airframe components*

Airframe noise is an ensemble of many different sources distributed across aerodynamic surfaces, appendages and cavities. Figure 1.1 of Chapter 1 shows the major sources of airframe noise at approach. Strategies for reducing airframe noise target the most dominant sources such as TEs, landing gear, slats and flaps. The importance of each source varies with the type of aircraft, but landing gear and high lift devices tend to dominate. Airframe noise is most important during approach when these devices are deployed causing flow separation and generation of high levels of turbulence. Reducing the airspeed helps reduce the source strengths, but there are obvious practical limits for low-speed flight. Also, since the certification metrics include duration, benefits from source reduction needs to exceed the offset from longer duration to reduce the EPNL.

(a)

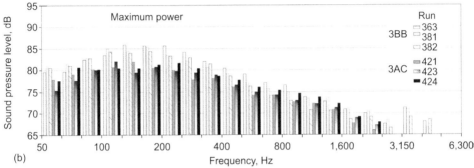

(b)

Figure 4.11.   Flight test for effect of chevrons on jet engine noise. (a) Chevron nozzles on Honeywell Falcon 20 flight test. (b) Flyover 1/3-octave sound pressure level without chevrons (3BB) and with chevrons (3AC). ((b) From Weir, D. S. 2004. Engine validation of noise reduction concepts—separate flow nozzles. AIAA–2004–188. Reprinted by permission of the American Institute of Aeronautics and Astronautics, Inc.)

Research on airframe-noise-reduction methods is difficult because of the large-scale test facilities required to mature the technologies. Even with very large anechoic wind tunnels, the scale factors are limited, causing issues with Reynolds number scaling. Flight tests are expensive, and it is difficult to measure noise from the airframe components because individual sound source levels can be low, although phased microphone arrays have been successfully used to distinguish between sources.

### 4.2.2.1   *Trailing edge noise*

Trailing edge noise is present for all flow conditions, but is particularly important when an airfoil is generating high lift. Adverse pressure gradients increase regions of turbulence and possible flow separation. Satisfying the Kutta condition (i.e., flow is tangential at the TE) can introduce periodic shedding of vortices. Vortex-shedding noise has a distinct broadband "hump" in the acoustic spectra that scales with the airfoil geometry and flow speed. Geometry characteristics of the TE, such as bluntness, are important. The directivity of TE noise is cardiod-like and it radiates above and below the TE, and is effectively a line source along the TE of the wing. A method for categorizing TE noise sources was developed by Brooks *et al.* (1989) based on experimental work.

Methods for reducing TE noise are tailored to the particular underlying noise generation mechanism. For example, it is known that noise can be produced when the boundary layer turbulence passes over the TE of the wing. The TE marks the location where the pressure discontinuity on the wing surface must give way to pressure continuity downstream. This sudden change in "boundary condition" causes some of the energy of turbulence to be converted to sound. Research has been done in laboratories on ways to reduce this source of TE noise by adding serrations or flexible wires/brushes at the TE. The goal is to make the transition in the boundary condition as smooth as possible so as to reduce the amount of sound generated. Figure 4.12 shows examples of such TE devices and the corresponding noise reduction from airfoil tests. TE noise can also be produced from flow separation at the TE of the wing. Flow-control methods such as plasma actuators have been shown to modify the boundary layer and

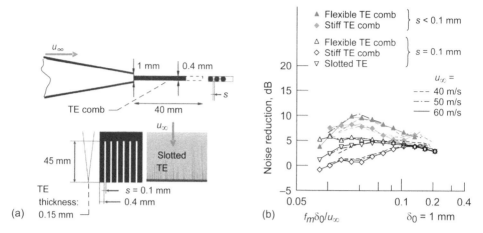

Figure 4.12.   "Comb" and slotted trailing edge (TE) extensions for a NACA 0012 airfoil test. $f_m\delta_0/u_\infty$ is the Strouhal number where $f_m$ is the one-third octave center frequency, $\delta_0$ is the characteristic length scale of the source region, and $u_\infty$ is the free stream velocity; $s$ is the thickness of the comb. (a) Schematic. (b) Noise reduction as a function of normalized frequency. ((a) From Dobrzynski, W. 2010. Almost 40 years of airframe noise research: what did we achieve? *J. Aircraft* 47:353–367. Reprinted by permission of the American Institute of Aeronautics and Astronautics, Inc. (b) From Herr, M. 2007. Design criteria for low-noise trailing-edges. AIAA 2007–3470. Copyright 2007 by Deutsches Zentrum für Luft- und Raumfahrt e.V. (DLR).)

reduce or eliminate flow separation. Although they have been shown to be effective, their implementation on aircraft has been limited because of practical concerns.

### 4.2.2.2   *Landing gear noise*

Landing gear noise is a dominant source of airframe noise when deployed during approach. It is a broadband source that results from highly irregular vortex shedding from multiple components of the landing gear system. The wake from one component can interact with another, such as landing gear struts. Holes and cavities can also contribute to the acoustic spectra as resonant tones. The shear layer from the upstream portion of the landing gear cavity can interact with landing gear components and the cavity itself. The noise radiates in all directions below the aircraft and is easily identified by phased microphone arrays.

A low-noise design strategy is to place the noise-producing components of the landing gear in regions of lower flow velocity, if feasible. Fairings that cover the struts and inner portion of the landing gear carriage have also been shown to significantly reduce noise. A fairing can eliminate exposure of noise-producing components to flow, and any shedding that occurs from the fairing that will be at a lower frequency. Both rigid and flexible fairings have been tested. Figure 4.13 shows an example of a "toboggan" fairing that was tested on a Boeing 777. Flow control methods have been proposed by introducing blowing and suction through porous materials, and plasma actuators to reduce separated flow and vortex shedding. Methods for reducing landing gear noise are difficult to implement because of the need for maintenance access and cooling of the brakes.

### 4.2.2.3   *Slat noise*

When the slats are deployed, there is a slot at the leading edge that emits broadband noise below the wing. Cavity resonances can generate tones but are rarely found in aircraft applications. The track and structural members that hold the slats are also a source of noise. Slat noise spans the low to middle frequencies of the audible range, where bracket noise is more likely in the mid- to high-range frequencies. Computations of the complex flow within the slat have helped identify source mechanisms, as shown in Figure 1.14 in Chapter 1.

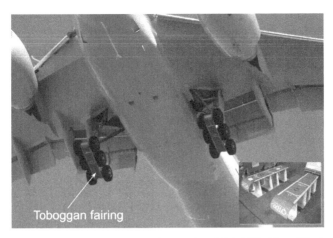

Figure 4.13. Toboggan landing gear fairings tested on Boeing 777-300ER. (From Elkoby, R., L. Brusniak, R. W. Stoker, *et al.*, 2007. Airframe noise results from the QTD II flight test program. AIAA 2007–3457. Copyright Boeing. With permission.)

Figure 4.14. Slat cove seal on Boeing 777 aircraft. (From Khorrami, M. R., and D. P. Lockard. 2010. Effects of geometric details on slat noise generation and propagation. *Int. J. Aeroacoust.* 9:655–678. Work of the U.S. Government.)

Good aerodynamic designs of slats that reduce separation and drag also help reduce slat noise. Decreasing the gap size and increasing the overlap of the slat with the wing helps reduce noise (Brooks *et al.*, 1989), but the region identified as the cove can still produce significant noise. Filler materials have been used to reduce the separation. Mixing devices added to the cusp and TE of the slat can reduce vortex shedding and the noise resulting from interaction with the slat TE. Leading edge trips located on the pressure side of the slat and just upstream of the gap can reduce tones from cavity resonances. Porous TEs on the slats have been found to reduce noise. Slat cove covers and seals used to improve aerodynamics can also help reduce the noise. For the seals used on the Boeing 777 (Fig. 4.14), the "blade seal" reduces the noise, as concluded from a computational study (Khorrami and Lockard, 2010). Acoustic treatment has been added to the slat, leading edge of the wing, and near the support tracks to reduce local noise sources. Many of these devices have not made their way into applications yet, but research has quantified the potential benefits for design trade studies. Noise reduction devices need to be integrated with other systems such as anti-icing and cannot negatively impact the aircraft aerodynamics. They also cannot interfere with slat retraction.

A method that has been implemented for reducing slat noise is to eliminate the slat and use a "drooped" leading edge that meets high lift requirements. Drooped-leading-edge designs extend the chord of the wing and minimize the curvature of the upper and lower surface to optimize for high lift and low noise. This technique was successfully implemented on the inboard section of the wing on the Airbus A380 but was an aerodynamic compromise to meet noise reduction goals.

### 4.2.2.4    *Flap (side-edge) noise*

Flaps are part of the high-lift system that creates noise when it is deployed. Flaps produce broadband noise with directivity peaks at multiple angles. There is increased noise from the flaps as they are fully extended. In addition to noise from the flap cove, the side edges of the flaps become a noise source as they are exposed to the tip vortices generated by the flap and interacting with the flap surface. The flap tracks that provide structural support are also a source of noise. For some aircraft configurations, the wakes from the landing gear impinge on the flaps and produce noise. An even stronger noise source occurs when the exhaust plume from the engine interacts with the flap.

A primary focus for flap noise has been to reduce the side-edge interaction source. Mitigation methods include making the edge of the flap from porous materials and brushes that reduce the unsteady aerodynamic response to the vortices and help absorb the noise. Adding fences at the tip of the flap can change the aerodynamics and the trajectory of the vortices to reduce the strength and location of the interactions, but they can have negative impacts on aerodynamic performance. Baffles have also been tested to introduce resonators within the flap tip. Finally, just as slat noise can be reduced with aerodynamic design, flap noise can be reduced by optimizing the overall lift and aircraft speed with the aim of reducing drag.

Wind tunnel tests have been conducted using phased microphone arrays to guide the decision on which flap-edge geometries and treatments should be tried for noise reduction investigations. Figure 4.15 shows the noise reduction measured in a 26%-scale model test of a Boeing 777 with porous flap edges.

Another method for reducing side-edge noise is to introduce a filler that eliminates the edges. A continuous mold-line flap that elastically stretches as the flap is deployed has been shown to significantly reduce noise (Fig. 4.16). The challenge for this concept is ensuring structural integrity and making sure there are no safety issues that would prevent flap deployment or retraction. This continues to be an area of research today that looks promising with advances in materials and structures.

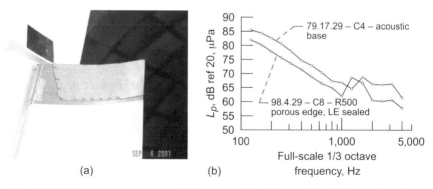

(a)                                (b)

Figure 4.15.    Flap noise reduction testing on 26%-scale model of Boeing 777. (a) Porous flap side-edge treatment. (b) Test results, where LE is leading edge and $L_p$ is sound pressure level. (From Horne, W. C., K. D. James, N. Burnside, *et al.*, 2005. Measurements of 26%-scale 777 airframe noise in the NASA Ames 40- by 80 Foot Wind Tunnel. AIAA 2005–2810. Work of the U.S. Government.)

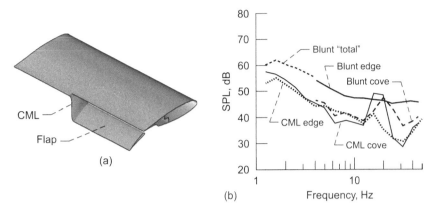

Figure 4.16. Technique to reduce side-edge airframe noise. (a) Continuous mold-line (CML) flap. (b) Sound pressure level (SPL) from phased microphone array measurements. (From Hutcheson, F. V., T. F. Brooks, and W. M. Humphreys, Jr. 2008. Noise radiation from a continuous mold-line link flap configuration. AIAA 2008–2966. Work of the U.S. Government.)

Methods for reducing the landing gear noise can also reduce the interaction noise with the flaps since the resulting turbulent field is reduced. There have been several aircraft with cutouts in the flaps to reduce engine plume interaction noise. However, there is usually a desire to make the flaps continuous for aerodynamic reasons.

### 4.2.2.5   *Other noise*

Wing spoilers located on the upper surface generate high noise levels, especially inside the cabin. Fortunately full deployment occurs after the aircraft has landed and does not create a community noise issue. Partial deployment of spoilers during approach tends to be a secondary, low-frequency noise source.

Any hole or cavity can be a noise source. If they are not anticipated, they are discovered during flight tests and quick fixes are sought. Vent holes for fuel systems can cause tones and one solution is to add vortex generators just upstream of the vent holes to reduce the added noise. Cavities that are exposed during flight can cause broadband noise. Newer aircraft designs tend to reduce the number of exposed cavities to reduce airframe noise.

Boundary layer noise from the fuselage and wing is considered the low-noise limit once the aircraft is flying as slow as possible and all other noise sources have been reduced. Newer aircraft have increased the amount of laminar flow on these surfaces to reduce drag. This will also help reduce this noise source, but it is expected that other noise sources will mask this benefit for community noise. Additionally, interior noise should be reduced if the fuselage can achieve higher portions of laminar flow.

### 4.3   FUTURE NOISE-REDUCTION TECHNOLOGIES

Implementation of noise-reduction technologies depends on many factors such as new aircraft and engine design opportunities, requirements, and trades across many objectives such as safety, cost, and performance. It is rare for noise-reduction technologies to be retro-fitted on aircraft or engines after they have been certified. It is more common to implement the technologies on new or derivative versions of the aircraft. Technologies need to buy their way into the design process by proving that they can be reliable and cost effective. The technology readiness level (TRL) scale introduced by National Aeronautics and Space Administration (NASA) has been used worldwide to help evaluate the maturity of technologies. The scale ranges from 1 for basic

Table 4.1.    Technology Readiness Level (TRL) definitions.

| TRL | Definition |
| --- | --- |
| 1 | Basic principles observed and reported |
| 2 | Technology concept and/or application formulated |
| 3 | Analytical and experimental critical function and/or characteristic proof-of-concept |
| 4 | Component and/or breadboard validation in laboratory environment |
| 5 | Component and/or breadboard validation in relevant environment |
| 6 | System/subsystem model or prototype demonstration in a relevant environment (ground or space) |
| 7 | System prototype demonstration in a space environment |
| 8 | Actual system completed and "flight qualified" through test and demonstration (ground or space) |
| 9 | Actual system "flight proven" through successful mission operations |

ideas to 9 for implementation on aircraft. Table 4.1 shows definition of the TRL scale as initially proposed by Mankins (1995). It was originally developed for space technology and later adapted for aeronautics.

Since 2008 the International Civil Aviation Organization has used two independent review panels to evaluate noise reduction technologies and their TRL. The panels were tasked to identify noise reduction technologies informed by industry and government studies. Aircraft noise reduction goals were defined for 10-year (mid-term) and 20-year (long-term) time horizons with consideration for novel aircraft designs. This was a thorough effort that represented the views of all major aircraft and engine manufacturers and research organizations throughout the world. Noise-reduction technologies were identified for each engine and aircraft component. Then they were evaluated using aircraft noise prediction tools for four reference aircraft categories: regional jet, small- and medium-range twin engine aircraft, long-range twin engine aircraft, and long-range four-engine aircraft. Reference noise levels were established, defining 2010 Best Practices using certification values, and were found to be 4 to 6 EPNdB cumulative below the ICAO "Chapter 4" standard (ICAO, 2008) (see Chapter 1 of this publication for details on noise certification). Noise reduction technologies were evaluated first as component effective perceived noise level and then combined for an aircraft system to predict overall noise levels for each certification point. This was done for mid-term (2020) and long-term (2030) by estimating the TRL for each technology. Technologies deemed ready for certification (TRL 8) within each time period were included in the aircraft system evaluations and were used to define aircraft noise goals.

Detailed reports are available through ICAO (ICAO, 2010; ICAO, 2014). A summary of the results are given in this section. Note that all noise reduction estimates are given in cumulative EPNdB.

### 4.3.1    Engine noise reduction

#### 4.3.1.1    Engine mid-term technologies

Fan blade sweep, swept and leaned stators, and fan speed optimization for higher bypass ratio engines are currently TRL 5 through 9 and are projected to be ready for production by 2020. The range of values recognizes that some engines have already implemented these technologies. The noise-reduction potential was estimated to be 2 to 4 dB for tones and 1 to 3 dB for fan broadband noise. Stator sweep was noted to have greater noise reduction potential on the order of 3 to 5 dB.

Zero splice inlet liners, scarf inlets, inlet nose lip liners, and aft cowl liners are future technologies with the potential of reducing the fan noise by 2.5 to 4.5 dB. Zero splice liners have already been implemented on aircraft such as the Boeing 787 and Airbus A380 and are expected to be commonly used on future engines. Scarf inlets have been tested on engines and found to reduce inlet noise, but work still needs to be done on their aerodynamic performance between takeoff and cruise. Inlet lip liners and aft cowl liners effectively increase the treatment area, but need to be integrated with other systems such as anti-icing and thrust reversers.

Jet noise reduction is strongly dependent on the BPR of the engine. For a BPR less than 8, jet noise reduction estimates are 1 to 3 dB using fixed and "variable" chevrons and advanced long-duct forced mixers. Variable-geometry chevrons are used to change the penetration angle to control the thrust loss and noise reduction. They have been tested on engines using shape memory alloy materials. Long-duct mixers are projected for applications on regional and corporate jets with BPRs of 4 to 6. For larger engines with BPRs exceeding 12 or 13, the jet noise reduction is estimated to be 3 to 4 dB because of reducing the exhaust velocities.

Core noise reduction technologies for the turbine have potential application of hot-stream liners and long-cowl common nozzle liners. Acoustic treatment on inner and outer sides of the turbine exit annulus can reduce noise by 2 dB for each unit of treatment length per nozzle radius. Optimizing the blade/vane counts and the spacing is already done on many engines, but an additional 2 to 4 dB for tones and 3 to 4 dB for broadband noise reduction is possible, especially if swept and leaned blades can be implemented. Combustion noise reduction using tailcone resonators can be tuned to reduce specific frequencies of narrowband spectra by 3 to 9 dB, but the impact on EPNL is strongly dependent on the specific engine application. Mixing devices used to reduce bleed valve noise are very effective, have already been shown to reduce noise by up to 10 dB, and can have a significant impact on the overall aircraft noise reduction when this source dominates. Combining all of the core noise reduction technologies, the ICAO review panel projected that 1 to 2 dB noise reduction is possible in the mid-term for the core EPNL component.

### 4.3.1.2 *Engine long-term technologies*

By 2030, additional noise reduction technologies are expected to be ready for applications. For fan noise, they include variable-area nozzles needed for the operation of low-pressure-ratio fans that also provide about 2 dB tone and broadband noise reduction. Placing acoustic treatment in alternative locations such as within the bypass duct stators and over the tip of the fan can offer additional noise reduction in the order of 1.5 to 3.0 dB. Active noise control has already been shown to reduce multiple tones with multiple-mode content, but it has proved too complex for practical applications to date. In addition, the importance of fan tones has diminished with lower speed fans placing more emphasis on broadband noise, which is even more of a challenge for active noise control. For fan noise, the review panel estimated that 1 to 3 dB additional noise reduction is possible beyond the mid-term estimates.

Advanced acoustic treatment candidates include adding half-wave resonators called Herschel-Quicke tubes within the liners to increase attenuation at specific frequencies. They have already been tested on an engine and could be matured by 2030. Zone liners designed with specific knowledge of the fan noise source characteristics can improve overall attenuation by optimizing the acoustic impedance. It may be also possible to increase the effective treatment area by extending liners into the aft cowl duct. Treated splitters located in the fan duct are another way to increase the treatment area and create a narrower annulus that makes the liners more effective. They have been tested in both the inlet and the aft bypass duct. While they can be very effective for noise reduction, performance losses and integration with other systems such as thrust reversers have made them impractical. Just as with active noise control, developments with enabling technologies will be needed before they will be ready for engine applications. Variable-impedance liners are also being developed, including active systems that change the impedance of the liner as the engine changes speed. Advanced manufacturing methods like three-dimensional printing are enabling complex liner designs, some of which are include bulk absorber material within the honeycomb resonators. When these technologies are applied to the fan noise component, an additional 2.5 dB of noise reduction is anticipated beyond the mid-term estimates.

Open rotors are not considered a noise-reduction technology. In fact they produce higher noise levels than turbofans. But the ICAO study included open rotors for noise evaluations since they are being considered for future aircraft because of their fuel efficiency advantage. "Open rotors" is a newer term being used for high-speed counter-rotating propellers similar to the General

Electric unducted fan (UDF) that was flown in the 1980s (Fig. 1.9(c), Chapter 1). Single-rotation propellers (turboprops) were discussed in Section 4.2.1.2. All of the noise-generation mechanisms for single-rotation propellers exist for counterrotating propellers, plus additional sources due to blade-to-blade interaction and pylon wake interaction with the blade rows. The propeller sound spectrum in Figure 1.11 of Chapter 1 is from a counterrotating open rotor. Mitigation methods include reducing the rotor speed, increasing the number of blades, clipping the aft rotor to reduce tip vortex interactions from the front rotor, using flow control to fill the pylon wakes, reducing the inflow distortions with optimized airframe installation, and increasing blade row spacing. Tremendous progress has been made since the first flight test of the UDF in the late 1980s. Noise levels can meet the ICAO Chapter 4 noise certification (ICAO, 2008) with significant margin (and will be discussed further in Sec. 4.3.3.2).

Long-term jet noise reduction will focus on engines with lower BPR. There will likely be an emphasis on other dominant noise sources as higher BPR engines enter service in the future, but smaller jets with high specific thrust requirements will still need jet noise reduction concepts. Beveled nozzles are analogous to scarf inlets and modify the directivity of the noise away from the community. Advanced methods like fluidic injection, microjets, and high-frequency excitation were reviewed by the ICAO panel. They are similar to each other in that they introduce a way to change the mixing process. The noise-reduction potential for all of these methods was estimated to be 1 to 3 dB.

The ICAO reports note that very little work has been done in long-term research for core noise reduction. There is an acknowledged need to understand the impact of low-emission combustors on noise. If the pressure ratio continues to increase and the combustors operate with lean injection, combustion noise is expected to increase. Ideas for reducing combustion noise include using aerated injectors instead of high-pressure injectors, if possible, and increasing the cross-sectional area of the combustion chamber. Turbine noise can benefit from high-temperature liners such as ceramic and metal foams. Over-the-rotor treatment being explored for fans (acoustic treatment in the nacelle directly over the fan) could also be implemented on multistage turbines. Active tip clearance control has been considered for compressors and may have acoustic benefits. The potential benefit of advanced technologies for long-term core noise reduction was not presented by the ICAO review due to the lack of quantitative information even at low TRL.

### 4.3.2   Airframe noise reduction

#### 4.3.2.1   Airframe mid-term technologies

Airframe noise-reduction technologies that can be implemented by 2020 focus on landing gear and high-lift noise sources. Landing gear fairings look promising and have matured to TRL 5. Some of the streamlining of bluff shapes and filling of voids has already been implemented. Careful placement of landing gear door positions and shaping have also been included in new aircraft designs. The review panel estimated that up to 5 dB of noise reduction is possible for landing gear.

Flap side-edge noise is expected to be addressed by treatments such as porous surfaces discussed in Section 4.2.2.4. Slat and slat track treatments are expected to be available in this time period. The combined benefit of all high-lift noise-reduction devices is expected to provide 3 to 4 dB noise reduction for this component.

#### 4.3.2.2   Airframe long-term technologies

For the long term, after fairings, caps, and design optimizations are utilized from the mid-term technologies, there is very little that can be done about landing gear noise beyond some type of flow control. The ICAO review panel estimated that the benefit would be less than 1 dB, which would make it difficult to justify the added complexity. An alternate approach is to change the aircraft design to use shorter landing gear such as those mounted under the fuselage for large military transports.

Continued development of methods for reducing slat and flap noise include more fillers as materials and structures technologies mature. Noise reduction up to 5 dB was estimated at low TRL of 1 to 2, which was deemed too low to provide credible noise reduction estimates at higher TRL for component and aircraft system predictions.

One idea that has emerged since the ICAO reviews is to add acoustic treatment on the wing or fuselage at locations that reduce engine noise as it reflects from these surfaces. This will require cost-effective liners that do not increase drag and can withstand the flow environment.

Aircraft noise has been reduced considerably maintaining a familiar "tube and wing" aircraft configuration. There is concern that further noise reduction will be limited without significant changes to the airframe and engine installation. Studies conducted and sponsored by NASA have identified so-called "N+2" and "N+3" configurations aimed at simultaneously meeting very aggressive noise, emission, and fuel-burn-reduction goals. Figure 1.8 of Chapter 1 shows two concept aircraft from these studies. The engines are mounted in locations where the wing can provide shielding of the propulsion noise. Prior to the NASA studies, Massachusetts Institute of Technology and Cambridge University developed a concept aircraft called the SAX-40 under the Silent Aircraft Initiative (Hileman *et al.*, 2010). A goal of this work was to design a functionally silent aircraft with minimal impact on other design parameters like fuel efficiency. All of these studies show that further noise reduction is possible. The ICAO review panel felt that introduction of these aircraft is well beyond the 2030 time frame for entry into service and thus were not included in the long-term goal assessments.

### 4.3.3 *Aircraft noise-reduction goals*

Using the component noise-reduction estimates described in the previous sections, the ICAO review panel, working with industry and government organizations, estimated aircraft noise levels. Each aircraft category was considered from regional jets to large four-engine aircraft. Allowable noise levels vary with aircraft takeoff weight, so the studies evaluated both nominal and expected maximum weights within each category. The aircraft noise levels typically increase with increasing weight and at a higher rate than the slope of the regulations, so considering maximum weights for growth versions of aircraft is important.

#### 4.3.3.1 *Mid-term goals*

Results from the mid-term aircraft noise predictions are shown in Table 4.2. The target BPR is shown for turbofan-powered aircraft. Aircraft noise reduction values were determined from systems analyses incorporating noise reduction technologies that can reach TRL 8 by 2020. The reference values show the 2010 baseline aircraft cumulative noise margins relative to the ICAO Chapter 4 standard (ICAO, 2008). The noise reduction goals listed in the last column show the anticipated noise levels of new aircraft that could be certified by 2020. These values can be compared with historical levels shown in Figure 1.6 of Chapter 1. An uncertainty analysis was included in these studies but not shown here for simplicity. In general, there is a ±4-EPNdB cum uncertainty in the noise predictions.

#### 4.3.3.2 *Long-term goals*

A similar analysis was done for long-term noise reduction technologies and is summarized in Table 4.3. The ICAO review panel felt that it was not possible to estimate long-term technologies at TRL 8 since many design trade studies need to be made for certification that do not involve just noise reduction technologies. The aircraft noise estimates were provided at TRL 6. Large turboprops were not part of this study, but aircraft in a Boeing 737/Airbus A320 size category (short medium-range twin) were compared using aft-mounted advanced turbofans and open rotors for the propulsion system. The uncertainty for the open rotor predictions is higher than turbofans since there is no current experience certifying an open-rotor propulsion system. The sensitivity of noise with increasing aircraft weight is greater for open-rotor systems compared to turbofans.

Table 4.2.  Mid-Term (2020) Aircraft Cumulative EPNdB Noise Goals at Technology Readiness Level 8.

| Aircraft category (weight) | Bypass ratio goal | Aircraft noise reduction | Reference Chapter 4 cumulative margin[a] | Noise reduction goal relative to Chapter 4[a] |
|---|---|---|---|---|
| Regional jet | | | | |
| 40 tonnes (nominal) | 7 | 9 | 4 | 13 |
| 50 tonnes (max.) | 7 | 9 | −0.5 | 8.5 |
| Large turboprops | | | | |
| 45 tonnes (nominal) | – | 9 | 3 | 12 |
| 53 tonnes (max.) | – | 9 | 0.5 | 9.5 |
| Short medium-range twin | | | | |
| 78 tonnes (nominal) | 9 | 16 | 5 | 21 |
| 98 tonnes (max.) | 9 | 16 | 1.5 | 17.5 |
| Long-range twin | | | | |
| 230 tonnes (nominal) | 10 | 14.5 | 6 | 20.5 |
| 290 tonnes (max.) | 10 | 14.5 | 2.5 | 17 |
| Long-range quad | | | | |
| 440 tonnes (nominal) | 9 | 16 | 5 | 21 |
| 550 tonnes (max.) | 9 | 16 | −1.5 | 14.5 |

[a]Referring to the standard presented in Chapter 4 of ICAO (2008).
*Source*: Mongeau, L. 2013. Noise technology goals—summary of the conclusions of the Second CAEP Noise Technology Independent Expert Panel (IEP2). Presented at the ICAO Symposium on Aviation and Climate Change, "Destination Green," ICAO Headquarters, Montreal. *http://www.icao.int/Meetings/Green/Documents/day%201pdf/session%202/2-Mongeau.pdf* (accessed March 22, 2017). With permission.

Table 4.3.  Long-term (2030) aircraft cumulative EPNdB noise goals at Technology Readiness Level 6.

| Aircraft category (weight) | Bypass ratio goal | Aircraft noise reduction | Reference Chapter 4 cumulative margin[a] | Noise reduction goal relative to Chapter 4[a] |
|---|---|---|---|---|
| Regional jet | | | | |
| 40 tonnes (nominal) | 9 | 17.5 | 4 | 21.5 |
| 50 tonnes (max.) | 9 | 17.5 | −0.5 | 17 |
| Short medium-range twin turbofans | | | | |
| 78 tonnes (nominal) | 13 | 25 | 5 | 30 |
| 98 tonnes (max.) | 13 | 25 | 1.5 | 26.5 |
| Open rotors | | | | |
| 78 tonnes (nominal) | – | 8.5 | 5 | 13.5 |
| 91 tonnes (max.) | – | 8.5 | 2 | 10.5 |
| Long-range twin | | | | |
| 230 tonnes (nominal) | 13 | 22 | 6 | 28 |
| 290 tonnes (max.) | 13 | 22 | 2.5 | 24.5 |
| Long-range quad | | | | |
| 440 tonnes (nominal) | 11 | 22 | 5 | 27 |
| 550 tonnes (max.) | 11 | 22 | −1.5 | 20.5 |

[a]Referring to the standard presented in Chapter 4 of ICAO (2008).
*Source*: Mongeau, L. 2013. Noise technology goals—summary of the conclusions of the Second CAEP Noise Technology Independent Expert Panel (IEP2). Presented at the ICAO Symposium on Aviation and Climate Change, "Destination Green," ICAO Headquarters, Montreal. *http://www.icao.int/Meetings/Green/Documents/day%201pdf/session%202/2-Mongeau.pdf* (accessed March 22, 2017). With permission.

## 4.4   SUMMARY

Sustained research on aircraft noise mitigation methods has significantly reduced the environmental impact of air transportation. Initial work focused on jet noise reduction since it was the dominant source for turbojets and low bypass ratio turbofans. Gradually, research expanded to other engine and airframe noise sources, and the evolution to higher BPR engines has reduced the jet and fan noise components. Incorporating noise-reduction technologies has led to quieter aircraft that meet regulations with sufficient margin. Since many of the noise sources are reaching parity in terms of noise levels, research for future aircraft will need to address a greater number of noise sources. Traditional "tube and wing" aircraft configurations are expected to reach a limit for further noise reduction, and it is expected that novel aircraft concepts that shield engine noise and also reduce airframe noise sources will further noise-reduction progress. Studies sponsored by the International Civil Aviation Organization show that 8 to 21 cumulative EPNdB noise reduction relative to their Chapter 4 regulations (ICAO, 2008) are possible for new aircraft entering service by 2020. Longer-term predictions for 2030 show additional noise reduction is possible, but it is difficult to estimate noise benefits at technology readiness level 8 because of design uncertainties. Actual noise levels depend greatly on specific aircraft, requirements, and design decisions.

## REFERENCES

Adib, M., F. Catalano, J. Hileman, *et al*. 2014. Novel aircraft-noise technology review and medium- and long-term noise reduction goals. Report to CAEP by the CAEP Noise Technology Independent Expert Panel, ICAO Doc 10017.

Brooks, T. F., D. S. Pope, and M. A. Marcolini. 1989. Airfoil self-noise and prediction. NASA RP–1218.

Dittmar, J. H., D. M. Elliott, and L. A. Bock. 1999. Some acoustic results from the Pratt & Whitney advanced ducted propulsor: fan 1. NASA/TM—1999-209049.

Dobrzynski, W. 2010. Almost 40 years of airframe noise research: what did we achieve? *J. Aircraft* 47: 353–367.

Dowty Propellers. 2016. R408 propeller system for the Bombardier Q400 Dash 8. *http://dowty.com/products/regional-airliners/* (accessed August 31, 2016).

Elkoby, R., L. Brusniak, R. W. Stoker, *et al*. 2007. Airframe noise results from the QTD II flight test program. AIAA 2007–3457.

Herr, M. 2007. Design criteria for low-noise trailing-edges. AIAA 2007–3470.

Hileman, J. I., Z. S. Spakovszky, M. Drela, *et al*. 2010. Airframe design for silent fuel-efficient aircraft. *J. Aircraft* 47:956–969.

Horne, W. C., K. D. James, N. Burnside, *et al*. 2005. Measurements of 26%-scale 777 airframe noise in the NASA Ames 40- by 80 Foot Wind Tunnel. AIAA 2005–2810.

Hutcheson, F. V., T. F. Brooks, and W. M. Humphreys, Jr. 2008. Noise radiation from a continuous mold-line link flap configuration. AIAA 2008–2966.

International Civil Aviation Organization. 2008. Annex 16 to the Convention on International Civil Aviation: Environmental Protection, Volume II—Aircraft Engine Emissions. Third edition, ICAO, Montréal, Quebec. *https://law.resource.org/pub/us/cfr/ibr/004/icao.annex.16.v2.2008.pdf* (accessed December 20, 2016).

International Civil Aviation Organization. 2010. Report to CAEP by the CAEP Noise Technology Independent Expert Panel: aircraft noise technology review and medium and long term noise reduction goals. Doc 9943, ISBN 978-92-9231-5p94-8.

International Civil Aviation Organization. 2014. Report to CAEP by the CAEP Noise Technology Independent Expert Panel: novel aircraft-noise technology review and medium- and long-term noise reduction goals. Doc 10017, ISBN 978-92-9249-401-8.

Jones, M. G., M. B. Tracy, W. R. Watson, *et al*. 2002. Effects of liner geometry on acoustic impedance. AIAA 2002–2446.

Khorrami, M. R., and D. P. Lockard. 2010. Effects of geometric details on slat noise generation and propagation. *Int. J. Aeroacoust.* 9:655–678.

Lighthill, M. J. 1952. On sound generated aerodynamically. I. General theory. *Proc. R. Soc. Lond. A* 211:564–587.

Majjigi, R. K., and P. R. Gliebe. 1994. Development of a rotor/wake-vortex model, Volume 1. NASA CR–174849.

Mankins, J. C. 1995. Technology readiness levels. NASA Office of Space Access and Technology White Paper. *https://www.hq.nasa.gov/office/codeq/trl/trl.pdf* (accessed August 31, 2016).

Mathews, D. C., N. F. Rekos, and R. T. Nagel. 1977. Combustion noise investigation. DOT/FAA Report No. FAA RD–77–3.

Mengle, V. G., V. D. Baker, and W. N. Dalton. 2002. Lobed mixer design for noise suppression: plume, aerodynamic and acoustic data. NASA/CR—2002-210823/VOL2.

Mongeau, L. 2013. Noise technology goals—summary of the conclusions of the Second CAEP Noise Technology Independent Expert Panel (IEP2). Presented at the ICAO Symposium on Aviation and Climate Change, "Destination Green," ICAO Headquarters, Montreal. *http://www.icao.int/Meetings/Green/Documents/day%201pdf/session%202/2-Mongeau.pdf* (accessed March 22, 2017).

Tyler, J., and T. Sofrin. 1962. Axial flow compressor noise studies. SAE Technical Paper 620532.

Weir, D. S. 2004. Engine validation of noise reduction concepts—separate flow nozzles. AIAA–2004–188.

Woodward, R. P., and C. E. Hughes. 2012. Acoustic performance of the GEAE UPS Research Fan in the NASA Glenn 9- by 15-Foot Low-Speed Wind Tunnel. NASA/TM—2012-217450.

# CHAPTER 5

## Advanced materials for green aviation

Ajay Misra

### 5.1 INTRODUCTION

Over the years materials advances have enabled a significant reduction in aircraft weight and an increase in the efficiency of gas turbine engines, both of which have contributed toward reducing fuel burn and emissions. It is expected that continuing advances in new materials will enable new aircraft and propulsion systems with significantly reduced weight and higher efficiency to meet future green aviation needs. This chapter provides an overview of various advanced materials that are under development for future aviation. There will be increasing use of composite materials in aircraft, with expanded use of polymer matrix composites (PMCs) in aircraft structures and in the cold sections of gas turbine engines, and with ceramic matrix composites (CMCs) replacing Ni-base superalloys in the hot section of gas turbine engines. The development of carbon nanotube (CNT) fibers with significantly higher strength than carbon fibers will lead to a new class of high-strength PMCs with superior strength. Ultra-lightweight cores based on ordered hierarchical cellular materials for sandwich structures will contribute to further weight reduction. Materials with a combination of load-bearing and other functional properties, along with innovative fabrication methods, will be used create multifunctional structures that will reduce weight and complexity. Smart materials based on a lightweight and simpler actuation system will enable the development of adaptive structures that can morph, or change shape, during different phases of flight. Advanced materials will enable high-power-density electrical components and a lightweight power transmission system for large commercial hybrid electric aircraft.

Over the past three decades, advances in materials have contributed significantly toward decreasing the emissions and noise of commercial and military aircraft. Materials contribute to green aviation in many ways. First, advanced lightweight materials such as PMCs lead to reduced aircraft weight, resulting in reduced fuel burn and $CO_2$ emissions. For example, for a single-aisle aircraft, a 12% reduction in structural weight can lead to a 7% reduction in fuel burn. Second, for the gas turbine engines used for aircraft propulsion, increases in thermal efficiency (and the associated reductions in fuel burn and $CO_2$ emissions) require an increase in the temperature capability of the materials used in the hot section components of engines. As high-temperature metallic alloys are reaching their use-temperature limits, CMCs are emerging as a new class of high-temperature materials for aircraft gas turbine engines. Third, smart materials can replace many of the heavy hydraulic-based actuation systems that are used for aerodynamic control of aircraft surfaces. Smart materials can benefit green aviation by (1) reducing the weight of heavy actuation systems and (2) increasing aircraft efficiency. In addition, smart materials can be enabling for reducing aircraft noise.

Advances in nanotechnology will have a significant impact on future aircraft materials. Of particular interest is the use of CNTs and graphene, both of which have unique thermal, electrical, and functional properties in addition to high strength. Nanomaterials offer the potential for multifunctionality in aircraft structures, enabling the structures to carry the load and perform other functions (such as thermal management, acoustic attenuation, energy storage, and health management, to name a few). Multifunctional structures offer the potential to reduce weight even further than what is possible with the use of composite materials.

There is growing interest in electric propulsion for future green aircraft. The ultimate goal for low-emission aircraft is an electric propulsion system that can significantly reduce emissions by using an energy storage system to power an electric motor that drives the fan. However, this improvement would require significant technological advances for large commercial aircraft, which may not mature for a long time. Near- and mid-term options for electric propulsion include turboelectric and hybrid electric propulsion, both of which use gas turbine engines. In turboelectric propulsion, the gas turbine is connected to an electric generator that generates power for the electric-motor-driven fan. In hybrid electric propulsion, a gas turbine is used in combination with an energy storage system to power an electric motor, which in turn drives the fan. The electric motor is a key component of electric propulsion systems. For aircraft applications, high power density is required for the electric motor system. Materials are enabling for achieving high power density in electric motor systems. Materials are also enabling for transmitting large amounts of power, on the order of megawatts, in electric aircraft. With the introduction of high-power-density components (e.g., electric motors and power electronics) in electric aircraft, a large amount of heat needs to be rejected, requiring an advanced thermal management system. Materials are also enabling for advanced thermal management systems in aircraft.

This chapter provides an overview of the advanced materials that are enabling for reducing aircraft emissions. It is divided into four sections: (1) lightweight materials, (2) smart materials, (3) high-temperature materials, and (4) materials for electric aircraft.

## 5.2    LIGHTWEIGHT MATERIALS

### 5.2.1    *Polymer matrix composites*

The most notable advances in lightweight materials over the last several years have involved the introduction of carbon-fiber-reinforced PMCs in the aircraft structure, replacing metallic components such as Al and Ti alloys. Composite materials are preferred in aircraft structures because of their high strength-to-weight ratio and excellent tailorable characteristics in comparison to metals. There has been steady increase in the extent of PMCs used in commercial aircraft, as shown in Figure 5.1. Today, PMCs constitute 50% of the structural weight for large commercial aircraft,

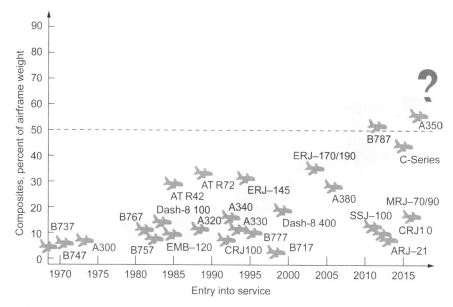

Figure 5.1.    Increasing use of composites in aircraft structures as a function of time. (Chris Red, Composites Forecasting and Consulting, LLC. With permission.)

such as the Boeing 787. Besides reductions in fuel burn and the weight of airframe structures, there are several other benefits associated with using composites in airframe structures. These include the ability of the composites to resist fatigue and corrosion and to withstand temperature extremes. These benefits translate to reduced maintenance costs because airframe inspection/ maintenance activities do not need to be as frequent as for the metallic aircraft. Another benefit of composites is that their strength can be tailored in any direction. By varying the orientation of the fibers, the strength of the composite material can be varied for different applied loads. The use of composites in airframe structures is expected to grow with time, with composite applications being extended to the Boeing 737 and Airbus A320 class of aircraft as well as regional jets.

PMCs are also finding increasing application in the cold sections of gas turbine engines, such as ducts and cowls, the fan containment system, and fan blades. Although PMCs have been used for noncritical components (e.g., ducts and piping) for a long time, there has been a steady growth in the application of PMCs for critical turbine engine components over the last decade or so, with the introduction of PMC fan blades and fan containment system (Fig. 5.2). GE was the first to introduce a PMC fan blade in the GE–90 engine for Boeing 777 (Mecham, 2012) and has continued to incorporate PMC fan blades in the subsequent versions for large engines, for example, GEnx (Boeing 787) and GE9X (Boeing 777X).

The PMC fan blade concept is being adopted in engines for single-aisle aircraft, such as the Leading Edge Aviation Propulsion (LEAP®) engine (CFM International, Cincinnati, Ohio) for Boeing 737 replacement. There is also growing trend toward using PMCs for composite fan cases, primarily to reduce weight. The GE GEnx engine used in the Boeing 787 uses a composite fan case with a damage-tolerant braided composite architecture. Other engine manufacturers besides GE have embraced the composite fan case concept, as evidenced by the introduction of a PMC fan case in the Pratt & Whitney PW1500G engine dedicated to Bombardier C-series aircraft and Williams International FJ44–4A engine now flying on the Cessna Citation CJ4 aircraft. The fan cases used in current gas turbine engines consist of braided architecture (Roberts *et al.*, 2009), as shown Figure 5.3. Future applications of PMCs in gas turbine engines could include advanced inlets with integrated acoustic treatment, structural vanes and frames, and first-stage compressor case.

A widely used fabrication method for composites consists of laying of fabric layers, or "plies," to form prepreg tape, which is the building block of a laminate stack. This is the process used for Boeing 787 composite structures. There is growing trend toward using a textile architecture consisting of woven fabrics to increase the damage tolerance of composites. An example of this is the use of a woven architecture for fan blades in the LEAP engine (Mecham, 2012) in which the fan blades are woven from carbon fiber strands rather than being formed of layered plies. This architecture increases the damage resistance of fan blades in the event of a bird strike. Another

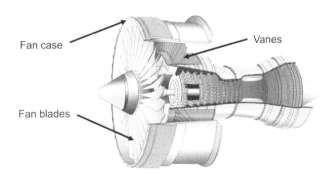

Figure 5.2.    Current use of polymer matrix composites (PMCs) in the load bearing sections of gas turbine engines.

Figure 5.3.   Braided architecture used to make the composite fan case. (From Roberts, G. D., J. M. Pereira, M. S. Braley, J. D. Dorer, and W. R. Watson. 2009. Design and testing of braided composite fan case materials and components. NASA/TM—2009-215811 (ISABE 2009–1201). Work of the U.S. Government.)

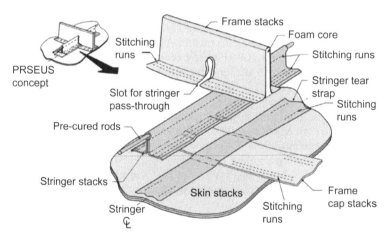

Figure 5.4.   Pultruded Rod Stitched Efficient Unitized Structure (PRSEUS) concept. (From Jegley, D., A. Przkeop, M. Rouse, *et al.*, 2015. Development of stitched composite structure for advanced aircraft. NF 16762-20691, Sept. 2015. *http://ntrs.nasa.gov/search.jsp?R=20160006276* (accessed August 12, 2016). Work of the U.S. Government.)

concept that is being developed jointly by NASA and Boeing is the Pultruded Rod Stitched Efficient Unitized Structure (PRSEUS) concept (Jegley *et al.*, 2015). The concept, shown in Figure 5.4, is an integrally stiffened panel design, in which elements are stitched together and designed to maintain residual load-carrying capabilities under a variety of damage scenarios. With the PRSEUS concept, through-the-thickness stitches are applied through dry fabric prior to resin infusion, and these stitches replace fasteners throughout each integral panel. Through-the-thickness reinforcement at discontinuities, such as along flange edges, has been shown to suppress delamination and turn cracks, which expands the design space and leads to lighter designs. Another concept that is receiving considerable attention is thin-ply composites (Shin *et al.*, 2007). The major advantages of thin-ply composites are suppression of microcracking and delamination, high strength, and greater damage tolerance.

   The strength and stiffness of PMCs depend on the fiber properties. Polyacrylonitrile (PAN)-based carbon fibers are currently being used in the aerospace industry. One of these, IM7 from Hexcel, which is widely used in the aerospace industry, has a tensile strength of 5.1 to 5.6 GPa and tensile modulus of 276 GPa (Hexcel Corporation, 2016a). There is continued development effort to increase the strength and modulus of carbon fibers. Two new commercial high-strength and high-modulus carbon fibers have recently been introduced: Toray 1100 G, with a strength of 6.6 GPa and modulus of 324 GPa (Toray Industries, Inc., 2014), and IM 10 fiber from Hexcel, with a strength of 6.69 GPa and modulus of 310 GPa (Hexcel Corporation, 2016b). Recently, an experimental fiber with tensile strength in the range of 5.5 to 5.8 GPa and modulus in the range of 354 to 375 GPa has been developed (Chae *et al.*, 2015). Both the tensile strength and modulus of available carbon fibers are still significantly lower than the theoretical values, and it is expected that carbon fibers with higher strength and modulus will be developed in the future.

### 5.2.2   Carbon nanotubes for structural applications

CNTs (Wikipedia, 2017) are tubular cylinders of carbon atoms that have extraordinary mechanical, electrical, thermal, optical, and chemical properties (Fig. 5.5). Single-walled CNTs (Fig. 5.5(a)) are seamless cylinders composed of a layer of graphene, whereas multi-walled CNTs (Fig. 5.5(b)) consist of multiple rolled layers of graphene. The strength of CNTs, shown in Table 5.1, can be an order of magnitude higher than the carbon fiber, and the elastic modulus can be nearly 30 times higher than that of carbon fibers. With such attractive mechanical properties, composites reinforced with CNTs offer the potential of significant

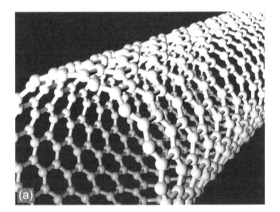 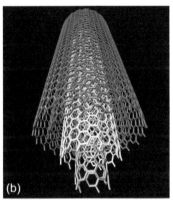

Figure 5.5.   Carbon nanotubes (CNTs). (a) Single-walled CNT. (b) Multiwalled CNT. ((b) From Rochefort, A. 2015. Nanotubes and buckyballs. Montreal, Quebec: Polytechnique Montreal. *http://www.nanotech-now. com/nanotube-buckyball-sites.htm* (accessed March 22, 2017). Image credit: A. Rochefort, Polytechnique Montreal.)

Table 5.1.   Comparison of strength and elastic modulus of carbon nanotubes (CNTs) with other structural material.

|  | Tensile strength, GPa | Elastic modulus, TPa |
|---|---|---|
| Single-walled CNT (experimental) | 13 to 53 | ~ 1 |
| Single-walled CNT (theoretical) | 94 to 126 | ~ 1 |
| Multiwalled CNT (experimental) | 11 to 150 | 0.2 to 0.95 |
| Stainless steel | 0.38 to 1.55 | 0.18 to 0.21 |
| Carbon fiber (IM10) | 6.7 | 0.31 |

*Source*: Wikipedia. 2017. Mechanical properties of carbon nanotubes. *https://en.wikipedia.org/wiki/Mechanical_ properties_of_carbon_nanotubes*. Creative Commons CC BY-SA 3.0 license, *https://creativecommons.org/licenses/ by-sa/3.0/legalcode* (accessed March 29, 2017).

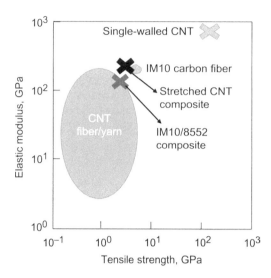

Figure 5.6.  Comparison of the properties of carbon nanotube (CNT) fibers and yarns with individual CNTs and the best available carbon fiber.

weight reduction for aircraft structures. However, CNTs cannot be used in structural applications unless they can be fabricated into useable forms such as fibers, yarns, or sheets. Although the development of high-strength CNT fibers and yarns has been the focus of intense research over the last several years (Behabtu *et al.*, 2008; Lu *et al.*, 2012; Wu and Chou, 2012), the strength and modulus of CNT fibers are still lower than that of carbon fiber. Typical values of the strength and elastic modulus of CNT fibers and yarns range from 0.1 to 3 GPa and 15 to 250 GPa, respectively, although values of 8.8 GPa strength and 397 GPa elastic modulus have been reported for short fibers. Nanocomp has produced CNT yarns on a commercial scale with a reported strength of 2.08 GPa (Gurau, 2014). These values are considerably lower than both the experimental and theoretical strength and modulus numbers for CNT.

One approach to increasing the strength and modulus of CNT fibers, yarns, and sheets is through post-processing treatments. Miller *et al.* (2014) showed that the specific strength and modulus of CNT sheets and yarns can be increased by a factor of 3 to 4 through a combination of functionalization and electron-beam irradiation treatments. Wang *et al.* (2013) employed a novel stretch winding process in which the CNT sheet was stretched by 12% before it was embedded in a PMC. The process resulted in a composite tensile strength up to 3.8 GPa and a modulus up to 293 GPa, which are higher values than for the conventional continuous-fiber-reinforced PMCs.

Figure 5.6 summarizes the properties of CNT fibers and yarns and compares them with commercial carbon fibers, composites, and a single-walled CNT. The measured strength and modulus of CNT fibers and yarns occupy a large band, with most of the data lying in the lower portion of the band. However, the strength and modulus of some CNT fibers and yarns approach that of the best carbon fiber, IM10. Also, the properties of CNT fibers and yarns are significantly lower than those of individual CNTs. There is continuing effort in many organizations to improve the properties of CNT yarns. There is a general consensus that the structure and properties of CNT yarns must be optimized at all hierarchical levels, from individual CNTs, to the bundle and network of CNTs, and to the macroscopic yarn structure. From the modeling of processing-microstructure-property relationships (Beese *et al.*, 2014), a potential path of improving strength and modulus of CNT fibers and yarns has been identified, which include (1) increasing bundle strength through the introduction of cross-links in the bundle network, (2) improving load transfer between adjacent CNTs, (3) developing fabrication methods for aligning straight CNTs and bundles, and (4) developing effective methods for compacting yarns to reduce porosity.

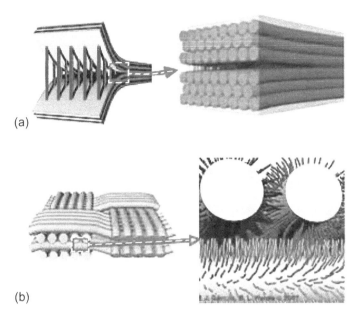

(a)

(b)

Figure 5.7.   Two concepts for increasing the interlaminar strength and fracture toughness of composites. (a) Carbon-fiber-reinforced polymer matrix composite (PMC) with aligned carbon nanotubes (CNTs) between laminates (Nanostitch™). (b) fuzzy fiber with the growth of aligned CNTs on the fibers. (From Wardle, B. L., *et al.*, 2013. Hierarchical nanoengineered composite (aerospace) structures: manufacturing and mechanical properties. Presented at Defense Manufacturing Conference (DMC) 2013, Orlando, FL. *http://web.mit.edu/aeroastro/labs/necstlab/documents/Wardle_NanoEngMIT-DMCInvitedDec2013sm.pdf* (accessed August 12, 2016). Figure courtesy of and Copyright MIT. Original images by Enrique Garcia (MIT).)

As CNT fibers with strength greater than 10 GPa are developed in the future, the specific strength of PMCs with CNT fiber reinforcement can be twice that of carbon-fiber-reinforced PMCs. Although it is expected that the properties of CNT fibers will improve significantly in the coming years, the strengths and moduli of carbon fibers are also likely to increase with further development. One approach for improving the strength of carbon fibers is to incorporate CNTs into the fiber, as shown by Kim *et al.* (2013), who increased the tensile strength of carbon fibers by more than 14% by the catalytic growth of CNTs on the surface of the fibers. It is likely that carbon fiber materials could be about 10 times stronger than they are presently.

Although the use of CNTs as load-bearing elements in CNT-reinforced polymer composites to replace carbon fibers is several years away, CNTs can be used effectively in carbon-fiber-reinforced PMCs to improve various properties such as interlaminar strength and fracture toughness. Wardle *et al.* (2013) developed two concepts known as Nanostitch™ (N12 Technologies, Inc., Cambridge, Massachusetts) and fuzzy fiber laminates (Fig. 5.7), in which CNTs are incorporated into carbon-fiber-reinforced composites to increase interlaminar strength and fracture toughness. In the nanostitched concept, aligned CNTs are introduced between the plies. In the fuzzy fiber concept, CNTs are grown on individual fibers, which creates fuzzy fibers. Both concepts have resulted in an increase in fracture toughness by a factor of 2 to 3. Fenner and Daniel (2014) have developed hybrid nanoreinforced carbon/epoxy composites in which the polymer matrix in a carbon fiber composite was reinforced with CNTs. With a CNT-reinforced matrix for carbon-fiber-reinforced PMCs, significant improvements were obtained in interlaminar shear strength (20%), fracture toughness (180%), and shear fatigue life (order of magnitude).

Incorporating CNTs in the polymer matrix of carbon-fiber-reinforced PMCs can dramatically increase the electrical conductivity of PMCs because of the high electrical conductivity of CNTs.

Such materials can protect the composite structures in aircraft from lightning strikes. Currently, the composite structures in aircraft are not electrically conductive, so the state of practice is to use heavy sacrificial embedded bronze or copper mesh in the outer layer of aircraft structures, which adds significant weight. A small addition (<2%) of CNT can significantly increase the electrical conductivity of a polymer matrix (Edelmann *et al.*, 2008; Russ *et al.*, 2011), which provides the benefit of significantly reducing the weight of the composite structure.

### 5.2.3   *Multifunctional materials*

In multifunctional material systems, the material constituents simultaneously and synergistically undertake two roles, which could be load bearing and thermal management, load bearing and acoustic attenuation, or load bearing and energy storage, to name a few. Such materials offer the potential for significant weight and volume reduction in a structural system or component/device. There has been considerable interest in developing multifunctional materials with energy storage capability, which would combine two separate components (structure for load bearing and energy storage component). Although multifunctional materials with energy storage and load bearing capability are important for many applications, including automobiles and unmanned aerial vehicles, they are of great importance for hybrid electric propulsion systems, which offer the potential for significant emissions reduction (Ashcraft *et al.*, 2011). The energy storage system (such as batteries and supercapacitors) is a critical component of hybrid electric propulsion system. Based on the current energy density of the state-of-the-art batteries, the energy system adds a significant weight penalty for aircraft, and it has been determined that energy storage systems with at  an increase in specific energy least 4 times would be required for large commercial hybrid electric aircraft. Clearly, multifunctional materials with load bearing and energy storage capability would enable use of the aircraft structure to store energy. The following paragraphs provide details for two multifunctional energy storage concepts employing multifunctional materials: supercapacitors with load bearing capability and batteries with load bearing capability.

   The supercapacitor concept, shown in Figure 5.8, has two plates or electrodes. The plates are made from a porous substance such as powdery, activated charcoal or carbon aerogel, which effectively gives them a high surface area that allows them to store much more charge. Both

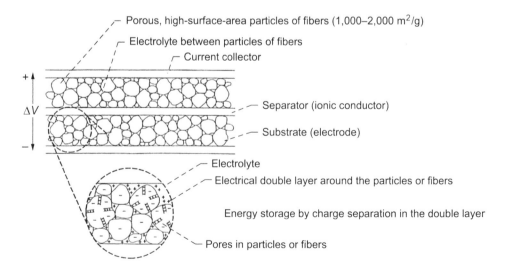

Figure 5.8.   Schematic of conventional supercapacitor. (Reprinted from Burke, A. 2000. Ultracapacitors: why, how, and where is the technology. *J. of Power Sources*, 91(1): 37-50. Copyright 2000, with permission from Elsevier.)

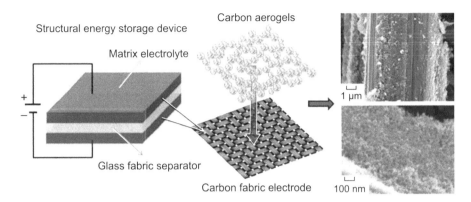

Figure 5.9.   Schematic of multifunctional structural supercapacitor composite. (Reprinted with permission from Qian, H., A. R. Kucernak, E. S. Greenhalgh, *et al.*, 2013. Multifunctional structural supercapacitor composites based on carbon aerogel modified high performance carbon fiber fabric. *ACS Appl. Mater. Interfaces* 5:6113–6122. Copyright 2013, American Chemical Society.)

plates are soaked in an electrolyte and separated by a very thin insulator (which might be made of carbon, paper, or plastic). The electrolyte is typically a liquid, although semi-solid and solid electrolytes are being developed. Individual layers of the conventional supercapacitor do not have the necessary strength to carry any load. Since carbon fiber is a structural element and carbon is a typical constituent for energy storage systems, adding carbon fiber to various elements of the supercapacitor is attractive for increasing the strength of the supercapacitor. Qian *et al.* (2013) developed a multifunctional structural supercapacitor composite based on carbon-aerogel-modified high-performance carbon fiber fabric. The multifunctional concept, shown in Figure 5.9, consisted of embedding structural carbon fiber fabrics in a continuous network of carbon aerogels. Incorporation of aerogels significantly increased the surface area of carbon fibers, which resulted in 100-fold increase in electrochemical performance. Glass fabric was used as a separator. An ionic liquid modified solid state polymer was used as a multifunctional matrix for the composite. In another concept developed by Lanzara and Basirico (2012), miniaturized CNT supercapacitors were prefabricated on a support and then embedded between layers of carbon-fiber-reinforced polymer composite.

In a conventional battery such as the Li-ion battery, each cell has essentially three components: a positive electrode (connected to the battery's positive, or +, terminal), a negative electrode (connected to the negative, or −, terminal), and a chemical called an electrolyte in between them. The positive electrode is typically is a complex oxide ceramic, such as lithium-cobalt oxide ($LiCoO_2$) and lithium iron phosphate ($LiFePO_4$). The negative electrode is generally made from carbon (graphite), and the electrolyte varies from one type of battery to another, being a liquid (mainly composed of lithium salts—$LiPF_6$, $LiClO_4$, and $LiBF_4$—and organic solvents). One of the approaches for strengthening various battery layers is through incorporation of structural fibers, an approach similar to that used for supercapacitors. Liu *et al.* (2009) developed a structural battery concept (Fig. 5.10) using a polyvinylidene fluoride (PVdF)-based electrolyte (more viscous). The active cathode and anode materials are nanofibers of $LiCoO_2$ and graphite, respectively. Carbon fibers are added to both anode and cathode layers to add structural capability as well as to provide electronic conduction pathways. CNT electrodes can also provide multifunctionality because of the high strength of CNTs. Evanoff *et al.* (2012) developed a scalable method utilizing CNT nonwoven-fabric-based technology to develop flexible, electrochemically stable battery anodes that can be produced on an industrial scale. Similar methods can be utilized for the formation of various cathode and anode composites with tunable strength and energy and power densities.

So far, multifunctional energy storage concepts have not demonstrated the combination of high strength and electrochemical storage capability needed for application in aircraft. Research

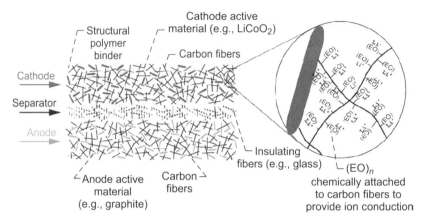

Figure 5.10.   Structural battery components reinforced with various active element fibers, including carbon fibers, to increase strength. Electrolytic oxide, $(EO)_n$. (Reprinted from Liu, P., E. Sherman, and A. Jacobsen. 2009. Design and fabrication of multifunctional structural batteries. *J. Power Sources,* 189:646–650. Copyright 2009, with permission from Elsevier.)

needs to continue to identify materials and material combinations that can provide both load bearing and energy storage capability. For example, a carbon fiber that has high strength may not have high electrochemical capacity, and it will have to be modified to have balanced properties. Nanotechnology will play an important role in the development of multifunctional materials. Nanofibers and nanoforests of active electrochemical materials need to be developed and incorporated into composite materials. A multifunctional property database for various materials of interest to multifunctional structures needs to be developed. Computational materials science modeling tools are needed to predict the different properties of materials so that new materials with optimized strength and functional (electrochemical, thermal, or sensing) properties can be developed.

### 5.2.4   *Cellular materials for sandwich structures*

Sandwich structures have been used in numerous aspects of aircraft construction. Typical sandwich structure are made up of thin, stiff, strong faces separated by a very lightweight cellular material known as the core. The core could be metallic (e.g., Al) or metallic (e.g., Nomex), and the facesheet could be composite or metallic. The current state of the art for the cores of conventional composite sandwich structures is honeycomb or foam cores. There could be several variation of the core material, such as foam-filled honeycomb or fiber-reinforced foam core. An class of core material that is receiving considerable attention uses a three-dimensional (3D), truss-like network of reinforcing fibers inside a lightweight foam core. One example of this emerging class of core materials include TYCOR® (Milliken & Company, Spartanburg, South Carolina) (TYCOR Products, 2011). The truss structure of the 3D fiber network provides added paths for load transfer and acts to impede crack propagation. Another class of new core material is the metallic lattice structure, in which 3D lattice truss structures replace honeycombs and foams to provide new combinations of stiffness, strength, robust performance, and multifunctionality (Sypeck, 2005). A schematic of a lattice block structure is shown in Figure 5.11. In their simplest form, lattice block panels are produced by casting fully triangulated truss-like configurations that provide strength and stiffness. Lattice block structures have been fabricated from a variety of materials, including high-temperature superalloys (Nathal *et al.*, 2004), and microlattice structures have been made from titanium alloys and stainless steel (Mines *et al.*, 2013).

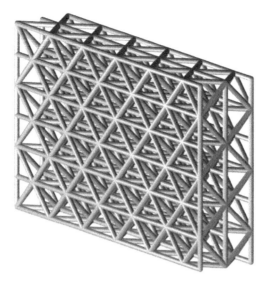

Figure 5.11.   Schematic of lattice block structure.

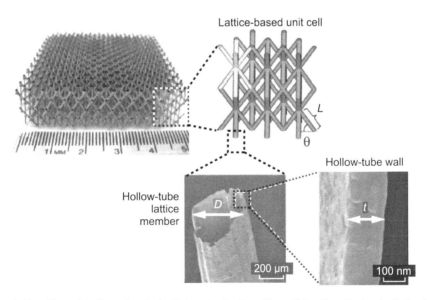

Figure 5.12.   Hierarchically ordered ultralight microlattice. (From Schaedler, T. A., A. J. Jacobson, A. Torrents, *et al.*, 2011. Ultralight metallic microlattices. *Science* 334:962–965. Reprinted with permission from AAAS.)

There is a continuing effort to develop ultra-lightweight core materials and structures that will substitute for honeycomb and foam cores. Schaedler *et al.* (2011) developed ultralight metallic microlattices with densities in the range of 0.9 to 10 mg/cm³. The material is 99.9% porous. The 0.1% of the material that is not air consists of a microlattice, or periodic array, of interconnected hollow nickel-phosphorous tubes with a wall thickness of 100 nm (Fig. 5.12). These tubes are angled to connect at nodes to form repeating, 3D asterisk-like cells. The microlattice material has excellent properties, including superior energy absorption properties with the ability to completely recover from compression exceeding 50% strain.

Microlattice structures with improved mechanical properties result from a material that contains microscale and nanoscale building blocks arranged in an ordered hierarchy. The principle of ordered hierarchy has been employed by Zheng *et al.* (2014) to develop an ultra-lightweight microarchitectured material with hollow tubes that have the same weight and density as aerogel, but with 10,000 times the stiffness. The material belongs to a class of materials called metamaterials whose mechanical properties are defined by their geometry rather than their composition. These microarchitectured materials were fabricated by microstereolithography, which allows rapid generation of materials with complex 3D microscale geometries with dimensions ranging from 40 nm to 20 mm. Superior properties were obtained with all classes of materials: metals, ceramics, and polymers.

Lightweight cellular core materials will be important for future aircraft design. For many aircraft components (such as the wing/empennage leading edge, fan blades, and fan casings) where impact resistance is important, cellular core materials provide high-energy absorption capability. Cellular core materials are attractive for multifunctional structures in which the open space can be filled with different materials to provide multifunctionality, such as a combination of load bearing and acoustic attenuation capability or a combination of load bearing and thermal management capability. A cellular core has been proposed as a key element of flexible skin for morphing aircraft (Olympio and Gandhi, 2010); it comprises high-strain-capable, low-modulus facesheets covering a cellular core, which has significantly higher strain capability compared with the virgin material. Although a cellular core made from aluminum can provide the strain capability required for a camber-morphing airfoil, new core materials with high strain capability are required for morphing wing span, chord, and planform. Ultralight microlattice cores have the potential to provide the high strain capability to enable a flexible skin for morphing aircraft wings.

## 5.3   SMART MATERIALS

Smart materials are enabling for morphing aircraft structures and components, also referred to as active structures or reconfigurable structures. In addition, smart materials can enable new concepts for noise reduction, flow control, and energy harvesting. There are several classes of smart materials, including shape memory alloys (SMAs), piezoelectric materials, shape memory polymers, magnetostrictive materials, and rheological fluids. Smart materials are differentiated by two primary attributes, strain and frequency, as shown in Figure 5.13. Two classes of materials, SMAs and piezoelectric materials, have received considerably more attention over the last decade or so compared with other classes of smart materials. SMAs find applications where high strain rate and low frequency response is required. Piezoelectric materials find applications where low strain rate and high frequency response are important. This section focuses on SMAs and piezoelectric materials.

### 5.3.1   *Shape memory alloys*

SMAs are alloys that have a "memory," which contributes to the name of the material. These materials have the ability to remember and recover their original shapes with load or temperature. This unusual property stems from the alloy undergoing a phase transformation from a high-temperature austenite to a low-temperature martensitic crystal structure. The SMA actuation concept is shown in simple form as a weight suspended from an SMA wire (Fig. 5.14). As heat is applied to the SMA wire, typically through self-resistance heating, the wire contracts because of the shape memory effect, and the weight is lifted, thus performing work. Work is defined by the mass of the weight times the distance moved and is a critical property for SMA development. Cooling the wire causes the weight to be returned to the lower position. Figure 5.15 displays the thermo-mechanical test employed that mimics actuator performance in a mechanical testing frame. A load is applied at low temperature, and it is maintained as the temperature is cycled through the martensite/austenite transformation. The applied stress multiplied by the transformation strain

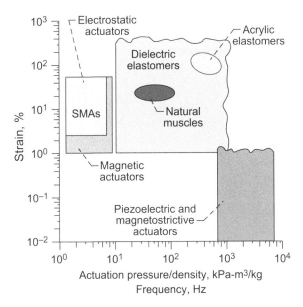

Figure 5.13.   Strain and operational frequency for various classes of smart materials. SMA, shape memory alloy. (From Ashley, S. 2003. Artificial muscles. *Sci. Am.* 289:55. Illustration credit: Sara Chen. With permission.)

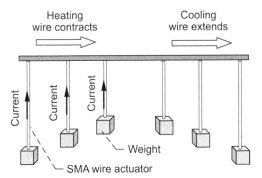

Figure 5.14.   Schematic of the use of a shape memory alloy (SMA) as an actuator. As heat is applied to the SMA, it contracts and performs work by lifting the weight.

defines the work capability of the alloy. Details on the principles and application of SMAs can be found in Otsuka and Wayman (1998).

   The most widely used SMA is a Ni-Ti alloy known as Nitinol. The actuation capability of SMA is of great interest for aircraft morphing structures, also known as smart structures and reconfigurable structures. Such morphing structures increase aircraft performance (such as reduced fuel burn and noise, and increased range and payload) by matching the structure shape and area to the operating conditions, thus avoiding compromise design points for traditional aircraft. Shape and area changes of aircraft structures are traditionally achieved through the deployment of hydraulic, electrohydraulic, and electromechanical actuation, which are heavy. SMAs offer lightweight actuation options for morphing/reconfigurable structures. An example of the weight benefit of an SMA actuator for a reconfigurable rotor blade is provided in Arbogast *et al.* (2008). For the reconfigurable rotor blade, the traditional actuation system consisting of electric motor and gear weighs 41 lb (18.6 kg), compared to only 0.99 lb (0.45 kg) for SMA-based actuation system.

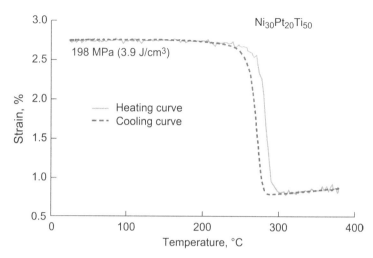

Figure 5.15. The thermo-mechanical test employed that mimics actuator performance in a mechanical testing frame. A load is applied at low temperature and maintained as the temperature is cycled through the martensite/austenite transformation. In this example, the experimental shape memory alloy (SMA) has an actuation temperature of about 250°C, and a work output of 3.9 J/cm³.

Figure 5.16.   Shape memory alloy (SMA)-actuated variable-area chevron flight-tested by Boeing. (From Mabe, J. H., F. T. Calkins, and G. W. Butler. 2006. Boeing's variable geometry chevron, morphing aero-structure for jet noise reduction. AIAA 2006–2142. Copyright, Boeing. With permission.)

SMA actuators offer many other benefits, including (1) high force per volume and weight, (2) compactness, (3) easy integration with existing systems, (4) frictionless and quiet operation, and (5) low maintenance.

A good overview of SMAs for morphing aerostructures can be found in Calkins and Mabe (2010). Applications include (1) a variable-geometry chevron (Fig. 5.16) on an engine exhaust that morphs the chevron between a shape optimized for noise reduction during takeoff and a shape at cruise that reduces shock cell noise without compromising engine performance, and (2) a variable-area jet engine fan nozzle with a large nozzle diameter at takeoff and approach, enabling the reduction of the jet velocity and resulting noise, and a small nozzle diameter during cruise to optimize fan loading and reduce fuel consumption. There is ongoing research to develop morphing structures for critical airframe components such as an adaptive trailing edge

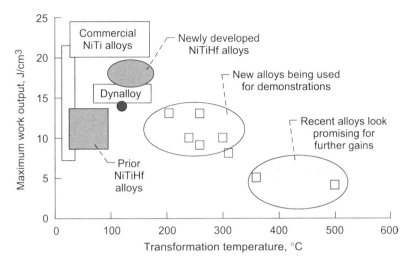

Figure 5.17   Summary of shape memory alloy (SMA) alloy development efforts, where the challenge is to increase temperature capability without sacrificing work output.

(Boeing, 2014) and a variable-camber continuous trailing edge flap (Urnes *et al.*, 2013) that would smoothly change the wing's shape continuously throughout the flight.

Commercially available Ni-Ti SMAs can be used at temperatures up to 100°C. However, many aircraft applications require application temperatures greater than 100°C. Gas turbine engine applications (such as blade shape control, vane shape and position control, and blade tip seals) require significantly higher temperature capability for SMAs. Several new high-temperature alloys have been developed at the NASA Glenn Research Center (Nathal and Stefko, 2013), as summarized in Figure 5.17. The NiTiHf series of alloys appears promising as a new class of SMA material. With higher temperature capability and high work output, NiTiHf alloys can prevent the accidental actuation of aircraft structures on hot summer days in certain regions of the world, when the ambient temperature exceeds temperature capability of commercially available Ni-Ti SMAs.

Although there has been continuing progress toward the development of SMAs for lightweight actuation systems, significant challenges still remain. These include (1) lack of long-term durability data, (2) an extensive database required for certification, (3) limited design tools, (4) thermal management and control techniques, (5) long-term dimensional stability, and (6) cost.

### 5.3.2   *Piezoelectric materials*

In piezoelectric materials, mechanical stress causes crystals to electrically polarize and vice versa. Application of electric current causes the piezoelectric material to deform, enabling the material to work as an actuator, shown in Figure 5.18. The most widely used piezoelectric material is lead zirconate titanate (PZT) ceramic, $Pb(Ti, Zr)O_3$. The displacement of a single layer of piezoceramic is too small for large stroke driving. Therefore several piezoelectric pieces are stacked together to provide large stroke. Two of the most popular actuator designs are multi-layers and bimorphs (Fig. 5.19). In a multilayer design, multiple thin piezoceramic sheets are stacked together. The multilayer design has the advantage of low driving voltage, quick response (10 μs), and high generative force (1000 N). However, the multilayer design has low displacement (on the order of 10 μm), which may not be suitable for some applications. In the bimorph design, multiple piezoelectric and elastic plates are bonded together to generate a large bending displacement of several hundred microns. The drawback of the bimorph design is relatively

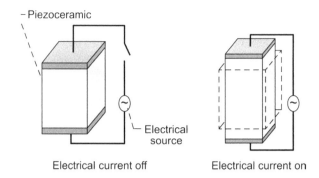

Figure 5.18.    Schematic showing the principle of actuation for a piezoelectric material.

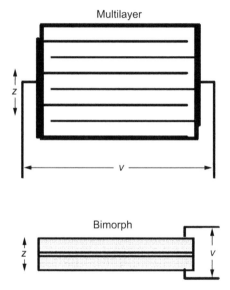

Figure 5.19.    Schematic of different types of piezoelectric actuators.

low response time (on the order of 1 ms) and small generative force (on the order of 1 N). Applications of piezoceramic materials for green aviation include noise reduction, vibration reduction, and active flow control to reduce emissions and improve aerodynamic performance.

Active noise plus vibration control along with improvement in aerodynamic efficiency have been demonstrated in wind tunnel tests for a smart helicopter rotor (Straub *et al.*, 2009) using a piezoceramic-actuated trailing edge flap on each blade (Fig. 5.20). The test demonstrated on-blade smart material control of flaps on a full-scale rotor for the first time in a wind tunnel, with conclusive demonstration of the effectiveness of active flap control for noise and vibration reduction.

The use of piezoceramic actuators for noise and vibration reduction in a gas turbine engine fan blade is an active area of research. Resonant vibrations of aircraft engine blades cause blade fatigue problem in engines, leads to thicker and aerodynamically lower performing blade designs, resulting in increases in engine weight and fuel burn. Min *et al.* (2012) investigated active piezoelectric vibration control in rotating fan blades, demonstrating the effectiveness of piezoceramic patches in reducing the vibration of a subscale rotating PMC fan blade (Fig. 5.21) through tests in a dynamic spin rig. Piezoelectric layers on static fan blades also are being considered for noise reduction in a European research effort (Leylekian *et al.*, 2014).

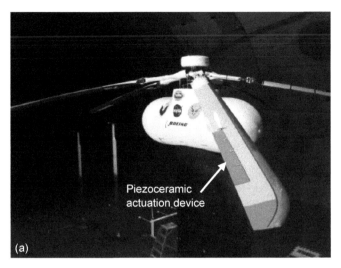

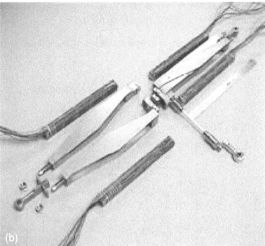

Figure 5.20. Active helicopter rotor blade using piezoelectric actuation. (a) Rotor blade wind tunnel test at the NASA Ames Research Center. (b) Details of piezoelectric actuation devices.

Piezoceramic actuators can enable fuel modulation in active combustion control (Schiller *et al.*, 2006) with the benefit of higher gas turbine engine combustor efficiency and increased combustion stability for low nitrogen oxide ($NO_x$) combustor concepts. In active control combustion, instabilities can be controlled by modulating the fuel out of phase with the thermo-acoustic vibration, thus canceling out the vibration. High-bandwidth fuel modulation is currently one of the most promising methods for active combustion control. To attenuate the large pressure oscillations in the combustion chamber, the fuel is pulsed so that the heat release rate fluctuations damp the pressure oscillations in the combustor. A key component of the fuel control valve with high-bandwidth fuel modulation is the piezoelectric actuator, which is capable of operating at the high frequencies and operating temperatures of the fuel nozzle.

Piezoelectric actuators are also attractive for flow control applications in aircraft with benefits for enhanced aerodynamic performance and reduced noise and fuel burn. Active flow control has been demonstrated to be effective in reattaching flow on an airfoil that would otherwise be stalled apart. Synthetic jets are an attractive option for active flow control, depicted in Figure 5.22.

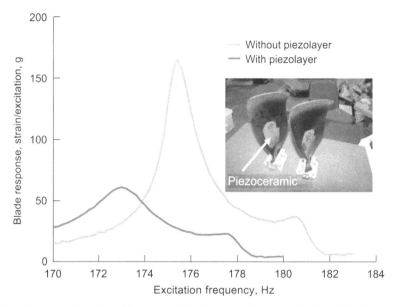

Figure 5.21.   Demonstration of the effectiveness of a piezoelectric layer on reduction of vibration in a rotating polymer matrix composite (PMC) fan blade.

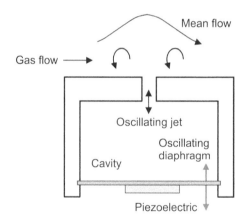

Figure 5.22.   Schematic of synthetic jet flow control using piezoelectric actuators (used to oscillate the diaphragm).

Synthetic jet systems have been produced with speakers, compressed air, air pumps, and bimorph diaphragms, among other components. All of these techniques add weight, require real estate, and add complexity to an airplane, making these options impractical. Piezoelectric actuators are an attractive solution because of their light weight and fast response. Mossi and Bryant (2006) have evaluated several piezoelectric actuators for synthetic jet applications. Despite the favorable attributes of piezoelectric actuators for synthetic jet application, issue of durability, system weight, size, and power have limited the use of these actuators outside of a laboratory.

Although piezoelectric materials are attractive for reducing noise and emissions in aircraft, significant challenges remain for commercializing actuation devices using piezoelectric materials. These include increasing strain capability, higher temperature capability, low weight and volume, integration with structure, long-term durability, and easy availability of a power source for operating piezoceramic devices. Work is in progress to address these challenges.

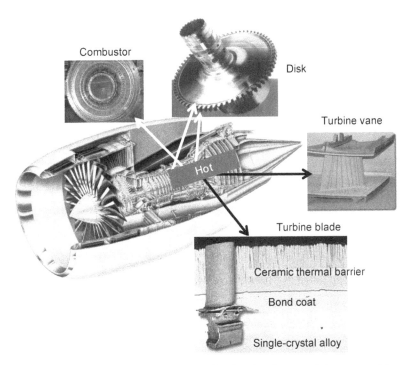

Figure 5.23.    Cross section of aircraft gas turbine engine showing the state-of-the-art materials.

## 5.4    HIGH–TEMPERATURE MATERIALS

High-temperature materials are used in the hot section of gas turbine engines, a cross section of which is shown in Figure 5.23. Nickel-base superalloys are the state-of-the-art (SOA) high-temperature materials in commercial gas turbine engines and are used in the later stages of the high-pressure compressor, combustor, and high-pressure turbine (HPT), and in the first stages of the low-pressure turbine (LPT). The SOA material for the HPT blade is cooled single-crystal Ni-base superalloy with a thermal barrier coating (TBC). The SOA disk alloy is a powder metallurgy Ni-base superalloy. The SOA combustor liner material is a Ni-base alloy with TBC. Titanium alloys find use in the initial stages of the HPC and the later stages of the LPT. The thermal efficiency of gas turbine engine is a direct function of the material temperature capability. Increasing the material temperature capability results in (1) an increase in turbine inlet temperature, which is required for high-efficiency engines, and (2) a reduction in cooling needed for the same turbine inlet temperature, with the benefit of increased engine efficiency.

### 5.4.1    *High-temperature Ni-base superalloys*

Nickel-base superalloys are extensively used in the hot section of commercial gas turbine engines. For the HPT blade and vane components, which have the highest temperature and are subjected to high thermal and mechanical loads, single-crystal Ni-base superalloys with a TBC are used. There has been steady increase in the temperature capability of Ni-base superalloys with TBCs over the last 70 years, as shown in Figure 5.24. Nickel-base superalloys are currently being used at close to 90% of their melting points, and further increases in temperature capability are expected to be incremental, with the possibility for another 25 to 30°C increase in temperature capability.

A further increase in the temperature capability of coated single-crystal Ni-base alloys can be achieved through decreasing the thermal conductivity of TBCs. Zhu and Miller (2004)

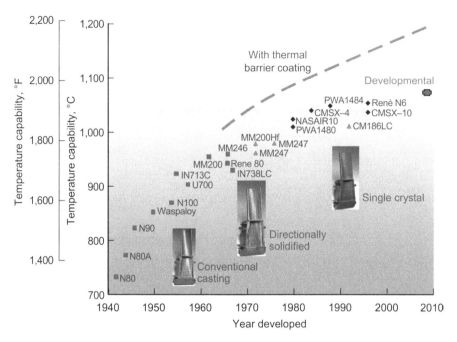

Figure 5.24.    Increase in temperature capability of nickel-base blade alloys over the last 70+ years.

developed low-conductivity TBCs with one-half to one-third of the thermal conductivity of the SOA yttria stabilized zirconia (YSZ) TBC. However, these coatings are not fully optimized with respect to different properties (e.g., fracture toughness, impact resistance, and erosion resistance) that are required for achieving long-term durability. An excellent review of the current approach to develop low-conductivity thermal barrier coatings can be found in Pan *et al.* (2012). Computational materials science tools, along with a carefully controlled experiment for a wide range of oxide chemistries, will be required to identify viable low-conductivity coatings. So far, the full potential for TBCs has not been realized in gas turbine engines because of a lack of high-fidelity, physics-based design and life prediction tools. With the increasing emphasis on an integrated computational materials engineering approach for the design and use of materials, the temperature capability of TBCs is expected to increase further.

The disk is another critical hot section component in the gas turbine engine. This component is subject to very high centrifugal stresses induced by the high rotational speed, plus additional loads on the rim imposed by the blades. In addition high thermal stresses are induced in the bulk of the disk by the temperature gradients during engine start and shut down. The current generation of disk alloys are predominantly powder metallurgy Ni-base alloys. There has been a steady increase in the temperature capability of disk alloys (Fig. 5.25), with a new generation of disk alloy entering into service every 10 to 15 years. The temperature capability of the current SOA disk alloys is on the order of 1,300°F (704°C), which has been achieved by careful control of the alloy microstructure to get balanced properties for both hub and rim regions. The hub region consists of a fine-grained microstructure for achieving good tensile strength and fatigue resistance. The rim region (the hotter region) consists of a coarse-grained microstructure for achieving good creep resistance.

Future gas turbine engines with very high thermal efficiency will require a very high overall pressure ratio (OPR), on the order of 50 or higher. Achieving such a high OPR will require 1,500°F (815°C) temperature capability for the disk, which is, beyond the projected temperature capability for powder metallurgy alloys. With alloying and innovative processing, the maximum temperature capability for Ni-base powder metallurgy disk alloys is projected to be ~1,450°F (787°C).

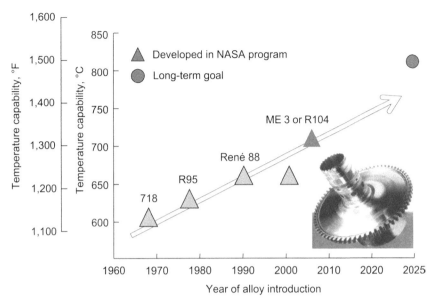

Figure 5.25.    Increase in temperature capability of gas turbine engine disk alloy over the last 50 years.

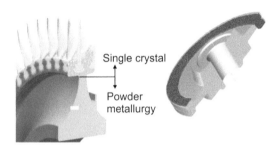

Figure 5.26.    Hybrid disk concept with powder metallurgy disk alloy bonded to a single-crystal rim to achieve 1500°F.

In order to achieve a disk temperature capability of 1,500°F (815°C), innovative approaches will be required. One approach that is currently being developed is the hybrid disk approach (Fig. 5.26), in which a disk hub, fabricated from a powder metallurgy disk alloy with 1,400 to 1,450°F (760 to 787°C) capability, is bonded to a single crystal Ni-base rim alloy. Since the rim regions have the highest temperatures in the disk, the single crystal alloy would provide the required high temperature in the rim.

### 5.4.2    *Ceramic matrix composites*

With Ni-base superalloys reaching their temperature limits, the emphasis for advanced materials for the hot section of gas turbine engines will be on CMCs. Although there are many types of CMCs, the one that is suitable for the gas turbine engine hot section is the silicon-carbide-fiber-reinforced silicon carbide (SiC/SiC) CMC, which weighs only a third of what Ni-based super-alloys weigh. The SiC/SiC CMC has significantly higher creep resistance (DiCarlo, 2013) than Ni-base superalloys and oxide/oxide CMCs, as shown in Figure 5.27. The temperature increase offered by SiC/SiC CMCs in comparison to SOA Ni-base superalloys could range from 200 to 500°F (128 to 260°C).

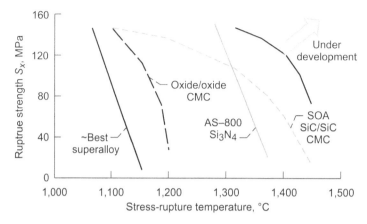

Figure 5.27.    Comparison of rupture strength versus temperature for different material systems. CMC, ceramic matrix composite; SOA, state of the art. (Adapted from DiCarlo, J. A. 2013. Advances in SiC/SiC composites for aero-propulsion. NASA/TM—2013-217889. Work of the U.S. Government.)

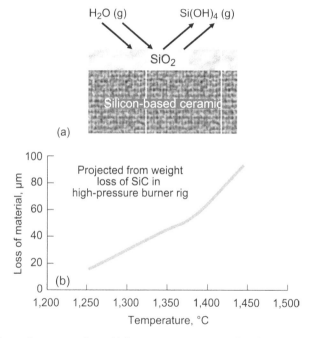

Figure 5.28.    SiC experiences recession at high use temperatures. (a) Silica film formed on the SiC surface reacts with water in the combustion environment, forming a gaseous product that is swept away; the film reforms, and the cycle repeats. (b) Loss of material after 100 h as a function of temperature.

The SiC/SiC CMCs react with the moisture in gas turbine engines (Smialek *et al.*, 1999), resulting in recession of the surface in a gas turbine engine environment (Fig. 5.28). Therefore, environmental barrier coatings (EBCs) are required to protect SiC/SiC CMCs from gas turbine engine environments. The first generation of EBCs (Lee *et al.*, 2005) capable of operating at 2,400°F (1,315°C), shown in Figure 5.29, consists of three layers: Si bondcoat, mullite (aluminosilicate) intermediate layer, and barium strontium aluminum silicate (BSAS) topcoat. The design of the EBC needs to take into consideration multiple factors, including chemical reactivity, diffusion of oxygen, mechanical integrity, and interdiffusion between various layers.

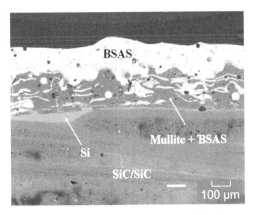

Figure 5.29.  Scanning electron micrograph showing three layers for the state-of-the-art environmental barrier coating (EBC). BSAS, barium strontium aluminum silicate.

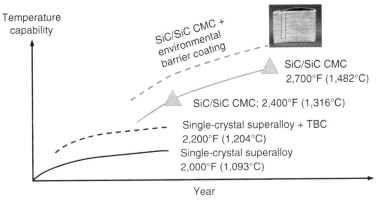

Figure 5.30.  Increase in temperature capability offered by SiC/SiC composite matrix composites (CMCs) over state-of-the-art nickel-base superalloys. TBC, thermal barrier coating.

The introduction of CMCs in the hot section of gas turbine engines will be game-changing for the gas turbine industry. As seen in Figure 5.30, CMCs offer significant increase in temperature capability beyond that of the Ni-base superalloys. The current temperature capability of SiC/SiC CMC is on the order of 2,400°F (1,315°C), which is 200°F (93°C) higher than that of the SOA coated Ni-base single crystal alloys. The first introduction of a SiC/SiC CMC in a commercial gas turbine engine was in 2016, with the use of a CMC shroud in the LEAP engine (GE Aviation and CFM International). It is expected that this will be followed by introduction of CMCs in LPT blades, combustor liners, and second-stage turbine vanes. The introduction of CMCs in HPT blades will require a significant increase in the load-carrying capability of CMCs. Ultimately, CMCs are expected to be used in all hot section components in gas turbine engines except for the disk. While SiC/SiC CMCs with 2,400°F (1,315°C) temperature capability are beginning to be introduced in commercial gas turbine engines, there is ongoing research to increase the temperature capability of these CMCs to 2,700°F (1,482°C).

System analysis studies (Grady, 2013) indicate that fuel burn reductions of up to 6% can be achieved in gas turbine engines by using CMCs with 2,700°F (1,482°C) temperature capability in the HPT vanes and blades and in the combustor as well as in the LPT vane and blade. Two factors contribute to the reduction in fuel burn. First, the higher temperature capability of CMCs in comparison to metals reduces or eliminates the cooling of hot section components. Second,

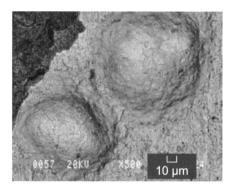

Figure 5.31.   Pitting observed on higher temperature disk alloy. (From Gabb, T. P., J. Telesman, B. Hazel, *et al.*, 2009. The effects of hot corrosion pits on the fatigue resistance of a disk superalloy. NASA/TM—2009-215629. Work of the U.S. Government.)

the weight of CMCs is one-third that of high-temperature superalloys, with the weight savings multiplier effect being much more than 3:1 because everything down the chain is affected as well. For example, it is estimated that incorporating CMC turbine blades on a GE90-sized engine could reduce the overall weight by about 455 kg (1,000 lb), which represents about 6% of 7,550 kg (~16,600 lb) dry weight of the full-sized GE90–115 (Trimble, 2010).

The key challenges for increasing the temperature capability of SiC/SiC CMCs are (1) the temperature capability of the fiber, (2) fabricating a dense composite without Si (the first generation of SiC/SiC CMCs contain free Si, which melts at 1410°C), and (3) developing an EBC with higher temperature capability, particularly developing a Si-free bondcoat.

### 5.4.3   *Environmental degradation challenges with increase in turbine operating temperatures*

With the increase in turbine operating temperatures, new environmental effects resulting from the ingestion of airborne contaminants (such as dust, sand, volcanic ash, and runway debris) are affecting the durability of turbine engine hot section components. Ingestion of airborne contaminants is resulting in the deposition of sulfate melts and glassy silicate melts on the surfaces of hot section components.

Recently, as disk temperatures have increased from ~648°C to ~704°C, hot corrosion has surfaced as a new environmental degradation mode (Gabb *et al.*, 2009). Hot corrosion is due to the deposition of salts—multicomponent sulfates—on the disk alloy surface. Figure 5.31 shows a hot corrosion pit that has been observed in service. The pitting morphology in the figure resembles typical Type II hot corrosion, which was identified as a new corrosion phenomenon in the 1970s for marine gas turbine engines (Luthra, 1982). There is a significant reduction in fatigue life due to the formation of hot corrosion pits, as shown in Figure 5.32. One approach to mitigating the effect of hot corrosion is to apply a protective coating. The key challenge for any coating is that it must be thin and ductile so that the fatigue life of the disk is not adversely affected by the coating.

Another mode of environmental degradation resulting from the ingestion of airborne contaminants is the deposition of glassy melts of calcium-magnesium aluminosilicate (CMAS) when the surface temperatures exceed ~1,200°C (Borom *et al.*, 1996; Kramer *et al.*, 2006). Single crystal Ni-base alloys with TBCs are susceptible to degradation by CMAS, resulting in delamination of the TBC. Several mechanisms have been postulated for the degradation of TBCs due to interaction with CMAS (Bacos *et al.*, 2011; Levi *et al.*, 2012; Mercer *et al.*, 2005; Wu *et al.*, 2010). The mechanisms include (1) a chemical reaction leading to the loss of strain tolerance in the coating and an increase in the stiffness of the coating; (2) residual stresses induced by the presence of

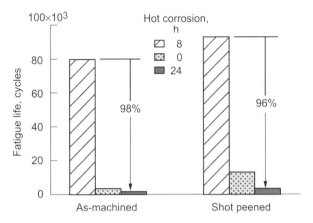

Figure 5.32.   Fatigue life degradation due to hot corrosion of ME3 alloy.

a CMAS layer, resulting in the initiation of in-plane and vertical cracks leading to partial removal of the TBC layer; (3) penetration of the CMAS liquid and interaction with the thermally grown oxide on the bondcoat, resulting in delamination of the coating along the TGO; and (4) CMAS promoting creep cavitation of the bond coat, resulting in the propagation of the lamination crack path within the metal. It is likely that new degradation mechanisms will emerge with further increases in temperature.

Several potential solutions are being considered to improve the resistance of TBCs to CMAS-induced degradation (Aygun *et al.*, 2007; Bacos *et al.*, 2011; Levi *et al.*, 2012). Potential solutions include (1) creating an impermeable layer on the top of the TBC that prevents infiltration of CMAS, (2) adding a sacrificial coating that reacts with CMAS to increase the melting temperature or viscosity so that the melt cannot infiltrate the TBC, (3) adding a nonwetting layer that minimizes contact between the coating and the molten deposit, and (4) manipulating the chemical reaction between the TBC and the melt to immobilize the melt by capturing the main constituents into crystalline phases and generating enough volume of reprecipitated products to fill the pore spaces and block access of any residual melt to the remaining TBC. So far, there have been some successes in improving the resistance of TBCs to CMAS attack; however, much more remains to be done to improve the durability of TBCs in the presence of CMAS.

With CMC hot section components coated with EBCs, the temperature of the EBC surface is expected to be greater than 1,315°C, and CMAS-induced degradation of EBCs will pose a major durability challenge for EBCs. The degradation of EBCs due to molten CMAS deposits is an emerging area of research (Grant *et al.*, 2007) and will continue to be studied in the near future.

## 5.5   MATERIALS FOR ELECTRIC AIRCRAFT

There is considerable interest in developing hybrid-electric and all-electric aircraft for the future, both of which offer significant potential for reducing fuel burn and greenhouse gas emissions. Several potential options are being considered: (1) an all-electric option with batteries or some other energy storage device driving an electric motor which in turn drives the fan propelling the aircraft, (2) a hybrid-electric option combining a gas turbine engine with an energy storage system with the energy storage system driving an electric motor and the fan in certain portions of the flight cycle, and (3) a turboelectric system in which the gas turbine is used to generate power through a generator, which drives the electric motor and the fan. Irrespective of the system, there are many common technical challenges associated with the electrical components, including

- High-power-density electrical motor
- High-power-density power electronics
- Lightweight power transmission system
- Lightweight thermal management
- Energy storage system with high specific energy

Materials are enabling to address each of the technical challenges. Various material classes of interest for electric aircraft are described in the following subsections.

### 5.5.1   *Advanced magnetic materials*

Advanced permanent magnets with a high amount of stored energy (measured by the energy product called $BH_{max}$) are enabling for high-power-density motors. Over the last 50 years, the $BH$ of permanent magnets has increased from 10 MGOe for Sm-Co alloys to 50 to 60 $MGO_e$ for Nd-Fe-B alloys. Further improvements in the energy product require the development of nanocomposites combining the magnetic hardness of rare earth compounds with the high magnetization of soft magnetic materials (Skomski and Coey, 1993). Theoretical predictions indicate that nanocomposites would have a $BH_{max}$ of 100 $MGO_e$. However, such high $BH_{max}$ values have not been achieved yet because of challenges associated with the fabrication of nanocomposites.

High-temperature magnets can enable increases in the operating temperature of electric motors, with the benefit of reduced cooling needs and reduced weight. The best commercially available high-temperature permanent magnets are based on the Sm-Co alloy system, which has a temperature capability on the order of 400°C. There has not been any significant progress over the last few decades in increasing the temperature capability of permanent magnets. New approaches based on computational material design and nanocomposites will be required.

High-power-density power converters (also known as power electronics) with reduced size and weight require an increase in operational frequency to above 1 MHz. It is estimated that the volume and weight of a converter operated at 2 MHz are 50 and 5 to 10 times less, respectively, than those of a converter working at 20 kHz. Passive components, such as inductors and transformers, make significant contributions toward the weight and volume of power converters. Therefore, soft magnetic materials with high operating frequency, high permeability, and high power capability need to be developed. Amorphous and nanocrystalline alloys are promising (Leary *et al.*, 2012) and have demonstrated low losses at high frequencies.

### 5.5.2   *Materials with high electrical conductivity*

Electrical cables contribute to significant weight in commercial aircraft. For example, the Boeing 747 has 140 mi of Cu electrical wiring contributing 3,500 lb (1,588 kg) of weight. Large commercial hybrid-electric aircraft require the transmission of megawatts of power, which will require significantly large diameter Cu cables with significant weight penalty. The weight of power cables can be reduced by developing materials with electrical conductivity higher than that of Cu or by deploying a high-voltage transmission system, both with significant materials challenges. Carbon nanotubes, in theory, can have an order of magnitude higher electrical conductivity than Cu and can carry an electric current density of $4 \times 10^9$ A/cm$^2$, which is more than 1,000 times greater than that of Cu (Hong and Myung, 2007). CNTs are therefore ideal candidates as power transmission cables. However, the present production method always produces a mixture of metallic and semiconducting nanotubes, which need to be separated to produce metallic nanotubes. Furthermore, the desired electrical properties need to be achieved in usable forms, such as fibers and yarns. Behabtu *et al.* (2013) recently developed CNT fibers whose specific conductivity (electrical conductivity/density) is slightly lower than that of Cu. The electrical conductivity of CNT fibers developed so far makes them suitable for data cables; however, further improvements in electrical conductivity will be required for power cable applications. CNT fibers with higher electrical conductivity can also replace Cu coils in electric motors, offering significant increases in power density and eliminating the need for superconducting motors that require cryogenic cooling.

Superconducting power transmission cables are another option for decreasing the weight of power cables for transferring megawatts of power. The use of high-temperature superconductor tapes enables power cable operation with a liquid nitrogen cooling system. Several types of superconducting cables have been demonstrated by Schmidt and Allais (2005).

### 5.5.3   *Advanced insulation materials*

Electrical insulation systems are critical for electric motors and generators and power transmission cables. Insulation components ensure that an electrical short does not occur and that the heat from the conductor losses is transmitted to a heat sink. Advanced insulation materials are enabling for reducing the thickness of insulation systems, with the benefits of higher power density for electric motors/generators and lightweight electrical power transmission systems.

There are several requirements for advanced insulation materials:

1. With the growing trend toward higher operating voltages, the electrical insulation material must be able to handle higher voltages. The improvement in dielectric breakdown strength, long-term durability under high-voltage operation, partial discharge resistance, and low dissipation factor will result in a thinner insulation system.
2. Higher thermal stability will enable the operation of the electrical insulation system at higher temperatures, which will reduce cooling requirements.
3. Higher thermal conductivity can enable faster heat transfer, enabling smaller size, higher speed, and higher power density for electrical motors and power electronics components. Polymer nanocomposites with nanofillers that are electrically insulating but have high thermal conductivity are attractive as insulation materials. One promising nanocomposite—epoxy with boron nitride fillers—has the potential for >3 times increase in thermal conductivity compared with epoxy only (Huang *et al.*, 2011).

### 5.5.4   *Advanced capacitors for power electronics*

Capacitors are key components of the power electronics that are used as part of the electric motor drive system. In state-of-the-art inverters,[1] the capacitors are the largest component and essentially determine the inverter volume. Consequently, an increase of the energy storage density of capacitors would have a huge impact on the size and power density of power electronics. Nanodielectric composites are particularly attractive for increasing the energy storage capacity of capacitors. The temperature capability of power electronics devices continues to increase with the use of SiC semiconductors instead of Si. This requires that the temperature capability of the capacitors be increased to take advantage of the SiC-based power electronics systems.

## 5.6   SUMMARY

Advanced materials will continue to be enabling for green aviation. Building upon the progress made so far, new materials will continue to be developed to meet the future needs of green aviation. Following the trend of the last decade or so, the use of polymer matrix composites in aircraft structures will continue to increase. With advances in manufacturing techniques and computational modeling tools, new composite architectures (such as three-dimensional and braided) will replace traditional two-dimensional  architectures. Nanotechnology is expected to play a key role in improving various polymer matrix composite properties (such as electrical conductivity, thermal conductivity, and interlaminar strength), thereby providing multifunctionality.

---

[1] Inverters convert electric currents from direct current (DC) to alternating current (AC).

The development of carbon nanotube-based fibers with significantly higher strength than commercially available carbon fibers will have a significant impact in reducing the overall weight of the aircraft. Ultra-lightweight cores utilizing ordered and hierarchical cellular materials will have a significant benefit for reducing the weight of the sandwich structures deployed in various parts of an aircraft.

Multifunctional and adaptive aircraft structures can provide significant weight reduction and aerodynamic efficiency benefits. Materials with high strength and energy storage capability will enable the development of multifunctional structures that store energy. Fibers with high thermal conductivity will serve as the building blocks for multifunctional structures with load bearing and thermal management capability. Continuous-fiber-reinforced PMCs with electrically conducting carbon nanotubes can provide lightning protection for aircraft structures. Adaptive capability has been demonstrated for several aircraft structures in various tests with the use of smart materials such as shape memory alloys. Although the potential for adaptive aircraft structures using shape memory alloys has been demonstrated, adaptive capability for critical aircraft components will require a demonstration of the durability of smart-material-based actuation systems, improvement in smart material capability, and computational modeling tools for three-dimensional actuation.

The operating temperature of gas turbines is expected to continue to rise to increase thermal efficiency, which will require an increase in the temperature capability of hot section materials. The temperature capability of Ni-base superalloys has been increasing steadily over the last 50 years. However, the use temperatures for these alloys are reaching close to the alloys' melting points, which has resulted in the introduction of ceramic matrix composites in gas turbine engines. CMCs offer a 200 to 500°F (128 to 260°C) increase in temperature capability in comparison with superalloys. There will be widespread use of CMCs in gas turbine engines over the next 20 years, and it is expected that all hot section engine components except for the disk will be made from CMCs. With the increase in the operation temperature of gas turbine engines, new environmental degradation issues (such as the deposition of molten materials on hot section components resulting from the ingestion of airborne contaminants) have surfaced, and coatings need to be developed to protect components from environmental degradation.

Advanced materials are enabling for the development of hybrid-electric and all-electric aircraft. High-power-density electric motors and power electronic systems will be required for these aircraft along with lightweight power transmission systems for transferring megawatts of power. Achieving the power density and power cable weight goals will require advanced magnetic materials, advanced insulation materials, materials with higher electrical conductivity than Cu, and advanced capacitors.

## REFERENCES

Arbogast, D. J., R. T. Ruggeri, and R. C. Bussom. 2008. Development of a ¼-scale NiTinol actuator for reconfigurable structures. *Proc. SPIE* 6930:69300L.

Ashcraft, S. W., A. S. Padron, K. A. Pascioni, G. W. Stout, Jr., and D. L. Huff. 2011. Review of propulsion technologies for N+3 subsonic vehicle concepts. NASA/TM—2011-217239.

Ashley, S. 2003. Artificial muscles. *Sci. Am.* 289:55.

Aygun, A., A. L. Vasiliev, N. P. Padture, *et al.* 2007. Novel thermal barrier coatings that are resistant to high-temperature attack by glassy deposits. *Acta Mater.* 55:6734–6745.

Bacos, M.-P., J.-M. Dorvaux, O. Lavigne, *et al.* 2011. Performance and degradation mechanisms of thermal barrier coatings for turbine blades: a review of Onera activities. *Journal Aerospace Lab* 3:1–11.

Beese, A. M., X. Wei, S. Sarkar, *et al.* 2014. Key factors limiting carbon nanotube yarn strength: exploring processing-structure-property relationships. *ACS Nano* 8:11454–11466.

Behabtu, N., M. J. Green, and M. Pasquali. 2008. Carbon nanotube-based neat fibers. *Nanotoday* 3:24–34.

Behabtu, N., C. C. Young, D. E. Tsentalovich, *et al.* 2013. Strong, light, multifunctional fibers of carbon nanotubes with ultrahigh conductivity. *Science* 339:182–186.

Boeing. 2014. FAA Continuous Lower Energy, Emissions and Noise (CLEEN) technologies. Boeing Program Review, Presented at the CLEEN Consortium Public Session, Atlanta, GA. *https://www.faa.gov/about/office_org/headquarters_offices/apl/research/aircraft_technology/cleen/2014_consortium/media/boeing_cleen_projects_briefing_november_2014.pdf* (accessed March 29, 2016).

Borom, M. P., C. A. Johnson, and L. A. Peluso. 1996. Role of environmental deposits and operating surface temperature in spallation of air plasma sprayed thermal barrier coatings. *Surf. Coat. Technol.* 86–87:116–126.

Burke, A. 2000. Ultracapacitors: why, how, and where is the technology. *J. Power Sources* 91:37–50.

Calkins, F. T., and J. H. Mabe. 2010. Shape memory alloy based morphing aerostructures. *J. Mech. Des.* 132:111012-1–111012-7.

Chae, H. G., B. A. Newcomb, P. V. Gulgunje, *et al.* 2015. High strength and high modulus carbon fibers. *Carbon* 93:81–87.

DiCarlo, J. A. 2013. Advances in SiC/SiC composites for aero-propulsion. NASA/TM—2013-217889.

Edelmann, K., B. Rackers, and B. L. Farmer. 2008. Nanocomposites for future airbus airframes. Presented at Wissenschaftstag 2008, Nuremberg, Germany. *http://www.dlr.de/fa/Portaldata/17/Resources/dokumente/institut/wissenschaftstag_2008/Edelmann.pdf* (accessed March 29, 2016).

Evanoff, K., J. Benson, M. Schauer, *et al.* 2012. Ultra strong silicon-coated carbon nanotube nonwoven fabric as a multifunctional lithium-ion battey anode. *ACS Nano* 6:9837–9845.

Fenner, J. S., and I. M. Daniel. 2014. Hybrid nanoreinforced carbon/epoxy composites for enhanced damage tolerance and fatigue life. *Compos. Part A* 65:47–56.

Gabb, T. P., J. Telesman, B. Hazel, *et al.* 2009. The effects of hot corrosion pits on the fatigue resistance of a disk superalloy. NASA/TM—2009-215629.

Gabb, T. P., J. Telesman, B. Hazel, *et al.* 2009. The effects of hot corrosion pits on the fatigue resistance of disk superalloy. *J. Mater. Eng. Perform.* 19:77–89.

GE Aviation. 2014. GE Aviation and Turbocoating SPA form coating joint venture. AVIATIONPROS.com. *http://www.aviationpros.com/press_release/12022105/ge-aviation-and-turbocoating-spa-form-coating-joint-venture* Accessed Feb. 23, 2016.

GE Aviation. 2015. The GEnx commercial aircraft engine: the GEnx engine delivers proven performance for the Boeing 787 Dreamliner and Boeing 747–8. *http://www.geaviation.com/commercial/engines/genx/* (accessed February 23, 2016).

Grady, J. E. 2013. CMC technology advancements for gas turbine engine applications. Presented at the American Ceramic Society's 10th Pacific Rim Conference on Ceramics and Glass Technology, San Diego, CA.

Grant, K. M., S. Kramer, J. P. A. Lofvander, *et al.* 2007. CMAS degradation of environmental barrier coatings. *Surf. Coat. Technol.* 202:653–657.

Gurau, M. 2014. The world's first commercial all-CNT sheets, tape and yarns. Blog, Nanocomp Technologies, Inc. *http://www.nanocomptech.com/blog/need-to-know-part-2-sheets-tape-yarn* (accessed March 29, 2016).

Hexcel Corporation. 2016a. HexTow IM7 carbon fiber. Hexcel product data sheet. *http://www.hexcel.com/resources/datasheets/carbon-fiber-data-sheets/im7.pdf* (accessed March 29, 2016).

Hexcel Corporation. 2016b. HexTow IM10 carbon fiber. Hexcel product data sheet. *http://www.hexcel.com/resources/datasheets/carbon-fiber-data-sheets/im10.pdf* (accessed March 29, 2016).

Hong, S., and S. Myung. 2007. Nanotube electronics—a flexible approach to mobility. *Nat. Nanotechnol.* 2:207–208.

Huang, X., P. Jiang, and T. Tanaka. 2011. A review of dielectric polymer composites with high thermal conductivity. *IEEE Electr. Insul. M.* 27:8–16.

Jegley, D., A. Przkeop, M. Rouse, *et al.* 2015. Development of stitched composite structure for advanced aircraft. NF 16762-20691, Sept. 2015. *http://ntrs.nasa.gov/search.jsp?R=20160006276* (accessed August 12, 2016).

Kim, K. J., J. Kim, W.-R. Yu, *et al.* 2013. Improved tensile strength of carbon fibers undergoing catalytic growth of carbon nanotubes on their surface. *Carbon* 54:258–267.

Kramer, S., J. Yang, C. G. Levi, *et al.* 2006. Thermochemical interactions of thermal barrier coatings with molten CaO-MgO-Al2O3-SiO2 (CMAS) deposits. *J. Am. Ceram. Soc.* 89:3167–3175.

Lanzara, G., and L. Basirico. 2012. Self-rechargeable multifunctional carbon fiber composites with CNTs supercapacitors. AIAA 2012–1647.

Leary, A. M., P. R. Ohodnicki, and M. E. McHenry. 2012. Soft magnetic materials in high-frequency, high-power conversion applications. *JOM* 64:772–781.

Lee, K. N., D. S. Fox, and N. P. Bansal. 2005. Rare earth silicate environmental barrier coatings for SiC/SiC composite and Si3N4 ceramics. *J. Eur. Ceram. Soc.* 25:1705–1715.

Levi, C. G., J. W. Hutchinson, M.-H. Vidal-Setif, *et al.* 2012. Environmental degradation of thermal-barrier coatings by molten deposits. *MRS Bull.* 37:932–941.

Leylekian, L., M. Lebrun, and P. Lempereur. 2014. An overview of aircraft noise reduction technologies. *Journal Aerospace Lab* 7:1–15. *http://www.aerospacelab-journal.org/sites/www.aerospacelab-journal. org/files/AL07-01_0.pdf*.

Liu, P., E. Sherman, and A. Jacobsen. 2009. Design and fabrication of multifunctional structural batteries. *J. Power Sources,* 189:646–650.

Lu, W., M. Zu, J.-H. Byun, *et al.* 2012. State of the art carbon nanotube fibers: opportunities and challenges. *Adv. Mater.* 24:1805–1833.

Luthra, K. L. 1982. Low temperature hot corrosion of cobalt-base alloys: Part I: morphology of the reaction product. *Metall. Trans. A* 13:1843–1852.

Mabe, J. H., F. T. Calkins, and G. W. Butler. 2006. Boeing's variable geometry chevron, morphing aero-structure for jet noise reduction. AIAA 2006–2142.

Mecham, M. 2012. Patterned breakthrough. *Aviation Week & Space Technology* 74–77.

Mercer, C., S. Faulhaber, A. G. Evans, *et al.* 2005. A delamination mechanism for thermal barrier coatings subject to calcium-magnesium-alumino-silicate (CMAS) infiltration. *Acta Mater.* 53:1029–1039.

Miller, S. G., T. S. Williams, J. S. Baker, *et al.* 2014. Increased tensile strength of carbon nanotube yarns and sheets through chemical modification and electron beam irradiation. *ACS Appl. Mater. Interfaces* 6:6120–6126.

Min, J. B., K. P. Duffy, B. B. Choi, *et al.* 2012. Piezoelectric vibration damping study for rotating composite fan blades. NASA/TM—2012-217648 (AIAA 2012–1644).

Mines, R. A. W., S. Tsopanos, Y. Shen, *et al.* 2013. Drop weight impact behaviour of sandwich panels with metallic microlattice cores. *Int. J. Impact Eng.* 60:120–132.

Mossi, K., and R. Bryant. 2006. Piezoelectric actuators for synthetic jet applications. *MRS Proceedings* 785:D11.8.1–D11.8.6.

Nathal, M. V., and G. L. Stefko. 2013. Smart materials and active structures. *J. Aerospace Eng.* 26:491–499.

Nathal, M. V., J. D. Whittenberger, M. G. Hebsur, *et al.* 2004. Superalloy lattice block structures. In *Superalloys 2004*, ed. K. A. Green, T. M. Pollock, H. Harada, *et al.*, Warrendale, PA: The Minerals, Metals & Materials Society.

Olympio, K. R., and F. Gandhi. 2010. Flexible skins for morphing aircraft using cellular honeycomb cores. *J. Intel. Mat. Syst. Str.* 21:1719–1735.

Otsuka, K., and C. M. Wayman. 1998. *Shape memory materials.* Cambridge: Cambridge University Press.

Pan, W., S. R. Phillpot, C. Wan, A. Chernatynskiy, *et al.* 2012. Low thermal conductivity oxides. *MRS Bull.* 37:917–922.

Qian, H., A. R. Kucernak, E. S. Greenhalgh, *et al.* 2013. Multifunctional structural supercapacitor composites based on carbon aerogel modified high performance carbon fiber fabric. *ACS Appl. Mater. Interfaces* 5:6113–6122.

Roberts, G. D., J. M. Pereira, M. S. Braley, *et al.* 2009. Design and testing of braided composite fan case materials and components. NASA/TM—2009-215811 (ISABE 2009–1201).

Rochefort, A. 2015. Nanotubes and buckyballs. Montreal, Quebec: Polytechnique Montreal. *http://www. nanotech-now.com/nanotube-buckyball-sites.htm* (accessed March 22, 2017).

Russ, M., S. Rahatekar, K. Koziol, *et al.* 2011. Development of carbon nanotube/epoxy nanocomposites for lightning strike protection. Paper presented at 18th International Conference on Composite Materials, Jeju Island, Korea.

Schaedler, T. A., A. J. Jacobson, A. Torrents, *et al.* 2011. Ultralight metallic microlattices. *Science* 334:962–965.

Schiller, N. H., W. R. Saunders, W. A. Chishty, *et al.* 2006. Development of piezoelectric-actuated fuel modulation system for active combustion control. *J. Intel. Mat. Syst. Str.* 17:403–410.

Schmidt, F., and A. Allais. 2005. Superconducting cables for power transmission applications—a review. Proceedings of the Workshop on Accelerator Magnet Superconductors, Hanover, Germany. *https://cds. cern.ch/record/962751/files/p232.pdf* (accessed March 30, 2016).

Shin, S., R. Y. Kim, K. Kawabe, *et al.* 2007. Experimental studies of thin-ply laminated composites. *Compos. Sci. Technol.,* 67:996–1008.

Skomski, R., and J. M. D. Coey. 1993. Giant energy product in nanostructured two-phase magnets. *Phys. Rev. B* 48:15812–15816.

Smialek, J. L., R. C. Robinson, E. J. Opila, *et al.* 1999. SiC and Si3N4 recession due to SiO2 scale volatility under combustor conditions. NASA/TP—1999-208696.

Straub, F. K., V. R. Anand, T. S. Birchette, *et al.* 2009. SMART rotor development and wind tunnel test. Paper presented at the 35th European Rotorcraft Forum, Hamburg, Germany.

Sypeck, D. J. 2005. Cellular truss core sandwich structures. *Appl. Comp. Mat.* 12:229–246.

Toray Industries, Inc. 2014. Toray develops high tensile strength and modulus carbon fiber TORAYCA T1100G and high-performance TORAYCA prepreg. *http://cs2.toray.co.jp/news/toray/en/newsrrs02. nsf/0/B126D9D8433675EE49257D11001770C5* (accessed April 12, 2016).

Trimble, S. 2010. General Electric primes CMC for turbine blades. Flightglobal. *http://www.flightglobal. com/news/articles/general-electric-primes-cmc-for-turbine-blades- 349834/* (accessed March 30, 2016).

TYCOR Products. 2011. Milliken & Co. Foam core fiberglase reinforced compositie material. *http://www. webcoreonline.com/tycor%C2%AE-products* (accessed March 29, 2016).

Urnes, Sr., J., N. Nguyen, C. Ippolito, *et al.* 2013. A mission-adaptive variable camber flap control system to optimize high lift and cruise plift-to-drag ratios of future N+3 transport aircraft. AIAA 2013–0214. *http://ntrs.nasa.gov/archive/nasa/casi.ntrs.nasa.gov/20140006948.pdf* (accessed August 12, 2016).

Wang, X., Z. Z. Yong, Q. W. Li, *et al.* 2013. Ultrastrong, stiff and multifunctional carbon nanotube composites. *Mater. Res. Lett.* 1:19–25.

Wardle, B. L., *et al.* 2013. Hierarchical nanoengineered composite (aerospace) structures: manufacturing and mechanical properties. Presented at Defense Manufacturing Conference (DMC) 2013, Orlando, FL. *http://web.mit.edu/aeroastro/labs/necstlab/documents/Wardle_NanoEngMIT-DMCInvitedDec2013sm. pdf* (accessed August 12, 2016).

Wikipedia. 2017. Mechanical properties of carbon nanotubes. *https://en.wikipedia.org/wiki/Mechanical_ properties_of_carbon_nanotubes.* (accessed March 27, 2017).

Wu, A. S., and T.-W. Chou. 2012. Carbon nanotube fibers for advanced composites. *Mater. Today* 15:302–310.

Wu, R. T., M. Osawa, T. Yokokawa, *et al.* 2010. Degradation mechanisms of an advanced jet engine service-retired TBC component. *J. Solid Mechanics Materials Eng.* 4:119–130.

Zheng, X., H. Lee, T. H. Weisgraber, *et al.* 2014. Ultralight, ultrastiff mechanical metamaterials. *Science* 344: 1373–1377.

Zhu, D., and R. A. Miller. 2004. Development of advanced low conductivity thermal barrier coatings. *Int. J. Appl. Ceram. Tec.* 1:86–94 (also NASA/TM—2004-212961 (ARL TR 3259)).

# CHAPTER 6

## Clean combustion and emission control

Changlie Wey and Chi-Ming Lee

### 6.1  INTRODUCTION

This chapter outlines the characteristics of gaseous and particulate aircraft emissions as well as potential technology options to reduce them. Reductions in overall emissions can be obtained by reducing the total amount of fuel burned via more efficient engine technology. Advanced combustor designs and/or alternative fuels may also produce less pollutant per unit weight of fuel burned. However, there are trade-offs that need to be considered when targeting the reduction of a particular emission.

The goal for environmental protection is to reduce all pollutants. Current emphasis on combustor design has been to reduce emissions of nitrogen oxides ($NO_x$). However, other emissions are also of concern. Particulate emissions and related atmospheric effects are current areas for research efforts and potential future regulations. The key technical approaches, outcomes, and issues involved in the reduction of these emissions are discussed in this chapter.

### 6.2  PRODUCTS OF COMBUSTION

Emissions from aircraft jet engines include excess oxygen ($O_2$), excess nitrogen ($N_2$), carbon dioxide ($CO_2$), water vapor ($H_2O$), nitrogen oxides ($NO_x$), carbon monoxide (CO), sulfur oxides ($SO_x$), unburned hydrocarbons (UHC), particulate matter (PM), and other trace compounds.

$H_2O$ and $CO_2$ are greenhouse gases; that is, they impact climate change. Their emissions are directly related to the amount of fuel burned. Fuel consumption is largely self-controlled by the aviation industry, which is operating under the tight constraints imposed by economic pressures. Figure 6.1 shows equilibrium values of these emissions at one inlet temperature and four pressures. $SO_x$ impacts local air quality; consequently, they have been regulated by the U.S. Environmental Protection Agency (EPA). However, $SO_x$ are not a regulated aircraft engine emission. Instead, they are controlled through limitations on jet fuel sulfur content.

Aviation $NO_x$, CO, UHC, and PM (defined as smoke) emissions have various impacts on the environment and are currently controlled under internationally accepted standards. $NO_x$ produced by aeroengines are the byproducts of hydrocarbon fuel combustion in air at high pressure and high temperature in the aircraft combustor. $NO_x$ emissions impact both climate change and local air quality. $NO_x$ emissions have been continuously regulated, and limits are becoming more stringent (see Sec. 6.5). The technology approaches to comply with these standards are complex, and new combustor designs that are created to reduce $NO_x$ emissions are equally complex.

CO and volatile organic compounds impact local air quality and human health and are regulated by the EPA. Aircraft engine CO and UHC emissions are the result of incomplete combustion. For most gas turbine aircraft engines, combustion efficiency is very high except at idle or low power, where inlet temperatures and pressures are low. Areas of concern are as follows:

1. The combustion zone may be too lean or too rich (too much or too little air to mix with fuel).
2. The residence time of the combustion zone may be too short.

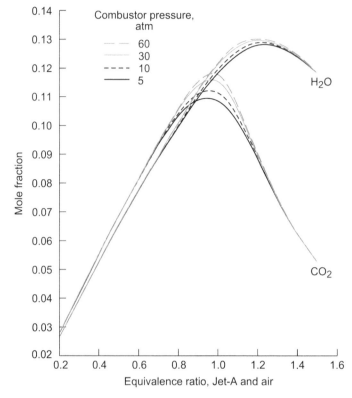

Figure 6.1.　Equilibrium $CO_2$ and $H_2O$ emissions versus equivalence ratio at 1,000°F and various combustor pressures, calculated by the NASA Chemical Equilibrium with Applications software program: maximum $CO_2$ emissions at equivalence ratio of 1.0 (stoichiometric, neither rich nor lean burn).

3. Fuel atomization and aerodynamic stabilization may be poor.
4. Reactions may be quenched by liner cooling.
5. Dilution (mixing air) may be too slow.

Since the combustor efficiency is essentially 100% at any condition except at idle (when it is very low—less than 7%—power), CO and UHC emissions mostly impact the local air quality around airports.

PM also impacts the local air quality and affects human health. Soot particles in the exhaust plume obscure light. To avoid visible smoke, the smoke number is defined as a measure of the concentration of particles in the exhaust plume. Plume visibility depends not only on the concentration of soot particles in the plume, but also the characteristic path length of light through the plume. For a fixed smoke number, the visibility will then be a function of the volume and aspect ratio of the plume and is related to the thrust of the engine, which is in turn a function of its power consumption, geometry and design. Given the desire to avoid visible jet engine exhaust plumes, these relationships can be used to formulate a limit on maximum smoke number. It is expressed as a function of engine thrust, and it is designed to ensure that the plume is not visible over a typical range of viewing angles. A plot of smoke number limit as a function of engine-rated thrust is shown in Figure 6.2.

Current aircraft engine standards are defined by the Committee on Aviation Environmental Protection (CAEP) of the International Civil Aviation Organization (ICAO), a United Nations specialized agency established for setting international civil aviation standards. Such standards have been adopted internationally. They regulate only emissions below 3,000 ft (about 914 m) in

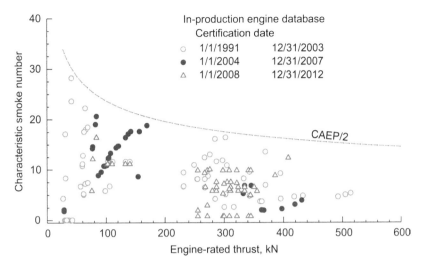

Figure 6.2.   Smoke number versus engine-rated thrust. Relationship of jet engine performance to smoke number limit set by the second meeting of the International Civil Aviation Organization's Committee on Aviation Environmental Protection (CAEP/2).

altitude, which covers the take-off, climb, descent, and taxiing/ground idle phases of the engine operation—the landing and takeoff (LTO) cycle.

### 6.2.1   *Fundamentals of NO$_x$ formation*

Aviation fuel typically does not contain significant amounts of nitrogen. The formation of NO$_x$ in aircraft gas turbines that consume aviation kerosene is overwhelmingly dominated by a thermal mechanism. Thermal NO$_x$ arises from the thermal dissociation of N$_2$ and O$_2$ molecules that are present in the air drawn into the combustor. At high temperatures, N$_2$ and O$_2$ dissociate into their atomic states, N and O, and react with N$_2$ and O$_2$ to form NO via the "Zeldovich mechanism":

$$N_2 + O \rightarrow NO + N \tag{6.1}$$

$$N + O_2 \rightarrow NO + O \tag{6.2}$$

Rates of these reactions are dependent upon the stoichiometric ratio (the fuel-to-air ratio allowing for complete combustion of the reactants) in the primary combustion zone, flame temperature, system pressure, and the time spent at the flame temperature (residence time).

Liquid fuel must be mixed with the oxidizer in the air on the molecular scale before it burns. Since the average residence time in the combustor is only around 5 ms, fuel-air mixing must be done in much shorter time scales. Converting liquid fuel into a combustible fuel-air mixture in a short time requires atomizing the fuel into small droplets that can vaporize quickly. Immediately after injection, the fuel-air distribution is typically non-uniform, with rich and lean pockets. As vaporization and turbulent mixing occurs, the distribution becomes progressively more uniform and trends toward the average equivalence ratio $\phi$, which is the ratio of the actual fuel-to-air ratio normalized by the stoichiometric fuel-to-air ratio. Figure 6.3 shows the fuel-air distribution downstream of a fuel injector becoming narrower with time (Chang, 2012).

Figure 6.4 illustrates the rate at which the emissions index for NO$_x$ (EINO$_x$) increases with inlet pressure as well as inlet temperature in combustors that did not feature any specific NO$_x$ reduction technology. NO is the primary NO$_x$ species produced in the flame. Subsequent reactions form

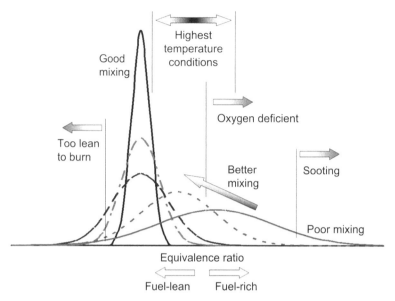

Figure 6.3.   Fuel-air mixture equivalence ratio distribution for various degrees of mixing, and their associated phenomena. (From Chang, C. T. 2012. Gas turbine engine combustors. In *Encyclopedia of aerospace engineering, online*, New York, NY. Copyright 2012 John Wiley & Sons, Ltd. With permission.)

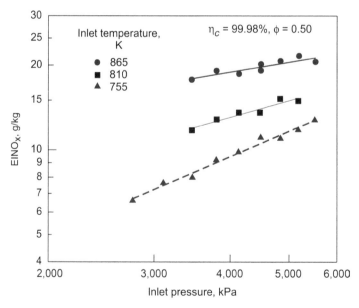

Figure 6.4.   $NO_x$ emissions index ($EINO_x$) increases with combustor inlet pressure and temperature. $\eta_c$ represents combustion efficiency, and $\phi$ is the equivalence ratio.

$NO_2$ throughout the turbine and jet-exhaust as well as in the ambient environment. This reaction is enhanced by relatively low gas temperatures and by traces of unburned fuel and CO. The range of possible temperature regimes, times, and trace gas concentrations within the engine and near plume can result in conversions of NO to $NO_2$ ranging from almost none to levels above 80%.

## 6.2.2 $NO_x$ emissions standards

The CAEP is a technical committee established by the ICAO Council in 1983, superseding the Committee on Aircraft Noise and the Committee on Aviation Engine Emissions (CAEE). CAEP assists the Council in formulating new policies and adopting new standards and recommended practices related to aircraft noise and emissions, and more generally to aviation environmental impact.

The first aircraft engine certification standards were adopted in 1981 by the CAEE, the predecessor of CAEP. ICAO emissions standards are summarized in Annex 16 of International Civil Aviation Organization (2008). At any power condition, regulations specify that aircraft engine particulate emissions must not exceed $83.6 \times (F_{oo})^{-.274}$ where $F_{oo}$ is engine-rated thrust, kN or a value of 50, whichever is lower. This standard applies to engines whose date of manufacture is on or after January 1, 1983.

At this time, regulatory levels for gaseous emissions apply only to engines whose rated output is greater than 26.7 kN and whose date of manufacture is on or after January 1, 1986. These standards control the engine characteristic, or $D_p/F_{oo}$, which is defined as the mass of emissions produced ($D_p$) during a static sea-level ($oo$) engine test for a simulated idealized LTO cycle normalized by the maximum engine thrust ($F_{oo}$). For UHC, $D_p/F_{oo}$ must not exceed 19.6 g/kg; for CO, 118 g/kg; and for $NO_x$, $40 + 2\pi_{oo}$ g/kg where $\pi_{oo}$ is engine pressure ratio, the ratio of the pressures at the nozzle exit and the combustor inlet.

The CAEP/2 meeting in 1992 amended the test procedures and made the $NO_x$ standard more stringent by 20% for any newly certified engine with the date of manufacture of the first individual production model on or after January 1, 2000. Regulatory levels for maximum allowable $NO_x$ then became

- $D_p/F_{oo} = 40 + 2\pi_{oo}$ for engines of a type or model of which the date of manufacture of the first individual production model was before January 1, 1996, and for which the date of manufacture of the individual engine was before January 1, 2000
- $D_p/F_{oo} = 32 + 1.6\pi_{oo}$ for engines of a type or model of which the date of manufacture of the first individual production model was on or after January 1, 1996, or for which the date of manufacture of the individual engine was on or after January 1, 2000

At the following CAEP meeting (CAEP/3) in 1995, the Committee recommended a further tightening of 16% on $NO_x$ emissions and additional test procedure amendment. This recommendation on $NO_x$ stringency was rejected by the ICAO Council in 1997, but the test procedure amendments were approved.

The next stringency increase was agreed upon at the CAEP/4 meeting in 1998 to be 16% lower than the CAEP/2 standard for the engines certified whose date of manufacture or the first individual production model was on or after January 1, 2004 (similar to CAEP/3 recommendation). This is the first time the standards were set with the consideration of the engine pressure ratio $\pi_{oo}$ as well as the rated thrust level $F_{oo}$. These standards are summarized in Table 6.1.

The $NO_x$ standard was again set at the CAEP/6 meeting in 2004, and it was made more stringent by 12% compared with that of CAEP/4 for engines of a type or model whose date of manufacture of the first individual production model was on or after January 1, 2008, or for those whose date of manufacture of the individual engine was on or after January 1, 2013.

The most recent updates in the CAEP/8 meeting in 2010 included the agreement of production-cut-off for engines not complying with the CAEP/6 $NO_x$ standard, and adoption of a more stringent $NO_x$ standard: an approximate 15% reduction in $NO_x$ emissions from CAEP/6. The CAEP/8 standards will be mandatory for the engines of a type or model whose date of manufacture of the first individual production model is on or after January 1, 2014.

Table 6.1 summarizes the progression of the $NO_x$ emission standards adopted by the CAEP since 1986.

Until CAEP/4, the standard was a simple straight line of permitted $NO_x$ rising with increasing engine overall pressure ratio (OPR). From CAEP/4 onwards an increased slope kink in this line

Table 6.1.   Summary of International Civil Aviation Organization (ICAO) $NO_x$ standards.

| Standard[a] (date) | Engine pressure ratio, $\pi_{oo}$[b] | Engine-rated thrust, $F_{oo}$,[c] kN | $NO_x$ emissions, $D_p/F_{oo}$,[d] g/kN |
|---|---|---|---|
| CAEE (1986) | All | >26.7 | $40 + 2\pi_{oo}$ |
| CAEP/2 (1996) | All | >26.7 | $32 + 1.6\pi_{oo}$ |
| CAEP/4 (2004) | <30 | 26.7 to 89.0 | $37.572 + 1.6\pi_{oo} - 0.2087F_{oo}$ |
| | <30 | >89.0 | $19 + 1.6\pi_{oo}$ |
| | 30 to 62.5 | 26.7 to 89.0 | $42.71 + 1.4286\pi_{oo} + 0.4013F_{oo} + 0.00642\pi_{oo} \times F_{oo}$ |
| | 30 to 62.5 | >89.0 | $7 + 2.0\pi_{oo}$ |
| | >62.5 | >26.7 | $32 + 1.6\pi_{oo}$ |
| CAEP/6 (2008) | <30 | 26.7 to 89.0 | $38.5486 + 1.6823\pi_{oo} + 0.2453F_{oo} - 0.00308\pi_{oo} \times F_{oo}$ |
| | <30 | >89.0 | $16.72 + 1.4080\pi_{oo}$ |
| | 30 to 82.6 | 26.7 to 89.0 | $46.1600 + 1.4286\pi_{oo} - 0.5303F_{oo} + 0.00642\pi_{oo} \times F_{oo}$ |
| | 30 to 82.6 | >89.0 | $-1.04 + 2.0\pi_{oo}$ |
| | >82.6 | >26.7 | $32 + 1.6\pi_{oo}$ |
| CAEP/8 (2014) | <30 | 26.7 to 89.0 | $40.052 + 1.5681\pi_{oo} - 0.3615F_{oo} + 0.0018\pi_{oo} \times F_{oo}$ |
| | <30 | >89.0 | $7.88 + 1.4080\pi_{oo}$ |
| | 30 to 104.7 | 26.7 to 89.0 | $41.9435 + 1.505\pi_{oo} + 0.5823F_{oo} + 0.005562\pi_{oo} \times F_{oo}$ |
| | 30 to 104.7 | >89.0 | $-9.88 + 2.0\pi_{oo}$ |
| | >104.7 | >26.7 | $32 + 1.6\pi_{oo}$ |

[a]CAEE, Committee on Aviation Engine Emissions; CAEP, Committee on Aviation Environmental Protection.
[b] $p_{oo}$ is the engine pressure ratio.
[c] $F_{oo}$ is the engine thrust at sea level.
[d] $D_p$ is mass of $NO_x$ emissions produced.

appeared at OPR 30, which permitted higher OPR engines to produce more $NO_x$ than would have been the case previously with a straight line. At CAEP/6 the slope of the line was again reduced somewhat for engines below OPR 30.

Figure 6.5 shows these various ICAO $NO_x$ standards together with nonattributed in-production engine data from the databank also highlighting the more recently certificated engines. It clearly shows that the CAEP/6 regulation varies linearly on either side of OPR = 30, with the $NO_x$ require-ment being less stringent with increasing OPR; that is, the requirements for the higher thrust class engine are slightly more rigorous than those of the lower thrust-class, lower OPR range. The figure also shows that newly certified engines to be introduced into service typically have much lower $NO_x$ emissions for the current regulation in anticipation of future more-stringent regulations.

Figures 6.6 and 6.7 show the ICAO standards for CO and hydrocarbon (HC) emissions, respectively. Again, the figures show that newly certified engines to be introduced into service typically have much lower CO and HC emissions than the current regulation in anticipation of more stringent regulations.

## 6.3   EMISSIONS CONTROL

Even though the aircraft is designed to be able to sustain flight for a period of time if an engine failure occurs, the combustor must function reliably during all phases of engine operation. To increase engine efficiency, higher inlet temperatures and pressures are desirable, but are limited by the durability of the combustor. $NO_x$ production in the engine is greatest where temperatures and pressures are highest (Dodds, 2002). Controlling peak temperature is a key component to reducing $NO_x$, but development of new low-$NO_x$ combustors requires maintaining all opera-tional aspects of the combustion system to ensure safety and reliability.

At the inlet, the combustor temperature can approach 700°C (1,300°F), and the pressure may be up to 45 atm. Within the combustor, the temperature can exceed 2,200°C (4,000°F).

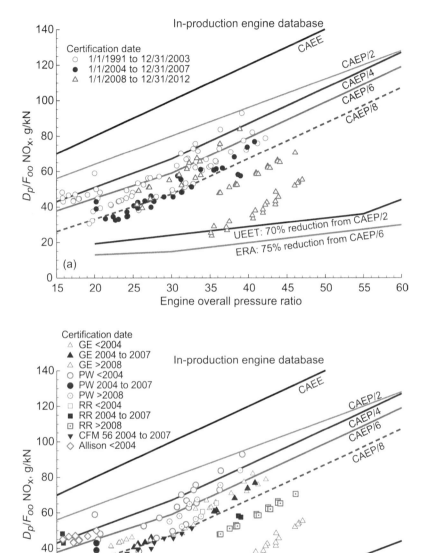

Figure 6.5.   Certified $NO_x$ emissions levels of current in-production engines versus International Civil Aviation Organization Committee on Aviation Environmental Protection (CAEP) $NO_x$ regulatory level, sorted by engine company and year. Levels also given for original Committee on Aviation Engine Emissions (CAEE) regulation and NASA aeronautics research projects Ultra Efficient Engine Technology (UEET) and Environmentally Responsible Aviation (ERA). (a) Sorted by year (CAEP meetings 2, 4, 6, and 8). (b) Sorted by engine company and year; $D_p/F_{oo}$ is the emissions regulatory parameter, defined as mass of emissions emitted in landing and takeoff (LTO) cycle divided by thrust. GE, GE Aviation; PW, Pratt & Whitney; RR, Rolls-Royce; CFM, CFM International; Allison, Allison Engine Company.

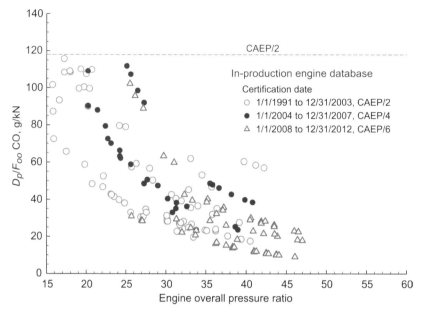

Figure 6.6.  Current in-production engines certified CO emissions level versus International Civil Aviation Organization Committee on Aviation Environmental Protection (CAEP) CO regulatory level, sorted by year (CAEP meetings 2, 4, and 6). Levels also given for original Committee on Aviation Engine Emissions (CAEE) regulation and NASA aeronautics research projects Ultra Efficient Engine Technology (UEET) and Environmentally Responsible Aviation (ERA). $D_p/F_{oo}$ is emissions regulatory parameter, defined as mass of emissions emitted in landing and takeoff (LTO) cycle divided by thrust.

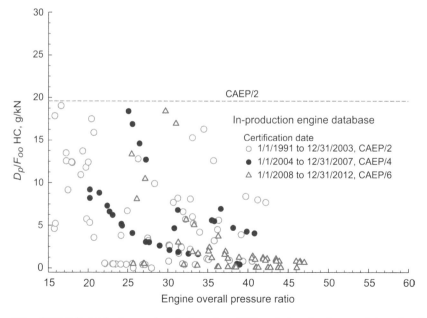

Figure 6.7.  Current in-production engines hydrocarbon (HC) emissions level versus International Civil Aviation Organization Committee on Aviation Environmental Protection (CAEP) HC regulatory level, sorted by year (CAEP meetings 2, 4, and 6). $D_p/F_{oo}$ is emissions regulatory parameter, defined as mass of emissions emitted in a landing and takeoff (LTO) cycle divided by thrust.

The temperature at the combustor exit is much reduced to values around 1,650°C (3,000°F). The survival of the combustor in this type of harsh environment is a major challenge, particularly since most metals melt around 1,350°C (2500°F) (Dodds, 2002).

Another major challenge in combustor design is flame stability, which may be the deciding factor against many potential technologies. Combustor burning efficiency must be essentially 99.9%+, at all conditions except very low power (idle) conditions, because any fall-off will cause the increase in $CO_2$ as well as other pollutants.

Combustion research sponsored by NASA's aeronautic research programs has focused on $NO_x$ reduction for more than three decades. As a result, $NO_x$ emissions have been reduced by about 50% with each new generation of aircraft, occurring at intervals of about 15 years. The aviation propulsion industry has manufactured real-world hardware using the NASA-sponsored combustor concepts initially developed under these research programs. Two notable examples, the GE90 and V2500 engines, were introduced into service a decade after the combustor concepts were developed under NASA's Experimental Clean Combustor Program (1974 to 1979) and its follow-on Energy Efficient Engine Program (1980 to 1984). Collaborative work during NASA's High Speed Research Program (1989 to 1994), the Advanced Subsonic Technology Program (1994 to 1999), and the Ultra Efficient Engine Technology (UEET) Program (2000 to 2004) generated valuable insight into the new combustor designs of Pratt & Whitney's (PW's) TALON combustor series and GE Aviation's twin-annular premixing swirler (TAPS) combustor for the GEnx engine (Lee *et al.*, 2013).

NASA's Environmentally Responsible Aviation (ERA) Project worked with the industry to develop the combustor technologies from 2010 to 2015 for a new generation of low-emissions combustors targeted for the 2020 timeframe. The goal for these new combustors is to reduce $NO_x$ emissions to half that of the current state-of-the-art combustors, while simultaneously reducing particulate emissions.

NASA aeronautics research programs always set goals of very low $NO_x$ emissions to provide a safety margin when developing the real hardware while fulfilling the safety and airworthiness requirements at the technology readiness level (TRL) 5 or 6 to TRL 9. After three decades, $NO_x$ levels have been considerably reduced and little room remains for further improvements (Fig. 6.8). Therefore, the ERA $NO_x$-reduction effort is even more challenging than it was under previous programs. $NO_x$ reduction at cruise level is also emphasized in the ERA Project.

In addition, ERA's system-level goals include a 50% fuel burn reduction for the platform as well as noise reduction (Table 6.2). Although much fuel consumption savings may be achieved by airframe drag reduction, a contribution is also required from the propulsion system, which requires increasing the engine overall pressure ratio to about 55 from the current state-of-the-art number of 45. The increased combustor pressure and temperature will increase the $NO_x$ formation rate, making the goal more difficult to achieve. In addition, the technology concept needs to be fuel flexible, capable of operating on a 50% blend of alternative non-petroleum-sourced hydrocarbon fuels. Therefore, fuel burn reduction technologies are needed that do not increase aircraft noise or $NO_x$ emissions.

The "N+1" timeframe represents the next generation of technology. The goal is to reach a TRL of 4 to 6 in 2015, supporting a 2020 entry into service. The "N+2" timeframe represents two technology generations into the future with the goal of reaching TRL 4 to 6 in the 2020 timeframe, supporting a 2025 entry into service. The "N+3" timeframe represents three technology generations into the future. Its focus is on technologies that will not meet the TRL 4 to 6 maturation level until 2025.

## 6.4   ENGINE $NO_x$ CONTROL STRATEGIES

The total amount of $NO_x$ produced by aircraft fuel burn is directly correlated to the residence time integral of the hot gas over the instantaneous $NO_x$ formation rate, which is exponentially related to flame temperature. The aggregate $NO_x$ level will be reduced when either the flame temperature or the residence time is reduced.

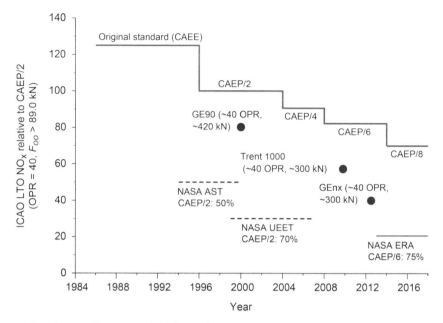

Figure 6.8.   History of International Civil Aviation Organization (ICAO) Committee on Aviation Environmental Protection (CAEP) and Committee on Aviation Engine Emissions (CAEE) regulations on $NO_x$ emissions for the landing and takeoff (LTO) cycle and overall engine pressure ratio (OPR) of 40 and engine-rated thrust $F_{oo}$ greater than 89 kN. Also shown are NASA aeronautics research program goals of the Advanced Subsonic Technology (AST) Program, Ultra Efficient Engine Technology (UEET) Program, and Environmentally Responsible Aviation (ERA) Project.

Table 6.2.   NASA's subsonic transport system-level metrics.

| Technology benefits | Technology generations (Technology readiness level = 4 to 6) | | |
| --- | --- | --- | --- |
| | N+1 (2015) | N+2 (2020) | N+3 (2025) |
| Noise (cumulative margin relative to Stage 4) | –32 dB | –42 dB | –71 dB |
| LTO $NO_x$ emissions (relative to CAEP/6)[a] | –60% | –75% | –80% |
| Cruise NOx emissions (relative to 2005 best in class) | –55% | –70% | –80% |
| Aircraft fuel consumption (relative to 2005 best in class) | –33% | –50% | –60% |

[a]LTO, landing and takeoff; CAEP/6, sixth meeting of the Committee on Aviation Environmental Protection.
*Source:* Del Rosario, R., J. Koudelka, R. Wahls, *et al.*, 2013. Technical progress and accomplishments of NASA's fixed wing project. Presented at 51st AIAA Aerospace Sciences Meeting, Grapevine, TX. *https://ntrs.nasa.gov/ archive/nasa/casi.ntrs.nasa.gov/20150009967.pdf* (accessed April 21, 2017). Work of the U.S. Government.

NOx formation rate is a function of the fuel-to-air ratio, flame temperature, and pressure. The total amount of $NO_x$ formed depends on the formation rate and the residence time. Potential designs to reduce $NO_x$ formation include burning rich (rich-burn, quick-quench, lean-burn, RQL), burning lean (lean-staged), and reducing combustor volume.

There are two primary $NO_x$-controlling combustion modes that could be applicable to gas turbines: RQL, and lean burn. Details of each will be discussed below in detail. Various other $NO_x$ reduction technologies, such as catalytic combustion or flameless combustion, have proven to be unsuitable for aeroengines because of excessive weight, size, stability, and other characteristics of these designs.

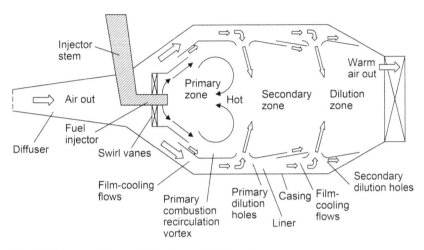

Figure 6.9.   Rich-burn, quick-quench, lean-burn (RQL) combustor.

### 6.4.1   *Rich-burn, quick-quench, lean-burn combustion*

The principles of RQL combustion include three stages, illustrated in Figure 6.9. In the first stage (rich burn), the fuel-air mixture is very rich (much more fuel than required for complete combustion) producing CO, $H_2$, and heat. Because of the low temperature and low concentrations of oxygen, the $NO_x$ formation rate is very low. In the second stage (quick quench), large amounts of additional air are rapidly added to transform the mixture into a lean condition (much more air than needed for complete combustion). This process will pass through stoichiometric conditions at some point between the first and second stages; this is characterized by the highest temperature and $NO_x$ formation rate. Very fast fuel-air mixing, minimizing the time at the stoichiometric condition, is critical for minimizing $NO_x$ formation. The third stage (lean burn) has a lean mixture, which completes the combustion process at reduced flame temperature and lower $NO_x$ formation rate.

There are many challenges in designing an RQL combustor:

1. In the fuel-rich burn zone (front end of the combustor), uniformity of the fuel-air mixture is mandatory to avoid rich pockets that produce a significant amount of soot particles.
2. In the quick-quench zone, advanced designs are needed that can introduce large volumes of air with rapid mixing to achieve uniformity in the fuel-air concentration.
3. Balancing the total emission content at all power conditions, ranging from idle to full thrust, can be challenging because it requires that tradeoffs be made; for example, increasing the production of soot particles in order to reduce $NO_x$ formation.

Before environmental awareness of the consequences of aviation emissions, combustors were generally designed to run with a rich burn and lean quench. It was therefore natural to start the work of emission reduction by addressing smoke generation, idling emissions and $NO_x$ formation. The current RQL designs are sophisticated versions of the original designs. They feature excellent control of fuel preparation, air/fuel ratios, internal aerodynamics, and residence times. Modern designs owe much to the investment in computational fluid dynamics (CFD) and combustion chemistry models over the last three decades.

Numerous improvements and new designs have been implemented to improve emissions performance of RQL combustors. This has been accomplished using a number of approaches. The average equivalence ratio in the rich zone has been increased to 1.5 to 1.8 by reducing the amount of the air entering the front end of the combustor. The higher equivalence ratio lowers the combustion zone temperature, thus reducing the $NO_x$ formed in the front end by an order of

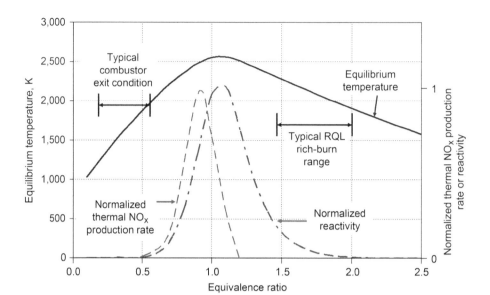

Figure 6.10.   The typical effect of combustion fuel-air equivalence ratio on gas temperature, normalized chemical reactivity, and normalized $NO_x$ formation rate. RQL is rich-burn, quick-quench, lean-burn. (From Chang, C. T. 2012. Gas turbine engine combustors. In *Encyclopedia of aerospace engineering,* New York, NY. Copyright 2012 John Wiley & Sons, Ltd.)

magnitude (see Fig. 6.10). To keep the soot emissions reasonably low, the fuel injector needs to mix well so that relatively few pockets with equivalence ratios above 2 exist.

The dilution jets in the legacy RQL combustor were modified to introduce the massive amount of fresh air all at once into the very fuel-rich gas from the primary zone and mix very quickly to bring the mixture to a lean equivalence ratio. This "quick-mix" process still goes through the stoichiometric high temperature condition, but the time spent at the high temperature is very short, thus producing less $NO_x$. Once it is mixed, the reacting gas temperature drops rapidly and the $NO_x$ formation rate rapidly decreases. As a result, the total $NO_x$ level is reduced.

The simplicity of the RQL concept belies the complexity of the implementation. The fuel-air mixing and turbulent jet-in-cross-flow quick mixing are the keys to low emissions. The other not-so-obvious issue is the possible liner burnout in the region immediately downstream of the dilution holes due to secondary flow from the horseshoe vortex, which tends to bring very hot gases in contact with the liner. In addition, the rich-burn front end also tends to produce larger soot particles and hence more visible smoke trails.

At cruise and approach, the primary zone may operate in or close to the high-$NO_x$-production band. Fortunately, the combustion air temperature and pressure are low at these conditions, and the design features built in for landing and takeoff $NO_x$ reduction (good fuel preparation and short residence time) reduce $NO_x$ (Faber, 2008).

Modern military engine combustors, such as GE Aviation's CF6SAC, Pratt & Whitney's TALON series, and the present Rolls-Royce (RR) Trent (Phase 5) combustors, all utilize some aspects of the RQL concept. The latest of these are good enough to satisfy the $NO_x$ limits established at the sixth meeting of the Committee on Aviation Environmental Protection (CAEP/6) (Collier, 2012), because the RQL's front end also can be run lean, and it has a large turn-down ratio and hence large operating range.

Two advanced combustor designs to control $NO_x$ formation are shown in Figure 6.11.

Low-emissions combustion for both single- and dual-annular combustors (SAC and DAC, respectively) have replaced standard combustors in new production by GE Aviation (Fig. 6.11(a)).

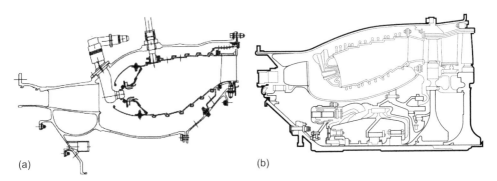

Figure 6.11. Advanced combustor designs to reduce NO$_x$ emissions. (a) GE Aviation CF6-80C low-emissions combustor. (b) Pratt & Whitney 4000-100 TALON II combustor. ((a) Copyright GE Aviation, used with permission.)

The most advanced technology is the optimization of the mixing that strikes a balance between NO$_x$ and smoke emissions to fulfill both regulations. These combustors reduce unburned hydrocarbon emissions by using technology to control fuel spray and cooling. These technologies had also been applied to GE90, CF34, and GP7000 engine families.

Similar technologies, namely optimized mixing for balancing NO$_x$ and smoke emissions, controlled fuel spray and cooling for UHC emissions, and rapid quench dilution to reduce NO$_x$, had been applied to the PW TALON combustor (see Fig. 6.11(b)). NO$_x$ emissions improved with the PW4000 TALON-II combustors. This combustor utilizes segmented Floatwall (Pratt & Whitney, East Hartford, Connecticut) liners to minimize cooling air. These technologies have also been applied to the JT8D.

Based on the principles of creating a rich primary zone for stability and a lean quick-quench zone to minimize NO$_x$ emissions, the RR Phase 5 combustor made the design choice of nonpremixed combustion and single-fuel staging. Applied technologies included using liner chutes to improve mixing, advanced wall cooling using "tiled" liners, and optimized zonal stoichiometry to reduce residence time.

These advanced concepts in combustor design have resulted in improvements in emissions output. The results for the PW4000 series are shown in Figure 6.12. NO$_x$ emissions improvements with RR Phase 5 combustors are clearly shown in Figure 6.13.

### 6.4.2   Lean-burn, premixed, prevaporized combustors

In the combustor industry language, "lean" and "rich" apply only to the front end of the combustor because all current combustor exits are lean. In the 1970s, research began on operating the front end of the combustor in a lean regime. The operating concept for lean-burn combustors is to entirely prevent the development of stoichiometric conditions, thus avoiding the high NO$_x$ formation rate associated with high temperature (Chang, 2012).

The principle of lean-burn, premixed, prevaporized technology (Fig. 6.14) is to completely mix and evaporate the fuel before combustion. In theory, the technique can produce extremely low NO$_x$ emissions, and it has been used successfully in gas-fired stationary power plant applications. In spite of considerable research activity in the 1980s and 1990s, the technology encountered numerous problems that appear to be insurmountable. Although NO$_x$ reductions of better than 90% were demonstrated at reduced combustor pressures, issues with autoignition, operability, and dynamics produced safety issues for aeroengines with this technology. The only way to avoid them was to use a nonpremixed pilot burner, which caused a significant penalty in NO$_x$ emissions (Faber, 2008).

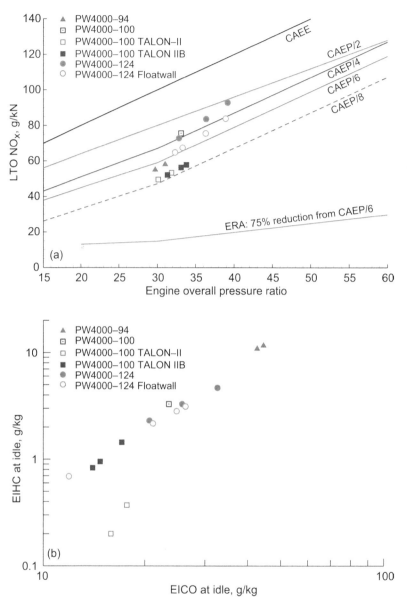

Figure 6.12.   Comparison of PW4000 engines with and without new technologies. (a) Landing and takeoff (LTO) $NO_x$ emissions versus engine overall pressure ratio for PW4000 engine. (b) Emissions index for hydrocarbons (EIHC) versus that of CO (EICO) emissions at idle condition.

### 6.4.3   *Lean-burn staged combustors*

In spite of successful significant reductions in $NO_x$ emissions that have been achieved by RQL technology, rising pressure ratio and combustion air temperature make further reduction of $NO_x$ increasingly difficult.

As much of the air as possible flows through the fuel injector to mix with the fuel in a lean-burn combustor; typically around 60% to 70% of the air flows through the front end. The dilution holes are greatly reduced, and the front flame-stabilizing zone is mostly lean. Lean-burn combustors are generally designed to be aerodynamically stable to ensure flame stability because

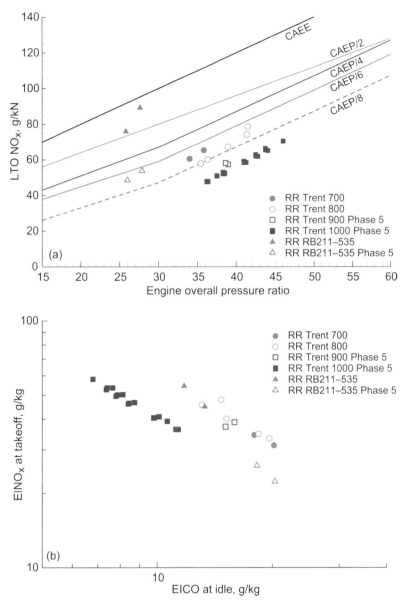

Figure 6.13.  Comparison of Rolls-Royce (RR) Trent engine with and without Phase 5 combustor. (a) landing and takeoff (LTO) NO$_x$ emissions versus engine overall pressure ratio. (b) Emissions index for NO$_x$ (EINO$_x$) at takeoff versus that for CO (EICO) at idle: NO$_x$ emissions are reduced with the Phase 5 combustors. Note that Trent 800 and RB211 engines had been out of production.

the flame temperature is lower, the average chemistry is slower, and only weaker jets from the primary dilution holes trap the recirculation vortex. Lean burning tends to produce smaller (sub-micron) soot particles, making the soot trail much less visible. Since there are fewer hot spots, the liners generally need less cooling air (Chang, 2012).

The lean-burn design approach requires that a very high percentage of the combustor air pass through the main fuel injector. The main fuel injector therefore tends to be large and complex, often introducing issues of cost, weight, and potential overheating. The atomizer needs to

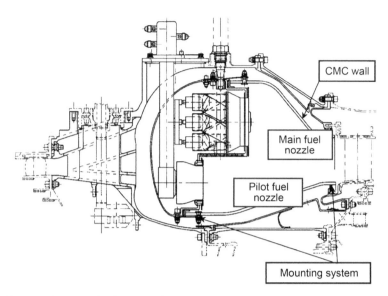

Figure 6.14.   Lean-burn premixed, prevaporized combustor.  CMC is ceramic-matrix composite. (From International Coordinating Council of Aerospace Industries Associations (ICCAIA). 2002. Emissions reduction technology workshop. Paris: CAEP SG. With permission from Kawasaki Heavy Industries.)

perform well in terms of fuel spray placement and quality so that the spray would be close to entirely vaporized and well mixed with the airflow prior to the flame zone.

Staged combustors incorporate multiple distinct combustion zones from respectively independent fuel injection systems. These fuel injectors are separated physically in space and can be optimized at different locations in the combustor. Different combinations and power levels of combustor stages can be used for various engine power settings. In addition, the combustor can be optimized to run lean for multiple flight conditions, giving better control over $NO_x$ production at all flight conditions (Lohman *et al.,* 1982). This is especially significant at the low power settings during which a significant portion of LTO $NO_x$ is formed. This feature also delivers the capability to reduce velocities in the near-injector region, reducing acoustic instabilities and improving lean blowout and ignition characteristics (Bruner *et al.,* 2010).

A combustor solely designed for lean combustion at high power will not light well nor burn stably at idle operating conditions. One potential solution is to have separate fuel systems, one for the "pilot zone" and the other for the "prime combustion zone." These systems can be controlled independently to perform "fuel staging" to have all fuel go to the pilot zone at low power and to have most fuel go to the prime combustion zone at high power. Lean-staged combustion still carries the same advantage of the lower flame temperature associated with lean burn. Also, significant $NO_x$ reduction can be achieved at high power with complete fuel vaporization and uniform fuel-air premixing.

Design challenges associated with these combustion systems include smooth control of the fuel staging: the additional complexity of separate fuel systems, which heightens cost and increases hardware weight; the additional fuel coking potential; possible fuel pre-ignition; and the unstable, wide range of dynamic pressures.

### 6.4.3.1   *Rolls-Royce lean-burn combustor*

The RR Advanced Low-Emissions Combustion Systems combines several approaches:

1. Lean-burn premixed combustion
2. Individually staged fueling for pilot and main an "advanced active control" system of valves to meter each fuel flow

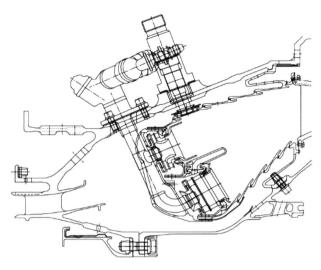

Figure 6.15.   GE Aviation's dual-annular combustor (DAC). (From Mongia, H., and W. Dodds. 2004. Low emissions propulsion engine combustor technology evolution past, present and future. ICAS 2004−6.9.2 (I.L.). Copyright GE Aviation, used with permission.)

3.  A rich pilot to assure stability at low-power conditions
4.  A lean main zone to minimize $NO_x$ Technology is likely to require considerable development to meet low-power efficiency and low-emissions targets, based on previous experiments with circumferential staging.

### 6.4.3.2   *GE Aviation's dual-annular combustor*

The dual annular combustor has a main burner, and the pilots are arranged in two rows of injectors in an annular configuration, shown in Figure 6.15. It consists of a pilot stage in the outer annulus of the burner, and a main stage in the inner annulus. Only the outer (pilot) stage was fueled during light-off and at low power. The pilot was designed to achieve good ignition and low CO and hydrocarbon emissions at low power. The main stage was designed with high airflow and high velocity at high-power conditions to provide a lean flame with minimal time for $NO_x$ formation.

The presence of the pilot row greatly expands the operating range of the combustor at the lower end. At very low power conditions, only a few of the pilot injectors need to operate, and they may operate lean but still within the flammability limit. As engine power increases, more fuel can be brought on line through all the pilot injectors. If more power is needed, this pilot could also be operated in an RQL mode. As more flow is needed, the main burners are brought on line. This could be considered a hybrid combustor combining the best features of rich-burn and lean-burn concepts (Chang, 2012).

The drawback to the DAC is its more complex fuel injector arrangement. Engine fuel injectors are normally inserted through the casing to fit onto the front end of the combustor. A dual-row configuration requires two injectors to be designed onto each fuel stem, complicating the mechanical structure. In addition, the larger wetted area inside the combustor requires correspondingly more cooling air. The cooler walls on the nonfueled main burner side also can produce more UHCs. As a result, the DAC produces slightly more CO and UHCs than the older SAC (Dodds, 2005). Other design challenges include emission at low-power conditions, combustor exit profile, and its impact on fuel burn. These issues led to the development of the next-generation low-emissions technology, a twin-annular premixed system, which emphasizes premixing (Mongia and Dodds, 2004).

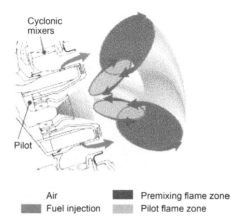

Figure 6.16.   GE Aviation's twin-annular premixing swirler (TAPS) operating principles for improved mixing. (From Chang C.-T, C. M. Lee, J. T. Herbon, *et al.*, 2013. NASA environmentally responsible aviation project develops next-generation low-emissions combustor technologies (Phase I). *J. Aeronaut. Aerospace Eng.*   2:116.   *https://www.omicsgroup.org/journals/nasa-environmentally-responsible-aviation-project-develops-next-generation-lowemissions-combustor-technologies-phase-i-2168-9792.1000116. php?aid=17626* (accessed April 6, 2016). Work of the U.S. Government.)

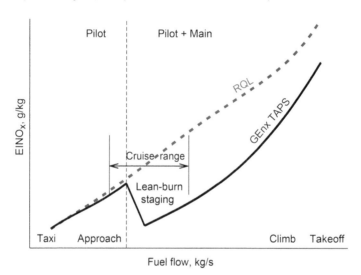

Figure 6.17.   Twin-annular premixing swirler (TAPS) fuel staging effects on $NO_x$ emissions index ($EINO_x$); comparison with rich-burn, quick-quench, lean-burn (RQL) combustor engine.

### 6.4.3.3   *GE Aviation's twin-annular premixed system combustor*

Based on two previously developed concepts, the premixing cyclone and the conventional pilot configuration, GE Aviation designed a new TAPS combustor. Shown in Figures 6.16 to 6.18, the TAPS consists of a premixing main swirler with a pilot swirler cup mixer located in the center. A majority of the combustion air flows through the mixer, which achieves significant premixing. By staging the fuel between the main swirler and the pilot, operation over all engine operating conditions is achieved. This allows a uniform temperature profile during pilot-only operation and adequate combustor exit temperature profiles at all operating conditions. The central pilot flame also provides stable operability at low-power conditions and improved control of the fuel spray, which reduces CO and UHC emissions. The multipoint injector and cyclone swirler system provides good premixing for the main flame at lower temperatures, resulting in low $NO_x$ emissions.

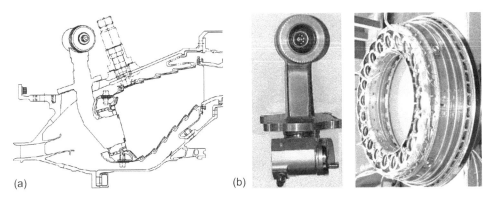

(a)                                    (b)

Figure 6.18.   Twin annular premixing swirler (TAPS) system. (a) Lean singular-annular combustor combustion system based on TAPS. (b) Tech56/CFM SAC TAPS fuel nozzle and combustor hardware. (From Mongia, H., and W. Dodds. 2004. Low emissions propulsion engine combustor technology evolution past, present and future. ICAS 2004–6.9.2 (I.L.). Copyright GE Aviation, used with permission.)

GE Aviation later consolidated the duel-row injector arrangement into a single row of coaxial complex nozzles for the GEnx in 2006, the engine used in the Boeing 787 and 747-8. The outer main injector is an airblast nozzle that produces extremely strong turbulence from two annular swirler rows, which atomizes and premixes most of the fuel before the mixture starts to burn.

The flame structure of the TAPS is compact and efficient. The pilot flame and the main burner flame form a pair of concentric flame cones. This combustor shares the similar squat cross section of the DAC, minimizing the hot gas residence time in the combustor. It also does not need the separating liner between the two rows in the DAC, thus freeing additional liner cooling air for better fuel mixing. The axis-symmetric arrangement allows unimpeded flame propagation around the injector azimuth, reducing the possibility of the fuel-air mixture being drawn into the cooler layer near the wall and producing incomplete combustion products.

### 6.4.4   *N+2 advanced low-NO$_x$ combustors*

Advanced concepts for reduced NO$_x$ combustors were developed under the Environmentally Responsible Aviation Project, described below.

#### 6.4.4.1   *GE Aviation technology*

In order to meet NASA's N+2 goals on NO$_x$ reduction and performance, GE Aviation started the combustor design, pushing for technologies that can advance low-NO$_x$-emissions capabilities based on the TAPS design, which was developed via multiple technologies and commercial programs including GEnx. The engine architecture, scale, and cycle were set by an engine-aircraft system analysis. A conceptual hybrid wing body aircraft and engine was the one that could meet the key N+2 objectives for NO$_x$, fuel burn, and noise reduction (Lee *et al.*, 2013).

The fundamental approach for the combustor design is to increase the fraction of air used for premixing in the front end of the combustor beyond the 70% used in previous TAPS designs (Foust *et al.*, 2012) (Figs. 6.19 and 6.20). Additional features that can further improve the fuel-air mixing are also under study. While pursuing better premixed conditions, significant challenges emerged regarding operability (efficiency and combustion dynamics) and durability (less cooling air for the combustor dome and liner).

High-temperature ceramic-matrix composite materials with advanced cooling are chosen for the combustion liners in order to meet the durability challenges. The new combustor design concepts have been benchmarked against data from previous successful development programs. A series of combustion tests ultimately provided the data needed to down select and further optimize the designs.

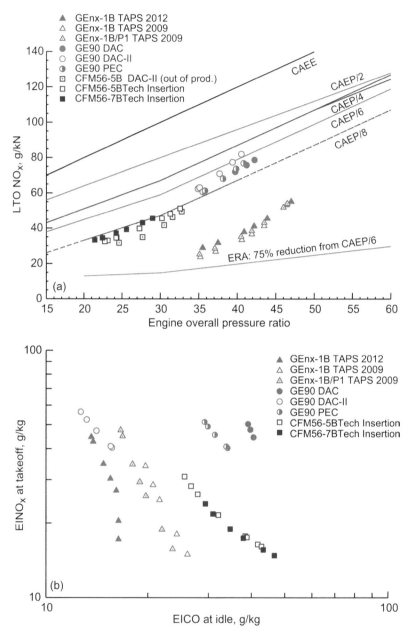

Figure 6.19.  Performance of GE Aviation and CFM International engines. (a) Landing-takeoff (LTO) $NO_x$ emissions index ($EINO_x$) versus engine overall pressure ratio. (b) $EINO_x$ at takeoff versus CO emissions index (EICO) at idle. TAPS, twin-annular premixing swirler; DAC, dual-annular combustor; PEC, performance-enhanced combustor; ERA, Environmentally Responsible Aviation Project.

One final configuration was tested in a new five-cup sector rig at the NASA Glenn Research Center (Fig. 6.21). Gaseous emissions, especially the $NO_x$ emissions at LTO and cruise conditions, and combustion efficiency were measured. Data showed that the GE Aviation N+2 combustor delivered an 81% reduction from (i.e., 19% of) the sixth meeting of the Committee on Aviation Environmental Protection (CAEP/6) $NO_x$ standards, surpassing the NASA ERA

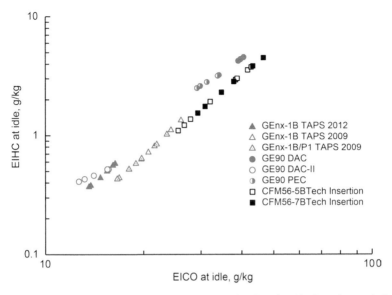

Figure 6.20.    Performance of GE Aviation and CFM International engines' hydrocarbon emissions index (EIHC) versus CO emissions index (EICO) at idle condition. TAPS, twin-annular premixing swirler; DAC, dual-annular combustor; PEC, performance-enhanced combustor.

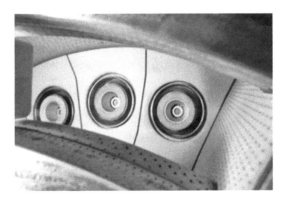

Figure 6.21.    GE Aviation N+2 five-cup ceramic matrix composite combustor sector rig. (From Chang C.-T, C. M. Lee, J. T. Herbon, *et al.*, 2013. NASA environmentally responsible aviation project develops next-generation low-emissions combustor technologies (Phase I). *J. Aeronaut. Aerospace Eng.* 2:116. *https://www.omicsgroup.org/journals/nasa-environmentally-responsible-aviation-project-develops-next-generation-lowemissions-combustor-technologies-phase-i-2168-9792.1000116.php?aid=17626* (accessed April 6, 2016). Work of the U.S. Government.)

Project's N+2 goal of 75% reduction from (25% of) the CAEP/6, while maintaining good combustion efficiencies and acceptable dynamic pressures. Data also showed 60% to 70% reduction in the $NO_x$ emissions index at cruise condition over the previous state of the art while maintaining better than 99.9% on the efficiency. Further development of this technology needs to focus on the thermal and mechanical durabilities, the manufacturability, and the optimization of the design to balance combustion efficiency and dynamics versus $NO_x$ emissions.

6.4.4.2   *Pratt & Whitney*

PW and United Technologies Research Corporation (UTRC) had developed various combustor concepts to reduce $NO_x$ emissions over the past four decades; for example, an axially staged combustion system was developed for the V2500 in the 1990s. PW had explored and developed several advanced concepts for the NASA-sponsored High Speed Civil Transport Program, the Advanced Subsonic Technology Program and the Ultra Efficient Engine Technology Program. Besides the RQL technology embodied as the TALON X, these technologies can be categorized as multipoint, fuel-nozzle radially staged, multidome, and various axially staged concepts. Each of these technologies faces different challenges.

The radially staged swirler faced the challenge of aerodynamically separating the pilot from the main burner in such a fashion to ensure good combustion efficiency at low-power, yet maintaining sufficient mixing for operational stability at all power conditions. Scaling the radially-staged swirler to a larger size would result in a challenge of uniform mixing. There would be mechanical challenges due to the much larger combined size of the multiple swirlers. As in all lean-staged systems, the additional challenge of coking was enhanced by the fact that there were a large number of fuel injection points and the need to distribute these points in such a way to allow for uniform mixing with air.

However, swirler mixing and pilot stability were improved by using previously developed concepts under the UEET Program and applying the lessons learned during TALON X development. Various swirler combinations and swirl distributions were conceptualized and analyzed using computational fluid dynamics (CFD) simulations. Based on the spray patterns predicted by CFD, the design was modified iteratively to produce successively refined versions of the design concept. The final design was then tested successfully in the UTRC Advanced Aeroengine Combustor rig.

In addition, as part of another NASA contract examining potential low-$NO_x$ concepts for supersonic engines, the PW radially staged swirler concept was tested at the NASA CE-5 single nozzle test facility. Data showed $EINO_x$ values were significantly less than 5, experimentally demonstrating the capability to meet ERA emissions goals. Combustion efficiency at the low-power condition was very good, demonstrating successful separation and stabilization of the pilot flame.

Axially staged combustors characteristically have good separation of the pilot with accompanying positive stability and efficiency. The challenge was in packaging and mixing of the main stage; that is, to implement it in a simpler fashion than the version developed for the V2500. Coking is also an issue for this technology. Based on the experience gained from the development of TALON combustors, the pilot stage was kept simple. Various concepts for the mains were conceptualized, analyzed, and explored with CFD. The key metric was to improve the degree of mixing, which was needed for the required low-$NO_x$ emissions was the key question. Various mixer designs were explored and then the final designs were evaluated through a series of idealized UTRC AAC rig tests.

The PW team chose the axially controlled stoichiometry concept to be the one tested at NASA Glenn Research Center. The separation of the pilot and the main provided good combustion efficiency low power condition, and operation stability at all power conditions. Based on the V2500 experience, mixing of the pilot and main is well controlled. This technology also distributes the heat release axially, reducing susceptibility to acoustics. Finally, $NO_x$ emissions data were the lowest of the tested PW configurations, which allowed significant developmental margin for any of the concepts.

The axially controlled stoichiometry concept was further rested in a three-sector arc rig, first at UTRC and then at NASA in the Advanced Subsonic Combustion Rig facility. Results between the two series of tests were consistent. Acoustic issues were only experienced at off-design conditions. Efficiency was above 99.9% at all fully staged high power points. Because of the conventional pilot zone, high efficiency was also achieved at idle and approach conditions. $EINO_x$ values at cruise were 2 and below. Emissions were measured at idle, takeoff, climb, and approach

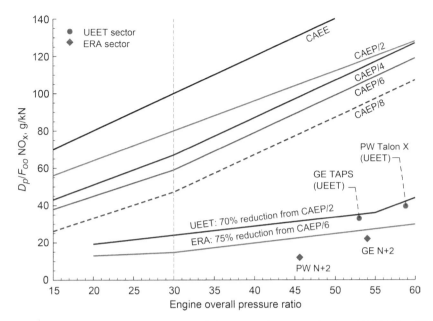

Figure 6.22.   Pratt & Whitney (PW) and GE Aviation sector rig test results from NASA Ultra-Efficient Engine Technology (UEET) Program and Environmentally Responsible Aviation (ERA) Project for $NO_x$ emissions levels ($D_p/F_{oo}$) at landing and takeoff (LTO) versus engine overall pressure ratio compared to International Civil Aviation Organization (ICAO) Committee on Aviation Engine Emissions (CAEE) and Committee on Aviation Environmental Protection (CAEP) regulatory levels.

conditions. Based on an advanced geared turbofan (GTF) engine (the Pratt & Whitney Geared Turbofan™, (East Hartford, Connecticut)), the mass of emissions produced (normalized by the engine-rated thrust) $D_p/F_{oo}$ for $NO_x$ of 88% below CAEP/6 was calculated using an N+2 cycle. Performance with respect to all CAEP-regulated emissions are shown in Figure 6.22.

Further development will include the way to simplify the packaging and to make it more product-ready: it must fit into the envelope available in the current and planned GTF engines as well as be verified and matured in a full-annular design. The evaluation and testing during the ERA Phase I efforts provided an excellent basis for the continued development of this concept.

In conjunction with the NASA ERA Project, PW continues to develop the TALON X combustor. In particular, swirler and front-end modifications were explored further. These approaches demonstrated the potential for low smoke and $NO_x$ emissions in the TALON X combustors, which were developed and tested in the GTF family of engines.

Experiment approaches for further advanced swirler designs were analyzed, evaluated using CFD, and tested in the spray facility at UTRC to determine critical uniformity parameters. A five-sector rig was designed and fabricated to simulate the results of the full-annular tests and the engine emissions. Data showed that lower smoke numbers and $NO_x$ levels were obtained than the swirlers currently used in those engines. Using PW experienced-based methods, with the same advanced GTF cycle a $NO_x$ level of 72% margin to CAEP/6 is projected. This is close to the NASA ERA Project N+2 goal of 75% $NO_x$ reduction to CAEP/6; therefore, the TALON X remains a viable option for the current generation of aeroengines.

## 6.5   TRADEOFFS INVOLVED IN REDUCING $NO_x$ EMISSIONS

Engine design involves making tradeoffs among many requirements. The most important tradeoffs at the engine level are those between $NO_x$ and $CO_2$ emissions, as well as the tradeoff between $NO_x$ emissions and noise. It is also important to recognize that tradeoffs occur not only at the engine level but also at the aircraft level, where the pollutants are eventually emitted. At the level of the whole aircraft, tradeoffs are broader than in the case of the engine alone.

For the past four decades, the natural desire to save fuel (improve fuel burn) and to reduce noise leaned towards two key technology trends: increasing the bypass ratio (BPR) and increasing the overall pressure ratio. BPR is the proportion of the total air moved by the front fan of the engine to the amount of air passing through the hot core of the engine (compressor, combustor, and turbine).

$CO_2$ emissions reduction is more dependent on the engine cycle during operation than on improvements to combustor design. The role of the combustor is to ensure that the demands of a more fuel-efficient cycle are met without compromising engine performance. Higher OPR and BPR will improve fuel efficiency (reduce fuel consumption) and reduce $CO_2$. Higher pressure ratio requires higher flame temperature and therefore higher temperatures within the combustor, resulting in more efficient combustion with associated reductions in CO and unburnt hydrocarbons emissions. However, as $CO_2$, CO, and UHC emissions are reduced, $NO_x$ emissions are increased because of the higher $NO_x$ formation rate associated with the higher pressure and temperature in the combustor. If engine core temperatures and pressures are reduced in an effort to reduce $NO_x$ emissions, then engine thermodynamic efficiency will be reduced. In order to maintain same thrust, additional fuel must be burned, resulting in additional $CO_2$ emissions. Therefore, advanced $NO_x$ technology is needed to minimize $NO_x$ emissions while allowing operation at high temperature and pressure.

In short, there is an inverse relation between the production of $CO_2$ and $NO_x$ at the fundamental level (Fig. 6.23). On the one side, regulations and standards demand $NO_x$ reduction; on the other side, economic pressures to minimize fuel use and maximize payload and range argue for $CO_2$ reduction. Therefore if $CO_2$ reduction is given greater priority, then $NO_x$ emissions will tend to rise, and vice versa. The competition between the conflicting drives for $CO_2$ and $NO_x$ reduction needs to be carefully balanced to achieve the most environmentally and economically desirable outcome.

In current designs, the permissible BPR value has increased from around 1:1 (equal flows through the fan and the core) to 10:1 (10 times amount of air flows through the fan than through the core). Improved fuel burn (reduced $CO_2$ emissions) has been achieved through improved propulsive efficiency provided by the front fan coupled with the improvements in thermal efficiency achieved by making the engine core work harder. Naturally engine diameter increases with BPR in order for the fan to pull sufficient quantities of air. Considering the aircraft mission, engine

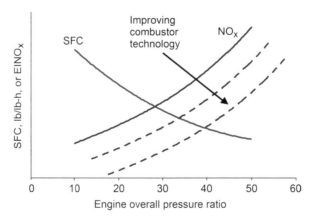

Figure 6.23.   $CO_2$-$NO_x$ tradeoff for varying combustor technology standards: Increasing engine overall pressure ratio (OPR) to reduce specific fuel consumption (SFC) will increase $NO_x$; hence, advanced combustion technology is required.

performance, and physical size, the highest BPR engines tend to be on long-haul wide-bodied aircraft. The modern narrow-bodied fleet tends to have engines with BPRs close to 5 or 6.

More radical changes can be considered if the fuel price significantly increases or the source of fuel is restricted: the geared turbofan engine. The GTF essentially is a conventional turbofan engine with the addition of a gearbox between the front fan and the turbine driving the fan. For narrow-bodied aircraft, this will open the possibility of increasing BPR to a value that is potentially equivalent to today's most efficient wide-bodied aircraft engine.

History has shown that reduced fuel burn and reduced $NO_x$ production have both been achieved by pursuing concurrent fuel-burn and $NO_x$-reduction technologies. If $NO_x$-reduction technologies can be further developed, then improvements in $CO_2$ emissions can continue. New materials and cooling technologies have also been introduced into new engines over the years, supporting the increase of engine pressure to improve thermodynamic efficiency.

Decades of improvements in fuel-air mixing processes, liner cooling techniques, and advanced materials have resulted in current combustion systems, which achieve essentially 100% combustion efficiency at all power settings, except for the lowest (idle). Hence levels of incomplete combustion products, such as CO, UHC, and visible smoke are all very low, only worthy of notice at low-power conditions in the vicinity of the airport.

Emission tradeoffs within the combustor include

1. A rich reaction zone that reduces $NO_x$ formation but tends to increase soot particles
2. A leaner reaction zone that reduces $NO_x$ and soot formation but tends to increase CO and hydrocarbon, which also reduces combustion stability
3. A reduced combustion chamber volume that reduces $NO_x$ but tends to increase CO and HC, which also tends to reduce altitude relight capability

Potential significant $NO_x$ reduction of a future direct-lean–injection-type combustor may cause additional engine weight because of the more complex fuel control system, combustor design, and manifolds, but the additional weight is not significant at the scale of the whole aircraft, therefore bringing a minimum $CO_2$ penalty.

Considering the regulatory perspective, "technical feasibility" and "economic reasonableness" are the two conditions to be met before $NO_x$-emission regulations can be any more stringent. Ongoing CAEP studies on the cost-effectiveness of modifying noncompliant engines by fitting modern low-$NO_x$ combustor systems only assume a $CO_2$ penalty of 0% to 0.5%, with 0.25% being viewed as a reasonable average value.

Improvements in fuel burn to date have been primarily constrained by issues relating to materials, such as temperature limits or cooling requirements, rather than technologies for reducing $NO_x$ emissions.

## 6.6    SUMMARY

Gaseous and particulate emissions from the aircraft directly impact the environment in terms of climate change and have a detrimental impact on human health. Gases such as carbon dioxide and water vapor are greenhouse gases and affect the global climate. Nitrogen oxides and particulate matter, especially fine-sized (submicron) particles, affect air quality in and around the airport, resulting in adverse health issues for the local residents. $NO_x$ emissions are currently regulated during take-off and landing to meet the local air quality standards set by the International Civil Aviation Organization (ICAO), a United Nations specialized agency established for setting international civil aviation standards. Particulate emissions are likely be regulated for local air quality in the very near future because of increased evidence of adverse health impacts associated with the ingestion of fine particles.

The amount of $NO_x$ produced by aircraft fuel burn directly depends on the residence time integral of the hot gas over the instantaneous $NO_x$ formation rate, which varies exponentially with flame temperature. The technologies for reducing $NO_x$ level in the combustor are aimed at reducing either the flame temperature, the residence time, or both. There are two primary $NO_x$-controlling strategies that have been applied to gas turbines. One is rich-burn, quick-quench, lean-burn and the other is burning lean (lean-staged). Both schemes have been used over the last three decades and resulted in $NO_x$ emissions reduction of about 50% with each new generation of aircraft, occurring at intervals of about 15 years. These technologies were initially considered by NASA for advancement through early- and mid-TRL (Technology Readiness Level) stages, which are typically associated with high risk and consequently not conducive to investment of resources by the industry until later stages of TRL, when the technology can be linked to a product.

Among the gaseous emissions, $NO_x$ and $CO_2$ have an inverse relationship in their production at the engine design level. $CO_2$ reduction, which is possible only from an increase of engine fuel efficiency, tends to increase $NO_x$ production because of the increase in combustion temperature needed for increasing cycle efficiency. Over the last few decades of technology advancements in fuel-air mixing processes and liner cooling techniques, it has been shown that reduced $NO_x$ production can be achieved at higher cycle temperatures, making concurrent reduction of $NO_x$ and $CO_2$ possible. Dramatic reductions of $NO_x$ and $CO_2$ emissions have been achieved through successful partnership of government agencies such as NASA with industry and academia. This is expected to continue in the future and play a key role in advancing the technology to meet the performance and increasingly stringent environmental compatibility requirements of the future aviation propulsion systems.

## REFERENCES

Bruner, S., S. Baber, C. Harris, *et al.* 2010. NASA N+3 subsonic fixed wing silent efficient low-emissions commercial transport (SELECT) vehicle study. NASA/CR–2010-216798.

Chang, C. T. 2012. Gas turbine engine combustors. In *Encyclopedia of Aerospace Engineering*, New York, NY: John Wiley & Sons, Ltd.

Chang C.-T., C. M. Lee, J. T. Herbon, *et al.* 2013. NASA environmentally responsible aviation project develops next-generation low-emissions combustor technologies (Phase I). *J. Aeronaut. Aerospace Eng.* 2:116. *https://www.omicsgroup.org/journals/nasa-environmentally-responsible-aviation-project-develops-next-generation-lowemissions-combustor-technologies-phase-i-2168-9792.1000116. php?aid=17626* (accessed April 6, 2016).

Collier, F. 2012. NASA aeronautics—environmentally responsible aviation project solutions for environmental challenges facing aviation. Paper presented at the 50th AIAA Aerospace Sciences Meeting, Nashville, TN.

Del Rosario, R., J. Koudelka, R. Wahls, *et al.* 2013. Technical progress and accomplishments of NASA's fixed wing project. Presented at 51st AIAA Aerospace Sciences Meeting, Grapevine, TX. *https://ntrs. nasa.gov/archive/nasa/casi.ntrs.nasa.gov/20150009967.pdf* (accessed April 21, 2017).

Dodds, W. 2002. Engine and aircraft technologies to reduce emissions. Paper presented at the UC Technology Transfer Symposium Dreams of Flight, San Diego, CA.

Dodds, W. 2005. Twin Annular Premixing Swirler (TAPS) combustor. In *Roaring 20th Aviation Noise & Air Quality Symposium*, Palm Springs, CA: University of California, Berkeley.

European Aviation Safety Agency (EASA). 2014. ICAO Aircraft Engine Emissions Databank. *http://easa. europa.eu/document-library/icao-aircraft-engine-emissions-databank* (accessed January 5, 2015).

Faber, J., D. Greenwood, D. Lee, *et al.* 2008. Lower $NO_x$ at higher altitudes—policies to reduce the climate impact of aviation $NO_x$ emission. Commissioned by the European Commission, DG Energy and Transport under contract TREN/07/F3/S07.78699. Delft: CE Delft.

Foust, M. J., D. Thomsen, R. Stickles, *et al.* 2012. Development of the GE Aviation low emissions TAPS combustor for next generation aircraft engines. AIAA 2012–0936.

International Civil Aviation Organization. 2008. International standards and recommended practices. In *Annex 16—Environmental protection*, Volume II, Third ed.

International Coordinating Council of Aerospace Industries Associations. 2002. Emissions reduction technology workshop. Paris: CAEP SG.

Lee, C.-M., C. Chang, S. Kramer, *et al.* 2013. NASA project develops next generation low-emissions combustor technologies. AIAA 2013–0540.

Lohmann, R. P., and J. S. Fear. 1982. NASA broad specification fuels combustion technology program—Pratt & Whitney Aircraft phase I results and status. AIAA−82−1088.

Mongia, H., and W. Dodds. 2004. Low emissions propulsion engine combustor technology evolution past, present and future. ICAS 2004−6.9.2 (I.L.).

Moran, J. 2007. Engine technology development to address local air quality concerns. Presented at the ICAO Colloquium on Aviation Emissions, Montreal, Quebec.

# CHAPTER 7

## Airspace systems technologies

Banavar Sridhar

## 7.1   INTRODUCTION

Civil aviation is a vital sector of the U.S. economy. In 2009, air transportation and related industries generated $1.3 trillion and employed 10 million people in the United States (Federal Aviation Administration, 2014). The U.S. National Airspace System (NAS), which provides the infrastructure needed for the operation of civil aviation in the United States, refers to all the hardware, software, and people involved in managing air traffic. At any given moment, as many as 5,000 aircraft may be flying in U.S. skies, and in 2011, the NAS managed the progress of nearly 10 million flights. However, the capacity of the U.S. airspace is limited by the ability of controllers to detect conflicts between aircraft and resolve them in a safe manner when traffic density is high. As a result, the air transportation system often experiences significant delays and lost productivity, and it produces greater amounts of noise pollution, carbon dioxide ($CO_2$), and other greenhouse gas (GHG) emissions than it would if operations were more efficient. In 2005, the Air Transport Association, a group representing airlines, estimated the cost of delays to airlines at $5.9 billion (Borener *et al.*, 2006).

Although air traffic has not increased in the United States over the last 10 years, the U.S. Federal Aviation Administration (FAA) predicts that air traffic demand will double in the next 20 years, with an approximate annual growth of 3.5% (FAA, 2011). This increase in demand could further strain airports and the airspace—resulting in large delays, a breakdown of airline schedules, and an adverse impact on the environment.

Incremental increases in the current capacity may not be sufficient to meet the future demands for air transportation. Major changes are needed, including increased levels of automation in an environment highly oriented toward safety, and new roles for automation, controllers, and pilots. The United States has created a multiagency Joint Planning and Development Office (JPDO) to lead the transformations required in the air transportation system. To facilitate this transformation, the JPDO developed a comprehensive Concept of Operations (Joint Planning and Development Office, 2011) that advocates operations based on four-dimensional aircraft trajectories. The National Aeronautics and Space Administration (NASA) is collaborating with the FAA and other industry partners to develop several advanced automation tools that will provide air traffic controllers, pilots, and other airspace users with more accurate real-time information about the nation's traffic flow, weather, and routing. The greater precision of this information will be a key enabler of the Next Generation Air Transportation System (referred to as "NextGen"). NextGen—a comprehensive transformation of the NAS—will be safer, more reliable, and more efficient. It also will reduce the impact of aviation on the environment. Transitioning to NextGen is vital to improving system performance, meeting continued growth in air traffic, and increasing the nation's mobility to support economic progress.

Section 7.2 provides an overview of current airspace operations, introduces primary air traffic control (ATC) and traffic flow management (TFM) activities, and describes issues facing the current system. Section 7.3 describes new concepts in air traffic management (ATM) and the key challenges in designing the future system. Section 7.4 describes the implementation of the advanced concepts in NextGen. Section 7.5 provides concluding remarks.

## 7.2   CURRENT AIRSPACE OPERATIONS

Airspace operations, as practiced today, can be viewed as a distributed, hierarchical process with multiple decisionmakers. An aircraft in flight is under the command of a pilot with assistance from air traffic controllers and flight dispatchers on the ground. Figure 7.1 shows this triad, which is responsible for the safe, smooth management of airspace operations. In the United States, the airspace is divided into 20 centers plus Alaska and Hawaii, with an Air Traffic Control System Command Center (ATCSCC) in Warrenton, Virginia. Figure 7.2 shows the 20 centers. The centers are subdivided into sectors (Nolan, 2003). For example, the high-altitude sectors in Oakland Center (ZOA) are depicted in the inset on the left side of Figure 7.2.

The two major functions associated with air traffic management are air traffic control, the ability to keep aircraft separated from each other, and traffic flow management, maintaining the efficient flow of traffic. These two functions have to be performed while (1) satisfying the demand for airport and airspace capacity and (2) reducing delay and the impact of aircraft emissions on the environment. The aircraft conflict detection and resolution task—separation assurance (SA)—is performed by air traffic controllers in sectors with the help of decision tools.

TFM involves planning air traffic to avoid exceeding airport and airspace capacity, while making effective use of available capacity. At the top level, the ATCSCC uses traffic predictions to form a strategic TFM plan. Under forecasted severe weather conditions, the ATCSCC may delay some aircraft at airports and/or reroute others. Regional adjustments to these plans are developed by the different centers. Dispatchers and air traffic controllers at airlines respond to changes to the traffic flow by rerouting, rescheduling, and canceling flights, thus changing flow patterns. Schedules and route preferences from airlines and other users of the system are factored into the development of the TFM strategy through the collaborative decisionmaking (CDM) process (Ball *et al.*, 2001).

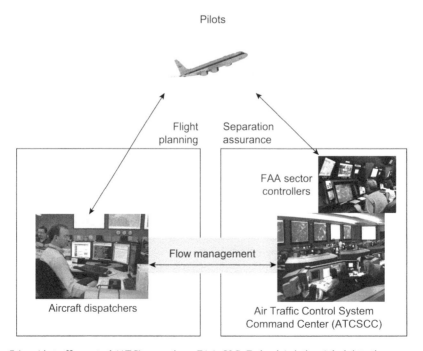

Figure 7.1.   Air traffic control (ATC) operations. FAA, U.S. Federal Aviation Administration.

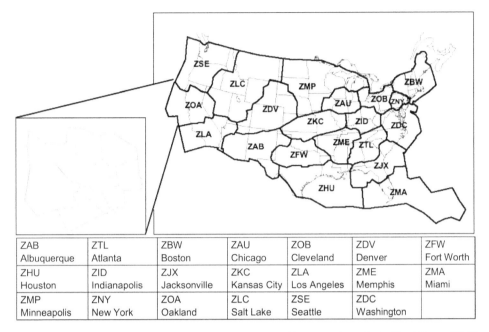

| ZAB | ZTL | ZBW | ZAU | ZOB | ZDV | ZFW |
|-----|-----|-----|-----|-----|-----|-----|
| Albuquerque | Atlanta | Boston | Chicago | Cleveland | Denver | Fort Worth |
| ZHU | ZID | ZJX | ZKC | ZLA | ZME | ZMA |
| Houston | Indianapolis | Jacksonville | Kansas City | Los Angeles | Memphis | Miami |
| ZMP | ZNY | ZOA | ZLC | ZSE | ZDC | |
| Minneapolis | New York | Oakland | Salt Lake | Seattle | Washington | |

Figure 7.2.  Airspace centers in the continental United States, with sectors shown for Oakland Center (ZOA). (From Sridhar, B., S. R. Grabbe, and A. Mukherjee. 2008. Modeling and optimization in traffic flow management. *Proc. IEEE* 6:1–29. *https://www.aviationsystems.arc.nasa.gov/publications/2008/AF2008077.pdf.* Work of the U.S. Government.)

### 7.2.1   *Separation assurance*

Air traffic controllers are responsible for separating aircraft operating within the region, or sector, of airspace under their control. Today, air traffic controllers provide SA by visual and cognitive analysis of a traffic display and by issuing control clearances to pilots via voice communication. The decision support tools deployed in recent years provide trajectory-based advisory information to assist controllers with conflict detection and resolution (Erzberger. 2006, 2004), coordinating departures and arrivals, and other tasks. Although decision support tools have reduced delays, cognitive abilities limit the number of aircraft that a human controller can safely handle to no more than 15 to 20—well below the capacity of the airspace when constrained only by legal separation.

   Because current airspace capacity is limited by controller abilities (Sridhar, 2011), a fundamental transformation of the way SA is provided is needed to achieve the capacity gains necessary to meet future demands.

### 7.2.2   *Traffic flow management*

The complexity of the TFM problem and the duration of the planning interval has led to a natural decomposition of the problem into national TFM decisions performed at the ATCSCC, which provide guidance to centers or groups of centers (regions), and regional TFM decisions performed in the regions. Regional TFM plans are local, more detailed, for shorter durations, and based on more accurate traffic and weather information.

#### 7.2.2.1   *National traffic flow management*

The goal of national TFM is to accommodate user-preferred gate-to-gate trajectory preferences through the management and allocation of National Airspace System resources when demand approaches or exceeds supply. The demand and supply situation is exacerbated during severe weather conditions, which can reduce both airspace and airport capacity. The ATCSCC uses traffic predictions to form a strategic plan over a 1- to 6-h time horizon (FAA, 2008). Depending on the expected weather conditions and demand in the different regions of the airspace and airports, the ATCSCC may delay some aircraft at airports and/or reroute others. The tools available to manage traffic in the presence of excess demand are the Airspace Flow Program (AFP), a Ground Stop (GS), the Ground Delay (GD) Program, the *National Severe Weather Playbook* (FAA, 2008), rerouting, and Miles-in-Trail (MIT). The AFP identifies flights scheduled to travel through capacity-limited regions of airspace, such as a region impacted by severe weather. The impacted flights are delayed at airports, or the airspace users are given the option to route around the constrained regions of airspace. AFPs are used to manage traffic flows due to en route constraints, whereas GD Programs and GSs are used for constraints impacting an airport. GSs hold all flights at their departure points that are destined for an affected airport for the duration of the GS initiative. Like the GS, the GD Program controls the flow of traffic to an airport where the forecasted demand is expected to exceed the airport's predicted acceptance rate.

Departure controls can be imposed on flights to regulate the flow of traffic into capacity-constrained regions of the NAS, or capacity-constrained regions can be routed around as a complementary control strategy. Under current operations, the U.S. Federal Aviation Administration relies on the *National Severe Weather Playbook* (FAA, 2008). It is a compendium of standardized alternative routes intended to avoid specific regions of airspace that are commonly impacted by severe weather during certain times of the year according to historically validated data. Figure 7.3 shows a planning template, known as "Green Bay," provided in the Playbook for rerouting

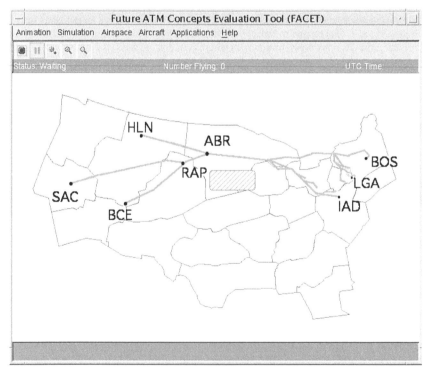

Figure 7.3.   Graphical representation of the Green Bay National Playbook route with a conceptual region of severe weather depicted by the hash-marked area. HLN, Helena Regional Airport (MT); ABR, Aberdeen Regional Airport (SD); BOS, Boston Logon International Airport (MA); SAC, Sacramento International Airport (CA); BCE, Bryce Canyon International Airport (UT); RAP, Rapid City Regional Airport (SD); IAD, Washington Dulles International Airport (DC); LGA, LaGuardia Airport (NY). (Federal Aviation Administration. 2008. Air Traffic System Command Center, National severe weather playbook. Figure from 2008 version; current version at *http://www.fly.faa.gov/PLAYBOOK/pbindex.html*. Work of the U.S. Government.)

eastbound traffic through Minneapolis Center (ZMP) when a large portion of airspace in the Midwest is affected by weather. The large hash-marked area in the southern portion of ZMP represents a predicted severe weather region. The routes represented by a solid line in Figure 7.3 represent alternative routes for aircraft originating on the West Coast and traveling to select East Coast destinations, such as Boston Logan International Airport (BOS), LaGuardia Airport (New York City, LGA), and Washington Dulles International Airport (IAD).

### 7.2.2.2   *Regional traffic flow management*

Regional TFM, which operates on a forecasted time horizon of roughly 20 min to 2 h, provides a tactical control loop for adjusting the control strategies generated by national TFM. Regional TFM is based on improved aircraft demand, airspace capacity, and weather intent information. These adjustments are implemented by rerouting aircraft locally or by spacing aircraft in a stream, referred to as MIT. The number of aircraft entering a region is inversely proportional to MIT. MITs are used in increments of 5 mi, and a spacing of 10 to 30 mi is routinely used to reduce congestion. The current TFM has a hierarchical and distributed control structure. Dispatchers and air traffic coordinators at airlines respond to these flow control actions by rescheduling and canceling flights, thus changing flow patterns. Schedules and route preferences from airlines and other users of the system are factored in the development of the TFM strategy through the CDM process (Ball *et al.*, 2001).

Traffic management initiatives such as Playbook routes, GSs, GDs, AFPs, and MIT restrictions are based on attempts to solve particular problems. For example, Playbook routes are used to circumvent severe weather, GSs and GDs are used to control demand at airports, and MITs are used to control workload in individual sectors. The various TFM actions are based on experience and are imposed independently. The interactions between different actions may not always be accounted for during the decisionmaking process.

Figure 7.4 shows the impact of these actions on aircraft schedules on a day with severe weather. Each dot in the display indicates an aircraft flying in the NAS, with gray dots indicating aircraft that are on time, blue dots indicating aircraft that have been delayed between 15 min and 2 h, and red dots indicating aircraft that have been delayed by more than 2 h. The overall capacity of the NAS will increase as methods are developed to integrate and optimize ATM initiatives such as AFPs, Playbook routes, GSs, GD Programs, and MITs to produce a single cohesive plan that improves traffic throughput, reduces delay, reduces congestion, and provides predictability and flexibility for aircraft operators.

A major problem with the current system is the insufficient sharing of information among decisionmakers (Wambsganss, 1997). Information about schedule changes and cancellations are not available to the decision support tools in a timely manner, and airlines are not fully aware of the traffic conditions and the status of the NAS in making their routing decisions. Another source of uncertainty is that most of the trajectory predictions in the crucial climb phase are based on nominal weights of the aircraft and nominal climb procedures (Krozel *et al.*, 2002). The FAA and the aviation industry have worked hard in recent years to improve collaboration between users and service providers in decisionmaking. CDM is limited to strategic planning, and user participation in planning decreases as the planning interval becomes smaller (Ball *et al.*, 2001).

### 7.2.3   *Terminal area operations*

The average spacing or distances between aircraft that can be safely accommodated with human-centered control limit the capacity at our airports and within the en route airspace. The spacing is influenced by the separation standards and the ability of the service provider to precisely control to those standards. Both of these factors are influenced by the technologies and procedures for tactical (time horizons of 20 min to 2 h) ATM available to the pilot and controller. Improving the efficiency of the terminal area (TA), which is the volume of airspace surrounding

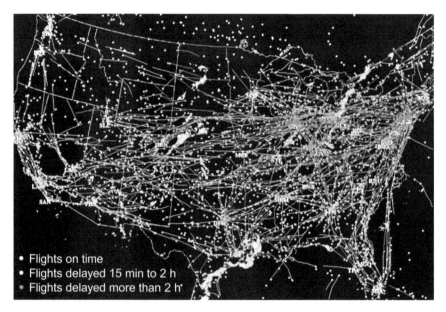

Figure 7.4.  Traffic display showing on-time and delayed flights. (From Sridhar, B., S. R. Grabbe, and A. Mukherjee. 2008. Modeling and optimization in traffic flow management. *Proc. IEEE* 6:1–29. *https://www.aviationsystems.arc.nasa.gov/publications/2008/AF2008077.pdf*. Work of the U.S. Government.)

airports to a radius of about 50 mi (80.47 km), is an especially complex task because of operating characteristics that are quite distinct from the en route environment. TA controllers manage both ascending and descending aircraft, more frequent turns, wider range of separation standards than in other areas, a mixture of equipment types and terrain, and increased traffic density within shorter time horizons.

In today's TA, as an aircraft transitions for landing, controllers track and guide the aircraft from cruise altitude to the runway using visual aids as well as their skills and judgment. They issue turn-by-turn instructions via radio communications. As aircraft approach runways, controllers manually merge aircraft and sequence them for arrival. Busy TA conditions often force the aircraft to fly inefficient arrival paths involving frequent changes in direction, altitude, and speed to maintain safe separation from other aircraft. Frequently, controllers must employ longer routes (known as path stretching) or holding patterns to tactically accommodate larger amounts of delay. The tactical nature of this manual approach leads to increased fuel burn and noise pollution, contributes to high controller workload, and exacerbates traffic congestion. Moreover, the imprecision of this current system creates greater uncertainty and forces controllers to add buffers to the separation required between aircraft—decreasing airspace capacity and leading to further delays.

Although more efficient arrival paths are achievable today, current technology limits their feasibility to light traffic conditions, such as during the middle of the night. During periods of high-density traffic, maintaining safe separation and throughput take precedence over achieving efficient operations. The technical challenge facing the aviation community is to make efficient arrival procedures common practice during heavy traffic when they are needed most, while still ensuring safety and throughput. Many of the inefficiencies associated with arrivals also apply to aircraft departing the TA and climbing to their en route cruising altitude.

Capacity at the busiest airports plays a key role in determining the efficiency and robustness of the NAS and ultimately defines the attainable growth in air traffic. Significant growth at the busiest airports as well as regional and smaller airports is needed to achieve NextGen capacity goals. The Joint Planning and Development Office envisions a combination of new technologies enabling significant growth at large airports and increased operations at underutilized airports to absorb the expected increase. Increasing capacity in the current architecture is not scalable to meet future needs.

### 7.2.4 *Surface traffic operations*

Current surface traffic operations lack advanced automation, and aircraft join the queue as they depart from the gate, resulting in taxiway waiting and stop-and-go situations. The management of aircraft on the surface affects both departing and arriving aircraft. Inefficient surface operations have repercussions throughout the NAS. Surface operations in the United States are a shared responsibility between airlines controlling the ramp (nonmovement) area and the FAA Air Traffic Control Tower (ATCT), which controls traffic on taxiways and runways (or movement area). Typically, airlines push back an aircraft from its gate as soon as the aircraft is ready, partly because of the scheduled gate push-back is a performance metric (FAA, 2014). Often these movements are uncoordinated, and during busy times they result in taxiway congestion and large runway queues (Balakrishnan and Jung, 2007; Malik *et al.*, 2010; Rathinam *et al.*, 2008). The taxi-out time is the time between the departure of an aircraft from the gate and the take-off time. The excess taxi-out time is measured as the difference between actual taxi-out time and the ideal unimpeded taxi-out time. A recent study estimated that during 2007 the excess time in the taxi-out phase in the United States was 7 min per departure for all airports and longer than 15 min for the three airports in New York City (Gulding *et al.*, 2009). An analysis of fuel consumption during taxi operations at the Dallas/Fort Worth International Airport (DFW) airport shows that approximately 7,000 gal/day (about 26,500 L/day) is wasted during stop-and-go operations (Gulding *et al.*, 2009).

The predeparture process may involve coordination between several operational positions within the ATCT and Traffic Management Units (TMUs) at both the associated Terminal Radar Approach Control (TRACON) and the Air Route Traffic Control Center (ARTCC) serving the departure airport. The amount of coordination needed depends on the complexity and activity level of the departure airport, the number of other airports within the TA, the merging of departures with overhead traffic flows, the presence of convective weather in the TA or *en route*, and constraints in the aircraft's destination airport. Currently the process is manual, and coordination relies on individual phone calls, manual data entry, and the use of restrictions such as MIT or Minutes-in-Trail (MINIT) for all flights crossing a particular departure fix (Doble *et al.*, 2009).

### 7.2.5 *Environmental operations*

Most of the current operational efforts in reducing the impact of aviation on the environment are focused on achieving sustainable growth by limiting community noise concerns, reducing the impact of aviation on air and water quality, and reducing $CO_2$ emissions by improving the fuel efficiency of aircraft trajectories.

## 7.3 ADVANCED AIRSPACE OPERATIONS CONCEPTS

Air traffic management is divided into several domains: as aircraft

1. Transit from gate to runway
2. Depart from the airport into cruise en route
3. Arrive in the terminal area
4. Land on the runway
5. Taxi back to the gate

Although each domain has special issues, some major deficiencies of the current system are excessive reliance on manual methods—some of which predate digital technology—and unwillingness to provide information and/or insufficient information. The two major challenges for future ATM systems are (1) determining the appropriate human/automation mix and (2) addressing the changing nature of the roles and responsibilities of the air traffic service provider and its agents (the Federal Aviation Administration, traffic flow management and other planners, and controllers) and the airspace users and their agents (airlines, airlines operations centers, pilots, and general aviation). As shown in Figure 7.5, the evolution from the current system, which is almost centralized with a small level of automation, to the future system depends on meeting those challenges.

Future ATM systems should be able to adapt to variations in the magnitude and distribution of the air traffic over the next several decades. It must accommodate aircraft with widely varying performance capabilities and different levels of equipment. The new systems will have to accommodate new types of vehicles like very light jets, unmanned air vehicles, and greater numbers of commercial space launch vehicles. Also, the traffic demand may become more variable in the future.

The future systems should have increased collaboration between users and service providers in all ways from strategic to tactical time frames. Another challenge will be to make good decisions in the presence of uncertainty in weather prediction. The effectiveness of probabilistic decisionmaking also should be factored into future ATMs. These criteria can only be achieved through increased levels of automation. As the level of automation increases, the design of fail-safe architecture is critical, and future ATMs should degrade gracefully under off-nominal conditions. All future designs must facilitate the transition from the current system to the ideal transformed system through a series of intermediate transitions. ATM users have widely varying and competitive mission objectives, and it will be a challenge to gain consensus among the different stakeholders as to the cost and benefits of various airspace technologies.

This section describes the major research issues and mentions a sample of proposed solutions. A comprehensive description of the current research areas and innovative solutions to address the problems can be found in the United States/Europe Air Traffic Management Research & Development Seminars; American Institute of Aeronautics and Astronautics (AIAA) Guidance, Navigation, and Control Conferences; AIAA Aviation Technology, Integration, and Operations Conferences; and journals specializing on aviation operations.

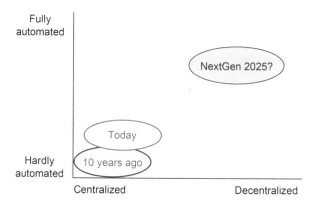

Figure 7.5.   Transformation of air traffic management (ATM) systems. NextGen, Next Generation Air Transportation System.

### 7.3.1   *Separation assurance*

The objective of SA research is to identify trajectory-based technologies and human/machine operating concepts capable of safely supporting a substantial increase in capacity (e.g., 50% more than current demand) under nominal and failure-recovery operations while accommodating airspace user preferences and favorable cost/benefit ratios. SA research focuses on three areas:

1. Automated SA technology development involving automatic conflict detection and resolution algorithms, trajectory analysis methods, and system architectural characteristics that together result in automated resolution trajectories that are safe, efficient, and robust under the huge variety of traffic conditions in the National Airspace System.
2. Functional allocation research aimed at developing human/machine air/ground allocations to provide integrated solutions for traffic conflicts, metering, and weather avoidance. This should include a series of human-in-the-loop simulations (HITLs) of increasing complexity with higher traffic densities, mixed equipage/operations in nominal and off-nominal conditions.
3. Human/automation operating concepts research analyzing cognitive workload, situational awareness, and performance under different service-provider-based concepts of operation; the roles and responsibilities of controllers and pilots; and a series of HITLs of increasing complexity and fidelity.
4. System safety and failure-recovery analysis research addressing the need to identify component failure and recovery modes for automated SA methods, including missed conflict alerts, data link failure, primary trajectory server failure, false read-back, human operator mistakes, and other factors.

### 7.3.2   *Traffic flow management*

Research in TFM (Sridhar *et al.*, 2008) is designed to accommodate future traffic growth while accounting for system uncertainties and accommodating user preferences. To accomplish this goal, research is organized into three focus areas: (1) traffic flow optimization, (2) Collaborative Traffic Flow Management (CTFM), and (3) weather impact assessment. Traffic flow optimization focuses on developing linear and nonlinear optimization techniques, as well as heuristic-based approaches and decomposition methods for effectively developing aircraft-level or aggregate flow control strategies in response to actual and predictive demand and capacity imbalances at the local, regional, and national levels. CTFM in TFM focuses on the development of methodologies for incorporating user preferences into TFM and improved intent information exchange. The outputs of this focus area are algorithms, procedures, and protocols for fully integrating CTFM into the TFM process. The weather impact assessment component of TFM develops metrics to predict and analyze the performance of the NAS with respect to observed or predicted weather, develops models to translate meteorological observations and forecasts into time-varying deterministic and probabilistic estimates of the available airspace and airport capacities, and defines requirements for Next Generation Air Transportation System ATM weather products. The output of the TFM focus area is a set of modeling, simulation, and optimization techniques that are designed to minimize or maximize a system performance measure (such as total delay) subject to airspace and airport capacity constraints, while accommodating weather uncertainty, user preferences, and predicted growth in demand.

### 7.3.3   *Terminal area*

TA operations development is focused on employing rapid prototyping and fast-time simulation to assess and iteratively refine the concept of operations based on the improved understanding of the fundamental challenges and the development of enabling technologies to address those challenges. The development of sequencing and deconfliction technologies focuses on advancing the state of the art beyond the current practices of modified first-come, first-served scheduling and tactical separation service. This research will lead to an understanding of the inherent uncertainty associated with the execution of precision trajectories in the terminal airspace together along with improvements in the optimization of multiobjective constraints for air traffic systems.

The development of precision spacing and merging technologies addresses the need to reduce the level of uncertainty inherent in aircraft operations in the terminal airspace and enables many aspects of equivalent visual operations, a key capability associated with NextGen, as defined by the Joint Planning and Development Office (JPDO, 2011). This research will produce procedures and technologies for precision merging and spacing in flight that are extended to meet multiple constraints and environmental considerations. TA operations research should include methods for managing precision and nonprecision operations in the same airspace. Concepts and technologies for runway balancing and assignments for arrival and departure need to be developed. As appropriate, these should be integrated with scheduling and surface management technologies. In particular, limitations due to wake, location, and strength must be considered for dynamic wake spacing.

### 7.3.4   *Surface operations*

Surface operation research is investigating new technologies and concepts to increase airport capacity by enhancing the flexibility and efficiency of surface operations. This research will result in evaluations of integrated automation technologies and procedures designed to provide the following capabilities:

1. Improving surface traffic planning through (a) balanced runway usage, (b) optimized taxi planning of departures and arrivals, (c) departure scheduling satisfying environmental constraints, dynamic wake vortex separation criteria, and constraints driven by other NAS domains, and (d) balanced runway usage and efficient runway configuration management through coordination with TA operations. Environmental impact will be considered as concepts are investigated.
2. Providing for trajectory-based surface operations by modeling aircraft surface trajectory prediction and synthesis, developing pilot display requirements and technologies for four-dimensional taxi clearances compliance, and taxi clearance conformance monitoring algorithms and procedures.
3. Maintaining safety in ground operations through the development of concepts and algorithms for both aircraft- and ground-based surface conflict detection and resolution and the integration of the two approaches. This research will develop surface traffic simulation capabilities (fast- and real-time HITL simulation) and a surface traffic data analysis tool. Then it will use these tools to evaluate integrated technologies. A software interface also will be developed to integrate the real-time surface traffic simulation with flight deck simulation capabilities.

### 7.3.5   *Environmentally friendly operations*

The main environmental constraints on the growth of aviation relate to noise, air quality, water quality, and climate. There is increased awareness of aviation-induced environmental impact affecting climate change (IPCC, 1999). Estimates show that aviation is responsible for 13% of transportation-related fossil fuel consumption and 2% of all anthropogenic $CO_2$ emissions (Brasseur and Gupta, 2010; Lee *et al.*, 2010). Although emission contributions from aviation are small, a large portion of these emissions occur at altitudes where the emissions remain in the atmosphere longer than they would if emitted at the surface. The desire to accommodate growing air traffic needs while limiting the impact of aviation on the environment has led to research in green aviation with the goals of better scientific understanding, utilization of alternative fuels, introduction of new aircraft technology, and rapid operational changes. There are many technologies (e.g., alternative fuels and new engine and airframe technologies) to reduce the impact of aviation, and these are described in Chapters 4 through 11. This section is limited to the manner in which aircraft operations can be tailored to minimize the aviation impact.

Aviation operations affect the climate in several ways. The climate impact of aviation is expressed in terms of "radiative forcing" (RF). RF is a perturbation to the balance between incoming solar radiation and outgoing infrared radiation at the top of the troposphere, and the amount of outgoing infrared radiation depends on the concentration of atmospheric greenhouse gases. The RF associated with each type of emission has an approximately linear relationship with the global mean surface temperature change. $CO_2$, water vapor, and other gases are unavoidable byproducts of the combustion of fossil fuel. Of these, $CO_2$ and water vapor are GHGs resulting in a positive RF. Because of its abundance and long lifetime, $CO_2$ has a long-term effect on climate change; the non-$CO_2$ emissions have a short-term effect on climate change. The important non-$CO_2$ impacts associated with aviation are water vapor, nitrogen oxides ($NO_x$), condensation trails (contrails), and cirrus clouds due to air traffic. Contrails are clouds that are visible trails of water vapor made by the exhaust of aircraft engines (Duda *et al.*, 2003). Contrails can lead to the formation of cirrus clouds that have a radiative impact on the climate. The latest estimates indicate that contrails caused by aircraft may be causing more climate warming today than all the residual $CO_2$ emitted by aircraft (Boucher, 2011).

Several new operational ATM strategies have been proposed that have the potential to mitigate the impact of persistent contrails on climate change. These strategies include adjusting cruise altitude and rerouting aircraft around regions of airspace that facilitate the formation of persistent contrails (Williams *et al.*, 2002). Such changes could result in longer travel times and increase fuel usage and $CO_2$ emissions.

None of the current methods for avoiding contrails consider the effect of wind on the aircraft trajectory and therefore neglect the potential fuel savings that aircraft can gain when flying wind-optimal routes. A new simulation capability (Sridhar *et al.*, 2013) analyzes the relationship between air traffic operations and their impact on the environment. The simulation integrates the positions of all air traffic in the United States based on flight plans, aircraft trajectory calculations based on predicted wind data, contrail calculations based on predicted temperature and humidity data, a common metric to combine the effects of different types of emissions, and algorithms to generate alternative trajectories for aircraft traveling between city pairs. Figure 7.6 shows the various components of the simulation, which can provide both $CO_2$ and non-$CO_2$ emissions resulting from different current and future operational scenarios. The integrated simulation is used to evaluate the energy efficiency of contrail reduction strategies as well as the tradeoff

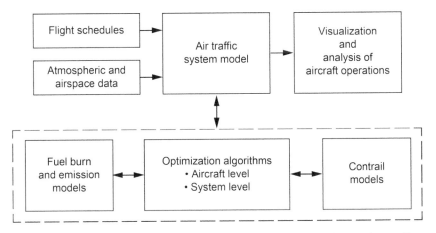

Figure 7.6.   Simulation capability to evaluate the impact of aviation emissions and contrails on the environment.

between the mitigation of persistent contrails and the increased consumption of fuel. The results are used by policy makers as they set aviation operation guidelines.

## 7.4   NEXT GENERATION AIR TRANSPORTATION SYSTEM TECHNOLOGIES

NextGen consists of a combination of projects to improve infrastructure and to introduce new technologies and procedures. This section describes some of the matured concepts being considered for implementation in the Federal Aviation Administration's Next Generation Air Transportation System  projects (FAA, 2012): Automatic Dependent Surveillance—Broadcast (ADS–B), Performance-Based Navigation (PBN), weather integration, and data communications.

### 7.4.1   *Automatic dependent surveillance—broadcast*

ADS–B (FAA, 2016) is a technology to transform air traffic control from the current radar-based system to a satellite-based system. ADS–B brings the precision and reliability of satellite-based surveillance and provides common situation awareness between pilots and controllers. ADS–B is expected to reduce separation margins between aircraft, leading to increases in airspace capacity. It also provides surveillance in remote areas currently without radar coverage and enables aircraft to fly more direct and more efficient routes.

Typically, an ADS–B-capable aircraft derives its position from the Global Positioning System (GPS) of satellites and may combine that position with any number of aircraft variables, such as speed, heading, altitude, and flight number. This information can be shared with other ADS–B-capable aircraft and ATC centers in real time. ADS–B provides two different services: ADS–B-out and ADS–B-in. ADS–B-out periodically broadcasts information about the aircraft itself—such as an aircraft's identification, current position, altitude, and velocity—through an onboard transmitter. It displays the aircraft's location to controllers on the ground or to pilots in the cockpits of aircraft equipped with ADS–B-in. ADS–B-in allows aircraft to receive traffic and weather information data and other ADS–B data, such as direct communication from nearby aircraft.

The FAA plans to implement ADS–B in stages. By 2011, more than 300 FAA-installed radio stations were already providing coverage of the East, West, and Gulf Coasts and of most of the areas near the U.S. border with Canada. The FAA expected the total complement of radio stations to number about 700. It is working with several airlines to obtain ADS–B data to validate the business case for early adoption of new equipment. The FAA has mandated the use of ADS–B-out in Class A, B, and C airspace by January 1, 2020. Currently, there is no mandate for ADS–B-in, and the airspace regularly used by general aviation is exempt from ADS–B-in requirements.

### 7.4.2   *Performance-based navigation*

Area Navigation (RNAV) is a method of navigation that enables an aircraft to fly along a desired flight path within the coverage of the navigational aids or within the limits of the aircraft, or a combination of both. The safety along an RNAV route is ensured through a combination of aircraft navigation accuracy, route separation, and ATC radar monitoring and communications. Required Navigation Performance (RNP) is RNAV operations with aircraft onboard equipment for performance monitoring and alerting. The PBN concept assumes that the navigation specification will be met through a combination of ground-based, satellite-based, and aircraft-based hardware and software. RNAV- and RNP-equipped aircraft can fly direct trajectories between points in the airspace, and RNAV and RNP specify the cross-track accuracy between the desired and actual trajectory of the aircraft. As shown in Figure 7.7, an aircraft with RNP 2 capability will be able to follow the desired trajectory with a cross-track accuracy of 2 nautical miles (3.7 km) 95% of the time and within a lateral containment region of 4 nautical miles (7.4 km) all (99.999%) the time. PBN varies from RNP 10 to the 0.1-nautical-mile

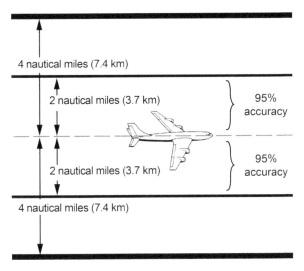

4 nautical miles (7.4 km)

2 nautical miles (3.7 km)    } 95% accuracy

2 nautical miles (3.7 km)    } 95% accuracy

4 nautical miles (7.4 km)

Figure 7.7.   Aircraft with Required Navigation Performance 2 (RNP 2).

Table 7.1.   Performance-based navigation (PBN) requirements during different phases of flight of an aircraft.

| | |
|---|---|
| Oceanic | Required Navigation Performance (RNP) 10 (RNP 10), RNP 4 |
| U.S. en route | Area Navigation 2 (RNAV 2) |
| Terminal area (TA) | RNAV 2, RNP 1 |
| Approach (instrument meteorological conditions) | RNP 0.3 |

precision and the curved paths of the RNP 0.1 Authorization Required approaches. Table 7.1 shows some of the commonly used performance and the functional requirements during the different phases of flight of an aircraft today.

Q routes are RNP 2 routes usable by RNAV-equipped aircraft from flight level 180 (FL180) to FL450. All current PBN terminal operations are RNAV and do not rely on onboard monitoring and alerting as elements of the overall safety of the operations. T routes are RNAV routes below FL180. The *en route* PBN provides more efficient routes around convective weather, increases airspace capacity using multiple Q routes, provides direct routes between city-terminal area pairs, and reduces delays.

The benefits of PBN in the TA include integrated Standard Terminal Arrival Route and RNP approaches, optimized profiles, decoupling of flows between primary and satellite airports, increased capacity in the TA, and reduction in noise and emissions. The realization of the benefits of PBN depends on how aircraft are equipped (Devlin *et al.*, 2009).

### 7.4.3   *Weather integration*

Severe weather has been identified as the source of 70% of the air traffic delays in the United States. NextGen Network Enabled Weather provides a common weather picture to all the users in the National Airspace System to enable dynamic and collaborative planning in the presence of severe weather. The National Weather Service is responsible for populating the National Oceanic and Atmospheric Administration (NOAA) NextGen Weather 4–D Cube (aka 4–D Data Cube). NextGen Network Enabled Weather interfaces with the 4–D Data Cube and provides a single authoritative source of weather information for air traffic management planning. Weather may reduce the arrival and departure capacity at airports because of wind, icing, convection, and visibility conditions. En route convective weather and turbulence may reduce the number of air-craft that can travel through a given airspace. The weather information is translated into airspace and airport constraints (Krozel et al., 2011) for use in advanced planning algorithms.

### 7.4.4   Data communication

Currently, voice is the primary communications method used to exchange critical information between the cockpit and ground-based controllers. However, voice communication is labor intensive, time consuming, and error prone, and it limits the ability of the NAS to meet future traffic demands. The availability of digital data communications improves system safety by relieving both pilots and controllers from routine tasks. Pilots and controllers are enabled to focus on strategic and critical tasks such as providing more preferred and direct routes and altitudes—saving fuel and time. In areas with dense traffic, fewer voice communications also will reduce radio-frequency congestion and spoken miscommunication.

Data communications were first introduced in operations as part of the Future Air Navigation System (FANS) program. FANS 1 and FANS A, developed by Boeing and Airbus respectively, provide the ability to autonomously send some data from the aircraft to the ATC system through Automatic Dependent Surveillance—Contract (ADS–C). The introduction of these capabilities in the oceanic airspace enabled safe separation distance between aircraft to be reduced from 100 to 50 nautical miles (185 to 93 km). In the continental United States, FANS will be modified for greater traffic density and available surveillance as FANS 1/A+, which will use the very-high-frequency (VHF) Digital Link (VDL) mode 2. A new data communication standard harmonizing the global needs of civil aviation—Aeronautical Telecommunications Network Baseline 2—is being developed by the International Civil Aviation Organization (ICAO; FAA, 2016).

### 7.4.5   Operations

The inefficiencies in aircraft trajectories have been recognized in air traffic operations (Reynolds, 2008), and this section describes efforts to improve trajectories during different phases of flight.

#### 7.4.5.1   Optimal descent trajectories

Optimal descent trajectory (ODT) tools provide ground-based trajectory advisories to air traffic controllers (Green and Vivona, 1996). They strive to enable continuous descents at near-idle thrust, conform to arrival schedule constraints for maximum throughput, avoid traffic and air-space constraints along the arrival path, allow clearance delivery by voice or data link, and lever-age existing flight deck capabilities for precision guidance and control. There are variations in the implementation of the ODT tools, and several versions, referred to as Efficient Descent Advisor (EDA; Coppenbarger et al., 2009), Continuous Descent Approach (Clarke, 2006; Ren and Clarke, 2007), and Optimized Profile Descent (OPD) (Shresta et al., 2009)—are in field evaluations. These tests have demonstrated that ODT technology can be implemented with fuel and emissions savings while avoiding conflicts with other aircraft in busy airports (Shresta et al., 2009).

Tailored arrivals to San Francisco is an operational trial to test the feasibility of issuing three-dimensional trajectory clearances over data link for automated guidance and control using an onboard flight management system. Tests of the NASA-developed technology were con-ducted by a partnership of the FAA, Boeing, and United Airlines, Inc., during 2006 and 2007. Boeing estimates that tailored arrivals can save between 400 to 800 lb (about 181 to 363 kg) of fuel per arrival. In 2012 NASA transferred the results of research to define and validate the EDA concept to the FAA for further evaluation and potential operational use as part of NextGen (Coppenbarger et al., 2009).

#### 7.4.5.2   Wind-optimal and user-preferred routes

The cruise phase uses the majority of the fuel consumed by the aircraft, and airline operations have focused on reducing the cost of fuel and crew time during cruise. The development of the ground support system has not kept up with the advances in aircraft avionics. Currently, aircraft cruise along a horizontal route following a predetermined altitude and speed profile. The horizontal route and altitude and speed profiles are selected to accommodate several factors like TA constraints, congested airspace, restricted airspace, and weather disturbances. The resulting aircraft trajectory consumes more fuel and produces more emissions than optimal four-dimensional trajectories. Several studies have estimated the inefficiencies of the current routing structure along with the benefits that could be realized with technology for wind-optimal routes (Gulding *et al.*, 2005; Howell *et al.*, 2003; Kettunen *et al.*, 2005; Reynolds, 2008). A recent study using air traffic data covering flights to and from the top 35 airports in the continental United States during 2007 estimated that the routes used by aircraft were 2.9% higher than the direct routes between these city pairs. The corresponding figure for traffic between the top 34 city pairs in Europe was 4% (Ren and Clarke, 2007). The extra distance traveled over direct routes is significantly higher over U.S.-to-Europe oceanic airspace because of the lack of radar surveillance, lack of VHF radio communication coverage, and general reliance on procedural separation. Similarly flights from Europe to Asia have large excess track distances because of large restricted airspace, strict entry points, and terrain.

### 7.4.5.3 *Surface*

Airport Surface Detection Equipment—Model X (ASDE–X) in 35 major airports will provide information about aircraft movements on the surface and share that data among air traffic controllers, traffic managers, flight operations centers, ramp operators, and airports. Data sharing enhances safety and traffic flow on runways, taxiways, and some ramps. It improves collaborative decisionmaking. The FAA also will provide surface data sharing at another nine busy and complex airports using Airport Surface Surveillance Capability (ASSC). Whereas ASDE–X tracks surface movement using radar, multilateration and ADS–B, ASSC collects data from multilateration and ADS–B only. Between 2014 and 2017, ASSC will begin to track transponder-equipped aircraft and ADS–B-equipped ground vehicles on the surface and aircraft flying within 5 nautical miles (about 9.25 km) of airports at Portland, Oregon; Anchorage, Alaska; Kansas City, Missouri; New Orleans, Louisiana; Pittsburgh, Pennsylvania; San Francisco, California; Cincinnati, Ohio; Cleveland, Ohio, and Andrews Air Force Base, Maryland.

### 7.4.6 *Integrated technologies*

NASA and the FAA are collaborating to develop and demonstrate an integrated set of NextGen technologies that provide an efficient arrival solution for managing aircraft beginning from just prior to the top of descent and continuing to the runway. These technologies are ADS–B, RNAV arrival routes, OPD procedures, terminal metering, Flight Deck Interval Management, and Controller-Managed Spacing tools (Thipphavong *et al.*, 2013). Thus far, these ATM technologies have been tested separately, and each has demonstrated throughput, delay, and/or fuel-efficiency benefits. The integration of these terminal arrival tools will allow arrival aircraft to safely fly closer together on more fuel-efficient routes to increase capacity and reduce delay, and minimizing fuel burn, noise, and greenhouse gas emissions.

### 7.4.7 *Global harmonization*

The FAA works with other international air navigation service providers (ANSPs) to ensure that the technology developments in the United States are synchronized with the developments in other countries. The ICAO organizes this activity. Interoperability of technology developments in the United States is improved by collaboration with Single European Sky ATM Research (SESAR).

The Asia and Pacific Initiative to Reduce Emissions (ASPIRE, 2011) is a joint collaboration between the FAA and ANSPs in Australia, New Zealand, Singapore, Japan, and Thailand. ASPIRE conducted a series of flights from 2008 through 2011 to successfully demonstrate the potential for fuel and emissions savings in the region. These flights made several changes to gate-to-gate operations, including reduced separation, more efficient flight profiles, and tailored arrivals. The best practices from these flights will be made available daily to all equipped aircraft on some city pairs between the United States and the Asia Pacific region.

During 2011, the FAA, the European Commission, several European ANSPs, and 40 European airlines participated in an effort to demonstrate NextGen and SESAR capabilities on trans-Atlantic flights. For the Atlantic Interoperability Initiative to Reduce Emissions (McDaniel, 2007; Sprong et al., 2008), Air France flew several flights between John F. Kennedy International Airport (JFK, New York City) and Paris Charles de Gaulle Airport (CDG) as well as between Paris Orly Airport (ORY) and Pointe-à-Pitre International Airport (PTP, Guadeloupe) during 2010 and 2011. The flights used procedures designed to reduce environmental impact with no special equipage. Improvements to taxiing, real-time lateral and vertical optimization, fuel-saving RNAV approaches, and other processes produced savings ranging from 200 to 300 gal (about 750 to 1,100 L) of fuel per flight.

## 7.5   CONCLUSIONS

This chapter provided an overview of the importance of aviation to the economy of the United States; a brief review of current aviation operations, research, and technology; and infrastructure upgrades being deployed to enable the current aviation system to meet the needs of future aviation systems. Improvements are being made in all phases of flight to achieve more efficiency, meet demand, and reduce emissions while maintaining safety. Air traffic operations need to be harmonized across all parts of the globe to achieve standardization and efficiency of operations. With investment in new aircraft and with higher oil prices and labor costs squeezing airline profits, greater automation and decentralization will be needed to sustain aviation systems. A fundamental challenge will be to develop the functional relationship between pilots, air traffic controllers, and other human decisionmakers in the presence of greater automation. The benefits of new technologies have to be demonstrated to speed up adaptation by industry.

## REFERENCES

ASPIRE. 2011. Asia and Pacific Initiative To Reduce Emissions. *http://www.aspire-green.com/default.asp* (accessed November 3, 2014).

Balakrishnan, H., and Y. Jung. 2007. A framework for coordinated surface operations planning at Dallas-Fort Worth International Airport. AIAA 2007–6553.

Ball, M. O., R. L. Hoffman, C.-Y. Chen, *et al.* Collaborative decision making in air traffic management: current and future research directions. In *New Concepts and Methods in Air Traffic Management*, ed. L. Bianco, P. Dell'Olmo, and A. R. Odoni, 17–30. Heidelberg: Springer Berlin Heidelberg.

Borener, S., G. Carr, D. Ballard, *et al.* 2006. Can NGATS meet the demands of the future? *Journal of Air Traffic Control* 48:34–38.

Boucher, O. 2011. Atmospheric science: Seeing through contrails. *Nat. Clim. Change* 1:24–25.

Brasseur, G. P., and M. Gupta. 2010. Impact of aviation on climate: research priorities. *Amer. Meteor. Soc.* 91:461–463.

Clarke, J.-P., Bennett, D., Elemer, K., *et al.* 2006. Development, design, and flight test evaluation of a continuous descent approach procedure for nighttime operation at Louisville International Airport. PARTNER Continuous Descent Approach Development Team Report No. PARTNER–COE–2005–002.

Coppenbarger, R. A., R. W. Mead, *et al.* 2009. Field evaluation of the tailored arrivals concept for datalink-enabled continuous descent approach. *J. Aircraft* 46:1200–1209.

Devlin, C. J., A. A. Herndon, S. F. McCourt, *et al.* 2009. Performance-based navigation fleet equipage evolution. Paper presented at the 28th Digital Avionics Systems Conference, Orlando, FL.

Doble, N. A., J. Timmerman, T. Carniol, *et al.* 2009. Linking traffic management to the airport surface: departure flow management and beyond. Paper presented at the 8th USA/Europe Air Traffic Management R&D Seminar, Napa, CA.

Duda, D. P., P. Minnis, P. K. Costulis, *et al.* 2003. CONUS contrail frequency estimated from RUC and flight track data. Presented at the European Conference on Aviation, Atmosphere, and Climate. Friedrichschafen at Lake Constance, Germany.

Erzberger, H. 2004. Transforming the NAS: the next generation air traffic control system. Proceedings of the 24th International Congress of the Aeronautical Sciences (NASA/TP—2004-212828), Yokohama, Japan.

Erzberger, H. 2006. Automated conflict resolution for air traffic control. Proceedings of the 25th International Congress of the Aeronautical Sciences (ICAS), Hamburg, Germany.

Federal Aviation Administration. 2008. Air Traffic System Command Center, National severe weather playbook. Current version at *http://www.fly.faa.gov/PLAYBOOK/pbindex.html* (accessed February 14, 2014).

Federal Aviation Administration. 2011. FAA aerospace forecast, fiscal years 2011–2031. *http://www.faa.gov/about/office_org/headquarters_offices/apl/aviation_forecasts/aerospace_forecasts/2011-2031/media/2011%20Forecast%20Doc.pdf* (accessed November 18, 2014).

Federal Aviation Administration. 2012. NextGEN implementation plan. *http://www.faa.gov/nextgen/library/media/nextgen_implementation_plan_2012.pdf* (accessed November 18, 2014).

Federal Aviation Administration. 2013. RTCA SC–214/EUROCAE WG–78 standards for air traffic data communication services. *http://www.faa.gov/about/office_org/headquarters_offices/ato/service_units/techops/atc_comms_services/sc214/* (accessed August 8, 2013).

Federal Aviation Administration. 2014. The economic impact of civil aviation on the U.S. economy. *http://www.faa.gov/air_traffic/publications/media/2014-economic-impact-report.pdf* (accessed November 18, 2014).

Federal Aviation Administration. 2016. Automatic Dependent Surveillance-Broadcast (ADS–B). *https://www.faa.gov/nextgen/programs/adsb/* (accessed January 30, 2017).

Green, S., and R. Vivona. 1996. Field evaluation of descent advisor trajectory prediction accuracy. AIAA–2001–4114.

Gulding, J., D. Knorr, M. Rose, *et al.* 2005. US/Europe comparison of ATM-related operational performance. *Air Traffic Control Quarterly* 18:5–27.

Gulding, J., D. Knorr, M. Rose, *et al.* 2009. US/Europe comparison of ATM-related operational performance. Paper presented at the 8th USA/Europe Air Traffic Management Research and Development Seminars, Napa, CA.

Howell, D., M. Bennett, J. Bonn, *et al.* 2003. Estimating the en route efficiency benefits pool. Paper presented at the 5th USA/Europe Air Traffic Management Seminar, Budapest, Hungary.

Intergovernmental Panel on Climate Change (IPCC). 1999. Aviation and the global atmosphere, ed. J. E. Penner, D. H. Lister, D. J. Griggs, *et al. http://www.ipcc.ch/ipccreports/sres/aviation/index.php?idp=3* (accessed November 19, 2014).

Joint Planning and Development Office (JPDO). 2011. Concept of operations for the next generation air transportation system. Version 3.2, Washington, DC. *http://www.dtic.mil/dtic/tr/fulltext/u2/a535795.pdf* (accessed November 18, 2014).

Kettunen, T., J.-C. Hustache, I. Fuller, *et al.* 2005. Flight efficiency studies in Europe and the United States. Paper presented at the 6th USA/Europe ATM 2005 R&D Seminar, Baltimore, MD.

Krozel, J., D. Rosman, and S. R. Grabbe. 2002. Analysis of en route sector demand error sources. AIAA–5016.

Krozel, J., R. Kicinger, and M. Andrews. 2011. Classification of weather translation models for NextGen. Paper presented at the 15th Conference on Aviation, Range & Aerospace Meteorology, Los Angeles, CA.

Lee, D. S., G. Pitari, V. Grewe, *et al.* 2010. Transport impacts on atmosphere and climate: aviation. *Atmos. Environ.* 44:4678–4734.

Malik, W., G. Gupta, and Y. C. Jung. 2010. Managing departure aircraft release for efficient airport surface operations. AIAA 2010–7696.

McDaniel, J. 2007. *Atlantic interoperability initiative to reduce emissions (AIRE) industry kick-off meeting.* Washington, DC: Federal Aviation Administration.

Nolan, M. S. 2003. *Fundamentals of Air Traffic Control.* 4th ed., Belmont, CA: Thomson Brooks/Cole.

Rathinam, S., J. Montoya, and Y. Jung. 2008. An optimization model for reducing taxi times at the Dallas-Fort Worth International Airport. International Congress of the Aeronautical Sciences paper ICAS 2008–8.6.1.

Ren, L., and J.-P. Clarke. 2007. Flight demonstration of the separation analysis methodology for continuous descent arrival. Draft paper for the 7th USA/Europe ATM 2007 R&D Seminar, Barcelona, Spain.

Reynolds, T. G. 2008. Analysis of lateral flight inefficiency in global air traffic management. AIAA 2008–8865.

RTCA, Inc. 2009. CTF–5 RTCA Task Force 5 NextGen mid-term implementation report. Document no. CTF–5.

Shresta, S., D. Neskovic, and S. S. Williams. 2009. Analysis of continuous descent benefits and impacts during daytime operations. Paper presented at the 8th USA/Europe Air Traffic Management Research and Development Seminars, Napa, CA.

Sprong, K. R., K. A. Klein, C. Shiotsuki, *et al.* 2008. Analysis of Atlantic Interoperability Initiative to reduce emissions continuous descent arrival operations at Atlanta and Miami. In *Digital Avionics Systems Conference, 2008. DASC 2008. IEEE/AIAA 27th.* New York: IEEE.

Sridhar, B., S. R. Grabbe, and A. Mukherjee. 2008. Modeling and optimization in traffic flow management: New approaches to achieving, assessing, and optimizing safe and efficient management of our ever-growing civil aircraft traffic aim to improve traffic flow and reduce costs. *Proc. IEEE* 96:1–29. *https://www.aviationsystems.arc.nasa.gov/publications/2008/AF2008077.pdf* (accessed February 8, 2017).

Sridhar, B., D. Kulkarni, and K. Sheth. 2011. Impact of uncertainty on the prediction of airspace complexity of congested sectors. *ATC Quarterly* 19.

Sridhar, B., N. Y. Chen, and H. K. Ng. 2013. Energy efficient contrail mitigation strategies for reducing the environmental impact of aviation. In *Tenth USA/Europe Air Traffic Management Research and Development Seminar (ATM2013).* Chicago, IL: Air Traffic Management.

Thipphavong, J., J. Jung, H. N. Swenson, *et al.* 2013. Evaluation of the controller-managed spacing tools, flight-deck interval management and terminal area metering capabilities for the ATM technology demonstration #1. Paper presented at the *Tenth USA/Europe Air Traffic Management Research and Development Seminar (ATM2013),* Chicago, IL.

Wambsganss, M. 1997. Collaborative decision making through dynamic information transfer. *Air Traffic Control Quarterly* 4:107–123.

Williams, V., R. B. Noland, and R. Toumi. 2002. Reducing the climate change impacts of aviation by restricting cruise altitudes. *Transportation Research Part D Transport and Environment* 7:451–464.

# CHAPTER 8

## Alternative fuels and green aviation[1]

Emily S. Nelson

### 8.1 INTRODUCTION

For engineers and scientists who are new to the field of biofuel production for aviation, it can be bewildering to come to terms with the conflicting conclusions drawn in the literature or evaluate the claims found in a casual Web search. It is a multidisciplinary field that spans chemical, petroleum, environmental, mechanical, aerospace, materials, and industrial engineering. The field is also overlaid by a web of geographically varying public policies that profoundly impact economic feasibility and the pace and directions in technology development. It is a field that has grown rapidly in all of these areas, so that there is a vast, rich literature base to peruse. Finally, there are passionate and articulate advocates for every conceivable viewpoint. This can make the leap across the interdisciplinary aspects rather challenging. Some of the most embedded sources of confusion arise from the following:

- Terminology may be conflicting or otherwise difficult to understand. For example, the naming of the same set of hydrocarbon compounds in petroleum engineering differs from the conventions in organic chemistry.
- The standard units in which values are reported vary widely. Often this is because conventions had already been established in various disciplines and geographical regions.
- There may be no clear standards or protocols for measurements, which leads to difficulty in comparing values; for example, the manner in which to quantify biomass yield from algae.
- Some studies do not clearly state their assumptions or identify their uncertainties, which are critical to evaluating its conclusions.
- In order to create assessments for global applications, the numbers used to describe the physics may be derived from laboratory results; thus, the uncertainties/error present in the small-scale data (e.g., producing 500 mL of crude algal oil over a few days in a laboratory) become grossly magnified when the process is scaled up (e.g., producing millions of barrels per hectare per year (MMbbl/ha·y) of biodiesel).

With regard to scaleup, a number of factors come into play: To be useful, the lab-scale experiment must meticulously keep track of all nutrients and additives, gas exchange, and energy input. Sometimes the processes or equipment that are useful for lab experiments for mixing, separation, or combustion do not translate up to industrial-scale facilities. For example, centrifuging to remove excess water from algal broth is both simple and effective for small quantities of fuel, but it is too expensive for commercial production. Finally, the lab has the luxury of a controlled environment and may run its algal growth tests over a few days, whereas commercial producers and refiners will be subject to the vagaries of seasonal and diurnal temperature swings, cloudy days, too much or too little rainfall, and the intermittent interest of predators or competitors.

The next generation of engineers, scientists, and policymakers must find a way to bridge these incompatibilities in order to provide the most useful, comprehensive, evidence-based data on current technologies and new strategies to make those technologies commercially viable. The

---

[1]This chapter originally appeared as chapter 11 in Dahlquist, E., ed., 2013. Biomass utilization—combustion, gasification, torrification and fermentation. Boca Raton, FL: CRC Press.

purpose of this chapter is to present basic concepts in the key areas of alternate-fuel production for green aviation and provide a foundation for critical thinking. Where available, uncertainties in the data are given. Such uncertainties can arise from measurement errors or from incomplete access to all the relevant data or processes, or they can be because of real statistical variation (e.g., variation in fuel sulfur content depending on whether the fuel was extracted from shale or more conventional petroleum reservoirs). Either way, it is useful to develop an understanding of how much variation can be expected in any given factor.

Over the last several decades, air traffic has increased by about 5% annually. As shown in Figure 8.1, its momentum was slowed only briefly through 2006 by major events such as the attack on the World Trade Center in 2001, wars, the SARS (severe acute respiratory syndrome) epidemic, and gloomy economic conditions. Since the global economic crisis in 2008 and 2009, the airline industry has recovered to prerecession levels and is expected to grow at or above historical levels (Boeing, 2011). However, even when accounting for technology advancements (dashed line) and logistics optimization (dotted line), jet fuel demand is expected to increase for the foreseeable future. In other studies that attempt to predict energy demand for the aviation industry during the coming decades, the details may vary somewhat, but the fundamental conclusion is that the demand for jet fuel will continue to increase. Better air traffic management, such as separation distance control and speed optimization, could provide a potential fuel consumption savings up to 15% (Blakey *et al.*, 2011). With the right infrastructure on the ground, continuous descent approaches could improve fuel efficiency over the current stepped approaches. Although there is some ongoing work in alternative modes of propulsion, such as fuel cells (Novillo *et al.*, 2011; Renouard-Vallet, *et al.*, 2012), this technology will not be commercially viable for decades, at best. The major efforts at reducing fuel consumption in engine and airframe design include increased efficiency through new turbofan designs (e.g., high-pressure-ratio cores and super-high-bypass-ratio fans), reduced drag airframes, and advanced materials in ceramic composites. The recently unveiled Boeing 787 delivers a 20% fuel reduction relative to similarly sized planes through the use of advanced materials, new engine design, and improved fuselage integration. Aircraft lifetimes are typically in the range of 20 to 30 years or more (Moavenzadeh *et al.*, 2011), so that it will be some time before there is a complete turnover to more efficient aircraft designs. To work with the fleet

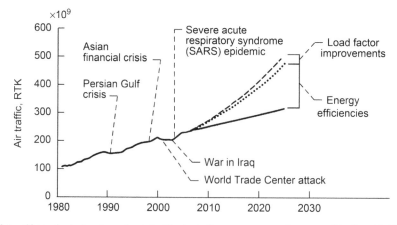

Figure 8.1    Chronological patterns in air traffic, including passenger and cargo traffic, from 1960 to 2006, overlaid with relevant events. Air traffic is shown in terms of revenue tons per kilometer. Predicted growth in jet fuel demand is shown from 2006 through 2025 (solid line), which includes expected efficiency gains from technology development (dotted line) and logistics improvements (dashed line). (Adapted from Chèze, B., P. Gastineau, and J. Chevallier. 2011. Forecasting world and regional aviation jet fuel demands to the mid-term (2025). *Energy Policy* 39:5147–5158. *http://www.sciencedirect.com/science/article/pii/S0301421511004496* (accessed April 17, 2017). With permission.)

as it exists, some aircraft can be retrofit with winglets to improve aerodynamics, but this is only an aid—not a fix—to the problem. In addition, airport supply chains and infrastructure are currently set up for petroleum-based fuels. So unlike the automobile industry, which is well suited to take advantage of all-electric and hybrid engines to reduce overall fuel consumption, the aviation sector has no practical alternative to the internal combustion engine in the coming decades. Thus, despite technological and logistical improvements, jet fuel demand will continue to grow.

At the same time, the availability of suitable energy sources to meet this demand will depend on strategic and innovative actions. To date, economic growth in developed and developing nations has depended upon the availability of cheap oil. According to the International Energy Agency (IEA), the global peak of conventional oil reserves was reached in 2006 (International Energy Agency, 2010). Historical data indicate that production in oil reservoirs have declined sharply after the peak is reached (Hirsch, 2005), which is also predicted in the IEA report. The sharp decline in output can be at least somewhat mitigated through development of new oil reservoirs, advanced techniques for recovering oil from unconventional sources such as oil shale, and increased reliance on coal, natural gas, and alternative fuels. However, a large environmental impact could result without appropriate clean coal technologies, safe recovery methods particularly for natural gas, and sustainable practices in alternative fuel manufacture. There is a real sense of urgency, since there are consequences associated with waiting too long to put mitigation strategies into place (Vaughan *et al.*, 2009). The rapidly fluctuating price of oil puts pressure on the global economy that a stable source of energy could help to alleviate. One thing that all sides should agree on is that it is crucial to make informed public policy decisions that are based on high-quality, up-to-date science.

We are feverishly seeking a way to satisfy our thirst for energy in a way that respects our planet and the life that depends on it. This will require the creative development of new energy sources, and biofuels are widely considered to be the best, if not the only, solution. Aviation is responsible for ~10% of global transport energy consumption (Moavenzadeh *et al.*, 2011). Air transport dumps hundreds of millions of metric tons of greenhouse gases (GHGs) into the atmosphere annually, currently accounting for ~3.5% of all GHG emissions. Legitimate concern over this contributor to climate change has led to a deluge of environmental and economic analyses that attempt to provide guidance to a rational way to move forward on the global scale. The life cycle assessment (LCA) has emerged as a framework that is used to quantitatively evaluate local and worldwide biofuel approaches in terms of their environmental and/or economic impacts (see Sec. 8.6).

Until the biorefining industry becomes more established or another serious global oil crisis is at hand, it is difficult to make a purely commercial case for biofuel manufacture. Public policy can make all the difference in providing incentives. Despite a rocky beginning in 2005, the European Union (EU) has been the leader in instituting carbon trading mechanisms in response to the Kyoto Protocol (Gourlay *et al.*, 2011). To account for air traffic's contribution to climate change, the European Union will include air travel in its comprehensive carbon trading system in January 2012 (Convery, 2009).[2] All domestic and international flights that arrive or depart from the EU will be covered by the EU Emissions Trading System. Airlines will receive carbon credits through the use of biofuels; if their net output exceeds the specified targets, the airline must offset the overproduction through carbon trading.

Other countries have also developed goals for reducing dependence on petroleum-based fuels through the use of alternative fuels. China has set targets of reducing energy consumption by 16% and $CO_2$ emissions per unit of gross domestic product by 17% in 2015 from its baseline in 2010 (Li *et al.*, 2011).[3] The U.S. Energy Independence and Security Act of 2007 specifies a ramping up of biofuel production through 2012, which includes 36 billion gal ($136 \times 10^9$ L) of

---

[2]This change did occur in 2012. For more details, see *https://ec.europa.eu/clima/policies/transport/ aviation_en*.

[3]For more current information on global emissions status for all of the emitters discussed here, see *https:// ec.europa.eu/clima/policies/transport/aviation_en*.

renewable fuels by 2022, which must include 21 billion gal ($79.5 \times 10^9$ L) of next-generation biofuels that are not derived from corn ethanol. In addition, life cycle GHG emissions must be reduced by at least 50% relative to the 2005 output of petroleum-based transportation fuels (U.S. Department of Energy, 2010). The U.S. Air Force has targeted that half of its aircraft will use blends of conventional and alternative fuels by 2016 (Byron, 2011), and the U.S. Navy plans to build the "Great Green Fleet" of carrier ships by 2016 powered entirely by non-fossil fuels (Karpovitch, 2011).[4] The commitment to purchase large quantities of biofuel will help bridge the gap between a product that is viable in the laboratory or small-scale pilot plant and the establishment of large-scale biofuel production facilities that will significantly bring down the cost of manufacture.

Although biofuels are an attractive solution, whether developed from feedstocks such as vegetable oils, agricultural waste, or the much-anticipated algae, there are many issues that hinder wide availability. Most studies agree that the cost of the feedstock is by far the largest contributor to the final biodiesel cost, although the numbers vary from about 60% to 95% (Balat, 2011a; Razon, 2009). Embedded within that issue are considerations of the economics and logistics of manufacturing, economies of scale (Knothe, 2010a), sustainable land development, lack of refining infrastructure, and market considerations, particularly with regard to the fluctuating price of oil.

In Section 8.2, the international standards for jet fuel are outlined, along with a description of the most important physical characteristics for aviation fuel. Fuel composition and its corresponding effects on fuel properties are the subject of Section 8.3. The following sections discuss alternative fuel feedstocks (Sec. 8.4), biorefining techniques (Sec. 8.5), and an introduction to life cycle analysis for aviation fuel (Sec. 8.6). For those without a solid background in chemistry and petroleum engineering, see the appendix for a review of definitions and basic hydrocarbon chemistry.

## 8.2   AVIATION FUEL REQUIREMENTS

### 8.2.1   *Jet fuel specifications*

Aviation fuel must live up to carefully developed standards, which have evolved over time (Edwards, 2007), because the fuel must operate effectively even in extreme conditions. For example, if the temperature and pressure on the ground are 15°C and 101.3 kPa, respectively, then at a cruise altitude of 11,000 m, the external temperature is about –56.5°C and the pressure is ~22.6 kPa.

All aviation fuels are blends of various hydrocarbons. There are two basic types of jet fuel, which differ by the proportions of hydrocarbons present in the fuel. The carbon number is a measure used to indicate the number of carbon atoms present in a hydrocarbon molecule. For example, methane ($CH_4$) is assigned a carbon number of C1, and octane ($C_8H_{18}$) is C8. The most common type of jet fuel is a kerosene blend with carbon numbers from C8 to C12, and the less typical naphtha-kerosene blend is a "wide-cut" fuel with a broader range of about C5 to C12. By including the lighter hydrocarbons, the fuel's vapor pressure is reduced, and it has better cold-temperature properties. For civilian aircraft, the most common aviation fuels for powering jet and turboprop engines are:

- Jet A: a kerosene-grade fuel used throughout the United States, designed to operate under the demanding conditions of flight. Its freeze temperature must be ≤–40°C.
- Jet A-1: a kerosene-grade fuel widely available outside the United States. It has a lower freeze point of ≤–47°C, and there are other minor differences relative to Jet A.
- Jet B: a naphtha-kerosene blend, used primarily in cold climates such as northern Canada. Although it operates more effectively at lower temperatures, it is also more volatile, so it

---

[4]Although most aircraft of the U.S. Air Force are now certified for alternative fuel use, the high price of alternative fuel rendered this target unreachable. Similar budgetary constraints and opposition by the U.S. Congress halted deployment of the "Great Green Fleet." See *http://www.defensemedianetwork.com/stories/ military-alternative-fuels-research/*

exhibits greater evaporation loss at high altitude. In addition, it is a greater fire hazard on the ground, and it makes a plane crash less survivable.

The most common military fuels are

- Jet propellant 4 (JP-4): the military equivalent of Jet B with the addition of corrosion inhibitors and anti-icing additives. It used to be the primary fuel of the U.S. Air Force, but it was phased out in the 1990s because of safety concerns. Although still in use by other air forces around the world, it is in limited production.
- JP-5: a high flash-point, wide-cut kerosene fuel used by the U.S. Navy, primarily for aircraft carriers
- JP-8: the military equivalent of Jet A-1 with the addition of corrosion inhibitor and anti-icing additives

Other common additives include antioxidants to prevent gumming; antistatic agents to dissipate static electricity; metal deactivators to remediate the effects of trace materials in the fuel that affect thermal stability; and biocides to reduce the likelihood of microbial growth within the system, which could plug filters and produce corrosive metabolites (Raikos *et al.*, 2011). Many countries and organizations provide somewhat different sets of specifications for jet fuels. Some of the most broadly based provisions are given in Table 8.1.

In the governing standards identified in Table 8.1, there are many additional specifications that are not shown here. Some of them set minimum or maximum values for various parameters; others simply require reporting. For example, the standards for Jet A-1 defined in the UK DEF STAN 91-91 (2016) require that fuel manufacturers report the percentage of fuel by volume (%v/v) that has been hydrotreated. This has implications for synthetic fuels,

Table 8.1.   Key specifications for aviation fuels.

| Property | Jet A-1[a] | Jet A[b] | Jet B[c] | JP-4[d] | JP-5[d] | JP-8[e] |
|---|---|---|---|---|---|---|
| Density at 15°C, kg/m³ | 775–840 | 775–840 | 750–801 | 751–802 | 788–845 | 775–840 |
| Viscosity at −20°C, mm²/s | ≤8 | ≤8.0 | – | – | ≤8.5 | ≤8.0 |
| Flash point, °C | ≥38 | ≥38 | – | – | ≥60 | ≥38 |
| Freeze temperature, °C | ≤−47 | ≤−40 | ≤−51 | ≤−58 | ≤−46 | ≤−47 |
| Distillation end point, °C | ≤300 | ≤300 | – | ≤270 | ≤300 | ≤300 |
| Vapor pressure, kPa | – | – | <21 | 14–21[f] | – | – |
| Specific energy, MJ/kg | ≥42.8 | ≥42.8 | ≥42.8 | ≥42.8 | ≥42.6 | ≥42.8 |
| Lubricity: wear scar diameter, mm | <0.85 | – | – | – | – | – |
| Total acidity, mg KOH/g | ≤0.015 | ≤0.010 | ≤0.010 | ≤0.015 | ≤0.015 | ≤0.015 |
| Aromatics, %v/v | ≤25 | ≤25 | ≤25 | ≤25.0 | ≤25.0 | ≤25.0 |
| Sulfur, %m/m | ≤0.30 | ≤0.30 | ≤0.40 | ≤0.4 | ≤0.30 | ≤0.30 |
| Hydrogen, %m/m | – | – | – | ≤13.5 | ≤13.4 | ≤13.4 |

[a]Values from specification DEF STAN 91-91. 2016. Turbine Fuel, Aviation Kerosine Type, Jet A-1. Ministry of Defence, Westminster, London. With permission.

[b]Values from specification ASTM D 1655. 2016. Standard Specification for Aviation Turbine Fuels. ASTM International, West Conshohocken, PA. With permission.

[c]Values from specification CGSB-3-22. 2012. Wide-Cut Type Aviation Turbine Fuel (Grade Jet B). Canadian General Standards Board, Gatineau, Canada. Reproduced with the permission from the Minister of Public Works and Government Services Canada, authorized in March 2017.

[d]Values from specification MIL-DTL-5624U. 2004. Detail Specification: Turbine Fuel, Aviation, Grades JP–4 and JP–5. U.S. Department of Defense, Washington, DC. Work of the U.S. Government.

[e]Specification: MIL-DTL-83133E. 2004. Detail Specification: Turbine Fuel, Aviation, Kerosene Type, JP–8. U.S. Department of Defense, Washington, DC (work of the U.S. Government) or DEF STAN 91-87. 2002. Turbine Fuel, Aviation Kerosine Type, Containing Fuel System Icing Inhibitor. Ministry of Defence, Westminster, London (with permission).

[f]At 37.8°C.

since hydrotreatment is the method of choice used to process biologically derived oils (or "green crude") into aviation-grade fuel (see Sec. 8.5). Each specification is associated with one or more standard measurement protocols, since the method of measurement can have an impact on the detected value. For a more detailed outline of specifications for other fuel types and the standards of other countries, see ExxonMobil Aviation (2005). For an excellent review of aviation fuel and testing methods, see Chevron (2006).

Between 1993 and 2011, kerosene-type jet fuel comprised 9 to 11% of the crude oil content in U.S. refineries (U.S. Energy Information Administration, 2011). The presence of aromatics, sulfur, and other trace components are highly correlated to the geographical source of the extracted crude. In Figures 8.2 and 8.3, representative properties are shown, as identified from 56 samples of aviation fuel obtained from around the globe in the World Fuel Sampling Program (WFSP) (Hadaller and Johnson, 2006). Not all of these samples were known to have been used as jet fuel, and one sample failed its thermal stability testing. Furthermore, some of the compiled regional results (Fig. 8.2(a)) or fuel type results (Fig. 8.2(b)) were predicated on one or two samples and should not be considered statistically significant. The samples from South Africa

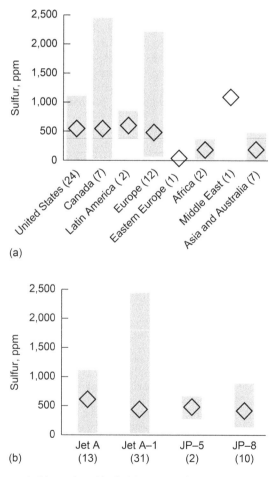

Figure 8.2.   Sulfur content in 56 samples of jet fuel from around the world. The number of samples in each category is shown in parentheses. Average values are marked with a diamond on a gray window representing minimum and maximum sample values. (a) Sulfur content by geographical region. (b) Sulfur content by jet fuel type. (Adapted from Hadaller, O. J., and J. M. Johnson. 2006. World fuel sampling program. Coordinating Research Council (CRC) Report No. 647.)

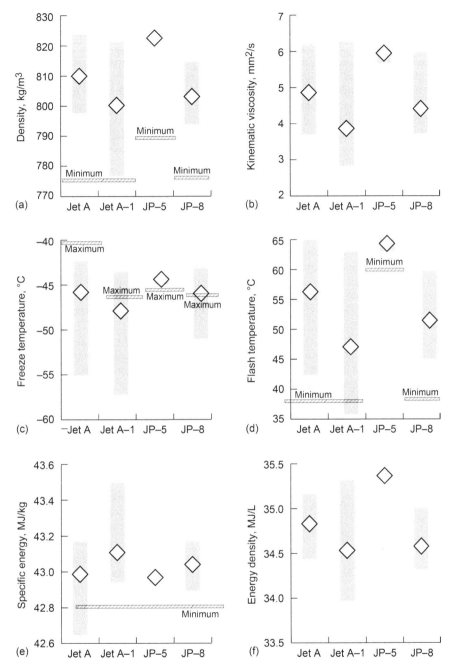

Figure 8.3.    Representative properties of aviation fuels Jet A, Jet A-1, JP-5, and JP-8 as found by the World Fuel Sampling Program. Maximum and minimum values per the specifications of Table 8.1 are identified by hashed lines. (a) Density. (b) Kinematic viscosity. (c) Freeze temperature. (d) Flash temperature (e) Specific energy. (f) Specific energy density. (Adapted from Hadaller, O. J., and J. M. Johnson. 2006. World fuel sampling program. Coordinating Research Council (CRC) Report No. 647.)

were either partially or completely composed of synthetic Fischer-Tropsch (FT) fuel. Three of the samples from Asia and Australia were extracted from oil shale. In Figure 8.3, the average values are marked by a diamond on a gray window representing the envelope of minimum and maximum values. Although a few samples of Jet A-1 exhibited relatively high values of sulfur, most of the samples had a sulfur content of about 500 ppm or less, which is significantly below the specified limit of 3,000 ppm (Chevron, 2006). Sulfur can be corrosive to the engine, and it is of concern in emissions, but extremely low levels of sulfur (below ~100 ppm) have also been correlated with increased engine wear.

The WFSP study presented results for a range of other fuel parameters as well, some of which are reproduced in Figure 8.3, and which will be discussed in the following paragraphs. The units for the parameters are identified by the symbols for mass ($M$), length ($L$), time ($T$), temperature ($K$), and amperes ($A$).

1. *Density*: Measured in $M/L$ and presented here in kilograms per cubic meter (kg/m$^3$), fuel density (Figs. 8.3(a) and 8.4(a)) represents the mass per unit volume of fuel. Density is a key parameter because increased fuel weight means that more energy must be supplied to move the loaded aircraft, but it is also correlated with other performance parameters discussed below, such as specific energy (heat of combustion).

Fuel injectors dispense their output of fuel by its volume, not its density. Consequently, when the fuel is injected into the combustion chamber, the density of the fuel will govern the fuel/air ratio. Thus, density is directly related to the thrust through the injected fuel volume and fuel reaction properties (Fazal *et al.*, 2010).

Since the density of liquid fuel decreases linearly with temperature (Fig. 8.4(a)), the standards in Table 8.1 require that the measurement be taken at a standardized reference temperature of 15°C. All of the fuel sampled in the WFSP met the minimum requirements for aviation fuel; none of the samples exceeded the maximum values of 845 kg/m$^2$ (JP-5) or 840 kg/m$^2$ (Jet A, Jet A-1 and JP-8).

2. *Kinematic viscosity at –20°C*: Measured in units of $L^2/t$ and presented here in square millimeters per second (mm$^2$/s), viscosity (Figs. 8.3(b) and 8.4(b)) is a measure of fluid resistance to shear stress. Consequently, lower viscosity fluids deform upon application of shear forces more readily, and they pour more easily. In aircraft operation, this property is important in cold

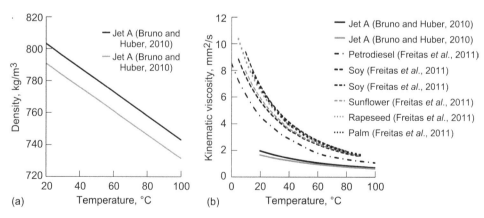

Figure 8.4.   Experimentally measured density and viscosity of jet fuel, petroleum diesel, and biodiesel. (a) Density of two samples of Jet A. (b) Viscosity of two samples of Jet A, petrodiesel and several biodiesels derived from various vegetable oils. (Data for jet fuel from Bruno, T. J., and M. L. Huber. 2010. Evaluation of the physicochemical authenticity of aviation kerosene surrogate mixtures. Part 2: Analysis and prediction of thermophysical properties. *Energy Fuels* 24:4277–4284. Data for petrodiesel and biodiesel from Freitas, S. V. D., M. J. Pratas, R. Ceriani, *et al.*, 2011. Evaluation of predictive models for the viscosity of biodiesel. *Energy Fuels* 25:352–358.)

starts, reignition at altitude, and the quality of lubrication and combustion. Higher viscosity causes larger pressure drops across the fuel lines, requiring the pumps to work harder to maintain a given flow rate. In addition, higher viscosity fuels have an impact on combustion quality: Liquid fuel enters the combustion chamber as an atomized spray. Higher viscosity fuels tend to cause larger droplets, and the spray pattern does not penetrate as deeply into the chamber. Incomplete combustion can result, with accompanying increases in exhaust smoke and emissions. Alternatively, low-viscosity fuels may not provide enough lubrication for moving engine components to work properly.

Viscosity decreases with increasing temperature, but the form of the functionality is not as simple as for density. Many mathematical descriptions for viscosity as a function of temperature have been proposed, and the form of that functionality differs among the models. To complicate matters further, there are two commonly used definitions for viscosity: kinematic viscosity $\nu$ measured in units of $L^2/t$ and dynamic (or absolute) viscosity (usually denoted as $\mu$ or $\eta$, depending on the discipline) measured in $ML/t$. The viscosities can be converted back and forth through the density $\rho$; that is, $\nu = \mu/\rho$. Strictly speaking, the dynamic viscosity is the factor that relates fluid response to shear stress, but the kinematic viscosity comes in handy since this compound parameter can simplify the form of the equations of fluid motion in many cases. From a physics standpoint, the dynamic viscosity of a typical Newtonian liquid should obey an Arrhenius-type expression in temperature $T$ (i.e., an exponential: $\mu \propto e^{-A/T}$). The constant $A$ is specific to the type of liquid under consideration. Since density is inversely proportional to temperature, $\mu \propto 1/T$, it follows that $\mu \propto T \cdot e^{-A/T}$. However, over small temperature ranges even a linear approximation may be sufficient. Near the freezing point, the viscosity behavior becomes more sensitive to temperature decreases, and the viscosity increases rapidly (Kerschbaum and Rinke, 2004). Freitas and coworkers evaluate several viscosity models against experimental measurements of diesel and biodiesel, shown in Figure 8.4(b), that vary substantially over a range of temperature and the fuel's hydrocarbon content (Freitas *et al.*, 2011).

In Figure 8.4(b), experimental viscosity data on two samples of Jet A (the two traces at the bottom) show that the temperature dependence is not linear between 20 and 100°C, although it is not grossly nonlinear over small temperature variations. However, aircraft operations occur over a broad range of temperatures, so the nonlinearity of viscosity is an issue, particularly at low temperatures. Note that in Figure 8.4(b) the petrodiesel (third from bottom) and biodiesels (clustered above the petrodiesel curve) all exhibit increasing sensitivity to lower temperatures. In this case, the petrodiesel was a commercial product suitable for automotive use, and the biodiesel comprised pure methyl esters (MEs) from a variety of vegetable oils. For jet fuels, the WFSP found that the temperature dependence became markedly nonlinear when the temperature approached –40°C and below (Hadaller and Johnson, 2006). For the purposes of fuel qualification, each of the fuels in the WFSP survey easily met the specification at the upper bound, 8 mm²/s at –20°C. For other discussions on the temperature dependence of diesel-grade fuel viscosity, see Freitas *et al.* (2011), Pratas *et al.* (2011a, 2011b), and Yuan *et al.* (2003, 2005, 2009).

3. *Freeze temperature*: Measured in units of $T$, and presented here in degrees Celsius (°C), freeze temperature (Fig 8.3(c)) is not a straight forward concept. Since jet fuels are a blend of different hydrocarbon compounds, they freeze over a temperature range rather than at a single temperature, like a pure liquid does. This is due to the fact that as the temperature is decreased, the heaviest hydrocarbons freeze into waxy crystals before the lighter components solidify. To create a systematic means of comparing the freezing properties of jet fuels, the term "freeze temperature" (not "the freezing temperature") is defined through the following procedure: the hydrocarbon fuel blend is cooled until wax crystals form. As the fuel is gradually warmed back up, the lowest temperature at which all of the wax crystals have melted is defined as the freeze temperature (also sometimes denoted the "freeze point"). Consequently, the freeze temperature is well above the temperature at which the fuel completely solidifies. A related term is "cloud point," which is the temperature at which wax crystals first start to form as the temperature is lowered. Roughly 10°C below the cloud point, the freezing fuel reaches the "pour point," at which the wax in the fuel has built up sufficient solid structure to prevent pouring. The combination of

viscosity and freezing point define the pumpability of a fuel; that is, the ease of pumping fluid through the fuel lines (Chevron, 2006).

4. *Flash temperature (flash point)*: Measured in units of $T$, and presented here in degrees Celsius (°C), the flash point (Fig. 8.3(d)) is the lowest temperature at which vaporized fuel above a flammable liquid will burn when exposed to an ignition source. Vapor burns only when the air/vapor mixture is in a certain range. Below the lower flammability limit, there is insufficient fuel in the mixture to combust. For kerosene-type jet fuel, the range is 0.6 to 4.7 volume percent (%v/v) vapor, while for wide-cut fuel, it is 1.3 to 8.0 %v/v (Chevron, 2006). The upper flammability limit is a function of the local temperature and pressure. The flash point of wide-cut fuels like Jet B is not specified, but is below 0°C (Chevron, 2006).

5. *Specific energy (heat of combustion)*: Measured in units of $L^2/T^2$ and presented here in megajoules per kilogram (MJ/kg), the specific energy (Fig. 8.3(e)) can be used to compare fuels by the relative energy content that a kilogram of fuel could release through complete and perfect combustion. The WFSP found that all but two samples reached the minimum required energy content.

6. *Energy density*: Measured in units of $M/LT^2$ and presented here in megajoules per liter (MJ/L), the energy density (Fig. 8.3(f)) measures the fuel energy content per unit volume and is the product of the specific energy and density. If these properties are measured by the units used for this work, that product must also be multiplied by the necessary factors to convert between liters and cubic meters (1 $m^3$ = 1,000 L). When a system is limited by volume, as in a completely filled fuel tank, a fuel with higher energy density can release more net energy and can be used to travel a longer distance.

7. *Aromatics content*: Measured as a nondimensional number, the aromatics content is the percentage of aromatic hydrocarbons in terms of volume per total fuel volume (%v/v). Aromatics are unsaturated (multiply bonded carbon-carbon bonds) hydrocarbon rings in a conjugated configuration, usually indicating alternating double and single bonds (see App. A). As the fuel combusts, carbonaceous particles form that become incandescent at the high temperatures and pressures in the combustor. The hot particles emit infrared radiation, which can heat up the surrounding walls and create hot spots. The result can be a loss in combustion efficiency, or even worse, a loss of structural integrity. Carbon that deposits on the wall can change the carefully designed flow pattern in the combustor by inhibiting the entrance of diluting air through combustor walls. If the particles are not completely consumed by the time they reach the turbine blades and the stator, they can damage these key engine components. Since aromatics tend to produce more of these carbonaceous particles, their content is limited to a maximum of 25 %v/v in aviation fuels. Nonetheless, the presence of aromatics can be critical for maintaining aircraft seals. Aromatics can cause seals and sealants to swell, developing a "set" to a particular swell level, which is a function of the aromatic content and exposure time. If the seals are subsequently exposed to jet fuel with either very low aromatics content or a sufficiently different mix of aromatics, the absorbed petroleum leaches out of the seal material, resulting in geometric shrinkage and possibly a leak (Hadaller and Johnson, 2006). An aging aircraft would be susceptible to this condition if it has used a wide-cut fuel like JP-4 for a long time, and it is switched over to a narrower cut like JP-8 or one of the synthetics. Synthetic fuels tend to have fewer aromatics than petroleum fuels and may require supplementation to meet jet fuel requirements (Corporan *et al.*, 2011). DEF STAN 91-91 (2016) requires a minimum of 8% aromatics content in the final blended product. In most of the synthetic FT fuels, up to 50% synthetic is blended with conventional jet fuel, with all of the aromatics coming from the petroleum stream (Moses, 2008).

The WFSP analyzed the hydrocarbon content of the sample fuels using a number of ASTM International protocols, some of which yielded conflicting results, particularly at low concentrations. Figure 8.5(a) shows the composition of the jet fuel analyzed *via* ASTM D1319 (2013). It is separated into categories of aromatics, saturates (alkanes, single-bonded carbon chains with hydrogens bonded to them) in the form of *n-*, iso-, and cycloparaffins), and olefins (alkenes, or unsaturated hydrocarbons with one or more carbon-carbon double bonds). Figure 8.5(b) shows a histogram of the aromatics content in all of the deliveries of Jet A to the U.S. military in 2004.

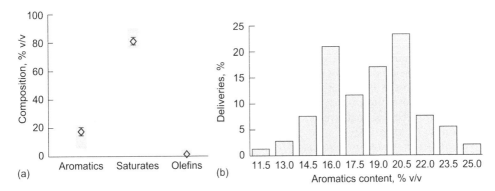

Figure 8.5.   Hydrocarbon content in jet fuels. (a) Composition by class. Data from the World Fuel Sampling Program. (b) Composition of all jet fuel deliveries to the U.S. military in 2004. ((a) Adapted from Hadaller, O. J., and J. M. Johnson. 2006. World fuel sampling program. Coordinating Research Council (CRC) Report No. 647. (b) From Colket, M., T. Edwards, S. Williams, *et al.,* 2007. Development of an experimental database and kinetic models for surrogate jet fuels. AIAA–2007–770. With Permission.)

Note that each of the numbers on the abscissa cover a 1.5% range in value: about 24% of the fuel deliveries had aromatics content in the range of 20.5±0.75 %v/v. On average, the aromatics content of fuels was in the range of 20%, although it varied from below 10% to about 24%. All of these fuels met the standards for aviation fuel, which indicates that, from a practical standpoint, the requirements allow substantial flexibility in composition. However, there are will be some differences in performance between the fuels with the highest and lowest aromatics content.

Naphthenes (cycloalkanes) are other fuel components of interest. They are also based on carbon rings, like aromatics, except that rather than some double bonds, they have only single carbon bonds; that is, they are saturated cyclic hydrocarbons.

8. *Hydrogen content*: Measured in nondimensional units as a percentage of hydrogen by mass to total fuel mass (%m/m), fuels with higher hydrogen content burn more cleanly and produce more energy per unit mass. The disadvantage of high-hydrogen-content fuels is the relatively lower energy content per unit volume. The WFSP found that hydrogen levels in jet fuel were in around 14 %m/m and that synthetic FT fuel had higher hydrogen content (Hadaller and Johnson, 2006).

9. *Lubricity*: Commonly measured by wear scar diameter in units of $L$ and presented here in millimeters (mm), lubricity refers to a fluid's capacity to reduce friction between moving parts. It is characterized by measuring the wear on a fixed steel ball after a specific length of time in contact with a rotating cylindrical ring that is partially immersed in fuel (Chevron, 2006). The depth of the wear scar increases with decreasing lubricity. This property is important for jet fuels, as aircraft components rely on the fuel itself to lubricate moving parts (e.g., fuel pumps and control units). For a discussion of wear scar testing on a wide range of samples, see Knothe (2008b).

10. *Thermal stability*: Commonly measured by the pressure drop $\Delta p$ across a filter in units of $L$ and presented here in millimeters of mercury (mmHg), thermal stability measurements utilize specialized test equipment, the JFTOT (jet fuel thermal oxidation tester), which exposes the fuel to a heated aluminum alloy tube in a controlled way and passes it through a filter to collect any particulates that have formed. After the test is complete, the pressure drop across the filter and a visual inspection of the aluminum tube for discoloration are used to evaluate thermal stability.[5] It is particularly important for aviation fuels because high-performance engines use the fuel as a

---

[5]It may seem unusual to measure a pressure drop in terms of length when it is actually a force per unit area. The height of the fluid column ($h$) is related to the pressure ($p$) through its density ($\rho$) and gravitational acceleration ($g$); specifically, $\Delta p = \rho g h$. By convention, standard atmospheric pressure at ground level is 760 mmHg = 101.325 kPa. This property represents the capacity for maintaining fuel properties due to thermal exposure.

mechanism for heat exchange; for example, for cooling certain engine components or hydraulic fluid. High temperatures can accelerate oxidation reactions in the fuel, leading to gum and particulate formation. Antioxidants are used as additives to improve thermal stability.

11. *Storage stability*: A fuel's capacity for retaining its essential properties while in storage, storage capacity is essential for reserve stocks that may be stored for several years, although a more typical storage time for aviation fuel is 6 months. Fuel properties can be altered by oxidation processes over time, which are a function of air exposure, elevated temperature, contaminants either in the fuel or absorbed from the container walls, and the presence of peroxides and antioxidants (Knothe, 2007).

12. *Electrical conductivity*: Measured in units of $T^3A^2/ML^2$ and presented here in picoSiemens per meter (pS/m), electrical conductivity is the fuel's capacity to conduct electricity, which is proportional to the rate at which electrical energy dissipates. A fuel made up of pure hydrocarbons would not be electrically conductive, but trace elements—specifically, the presence of polar molecules—could bestow the capability of holding a charge. During pumping operations, the fuel can be in contact with a variety of materials in hoses, pipes, fittings, and valves. Contact between dissimilar materials can cause the formation of an electrical charge in the fuel. Given time, the static electricity would dissipate. However, if the time constant for the dissipation is long, vaporized fuel could be at risk of burning should it come into contact with a stray ignition source. Naturally derived petroleum products have more contaminants than synthetic fuels and will tend to have a higher conductivity. Jet A-1 and JP-8 require the incorporation of a static dissipater additive, which increases electrical conductivity, thereby decreasing the time required for dissipation and reducing the risk of unplanned combustion.

13. *Vapor pressure*: Measured in units of $M/LT^2$ and presented here in kilopascals (kPa), vapor pressure is a property for a given substance at thermodynamic equilibrium in a closed system. Here, the liquid and gaseous phases (and solid phase, if any is present) have no driving force to change phase, so they will remain in the same proportions unless the system is perturbed. The proportionality is a function of temperature and pressure. At a fixed temperature, liquids will remain stable at all pressures above the vapor pressure, but they will boil when the pressure drops below the vapor pressure. Kerosene-type jet fuel, such as Jet A, has a vapor pressure of about 1 kPa at 38°C and carries no vapor pressure requirement. Wide-cut fuel, such as Jet B, is more volatile, has a higher vapor pressure, and has a maximum permissible vapor pressure of 21 kPa at 38°C.

14. *Heat capacity*: Measured in $L^2/T^2K$ and presented here in joules per gram per degree Celsius (J/(g·°C)), heat capacity is not specified in the requirements but is a property that can affect engine efficiency. More efficiency can be squeezed out of an engine if the fuel itself is used to cool engine components while it is simultaneously heated up en route to the combustor. If temperature is too high (over ~480°C), thermal or catalytic cracking can occur. In WFSP, the average value of the heat capacity at 0°C was 1.582 J/(g·°C) and showed a linear dependence on temperature.

15. *Cetane number*: A dimensionless term, the cetane number is related to the ignition delay time between the injection of the fuel and the onset of ignition. It is particularly important for piston-driven engines with periodic combustion spurts, for which precise control over combustion timing is critical. A higher cetane number correlates with a shorter ignition delay time; especially for automotive applications, a higher cetane number will result in cleaner, more complete combustion (Refaat, 2009).

For additional reading on jet fuel standards and testing methods, see Chevron (2006) and Knothe (2008b); for jet fuel properties and composition, see Hadaller and Johnson (2006), Moses (2008, 2009), and Moses and Roets (2009); and for thermal, oxidative, and storage stability, see Jain and Sharma (2010) and Knothe (2007).

### 8.2.2   *Alternative jet fuel specifications*

For the near term, alternative fuels must be "drop-in" fuels, which can be used directly in current aircraft turbine engines, usually by blending jet-grade petroleum and biojet fuel. Note that there are specifications for aviation gasoline (avgas) for small piston-driven aircraft. Since that represents a small percentage of aviation fuel, the focus here will be on the heavier aviation fuel designed for use in turbine-powered aircraft. A variety of alternative jet fuels have been thoroughly tested on the ground and have been used in demonstration flights by commercial airlines as well as the U.S. military. See Hendricks *et al.* (2011) and Kinder and Rahmes (2009) for some of this background and Edwards (2007) for a description of the historical evolution of aviation fuel standards. The challenge lies in making these fuels commercially viable, environmentally friendly, and broadly available. The basic fuel types for alternative fuels for aviation are:

- SPK: Synthetic paraffinic kerosene is derived from natural gas, such as Syntroleum's S-8, Shell's GTL, and Sasol's GTL-1 and GTL-2. (GTL stands for gas-to-liquid, a reference to its FT manufacturing process.) SPK fuels are functionally similar to Jet A, but they have a negligible aromatics content.
- IPK: Synthetic isoparaffinic kerosene is derived from coal and has properties that are comparable to SPK.
- FSJF: Fully synthetic jet fuel is a synthetic paraffinic kerosene developed by Sasol that consists of coal-derived fuel with synthetic aromatics. This is the only alternative fuel at this time that does not require blending with conventional fuels (Moses and Roets, 2009).
- HRJ/HEFA: Hydrotreated renewable jet fuel is derived from renewable fuel sources. Vegetable oils are one such source, but they require hydrotreatment to condition the oil to jet-fuel quality. This fuel has also been called "bio-SPK" to underscore its functional similarities to SPK. In 2011, it was renamed "HEFA" for its hydroprocessed esters and fatty acids composition.

An analysis by Hilemann *et al.* (2010) concludes that these fuels would reduce the net fuel energy consumed by the aviation industry, improving overall fleet-wide energy efficiency. With the exception of HEFA/HRJ, these fuels are derived from coal and natural gas. They are produced with a variant of the FT process, which is a liquefaction technique. In 2009, FT processing was the first manufacturing method to be approved by ASTM International under its alternative jet fuel specification, ASTM D7566 (2016). The SPKs were the first approved synthetic blending component under this standard in 2009. Revisions were made by ASTM in July 2011 for the use of HEFA as a jet fuel in a 50/50 blend.

"Neat" fuels are not mixed or diluted with other fuels. (For pure biofuels, the designation "B100" may also be used, in which the "100" indicates that 100% of the fuel is bioderived.) One study found that an array of SPK and IPK FT fuels met all jet fuel specifications except for density (Moses, 2008). To mitigate this property, a 50/50 blend of FT and conventional fuel was certified for commercial aviation. It has been widely tested on the ground and in civilian and military aircraft.

The composition of the FT synthetic fuels can vary in hydrocarbon content, but they tend to peak at a slightly lower carbon number than petroleum-based fuels. They also have a small aromatics content. The primary advantages and disadvantages of such synthetic fuels relative to conventional fuels are listed:

- *Advantages:* cleaner burn; reduced carbon monoxide (CO), sulfuric gases ($SO_x$), and particulate emissions; better thermal stability; and potentially carbon-neutral
- *Disadvantages:* lower energy density, poor lubricity, higher freeze point, higher viscosity, carbon capture and sequestration is required to be considered sustainable.

Commercial production of FT fuels is well established in South Africa, Qatar, and Malaysia. Because of the high temperatures and pressures required for the FT process, the refining process is energy-intensive and releases higher quantities of carbon dioxide ($CO_2$) during manufacture than petroleum refining. Consequently, although this fuel is the most commercially viable alternative fuel, it requires capture and sequestration of the carbon emissions in order to be considered a sustainable fuel (Bartis and Van Bibber, 2011; Kinsel, 2010). With sequestration, the environmental impact of FT fuels is comparable to that of petroleum (Bartis and Van Bibber, 2011). However, the capture and sequestration steps can add substantially to the investment cost for such fuels. Nevertheless, FT fuels are the most commercially ready solution for current energy needs. For the longer term, the $CO_2$ emissions of petroleum-based FT fuels render it less attractive than biofuels, which have the potential to be $CO_2$-negative. In this case, substituting biofuels for petroleum-based fuel would reduce the level of greenhouse gases in the atmosphere.

Whereas the molecules in petroleum-derived fuel are primarily composed of pure hydrocarbons, fuels from vegetable oil and biodiesel are composed of hydrogen, carbon, and oxygen. The long hydrocarbon chains that are typical of the paraffins are present in biodiesel as a component of a fatty acid, but they are coupled at one end of the chain to other functional groups. Condensation of an alcohol and a fatty acid produces an ester, which can be suitable for a biode-rived fuel blend. In biofuel production, the most commonly used alcohol is methanol, because it is inexpensive and provides good fuel properties. Ethyl esters, based on ethanol, are common in areas like Brazil, where ethanol is abundant. Since the ethanol in Brazil is largely derived from sugar cane, this type of biodiesel is completely based on renewable sources. In contrast, methanol is usually derived from natural gas. Butanol or propanol can also be used to form butyl and propyl esters, respectively, but this is much more expensive and less common. When the ester has only a single fatty acid, the biodiesel blend is composed of mono-alkyl esters; these are termed "fatty acid alkyl esters" (FAAEs) somewhat generically, the most common of which are fatty acid methyl esters (FAMEs). At one time, they were under consideration as a jet-fuel-blending component, but their poorer cold-weather properties and specific energy are inadequate for jet fuel. The new standards for bioderived jet fuel only include HEFA fuel (IRENA, 2017). In this work, vegetable oils and greases will be referred to as "green crude"; blends created from green crude, as "FAAE"; blends of FAAE and conventional diesel that conform to the standards of CEN EN 14214 (2014) and ASTM D6751 (2015), as "biodiesel"; and biofuel consisting of mixtures of pure hydrocarbons derived from renewable sources, as "HEFA."

Some requirements on biodiesel properties in its neat state are identified in Table 8.2. Note that these requirements are not targeted toward HEFA, but they are provided here as a reference.

Although the European EN 14214 specifications (CEN EN 14214, 2014) place a range limit on biodiesel density for FAME, and both standards identify a permissible range on viscosity, most of the requirements have to do with fuel composition. The presence of excess mono-,

Table 8.2.   Key fuel specifications for biofuels based on methyl esters for automotive applications.

| Property | Fatty acid methyl ester,[a] FAME | Fatty acid alkyl ester,[b] FAAE |
| --- | --- | --- |
| Density at 15°C, kg/m³ | 860 to 900 | – |
| Viscosity at 40°C, mm²/s | 3.5 to 5.0 | 1.9 to 6.0 |
| Methanol, %m/m | <0.20 | <0.20 or flash temperature <130°C |
| Sulfur, mg/kg | <10 | <15 |
| Total esters, %m/m | >96.5 | – |
| Monoglyceride, %m/m | <0.8 | – |
| Diglyceride, %m/m | <0.2 | – |
| Triglyceride, %m/m | <0.2 | – |
| Free glycerol, %m/m | <0.02 | <0.02 |
| Total glycerol, %m/m | 0.25 | <0.24 |
| Total acidity, mg KOH/g | <0.5 | <0.5 |
| Specification | EN 14214 | ASTM D 6751 |

[a]Values from specification CEN EN 14214. 2014. Liquid petroleum products—FAMEs for use in diesel engines and heating applications—requirements and test methods. Brussels: European Committee for Standardization. © CEN, reproduced with permission.
[b]Values from specification ASTM D6751-15ce1. 2015. Standard specification for biodiesel fuel blend stock (B100) for middle distillate fuels. West Conshohocken, PA: ASTM International. With permission.

di- and triglycerides, methanol, and glycerol would indicate an incomplete reaction in the refining process or flaws in the separation process or both. Relative to the specifications for jet fuel in Table 8.1, one important difference is that the standards for FAME and FAAE in Table 8.2 specify a range on kinematic viscosity at 40°C, whereas jet fuel specifications are referenced to a much lower temperature, −20°C. This will be discussed in more detail in the next section. Relative to petroleum fuel, biodiesel derived from FAAE have the following advantages and disadvantages:

- *Advantages*: cleaner burn; high lubricity; reduced carbon monoxide (CO), sulfuric gases ($SO_x$) and particulate emissions; better thermal stability; carbon-negative footprint (provided the manufacturing process is sustainable); and biodegradability
- *Disadvantages*: higher freeze point; lower energy density; higher $NO_x$ emissions; poorer oxidative and storage stability; and high cost and low availability of feedstock and refining plants.

Typically, green crude has a higher proportion of heavier carbon components than jet fuel, resulting in a fluid that is more viscous than diesel (Fig. 8.4(b)). Although vegetable oil can be blended with conventional fuel for use in some piston-driven diesel engines without treatment, it can form deposits, plugging up filters and injectors. Plant-derived fuels have very little sulfur or aromatic content, so, in general, their emissions have less environmental impact.

A recent review of material compatibility (Fazal *et al.*, 2010) finds that, relative to petroleum-based diesel, some biodiesel results in greater potential for plugged filters, sticking parts, and corrosion of some metals. The corrosion process can also degrade fuel properties such as density and viscosity. This study was primarily concerned with biodiesel for automotive applications, but its findings could also apply to aviation. In any event, the disadvantages of biodiesel can largely be mitigated by additives and blending with conventional fuels, with the exception of its inherently lower energy density.

Biorefining processes such as transesterification are used to break down the vegetable oils to alkyl esters to form FAAE. Vegetable oils, free fatty acids, and FAAEs can be used as precursors to HEFA. In order to meet freeze-point requirements and boost the specific energy, they must undergo hydroprocessing to remove the oxygen, become saturated with hydrogen, and reduce chain length (Sec. 8.5.2.2). The composition of HEFA fuel will be essentially the same as that of petroleum-derived fuels, but it will have much less cycloparaffin content and only traces of olefins, aromatics, and sulfur. As a result, it will have the following:

- *Advantages*: better stability and blending properties because of the absence of double bonds, oxygenated molecules, and heteroatoms (here, primarily nitrogen and sulfur); cleaner burn, lower sulfur and particulate emissions relative to petroleum jet fuel; and lower freeze point and higher specific energy than FAAE-diesel blends
- *Disadvantages*: lower lubricity than biodiesel, so it will likely require additives or blending. Since natural antioxidants may be lost during hydroprocessing, this fuel may also need additives for stability.

For further background on biodiesel, see Blakey *et al.* (2011), Knothe (2010a, 2010b); for SPK composition, properties, and standards, see Moser (2009).

## 8.3  FUEL PROPERTIES

### 8.3.1  *Effect of composition on fuel properties*

The hydrocarbon composition of any particular tank of jet fuel is dependent on the properties of the crude oil, refining operations, and the use of additives. Petroleum-based jet-grade fuels are mixtures of various hydrocarbons, including alkanes (paraffins) and cycloalkanes (naphthenes), aromatics (including small amounts of benzene and polycyclic aromatic hydrocarbons), and alkenes (olefins), as well as trace amounts of sulfur and other materials. Note that for the most part, the standards described in Table 8.1 did not specify hydrocarbon composition. Instead, the fuel properties are primarily specified in terms of performance. As a practical matter, the jet fuel performance specifications (Sec. 8.2.1) limit the carbon content of jet fuels to the carbon numbers from C8 to C16 (Edwards, 2010), but the molecules can appear as straight, branched, or linked chains. Most of the fuel components are saturated (i.e., only single carbon-to-carbon bonds appear in the mixture).

When crude oil is refined, it undergoes a distillation process to separate the hydrocarbon components by weight. As heat is applied, the lightest hydrocarbons will start to vaporize first and will move up a distillation column, followed later by heavier hydrocarbons. Jet B is a wide-cut fuel, which has a more diverse hydrocarbon blend with chain lengths from about C4 to C16. The presence of the light carbon components gives Jet B its cold-temperature properties. Although Jet A-1 has a lower freeze point than Jet A (Fig 8.3(c)) to makes it desirable for operation in colder climates, it can be less expensive to refine oil to Jet A standards, since a broader temperature range can be used in the distilling process to recover a wider range of hydrocarbons.

The thermophysical properties of a pure hydrocarbon fluid are dependent on its carbon number and its structural complexity. The fluid can be saturated (holding the maximum possible number of hydrogen atoms with only single bonds between all carbons), or unsaturated (replacing some of the carbon-carbon single bonds with double bonds). The most abundant forms are that of simple, straight-chain, saturated *n*-paraffins; branched, saturated isoparaffins; cyclic, saturated paraffins (cycloalkanes, or naphthenes); unsaturated olefins with at least one double bond; and unsaturated cyclic aromatics (see the appendix for a description of the underlying organic chemistry). Figure 8.6(a) shows that density increases with carbon number for saturated compounds. The densities of *n*-paraffins and isoparaffins at 20°C are indistinguishable, but further structural complexity and double bonds increase the density at a given carbon number. Discussion of density correlations as a function of composition can be found in Alptekin and Canakci (2008, 2009), Refaat (2009), and Saravanan and Nagarajan (2011).

Kinematic viscosity increases with chain length (Fig. 8.6(b)), degree of saturation, and amount of branching (Knothe, 2005; Refaat, 2009). The boiling point increases with increasing carbon number (Fig. 8.6(c)) and is dependent to a lesser extent on structural complexity. The specific energy decreases with increasing carbon number and the presence of naphthenes and aromatics (Fig. 8.6(d)), whereas the opposite holds true for the energy density (Fig. 8.6(e)). Note that the *n*-paraffins and isoparaffins share nearly the same dependence, while naphthenes are more strongly influenced by carbon number from C8 to C10. Since the energy density increases as the carbon number increases, in a situation with a fixed volume, such as a completely filled fuel tank, more energy can be released by fuels with higher carbon numbers. High carbon numbers are advantageous in this regard, although that must be balanced against the deleterious effects of increased viscosity (Fig. 8.6(b)) and increased freeze temperature (Fig. 8.6(f)).

Lighter, low-carbon fuels have a lower freeze temperature than high-density high-carbon fuels (Fig. 8.6(f)). This provides some insight as to the desirability of using a wide-cut blend for better cold-weather performance, since a wide blend tends to have more hydrocarbons with low carbon numbers. Also note that the only graph in Figure 8.6 in which *n*-paraffins substantially differ from isoparaffins is in freeze temperature. The freeze temperature decreases with increased structural complexity, such as increased branching and reduced level of saturation. Relative to straight-chain, ladder-like alkanes, such molecules exhibit reduced packing efficiency, so that lower temperatures are needed to reach an entropy level at which crystallization can occur (Refaat, 2009). The freeze temperature of naphthenes from C8 to C10 increases more strongly than the rate of increase for *n*-paraffins and isoparaffins. Considering the specific energy and freeze temperature, increasing isoparaffin content will produce better cold-weather properties, without a substantial hit on the available energy content. If the concentration of isoparaffins can be increased, even at the same carbon number, the fuel will have lower susceptibility to cold temperatures, as well as a lower viscosity (Moses, 2008). The presence of naphthenes may deliver

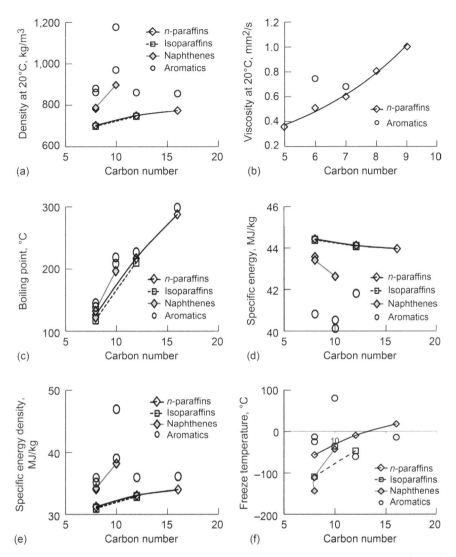

Figure 8.6. Compositional dependence of thermophysical properties of pure hydrocarbons. (a) Density at 20°C. (b) Kinematic viscosity at 20°C. (c) Boiling point. (d) Specific energy. (e) Specific energy density. and (f) Freeze temperature. (Data from ASTM DS4B. 1991. Physical constants of hydrocarbon and non-hydrocarbon compounds. ASTM International, West Conshohocken, PA; Touloukian, Y. S., S. C. Saxena, and P. Hestermans. 1975. Viscosity. New York, NY: IFI/Plenum Data Company; and TRC. 2011. TRC thermodynamic tables. College Station, TX: Thermodynamic Research Center, Texas A&M University.)

better specific energy density with increasing carbon number from C8 to C10, but it comes at the penalty of reduced tolerance to cold temperatures.

Other properties that are affected by carbon number include vapor pressure and lubricity. For alkanes, vapor pressure increases with decreasing carbon number. Low-density hydrocarbons, such as methane and propane, appear as vapors at standard pressure and temperature at ground level (by convention, 101.325 kPa and 20°C), whereas higher density hydrocarbons appear as liquids. In general, increasing carbon number correlates with better lubricity, although that may also be linked to the presence of certain polar molecules in the fuel (Refaat, 2009). Fuels with a bias toward lower carbon number will burn more cleanly. Consequently, there is less deposition of unburned hydrocarbons in the fuel system, which can improve engine combustion quality and simplify maintenance.

For precise control of an intermittent combustion process, such as that of a piston-driven engine, it is advantageous to have a fuel with a narrower carbon number distribution, rather than a wider one, so that all of the fuel components ignite almost simultaneously. A short, consistent ignition delay time allows for more efficient operation of such engines, as well as cleaner burns with fewer emissions. However, for turbine engines (which are based on a continuous combustion process), there may be difficulties in high-altitude relight if the distribution of components is too narrow (Blakey *et al.*, 2011).

In Figures 8.7 to 8.11, the composition of many fuel samples are shown as percentages of the fuel composition by carbon number, and, where available, structural class. Despite the obvious differences in the fuel blends, all of these fuels meet jet fuel specifications with one exception. Note that most of the graphs are presented in terms of mass fraction. However, the measurement for Jet A in Figure 8.7 is given as a molar percentage, although the units are still dimensionless. Fuel composition on a molar basis is useful for computational chemistry, because the equations for chemical kinetics are naturally expressed in those units. The mole fraction is related to the mass fraction through the atomic mass, which is expressed in grams per mole (g/mole). Mass-based carbon composition in a sample of JP-4 (the military equivalent of Jet B) is shown in Figure 8.8. The shale-derived Jet B in Figure 8.8(b) is much richer in the C7 to C9 range relative to the other petroleum-derived fuels, and it is strongly influenced by an abundance of isoparaffins in the fuel mixture.

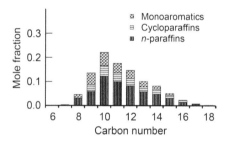

Figure 8.7.   Jet A fuel: composition by carbon number. (Adapted from LeClercq, P., and M. Aigner. 2009. Impact of alternative fuels physical properties on combustor performance. Presented at 11th Triennial International Conference on Liquid Atomization and Spray Systems, Vail, CO.)

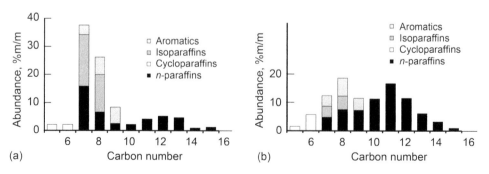

Figure 8.8.   Jet B (JP-4) fuel compositions by carbon number. (a) Derived from petroleum. (b) Derived from shale. (Adapted from Agency for Toxic Substances and Disease Registry. 1995. Toxicological profile for jet fuels JP–4 and JP–8. Atlanta, GA: U.S. Department of Health and Human Services. *http://www.atsdr.cdc. gov/ToxProfiles/TP.asp?id=773&tid=150* (accessed September 1, 2011). Work of the U.S. Government.)

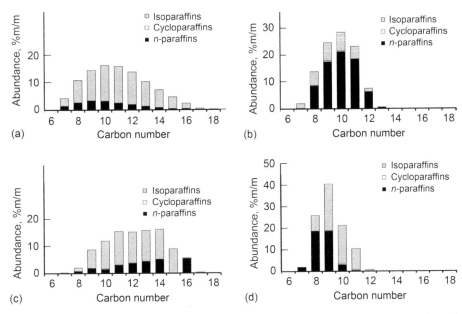

Figure 8.9.   Synthetic paraffinic kerosene (SPK) fuels derived from natural gas: composition by carbon number (a) S-8. (b) GTL-1 (gas to liquid). (c) GTL-2. (d) Shell GTL. (Adapted from Moses, C. A. 2008. Comparative evaluation of semi-synthetic jet fuels: final report. CRC Project No. AV-2-04a.)

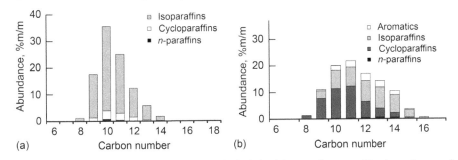

Figure 8.10.   Synthetic paraffinic kerosene (SPK) fuels derived from coal: composition by carbon number. (a) ITK. (b) Fully synthetic Fischer-Tropsch. ((a) Adapted from Moses, C. A. 2008. Comparative evaluation of semi-synthetic jet fuels: final report. CRC Project No. AV-2-04a. (b) Adapted from van der Westhuizen, R., M. Ajam, P. De Coning, *et al.*, 2011. Comprehensive two-dimensional gas chromatography for the analysis of synthetic and crude-derived jet fuels. *J. Chromatogr. A* 1218:4478–4486.)

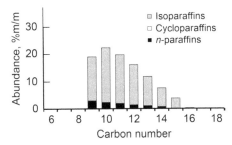

Figure 8.11.   Hydroprocessed esters and fatty acids (HEFA) fuel derived from oil extracted from *Jatropha* and algae: composition by carbon number. (Adapted from Kinder, J. D., and T. Rahmes. 2009. Evaluation of Bio-derived Synthetic Paraffinic Kerosenes (Bio-SPK). White paper, Sustainable Biofuels Research & Technology Program, The Boeing Company.)

Based on the discussion of hydrocarbon composition effects on fuel characteristics above, one should expect that these fuels would have good cold-weather properties, as is required for JP-4.

The synthetic paraffinic kerosene fuels in Figures 8.9 and 8.10 are manufactured from natural gas (Fig. 8.9) and coal (Fig. 8.10), in a process that compresses the fuel from the gaseous state into liquid. In order to meet the minimum density requirements and aromatics content for jet fuel, they must be blended in a 50-50 mixture with conventional petroleum jet fuel (Moses, 2008). The S-8 fuel (Fig. 8.9(a)) is a wide-cut fuel and was used in a 50-50 blend with JP-8 by the U.S. Air Force. The composition of the gas-to-liquid GTL-1 (Fig. 8.9(b)) was tuned for automotive diesel—not jet fuel—and it did not meet the freeze temperature requirements. (Note that this fuel is dominated by *n*-paraffins.) The GTL-1 fuel was further processed to create GTL-2 (Fig. 8.9(c)), which increased the isoparaffin content, although it also shifted and broadened the hydrocarbon distribution. This manipulation permitted GTL-2 to meet freeze temperature requirements. The Shell GTL fuel (Fig. 8.9(d)) is a natural-gas-derived synthetic fuel that exhibits significantly different composition from the prior fuels but still meets the requirements for jet fuel. The ITK Engineering GmbH fuel (Fig. 8.10(a)) is a coal-derived product, which has been used at the O.R. Tambo International Airport in Johannesburg since 1999. An example of the first fully synthetic Fischer-Tropsch fuel is shown in Figure 8.10(b). It is manufactured from coal and enhanced with synthetic aromatics in order to meet jet fuel standards. This is the only synthetic fuel that does not need to be blended with conventional fuels at this time.

One example of a HEFA fuel is shown in Figure 8.11, derived from oil extracted from *Jatropha* and algae by Boeing. This fuel has a relatively high percentage of isoparaffins, which would be expected to improve its cold-weather properties while retaining the advantageous energy density of higher carbon content. Boeing has also characterized other bioderived HEFA, for which the green crude was comprised of oils from *Camelina*, *Jatropha* and algae in varying proportions. All of these blends yielded similar hydrocarbon composition in terms of carbon number distribution and the ratio of *n*-paraffins to isoparaffins (Kinder and Rahmes, 2009).

While developing new aircraft designs and vetting new alternative fuels, it would be beneficial to improve the combustion efficiency through smart design of combustors, inlets, and nozzles and to minimize pollutant output. Computational fluid dynamics (CFD) has been used for decades to model both fluid flow and the chemical kinetics that govern fuel combustion in aircraft components. To properly model the chemical kinetics requires an understanding of the composition of the blended fuel components as well as appropriate chemical kinetic models that describe the cascade of chemical reactions governing the combustion process. Once the constituents of the fuel are determined, blending rules are applied to calculate the thermophysical properties of the hydrocarbon mixture, such as density, viscosity, specific energy, and others as needed. During a simulation, these properties can evolve over time as a result of chemical reaction, which changes the composition of the fuel, or through changes to the local temperature and pressure. Considering the variability shown in Figures 8.3, 8.5, and 8.7 to 8.11, it would be a disheartening task to attempt to comprehensively model the entire range of possible fuel mixtures. This leads to the desire for identifying representative fuel blends, or jet fuel surrogates (Anand *et al.*, 2011; Colket *et al.*, 2007; Dooley *et al.*, 2010; Herbinet *et al.*, 2010; Huber *et al.*, 2010; LeClercq and Aigner, 2009; Mensch *et al.*, 2010; Pitz and Mueller, 2011; Singh *et al.*, 2011; Westbrook *et al.*, 2009). Surrogate fuels provide a standardized reference composition for comparing different designs and operating conditions. Both experimental and numerical experiments have employed surrogate fuels to simplify the chemistry while maintaining the key elements of the combustion process to keep all of the essential physics in place. C16 (*n*-hexadecane) is a primary reference fuel for experimental work on diesel engines. The Lawrence Livermore National Laboratory (LLNL) developed a suite of chemical kinetic models for alkanes from *n*-octane to *n*-hexadecane for low and high temperatures (Westbrook *et al.*, 2009). Their work showed that all *n*-paraffins in this range of C8 to C16 exhibit nearly the same ignition behavior. If the right behavior can be appropriately captured in a numerical study by substituting a lower carbon number fuel for *n*-hexadecane, this vastly reduces the size of the chemical kinetic model. As a reference point, LLNL's model for the combustion of *n*-hexadecane has 8000 reactions

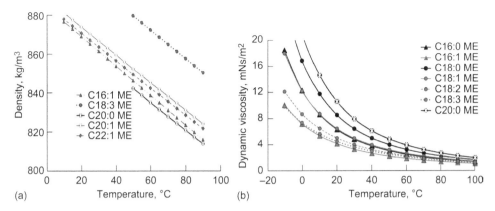

Figure 8.12. Temperature dependence of saturated and unsaturated methyl esters (MEs) for (a) density and (b) viscosity. (Data from Freitas, S. V. D., M. J. Pratas, R. Ceriani, *et al.*, 2011. Evaluation of predictive models for the viscosity of biodiesel. *Energy Fuels* 25:352–358; and Pratas, M. J., S. Freitas, M. B. Oliveira, *et al.*, 2011a. Densities and viscosities of minority fatty acid methyl and ethyl esters present in biodiesel. *J. Chem. Eng. Data* 56:2175–2180.

and 2100 species, in contrast to *n*-decane (C10), which has 3900 reactions and 950 species (Westbrook *et al.*, 2009).

The identification of the right blend of fuel components for surrogate fuels is currently an active area of research. Databases for petroleum fuel components have been laboriously built up over decades. In contrast, the characterization of the more complex chemistry of biofuels is relatively recent (Kohse-Hoinghaus *et al.*, 2010). Surrogates for biofuels (or surrogate blends) can be strategically developed to analyze a simpler, idealized combustion system. Although surrogates have a simpler chemistry, they are designed to share the essential combustion characteristics of commercial biofuels. For computational modeling of the chemical kinetics of biodiesel oxidation, Herbinet *et al.* (2008) used a virtual blend of *n*-heptane, iso-octane and methyl butanoate supplemented with methyl decanoate to provide more realistic response at high and low temperatures. More recent work has incorporated two large, unsaturated esters to show the influence of the double bond (Herbinet *et al.*, 2010). Another study used two of the five major components of biodiesel to examine toxic emissions (Kohse-Hoinghaus *et al.*, 2010). They studied methyl stearate and methyl esters using LLNL's pre-existing classes and rules for the chemical reactions, but added additional mechanisms (3,500 chemical species and >17,000 reactions) to represent some of the unique characteristics of MEs. For example, the monounsaturated compound, methyl oleate (also called oleic acid ME, with a lipid number[6] of C18:1 and a chemical formula of $C_{19}H_{34}O_2$) is slightly less reactive than methyl stearate (stearic acid, C18:0, $C_{19}H_{36}O_2$). The double carbon-carbon bond in methyl oleate inhibits some of the reaction pathways that produce chain branching at low temperatures (Naik *et al.*, 2011).

Reliable models for fuel properties such as density and viscosity as a function of temperature and composition are essential for computational analysis of aircraft design, such as calculating the power required to pump a given volume of fuel from its storage reservoirs to the combustor or the efficiency of combustion in a specific combustor design at a range of inlet temperatures. Figure 8.12 shows the effect of composition and temperature on properties for certain fatty acid esters that are typical of biodiesel. Figure 8.12(a) presents experimental data representing the temperature dependence of density for saturated MEs (C*x*:0) in the range of C16 to C22, as well as unsaturated MEs with one or three double bonds (C*x*:1, C*x*:3) (Pratas *et al.*, 2011a).

---

[6]The lipid number of the fuel's fatty acid subunit is written in the form of C*x*:*y*, in which *x* represents the carbon number and *y* represents the number of unsaturated carbon-to-carbon bonds. See the appendix for details.

As with the pure hydrocarbons, density decreases linearly as the temperature increases. The density also increases with increasing chain length and increasing levels of unsaturation (Refaat, 2009). The data in Figure 8.12(b) show the temperature dependence of viscosity, which were computed from exponential functions based on the revised Yuan model developed by Freitas *et al.* (2011). As the carbon number for the MEs increases, the viscosity also increases. As the level of saturation increases, however, the viscosity decreases for the MEs shown here. The curve for monounsaturated methyl oleate (C18:1 ME) is nearly indistinguishable from that of saturated methyl palmitate (C16:0 ME). This behavior can also be identified in the analysis of ME blends (Saravanan and Nagarajan, 2011).

To complicate matters further, the MEs shown in Figure 8.12 have a specific configuration of groups about their double bonds (a *cis* configuration). Some of these compounds also have isomers with a different configuration of groups about their double bonds (a *trans* configuration). A description of the *cis/trans* configurations is found in the appendix. Although *trans*-isomers are much less common in biodiesel, they can be introduced during the refining process through catalytic partial hydrogenation (Moser, 2009). Once the hydrocarbon chain increases beyond a handful of carbons, there are a number of possible structural configurations for unsaturated compounds depending on where the double carbon-to-carbon bond(s) are located and the orientation of groups about them. These variations can introduce changes to the ester's properties. Furthermore, the double bonds add rigidity to the molecule that can introduce kinks into the molecular structure, which can affect intermolecular interactions. For clarity, various forms of identification for the essential MEs characteristic of biofuels are shown in Table 8.3. The common name for the fatty acid alkyl ester is listed in the first column, followed by the lipid number associated with the fatty acid subunit. The chemical formula shows that the number of H atoms drops by two for each C=C double bond. All of these esters have two O atoms except for methyl ricinoleate with three O atoms. This compound is the primary fatty acid present in an important vegetable crude source, the castor bean, but it is rarely found in other vegetable oils.

Even when including the molecular weight, the first three identifiers in Table 8.3 are not specific enough to point to a single molecular structure due to the wide variety of possible structural configurations. The CAS number (of the Chemical Abstracts Service) denotes a system of identification that can reduce ambiguity when referring to these compounds. There are a number of other naming systems and conventions that have been developed for these molecules in different disciplines,[7] and two of the more common alternate names are found in the last column. The CAS identifier is associated with these and other synonyms for these esters.

Biodiesel from canola (rapeseed) and soy demonstrated differences in combustion properties that were related to the relative amounts of five ME components (Westbrook *et al.*, 2011). Specifically, the properties of the long-chain alkyl group—and not the carbon number—dictated the ignition delay time. An examination of thermophysical properties as a function of carbon number and level of saturation (Fig. 8.13) can illuminate a number of other trends. In Figure 8.13(a), the kinematic viscosity at 40°C clearly shows that viscosity increases with carbon number. It also decreases with increasing unsaturation: monounsaturated methyl palmitoleate (C16:1) is less viscous than the saturated methyl palmitate (C16:0). For the *cis*-type unsaturated compounds at C18, the viscosity reduces as saturation decreases. However, the viscosity of methyl elaidate (C18:1 *trans*) and methyl stearate (C18:0) are indistinguishable on this figure just below 6 mm²/s. The viscosity of methyl linoelaidate (C18:2 *trans*, 5.33 mm²/s) is substantially higher than methyl linoleate (C18:2 *cis*, 3.65 mm²/s). This provides evidence that the beneficial reduction of viscosity offered by *cis*-type unsaturated compounds may not exist for *trans*-type isomers.

The trends are different for other properties. For the specific energy (Fig. 8.13(b)), the value for methyl oleate (C18:1) falls cleanly on top of methyl stearate (C18:0) at about 40 MJ/kg, and the value for methyl palmitoleate (C16:1) is nearly the same as that of methyl palmitate (C16:0).

---

[7]Nomenclature of the International Union of Pure and Applied Chemistry is commonly used in chemistry instead of CAS nomenclature. Although it has the advantage of better precision by supplying more information about chemical structure, it comes at the expense of readability, particularly for complex molecules.

Table 8.3.   Fatty acid methyl esters (FAMEs) that are important in vegetable-oil-derived biofuels.

| FAME | Lipid number | MW,[a] g/mol | CAS number[b] | Alternate names |
|---|---|---|---|---|
| Methyl laurate $C_{13}H_{26}O_2$ | C12:0 | 214.34 | 111-82-0 | Lauric acid ME, methyl dodecanoate |
| Methyl myristate $C_{15}H_{30}O_2$ | C15:0 | 242.4 | 124-10-7 | Myristic acid ME, methyl tetradecanoate |
| Methyl palmitate $C_{17}H_{34}O_2$ | C16:0 | 270.45 | 112-39-0 | Palmitic acid ME, methyl hexadecanoate |
| Methyl palmitoleate $C_{17}H_{32}O_2$ (*cis* configuration) | C16:1 | 268.43 | 1120-25-8 | Palmitoleic acid ME, methyl *cis*-9-hexadecenoate |
| Methyl stearate $C_{19}H_{38}O_2$ | C18:0 | 298.5 | 112-61-8 | Stearic acid ME, methyl octadecanoate |
| Methyl oleate $C_{19}H_{36}O_2$ (*cis* configuration) | C18:1 | 296.49 | 112-62-9 | Oleic acid ME, methyl *cis*-9-octadecenoate |
| Methyl elaidate $C_{19}H_{36}O_2$ (*trans* configuration) | C18:1 | 296.49 | 1937-62-8 | Elaidic acid ME, methyl *trans*-9-octadecenoate |
| Methyl ricinoleate $C_{19}H_{36}O_3$ (*cis* confguration) | C18:1 | 312.49 | 141-24-2 | Ricinoleic acid ME, methyl 12-hydroxy-9-octadecenoate |
| Methyl linoleate $C_{19}H_{34}O_2$ (*cis* configuration) | C18:2 | 294.47 | 112-63-0 | Linoleic acid ME, methyl *cis*-9-*cis*-12-octadecadienoate |
| Methyl linoelaidate $C_{19}H_{34}O_2$ (*trans* configuration) | C18:2 | 294.47 | 2566-97-4 | Linolelaidic acid ME, methyl 9-*trans*-12-*trans*-octadecadienoate |
| Methyl linolenate $C_{19}H_{32}O_2$ (*cis* configuration) | C18:3 | 292.46 | 301-00-8 | Linolenic acid ME, methyl all-*cis*-9,12,15-octadecatrienoate |
| Methyl arachidate $C_{21}H_{42}O_2$ | C20:0 | 326.56 | 1120-28-1 | Eicosanoic acid ME, methyl eicosanoate |
| Methyl gadoleate $C_{21}H_{40}O_2$ (*cis* configuration) | C20:1 | 324.54 | 2390-09-2 | Cis-11-eicosanoic acid ME, methyl *cis*-11-eicosanoate |
| Methyl behenate $C_{23}H_{46}O_2$ (*cis* configuration) | C22:0 | 354.61 | 929-77-1 | Behenic acid ME, methyl docosanoate |
| Methyl erucate $C_{23}H_{44}O_2$ (*cis* configuration) | C22:1 | 352.09 | 1120-34-9 | Eruric acid ME, methyl *cis*-13-docosenoate |

[a]Molecular weight.
[b]Number assigned by the Chemical Abstracts Service (CAS) to identify substances.

Another feature in favor of MEs over ethyl esters for biofuel is that they exhibit greater lubricity (Holser and Harry-O'Kuru, 2006). In (Fig. 8.13(c)), the wear scar diameter decreases—indicating an increase in lubricity—with increasing chain length. The lubricity also increases with increasing saturation (i.e., wear scar diameter decreases for the C18:$y$ MEs as $y$ increases). At the molecular scale, oxygen-containing compounds like MEs can adsorb or react on surfaces

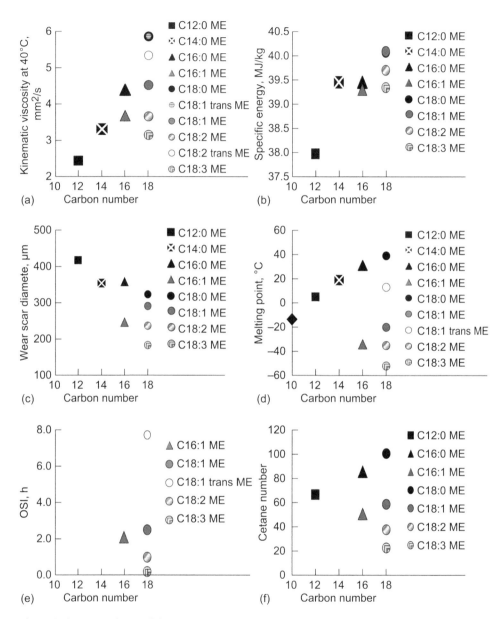

Figure 8.13.   Dependence of the thermophysical properties of methyl esters (MEs) on carbon number and saturation. (a) Kinematic viscosity. (b) Specific energy. (c) Wear scar diameter, a measure of lubricity; (d) Melting point. (e) Oxidative stability index (OSI). (f) Cetane number. (Melting point for C10:0 ME from Knothe, G. 2008a. "Designer" biodiesel: optimizing fatty ester composition to improve fuel properties. *Energy Fuels* 22:1358–1364. Remaining data from Moser, B. R. 2009. Biodiesel production, properties, and feedstocks. *In Vitro Cell. Dev. Biol. Plant* 45:229–266.)

that rub together, reducing the friction between asperities (microscale or smaller hills and valleys on a surface). This can reduce wear and the tendency toward seizure (Fazal *et al.*, 2010). This is, however, highly dependent on maintaining a water-free environment because of the potential for corrosion.

A review of material compatibility of diesel with respect to biodiesel (Fazal *et al.*, 2010) has identified some common themes among biodiesels: although they have better inherent lubricity than diesel (Holser and Harry-O'Kuru, 2006; Knothe and Steidley, 2005), they are hygroscopic and tend to absorb any moisture that may be present in their surroundings. This can promote the growth of microbes that produce corrosion-producing metabolic products, as well as render electrochemical corrosion more likely upon exposure to certain metals. In field testing for automotive applications, biodiesel generally produces similar or less wear than conventional diesel. However, elemental testing suggests that certain metals should not be paired with other particular metals. Biodiesel is, in general, more corrosive to copper than to ferrous compounds, but corrosion is dependent on both the parent feedstock and the type of metal (Fazal *et al.*, 2010).

Figure 8.13(d) shows that melting point of MEs increases with chain length and with increasing saturation for *cis*-type compounds. However, *trans*-type compounds exhibit reduced melting point relative to saturated compounds, and the melting point is substantially higher than their corresponding *cis*-type esters. Note that for most pure substances—and the MEs discussed here—the melting point is nearly equivalent to the freeze point: that is, the temperature at which solids and liquids coexist. Some substances can be supercooled below the freeze point when there are no nucleation sites, so the use of "melting point" may be considered more precise by some than "freeze point."

In any event, the high melting/freezing points of MEs become problematic for creating an adequate jet fuel blend, since the jet fuel specifications call for a freeze point of –40°C or below. The freeze point of a fuel was shown to be a function of its composition. Unlike pure hydrocarbons, some FAAEs may not exhibit a simple, linear dependence on blend ratio, making it more difficult to predict the freeze point of the final blend.

Examination of the melting point data brings out another issue with respect to jet fuel suitability. Jet fuel specifications use a reference temperature of –20°C for measurement of kinematic viscosity. At this temperature, Figure 8.13(d) shows that the saturated MEs from C10:0 and up are solids. Most vegetable oils are heavily weighted toward compounds in the range of C16 to C18 and consist of both saturated and unsaturated components. The *cis*-type unsaturated compounds at C16 to C18 may exist naturally as liquids at –20°C in their pure state, but the blends that are typical of vegetable oil will include solid components at this temperature. Even at –12.5°C, solids were identified in three commercial biodiesels because of freezing methyl palmitate (C16:0) and methyl stearate (C20:0) (Coutinho *et al.*, 2010). Unsurprisingly, most measurements of viscosity as a function of temperature for FAAE-based biodiesels stop short of –20°C, but they tend to exceed the 8 mm$^2$/s upper bound for jet fuel well above this temperature (e.g., see Freitas *et al.*, 2011).

The oxidative stability index (OSI), measured in hours, is used to quantify susceptibility to fuel degradation through exposure to oxygen. Increased OSI indicates increased oxidative stability (Fig. 8.13(e)). The OSI increases with chain length and decreases with unsaturation (i.e., the presence of double bonds). Finally, the cetane number (Fig. 8.13(f)), which increases as ignition delay time decreases, exhibits an increase with carbon number and a decrease with level of saturation. See Moser (2009) for further discussion.

The monounsaturated compounds (C*x*:1) thus present a marked advantage in terms of reduced viscosity and melting point with little or no penalty in terms of energy content. The polyunsaturated compounds (C*x*:2 and C*x*:3), however, although they may be beneficial in terms of reduced viscosity, they may increase the density to undesirable levels, as shown in Figures 8.13(a) and 8.12(a) for methyl linolenate (C18:3). Polyunsaturated esters are also less stable with a reduced storage stability (Ramos *et al.*, 2009).

For further reading on the effects of structural composition on biofuel properties, see Knothe (2005, 2008a), Moser (2009), and Yuan *et al.* (2003). For a good description of the analytical methods used for evaluating biodiesel composition, see Chapter 5 of Knothe *et al.* (2005) and also van der Westhuizen *et al.* (2011).

### 8.3.2   Emissions

The emissions from an airplane are a function of fuel properties, the amount of fuel used, operating conditions, and the combustion efficiency. To evaluate environmental impact, the following emissions are important for their effects on climate change and public health:

1. *Carbon dioxide (CO$_2$)* is a greenhouse gas that contributes to the warming of the planet. Airlines account for ~2.5% to 3% of global CO$_2$ emissions. Under conditions of complete combustion, the CO$_2$ emissions are ~3160±60 g/kg (Lee *et al.*, 2010). Carbon monoxide (CO) is also a regulated gaseous emission. Other greenhouse gases are also of concern with respect to climate change.

2. *Water vapor (H$_2$O)* is emitted in quantities of 1,230 ± 20 g/kg when fuel is completely burnt (Lee *et al.*, 2010). For supersonic aircraft, water vapor is the primary concern for gaseous emissions. At high altitudes, it can form contrails and promote formation of cirrus clouds. Contrails are formed when the hot, moist exhaust of the aircraft mix with sufficiently cold ambient air. Although the detailed mechanisms are not fully established at this time, contrail formation may exacerbate climate change. Other hydrogen-containing compounds of concern are hydroxyl radicals (•OH) and hydrogen peroxide (H$_2$O$_2$). The former is particularly important in chemical processes that produce sulfuric acid (H$_2$SO$_4$), NO$_x$, and ozone (O$_3$).

3. *Sulfur oxide (SO$_x$)* generation is directly proportional to sulfur content in the fuel. Sulfur dioxide (SO$_2$) is the most abundant sulfur-containing species. It is directly proportional to the sulfur content in the jet fuel and is mostly produced under the high-temperature conditions of the combustor (Lee *et al.*, 2010). These emissions may play a role in the development of acid rain. The specifications permit an upper limit on sulfur content to 3,000 ppm, although jet fuels are typically in the range of 500 to 1000 ppm currently (Chevron, 2006). Such compounds may also play a role through the formation of contrails and affect particulate emissions.

4. *Nitrogen oxide (NO$_x$)* is formed through reaction with atmospheric nitrogen under the high-temperature conditions in the combustor. Trace quantities of nitrogen bound to the fuel will also form NO$_x$. At ground level, NO$_x$ is of concern because it is toxic and is a precursor to chemical smog. It is linked to the creation of ozone in the troposphere (ground level to approximately 12,000 m). When emitted above the troposphere, NO$_x$ is implicated in the destruction of the stratospheric ozone layer.

5. *Particulate matter (PM) and unburned hydrocarbons (UHCs)* are the result of incomplete combustion. Emissions of this pollutant strongly depend on engine design and operating conditions, but fuel characteristics are also important. At altitude, PM can act as nucleation sites for contrail formation. Since the aromatic content in alternative fuels is nearly negligible, the generation of PM is also significantly reduced in terms of particle size, number density, and total PM mass (Timko *et al.*, 2011).

The effects of these pollutants on the atmosphere are compared through their impact on radiative forcing (RF) of the climate. The mean global surface temperature is directly related to RF. A positive RF corresponds to a warming effect, which is characteristic of CO$_2$ and soot emissions. A negative RF contributes to a cooling effect on the atmosphere, which is typical of sulfate particle emissions. NO$_x$ emissions result in a positive RF due to formation of tropospheric ozone, while it can induce a negative RF contribution from the destruction of ambient methane (CH$_4$). This framework provides a way to link the pollutants, so that the net effect on the atmosphere can be computed as a summation of all of the contributions.

Near ground level, air quality is an issue. At the high power conditions that are typical of airplane takeoff and landing, fuel is less likely to undergo complete combustion. This leads to higher emissions of CO, NO$_x$, UHCs, and PM, and these may induce smog formation and haze. PMs are also linked to respiratory distress and increased mutagen potential (Krahl *et al.*, 2005) that could be a factor in the development of lung cancer.

In testing for automotive emissions, the presence of FAAE reduces all types of emissions, except for NO$_x$. However, one study found that the release of this pollutant can be controlled by modification of injection timing (Krahl *et al.*, 2005). For aviation purposes, the presence of FAAE or FT fuel in a jet fuel blend reduced all categories of the emissions discussed here (Timko *et al.*, 2011).

For a thorough description of the complex interplay by which gaseous and particulate emissions influence the climate, see the excellent review by Lee *et al.* (2010). For a discussion of airborne particulate matter, see Kumar *et al.* (2010) and Timko *et al.* (2010); for biofuel combustion chemistry and the generation of toxic emissions, see Kohse-Hoinghaus *et al.* (2010) and Krahl *et al.* (2009); and for systems analysis of emissions reduction strategies for Europe, see Dray *et al.* (2010).

## 8.4   BIOFUEL FEEDSTOCKS FOR AVIATION FUELS

There are three commonly used categories of biofuels that indicate their readiness for commercialization. Although there is no universal agreement on the specifics, they will be categorized by the sophistication of the conversion technology as follows:

- *First-generation biofuels*: Are the easiest to bring to market using current technology. This category uses fermentation processes to produce bioethanol. Commonly used feedstocks include corn in the United States and sugar cane in Brazil. The corn-based bioethanol in particular is widely criticized for producing a larger environmental cost than petroleum fuels. The low energy density of bioethanol is inadequate for aviation fuel (Hileman *et al.*, 2010).
- *Second-generation biofuels*: Require process improvements in refining technology to commercialize. This category uses biomass or coal as feedstocks in gasification and/or liquefaction processes, such as Fischer-Tropsch, to produce biofuel. Biomass sources include switchgrass, agricultural waste, wood chips, and other forest residue.
- *Third-generation biofuels*: Require cost-effective, sustainable means of producing the feedstock, as well as efficient harvesting, oil extraction, and conversion. This category includes fuels from oils derived from vegetables and microbes, as well as greases from animal fat.

Second-generation biofuel processing is discussed elsewhere in this book and in other sources: For general background, see Kinsel (2010), Kreutz *et al.* (2008), Sims *et al.* (2010), and Sivakumar *et al.* (2010); for cellulosic genomics, see Rubin (2008); and for lignocellulosic coproducts, see Mtui (2009). This section will focus on feedstock production for first- and third-generation biofuels. The latter is an important area, since the high cost of feedstock has the biggest impact on the fuel price. For sustainability, the feedstock choice may change from one geographic region to another, depending on factors such as climate, land use, water availability, and the location of the nearest processing facilities.

### 8.4.1   *Crop production for oil from seeds*

Many plants have oil-rich seeds that can be converted to fuel, such as soybeans, *Camelina*, canola (rapeseed), and sunflower. Other feedstocks with good oil content are cotton seed, babassu, palm, and coconut. Much of the research to date on biodiesel has been performed on these crops, particularly soybeans (e.g., see Akbar *et al.*, 2009; Freitas *et al.*, 2011; Yuan *et al.*, 2005, 2009). The lipid composition of soy methyl esters is quite consistent across different studies, with a preference for C16 to C18 range, and a high proportion of monounsaturated and polyunsaturated C18 compounds, as shown in Figure 8.14(a). Yuan *et al.* (2005) presented data that showed genetically modified soy, which had a more desirable fatty acid profile with a higher concentration of monounsaturated methyl oleate (C18:1) (Fig. 8.14(b)). In Europe, canola has dominated the scene as a biofuel feedstock. It has a high lipid content of up to 50% and a composition that is mostly in unsaturated MEs at C18 and saturated MEs at C16 (Fig. 8.14(c)). Many vegetable oils exhibit a preference for the C16 to C18 range, such as cotton (Fig. 8.14(d)), *Jatropha* (Fig. 8.14(e)), sunflower (Fig. 8.14(f)), palm (Fig. 8.14(g)), milkweed, and *Camelina*. There are

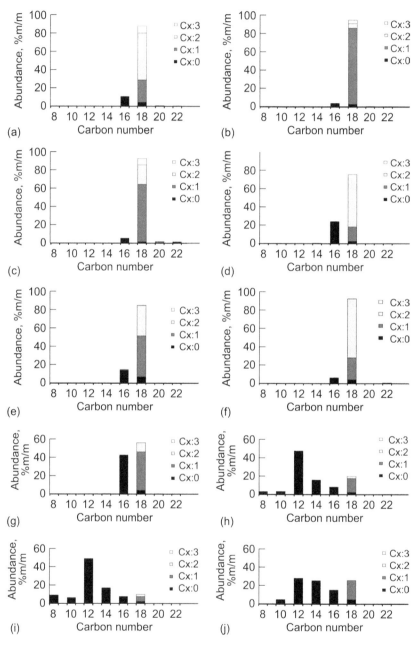

Figure 8.14.    Composition of vegetable oil methyl esters. (a) Soy. (b) Genetically modified soy. (c) Canola. (d) Cotton seed. (e) *Jatropha*. (f) Sunflower. (g) Palm. (h) Palm kernel. (i) Coconut. (j) Babassu. (Data were obtained from Akbar, E., Z. Yaakob, S. K. Kamarudin, *et al.*, 2009. Characteristic and composition of *Jatropha curcas* oil seed from Malaysia and its potential as biodiesel feedstock feedstock. *European J. Sci. Res.* 29:396–403; Freitas, S. V. D., M. J. Pratas, R. Ceriani, *et al.*, 2011. Evaluation of predictive models for the viscosity of biodiesel. *Energy Fuels* 25:352–358; Nogueira, Jr., C. A., F. X. Feitosa, F. A. N. Fernandes, *et al.*, 2010. Densities and viscosities of binary mixtures of babassu biodiesel + cotton seed or soybean biodiesel at different temperatures. *J. Chem. Eng. Data* 55:5305–5310; Yuan, W., A. C. Hansen, Q. Zhang, *et al.*, 2005. Temperature-dependent kinematic viscosity of selected biodiesel fuels and blends with diesel fuel. *J. Amer. Oil Chem. Soc.* 82:195–199; and Yuan, W., A. C. Hansen, and Q. Zhang. 2009. Predicting the temperature dependent viscosity of biodiesel fuels. *Fuel* 88:1120–1126.)

also some vegetable oils that present a broader range of carbon numbers, such as the palm kernel (Fig. 8.14(h)), which is the seed of the palm fruit, coconut (Fig. 8.14(i)), and babassu palm (Fig. 8.14(j)). For hydrotreatment of vegetable oil into renewable jet fuel, it is advantageous to have a carbon distribution that is heavily weighted in the C10 to C14 range, which is closer to conventional jet fuel. In this case, the biorefining process might be able to avoid the energy-intensive hydrocracking stage to break down long chain-length molecules, as is needed for C16 to C18 oils. However, most of the feedstocks shown in Figure 8.14 are not desirable feedstocks because they may be used for food and/or have limited availability. In addition, cultivation techniques, such as that used to create palm plantations, have been linked to deforestation.

For sustainability, avoiding competition with food crops or nutritional supplements is desirable, so much of the emphasis in feedstock development has shifted to other oil-producing crops that can be grown on marginal land that is unsuited for farming. The state of Montana in the United States has instituted a program to grow *Camelina* in intercropping; that is, in crop rotations when the land would otherwise have been left fallow. *Camelina* is related to canola, has 37 to 45% lipid content (~40% monounsaturated, ~50% polyunsaturated) and is a good source of ω3 fatty acid and vitamin E. It is one of the feedstocks that have been successfully used to create hydroprocessed esters and fatty acids fuel, and it has been tested extensively in civilian and military aircraft. However, it has already been approved as an animal feed supplement in the United States, and it may be approved for human consumption, which will limit its desirability as a biofuel feedstock. Two inedible beans that have generated much interest are *Jatropha curcas* (Figs. 8.11 and 8.14(e)) and castor beans.

A decade ago, *Jatropha curcas* appeared to be the perfect answer to sustainable biofuel production. *Jatropha* is a small tree (5 to 7 m tall) that produces fruit after its first year and matures at 3 to 5 years, with a lifetime of about 50 years. It is native to Mexico, Central America, and parts of South America. *Jatropha* seeds are oil rich, but inedible. It was believed that the trees were pest-, disease-, and drought-resistant and could thrive without irrigation on marginal land. Its large central tap root and shallow lateral roots are still widely believed to protect against wind and water erosion. Aggressive campaigns to increase land devoted to *Jatropha* production were instituted in India, Africa, and elsewhere.

Admittedly, *Jatropha* was known to have some undesirable properties that needed to be addressed. Some *Jatropha* varieties are quite toxic, and their seeds contain toxins, such as phorbol esters, curcin, trypsin inhibitors, lectins, and phylates. To mitigate this drawback, some have tested out the concept that the leftover meal after oil extraction could be detoxified with heat treatment and used for animal feed (Xiao *et al.*, 2011) or that the seed/kernel cake could be used as a fertilizer or as biomass in an anaerobic digester. Except for its role as biomass, the other uses have not yet been proven in a commercial setting.

The original optimism has been tarnished somewhat. Since it has not yet been domesticated, crop production is subject to wide variability. In a compilation of data from 1- to 9-year-old plants in South America, India, and Africa, the annual yields ranged widely from 313 to 12,000 kg oil/ha (Achten *et al.*, 2008). In general, trees produce seeds at a rate of 0.2 to 2 kg per tree (Achten *et al.*, 2008; Yang *et al.*, 2010), although there are reports of higher values. The oil content of the seeds is in the range of 27 to 44% oil by mass (Achten *et al.*, 2007). Under controlled conditions for 2-year-old trees developed from wild varieties throughout southern China, the maximum oil yield per tree per year was 15 times higher than that of the minimum oil yield (Yang *et al.*, 2010). (Note that it is not appropriate to extrapolate from biomass yield per tree to oil yield per hectare; the biomass yield depends on plant spacing, canopy management, and other production parameters, and the conversion between biomass and oil yield can vary substantially.)

Yields have not been impressive on marginal lands, and are highly dependent on rainfall, soil type, soil fertility, and genetics as well as plant age, spacing, and management methods. Although *Jatropha* can grow on a wide range of soils, for best biomass production, it requires an infusion of nitrogen and phosphorus as fertilizer (Foidl *et al.*, 1996) and water (Achten *et al.*, 2008; Yang *et al.*, 2010). Seeds do not mature all at once, which makes harvesting a labor-intensive process. Damage due to pests or disease have been noted in continuous monocultures in India (Achten *et al.*, 2008). In non-native localities, *Jatropha* may be an invasive species, and it has already escaped into the wild in Florida (Gordon *et al.*, 2011). In spite of these criticisms, there may be a role for *Jatropha* as a biofuel feedstock in some geographical locations, in part because plantations have already been established.

Castor beans are related to *Jatropha*, but are less toxic, and still have a lipid content of 40 to 60%. Castor-derived biodiesel blends may provide better lubricity than that of other vegetable oils, even at very low concentrations of less than 1% (Goodrum and Geller, 2005). This is likely because of the fact that castor oil contains high quantities of ricinoleic acid ($C_{18}H_{34}O_3$), an unsaturated ω9 fatty acid, and trace amounts of dihydroxystearic acid ($C_{18}H_{36}O_4$), both of which have more hydroxyl groups (1 and 2, respectively) than most vegetable oils (Refaat, 2009). Evogene Ltd., an Israeli company, is systematically breeding castor plants for good oil production potential. Biodiesel created from Evogene's castor has been produced by Honeywell's UOP LLC, with testing by the U.S. Air Force and NASA. Preliminary characterization of castor-based biodiesel indicates that the hydrocarbon composition has substantial content in the C9 to C11 range (Bruno and Baibourine, 2011) and looks promising for meeting the standards required for aviation fuel.

For further reading on other types of vegetable oils under consideration as biofuel feedstocks, see Balat (2011b), Holser and Harry-O'Kuru (2006), Kumar *et al.* (2010), Razon (2009), and Singh and Singh (2010).

### 8.4.2    *Crop production for oil from algae*

Third-generation feedstocks may be obtained from algae, cyanobacteria, and halophytes. Algae are photosynthetic organisms that span length scales from just a few microns (unicellular microalgae) up to 50 m (multicellular macroalgae, such as kelp). Cyanobacteria are also photosynthetic organisms, but, unlike algae, these microbes lack a membrane-bound nucleus. Halophytes are salt-tolerant plants, such as salt marsh grass, that can thrive in saltwater. There is a huge environmental benefit to growing saltwater-tolerant halophytes and microbes (Yang *et al.*, 2011), because they can be nourished from seawater or brackish water rather than freshwater. Algae are also an attractive crop because they can sequester carbon by using the flue gas from power plants as a nutrient source and remediate wastewater. In this section, we will limit the discussion to algae. For reading on cyanobacteria, see Bouriazos *et al.* (2010), Quintana *et al.* (2011), and Tan *et al.* (2011); and for halophytes, see Hendricks (2008), Hendricks *et al.* (2011), and McDowell Bomani *et al.* (2009).

Oleaginous (oil-producing) algae have been hailed as the most efficient producers of green crude over all feedstock types—potentially. The cost-effective growth of algae for conversion to biofuel in a production-scale setting has not yet been achieved, but there are many companies that are currently building such facilities. For a listing of commercial facilities for algae growth, see Singh and Gu (2010) as a starting point, but note that their data, based on a 2009 study, is already obsolete, and many more studies are in process.

From 1978 to 1996, the U.S. Department of Energy funded the Aquatic Species Program (ASP) to quantitatively explore the concept of producing biodiesel from algae. The program analyzed over 3,000 strains of microalgae and diatoms (algae with a cell wall of silica), which were narrowed down to the 300 most promising microbes. The intent was not only to understand which species were the best at oil production, but also their hardiness with respect to seasonal temperature variation, pH, and salinity, and the ability to outgrow wild competitors, all of which affect the stability of the culture. Algal growth in industrial-scale open ponds with 1,000 $m^2$ surface area was examined for feasibility of mass production in California, Hawaii, and New Mexico (Sheehan *et al.*, 1998). As is typical of open-pond aquaculture, the depth of the ponds was shallow to aid in light penetration; here 10 to 20 cm. The ASP determined that microalgae use far less water and land than oil-producing seed crops, estimating that 200,000 ha could produce significantly more energy than seed crops: about one quadrillion Btu energy ($\sim 1 \times 10^{18}$ J, or roughly 1% of global energy consumption). Nevertheless, the ASP concluded that biofuel from algae would not be cost-competitive with petroleum fuel. In their 1995 evaluation, they projected the cost of algae-based biofuel to be 59 to 186 U.S.$/bbl compared with petroleum at $20/bbl. Since then, the gap has likely narrowed because of adjustments on both types of fuel.

Biological productivity has been identified as the single largest influence on fuel cost (U.S. Department of Energy, 2010; Yang *et al.*, 2011). This includes such parameters as growth rate, metabolite production, tolerance to environmental variables, nutrient requirements, resistance to predators, and the culturing system. As with *Jatropha*, the reported yields for algal biomass have been wildly disparate because the technology for mass production is not mature.

The two major categories of growth facilities for photosynthetic production are (1) open ponds, typically in a racetrack configuration with flow driven by a paddlewheel and (2) closed, transparent photobioreactor systems that consist of an array of cylindrical tubes or an enclosure formed by two flat plates. Open ponds are the simplest design, and they are by far the least expensive to set up, operate, and clean after a growth cycle (about 2 weeks). However, they must contend with environmental conditions, such as large temperature swings, evaporative losses, and the threat of contamination by opportunistic wild species and predators. The algae selected for this type of operation must be hardy and fast-growing to outperform any competitors that are present in the environment. There are many examples of commercial success with this type of aquaculture for higher-value products, such as β-carotene and ω3 fatty acids. Photobioreactors are, in comparison, expensive to build, run, and maintain. However, they can operate at higher algal concentrations, so that the water that must be extracted at the end of the process is reduced. Depending on the design, the algae may have better access to light. They are also better equipped to handle less robust organisms, such as algae that have been genetically modified for improved oil production. Scaleup remains an issue; there is a limit as to the length of the cylindrical tubes if it is desirable to maintain a uniform temperature, pH, and dissolved gas content in the growth medium. Other design challenges are that heat and oxygen must be removed from the reactor, and a carbon source such as carbon dioxide must be replenished. The design must also have a protocol for preventing or removing biofilm buildup on the transparent surfaces. Both types of systems have strong advocates. For further discussion, see Brennan and Owende (2010), Carvalho *et al.* (2006), Cheng and Ogden (2011), and Jorquera *et al.* (2010). Within a few years, production facilities that are under construction or in development should provide more quantitative data, and there may well be a complementary role for open and closed systems.

In open-pond raceways that are supplied with nutrients from the flue gas of power plants, many strains of algae in an autotrophic growth regime can consistently increase its biomass on average by 20 grams (dry weight) per square meter of pond surface area per day ($g/m^2/d$). Overall biomass production from the ASP averaged 10 $g/m^2/d$, but at times achieved up to a maximum of 50 $g/m^2/d$ (Sheehan *et al.*, 1998). Theoretically, the average maximum yield could be substantially higher than that, but its attainment is elusive in a commercial-scale setting.

For algae that generate enough oil to be considered for biodiesel production, they are usually about 15 to 45% lipids when grown in the lab, depending on the strain and growth conditions, although some studies report lipid contents of up to 70%. Weyer and coworkers (2010) examined the theoretical maxima for oil production in many geographical locations with an optimistic algal lipid content of 50% and with efficiencies for photon transmission, photon conversion, and biomass conversion of 95, 50, and 50% respectively. With those assumptions, the annual oil yield ranged from the worst case in Kuala Lumpur of 40,700 L/ha/yr (4,350 gal/acre/yr) to the best case in Phoenix, Arizona, of 53,200 L/ha/yr (5,700 gal/acre/yr).

This work is consistent with another study that examined theoretical maxima. Cooney and coworkers (Cooney *et al.*, 2011) assumed that the maximum solar irradiance normal to the Earth's surface at ground level is 1,000 $W/m^2$, multiplied by correction factors that correspond to a sunny location with occasional clouds and account for the changing angle of the Sun during the day. Since algae can only use a specific spectrum of the incoming sunlight (about 45% of it), that number is also multiplied by a correction factor of 0.45. Then they used the highest photosynthetic conversion efficiencies to represent the maximum biomass that the algae could produce from that incoming energy. By relating the specific energy of the biomass constituents (protein and carbohydrates at 16.7 MJ/kg, lipids at 37.4 MJ/kg, and ash with 0 energy content), they could derive an expression for biomass production as a function of lipid and ash content. Assuming that the facility would grow algae 365 days a year (another optimistic assumption!), Figure 8.15(a) depicts the average maximum biomass yield per day, which ranges from 143 g/m$^2$/d (lipids = 0%, ash = 100%) to 63 g/m$^2$/d (lipids = 100%, ash = 0%). In these extreme cases, there are no proteins or carbohydrates produced, so the limits are not physically realistic, but in between those limits, all components (lipids, proteins carbohydrates, and ash) are represented. Interestingly enough, note that there is a tradeoff between lipid production and biomass yield. We will have more to say about this later. In an economic sense, whether or not a given number for lipid or biomass production is "good" depends on the value of the lipids and other coproducts from the biomass as well as the cost of oil extraction and refinement, and any other processing of coproducts that is needed.

Assuming that 100% of the oil can be recovered from the biomass, the oil yield is the product of the biomass yield and the lipid fraction (i.e., the lipid content on a percentage basis divided by 100). Figure 8.15(b) shows a family of curves representing annual oil yield that relates lipid content to biomass productivity. Weyer *et al.*'s (2010) annual oil yield of 4,350 to 5,700 gal/acre/yr at a fixed 50% lipid content corresponds to an average biomass yield of about 21 to 27 g/m$^2$/d in the Cooney *et al.* (2011) figure, shown as a vertical gray line. The gray box in Figure 8.15(b) represents an envelope that encompasses a generous estimate of current capabilities on a commercial scale. Some reviews critically evaluate the overly optimistic assumptions that are used in projections of algal oil output (Li *et al.*, 2008), based on understanding of the growth process. Others simply report the bloated figures without an astute appraisal of the claims. An understanding of the absolute upper limit that is physically possible can help assess claims that seem too good to be true.

Most algae can grow autotrophically (using photosynthesis and organic compounds to create food, also called phototrophically), heterotrophically (using complex organic compounds such as sugar for food without the need for sunlight), or mixotrophically (a combination of growth regimes). Contamination is a concern for heterotrophic growth since opportunistic bacteria also grow well on organic carbon. The type of growth regime can significantly affect the lipid (oil) composition. Under heterotrophic growth in the dark, the lipid content of *Tetraselmis* shifted dramatically toward lower carbon number and increased saturation. The abundance of palmitic acid (C16:0) increased from ~20% in mixotrophic/phototrophic growth to 79% for heterotrophic growth. The growth rate was somewhat slower, with a doubling time of 22 h for phototrophic and 25 h for heterotrophic growth (Day and Tsavalos, 1996). Figure 8.16 shows the effect of phototrophic versus heterotrophic growth regime for *Dunaliella tertiolecta* (Tang *et al.*, 2011). Less dramatic, but similar, compositional trends were reported for *Chlorella protothecoides* (Santos *et al.*, 2011). The ideal oil would be in the C10 to C14 range, which is closer to conventional jet, and would require less energy for refining. With this in mind, the shift toward lower carbon number is beneficial for the manufacture of jet-grade fuel, although the hydrocarbon chains are still longer than desirable. Since the C16 components are fully saturated, they will be more difficult to isomerize (Sec. 8.5.2.2) to bring down the freeze temperature to acceptable levels. Consequently, this oil may still require energy-intensive hydrocracking to bring it to jet-range quality.

When under stress such as conditions of nutrient deficiency, algae do not waste energy on cell division, so the growth rate decreases. Individual cells may become larger, and lipid production may increase, resulting in a net increase of lipids on a cell-wise basis. However, the ASP concluded that growing algae in a nutrient-deficient mode was generally counterproductive since the gain in lipid production was more than offset by the decrease in growth rate (Sheehan *et al.*, 1998). However, from a more holistic standpoint, it could reduce production cost: directly, since nutrients do not have to be supplied during the deprivation phase, and indirectly, from the reduction in biomass that must be processed. For example, one study found that after 4 days of nitrogen deprivation in the lab, *Dunaliella tertiolecta* cells accumulated five times more lipids than the control cells (Chen *et al.*, 2011). Similarly, *Nannochloropsis oculata* increased its lipid content from 35.0 ± 1.2% to 44.5 ± 2.2% after 4 days of nitrogen deprivation (Su *et al.*, 2011).

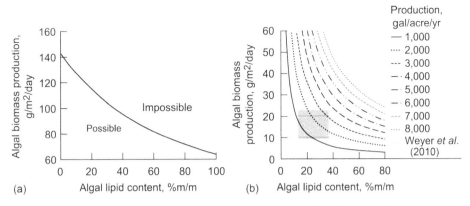

(a)

(b)

Figure 8.15. Maximum theoretical algal production. (a) Average daily biomass productivity as a function of lipid content; (b) Curves of constant annual oil yield as a function of lipid content and average biomass productivity, where 1,000 gal/acre/yr = 1,531.9 L/ha/yr. (Vertical line in (b) represents data derived from Weyer, K. M., D. R. Bush, A. Darzins, *et al.*, 2010. Theoretical maximum algal oil production. *Bioenergy Res.* 3:204–213. The gray box represents a rough envelope of current production capabilities. The remainder of the data is adapted from Cooney, M. J., G. Young, and R. Pate. 2011. Bio-oil from photosynthetic microalgae: case study. *Bioresource Technol.* 102:166–177, reprinted with permission from Elsevier, Inc.)

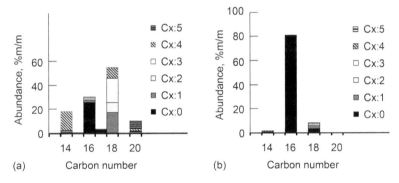

(a)

(b)

Figure. 8.16. Effect of sunlight on lipid content for *Dunaliella tertiolecta*. (a) Phototrophic growth. (b) Heterotrophic growth. (Data from Tang, H., N. Abunasser, M. E. D. Garcia, *et al.*, 2011. Potential of microalgae oil from *Dunaliella tertiolecta* as a feedstock for biodiesel. *Appl. Energy* 88:3324–3330.)

The lipid composition also exhibited a shift toward saturated compounds at C18, as shown in Figures 8.17. Many other studies have examined the utility of mixotrophic growth, initially setting growth conditions for high biomass production, followed by several days of nutrient deficiency to increase the lipid content (e.g., see Ben-Amotz, 1995; Levine *et al.*, 2010; Xiong *et al.*, 2010).

Another interesting concept is the growth of multispecies algal communities in open ponds. Recent work studied the impact of species diversity using cultures of 1 to 4 algal strains with 6 different species compositions that were drawn from a collection of 22 standard algal strains. The algae represented all major algal classes, including chlorophytes, cyanobacteria, cryptomonades, crysophytes, and diatoms. In addition, eight samples of naturally occurring phytoplankton with a shared evolutionary history were collected from the wild. This work suggests that such communities may store significantly more solar energy as lipids than monocultures in photobioreactors (Stockenreiter *et al.*, 2012). It would also be useful to know if naturally occurring communities have other beneficial properties, such as better stability than monocultures.

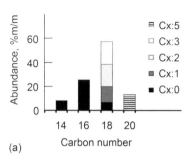 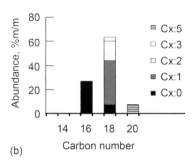

(a)                                    (b)

Figure 8.17.   Effect of nitrogen on lipid content for *Nannochloropsis oculata*. (a) Sufficient nitrogen. (b) Nitrogen deficiency. (Data from Su, C.-H., L.-J. Chien, J. Gomes, *et al.,* 2011. Factors affecting lipid accumulation by *Nannochloropsis oculata* in a two-stage cultivation process. *J. Appl. Phycol.* 23:903–908.)

Currently, open-pond design is based on deep familiarity and experience with such systems. Is there a way to move toward the theoretical maximum in a more scientific, cost-effective way? In high-density cell cultures, the cells nearest to the surface absorb most or all of the available light (Chisti, 2007) due to self-shading. An engineering solution is to improve the vertical mixing in the system, so that more cells have access to light. One design choice in pond construction is whether to use a rectangular cross section or an angled cross section with a wider top than bottom. Recent work at NASA Glenn Research Center showed that the addition of passive mixing devices is more effective in providing access to light than the cross-sectional geometry in dense suspensions. This knowledge permits the choice of channel geometry to be based on other factors, such as ease of construction or maintenance. Other, more sophisticated approaches are needed to examine in greater depth the stochastic effects of hydrodynamic mixing on cell growth. This requires better understanding of the time scales that are relevant to industrial production. There is a great deal of research on the short time scales associated with photosynthesis, such as the amount of time it takes a cell to absorb a photon, convert it into food, and be ready to absorb another one: on the order of milliseconds. There are similarly small hydrodynamic time scales associated with turbulence, as well as larger ones linked to pond traversal. In current designs, the primary locations for vertical mixing occur at the paddlewheel and at the circular bends at either end of the pond, so that time scales associated with the unmixed state are on the order of tens of minutes. Algae that are near the surface will be continuously in the light during that stage, while the those near the bottom may be in the dark. There are also photoadaptation time scales that appear to be on the order of hours. Exposure to too much light can lead to photoinhibition. There are growth time scales with doubling time on the order of days. The complex, species-specific interplay among all of these factors with the biokinetics is an area that is ripe for exploration and may provide a scientific basis for effective pond design.

See the Department of Energy's Algal Biotechnology Program Roadmap for a good description of all aspects of algae growth for biofuel (U.S. Department of Energy, 2010); for background on basic photosynthetic processes, see Nelson and Yocum (2006); for studies on bioprospecting/screening, see Araujo *et al.* (2011), Doan *et al.* (2011), Gouveia and Oliveira (2009), Griffiths and Harrison (2009), Lee *et al.* (2011), and Rodolfi *et al.* (2009); for screening for biodiesel and sewage remediation, see Sydney *et al.* (2011); and for algae and $CO_2$ remediation, see Ho *et al.* (2011).

## 8.5   MANUFACTURING STAGES

Extraction processes from vegetable oils, algae, and greases release oil that is high in triglycerides and fatty acids with hydrocarbon subunits that are primarily in the range of C16 to C18. The goal of biorefining is to convert this green crude (also called green diesel) into a less-viscous fuel with better cold-temperature properties.

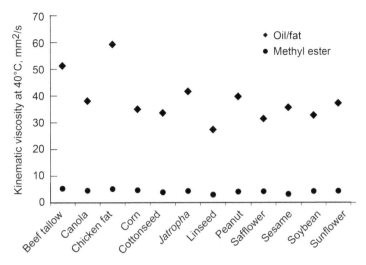

Figure 8.18.    Transesterification reduces viscosity of vegetable oils and greases by converting them to methyl esters. (Data obtained from compilations in Jingura, R. M., D. Musademba, and R. Matengaifa. 2010. An evaluation of utility of *Jatropha curcas L.* as a source of multiple energy carriers. *Int. J. Eng. Sci. Technol.* 2:115–122; Knothe, G. 2008a. "Designer" biodiesel: optimizing fatty ester composition to improve fuel properties. *Energy Fuels* 22:1358–1364; Lang, W., S. Sokhansanj, and F. W. Sosulski. 1992. Modelling the temperature dependence of kinematic viscosity for refined canola oil. *J. Amer. Oil Chem. Soc.* 69:1054–1055; and Refaat, A. A. 2009. Correlation between the chemical structure of biodiesel and its physical properties. *Int. J. Environ. Sci. Technol.* 6:677–694.)

Biodiesel is defined as fuel that is made up of mono-alkyl esters of long-chain fatty acids that are derived from vegetable oils or animal fats. In petroleum fuel, straight-chain, saturated hydrocarbons are called *n*-paraffins. For biodiesel, paraffinic subunits contained in fatty acids are combined with alcohols to form esters that can be blended with conventional diesel and used in vehicles (Sec. 8.5.2.1). This form of biofuel is referred to as fatty acid methyl esters, or equivalently fatty acid alkyl esters. Transesterification reduces the viscosity of the vegetable oil feedstock to a range that is acceptable for biodiesel, as shown in Figure 8.18.

For renewable jet fuel, the freeze point and cold-temperature properties must be improved further. To manufacture hydrotreated esters and fatty acids fuel, the fatty acids and triglycerides in these oil feedstocks undergo a different set of reactions to yield $CO_2$, $H_2O$, and long-chained *n*-paraffins. This is followed by a second reaction (hydrocracking) that breaks apart the dense hydrocarbon chains to form smaller, highly branched hydrocarbons, shorter chain *n*-paraffins, and a small amount of cycloparaffins (Sec. 8.5.2.2). As a result, the molecular makeup of hydroprocessed esters and fatty acids fuel is very similar to conventional jet fuel with a blend of pure hydrocarbons in the range of C9 to C15 (see Fig. 8.11).

## 8.5.1   *Dewatering, crude oil extraction, and preprocessing*

To prepare for refining, oil-rich beans or seeds are dried, cleaned, cracked, and compressed into flakes. The oil is extracted chemically by exposing it to a solvent or, less efficiently, mechanically by pressing. A combination of preheating and mechanical extractions can increase oil yield (Mahmoud *et al.*, 2011). For algae, the choices become more varied. The algal biomass can be dried, and oil extracted by chemical, biochemical, or thermochemical approaches, often with some mechanical assistance.

Dewatering can be an energy-intensive stage in oil production from algae. Even at the end of a growth cycle, the algal broth is about 5% to 17% biomass in industrial-scale open ponds.

Although the biomass concentration could be much higher when grown in photobioreactors, there is still significant water content. The biomass can be removed by a number of methods. The size and density of the algal cells are a primary consideration for choosing a separation technique:

1. *Sedimentation*. Most algae cells are heavier than water, so sedimentation is one possible, inexpensive strategy, but it is slow and is better for large, dense cells, roughly those larger than 50 to 100 μm. Many of the most promising algae for biodiesel production are about 1 to 30 μm in size. Flocculants reduce or neutralize the negative charge of microalgae so that they can form clumps, which effectively increases the particle size for better sedimentation. Flocculants, sometimes coupled with ultrasonic forcing, may be used as a pretreatment step to encourage clumping.

2. *Flotation*. In flotation, air is bubbled up through the culturing system. As algae (or clumps of algae) attach to the bubbles, they are carried up to the liquid surface, where they can be removed by skimming.

3. *Filtration*. Membrane microfiltration is highly efficient at separating water from biomass in small batches (Zhang *et al.*, 2010), but it is not a good fit for large-scale production due to high energy requirements for pumping and filters that require a lot of attention and maintenance.

4. *Centrifugation*. Centrifugation is very efficient in the lab, but it is prohibitive in an industrial setting because of the high energy costs of running this equipment. If the biomass needs to be dried for oil extraction, techniques range from the low-tech, slow process of drying in the sun to fast, expensive processes such as bed drying, freeze drying, drum drying, and spray drying.

Thermally assisted mechanical dewatering is discussed by Mahmoud *et al.* (2011).

Once the algal cells are isolated, the most common method for lipid extraction from them occurs by disrupting the cell walls, releasing algal contents into solution from which they can be separated. This can be accomplished chemically by solvent extraction. The most common solvent in the lab is *n*-hexane because it results in high oil yield, but it is also a slow process, which is not desirable for a large-scale manufacturing facility. Also, the separation and recovery of the solvent and lipids (e.g., through distillation) adds an additional step that requires significant energy input. Finally, there are fire and safety hazards that come with this method. Another option is enzymatic extraction. In this method, enzymes in a water solvent break down the cell walls; one such example is the enzyme alkaline protease for *Jatropha* (Achten *et al.*, 2008). Enzymatic extraction is a less effective technique than chemical extraction, however, and pretreatment may be helpful to assist the enzymes in breaking down the cell walls, such as ultrasonication. Assistance may also be given by other thermal or mechanical means, such as autoclaving, bead-beating, or microwaves, which are all viable in lab-scale processing. At large scale, mechanical pressing has been suggested as a cost-effective pretreatment for lipid extraction, perhaps in combination with solvent extraction (Gong and Jiang, 2011).

Some algae, such as *Dunaliella*, have a high extraction efficiency because they do not have a thick cell wall, but other species of interest, such as *Nannochloropsis* or *Chlorella*, are more challenging because of their hard cell walls. Diatoms also have hard cell walls. Other extraction methods may be more attractive for these microbes, such as chemically extracting lipids through the cell wall, or manipulating the hydrophobicity, which encourages algae to secrete the lipids in a process known as "milking." In the latter case, the algae remain viable after the extraction and can be returned to production as an active cell culture.

After the extraction step is complete, the oil must be filtered and cleaned before going on to the next manufacturing stage.

For further reading on dewatering of algae, see Uduman *et al.* (2010); for extraction processes, see Grima *et al.* (2003) and Mercer and Armenta (2011).

### 8.5.2   *Biorefining processes*

Processes such as transesterification and hydroprocessing are used to transform the molecular structure of these crude oils: hydrocarbon fuels are produced by removing oxygen and impurities from them.

#### 8.5.2.1   *Transesterification*

Transesterification is the classical method of producing biodiesel. Vegetable oils, or green crude, are highly viscous compounds based on saturated and unsaturated fatty acids, often concentrated in the range of C16 to C18. As such, their viscosity is high as is their freeze point, which make them inadequate except for use in some piston-driven diesel engines. The goal of transesterification is to transform the green crude to biodiesel, which has more desirable properties such as reduced viscosity and improved cold flow properties. This is accomplished by removing oxygen, increasing hydrogen content to saturation levels, and reducing hydrocarbon chain length. The best biodiesel blends will have high levels of monounsaturated and saturated fatty acid esters. They will also be low in polyunsaturated fatty acid esters, which lead to poor oxidative stability and high density (see Figs. 8.13(e) and 8.12(b), respectively)). These processes are effective strategies for production of biodiesel regardless of the feedstock.

   Most of the fatty acids in vegetable oils are bound up in triglyceride molecules (or triacylglycerols, TAGs). These compounds comprise a glycerol backbone that is connected to three fatty acids. For some algae, triglycerides are the main carbon storage molecule, but algal oil may also include free fatty acids.

   The fatty acids are stripped off the glycerol sequentially, until the result is three separate molecules of fatty acid esters and a glycerol molecule. In the finished product, the presence of monoglycerides or diglycerides (glycerol bound up with one or two fatty acids, respectively) is a sign of incomplete transesterification. In order for this reaction to occur, the triglyceride must be exposed to an alcohol, usually methanol, in the presence of a catalyst. In practice, the reaction is usually carried out with an excess of methanol. The liberated fatty acids react with the alcohol to form esters (in this case, FAMEs). This is depicted in Figure 8.19 in which $R_1$, $R_2$, and $R_3$ represent three alkyl groups, alkanes (paraffins) with one missing hydrogen at one end of the carbon chain. Esters are a molecular alliance of a carboxylic acid with an alcohol, in which the H of at least one of the acid's hydroxyl (OH) functional groups is replaced with a straight-chain R (alkyl) group. For mono-alkyl esters, there is exactly one replacement within the acid by an R group, as in Figure 8.19.

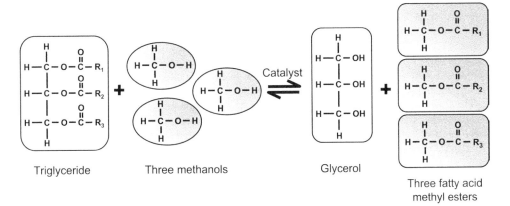

| Triglyceride | Three methanols | Glycerol | Three fatty acid methyl esters |

Figure 8.19.   Transesterification is the reaction of a triglyceride and alcohol to form glycerol and alkyl esters.

The process is reversible, in theory, but in actual practice it is unlikely since the glycerol is not miscible (i.e., it does not mix) with the product, although excess methanol can slow the separation. The result is a two-phase system, with the fatty acids floating to the top of the reaction vessel.

For biodesel, methanol is usually the alcohol of choice because of its low cost and ready availability, although ethanol is used in places such as Brazil in which that fuel is abundant. Some studies have examined the use of genetic modification of *E. coli* to produce higher chain alcohols, such as isobutanol (Atsumi *et al.*, 2008), but this process is not yet ready for commercialization.

A catalyst is a substance that speeds up, or otherwise assists, the speed or likelihood of a chemical reaction, without itself being altered. The catalyst can be alkaline (such as sodium hydroxide and sodium methylate), acidic (such as sulfuric acid), or enzymatic. Strong acids donate a proton to the carbonyl (–CO) group, whereas bases remove a proton from the alcohol (Cordeiro *et al.*, 2011). Acidic catalysts usually provide the most complete reaction, although they are excessively slower than alkaline catalysts, and they require higher temperatures. If free fatty acids are present in the reactant (roughly >0.5 to 1%), reaction with alkaline (base) catalysts will lead to a saponification reaction that turns the triglycerides into soaps rather than alkyl esters. In this case, a two-step procedure is needed to use the quicker base catalysts. The first step is a pretreatment with an acid catalyst, which converts both free fatty acids and triglycerides to alkyl esters. After this reaction is neutralized, it is followed by another step with a base catalyst to convert the remaining triglycerides. The additional energy and time requirements of acidic catalysis may be more economical than the extra steps that are needed in a two-stage procedure. Lipase enzymatic catalysts are more expensive and usually require longer reaction time, although they are more environmentally friendly. Alkaline protease was used in aqueous enzymatic oil extraction for best results from *Jatropha* (Achten *et al.*, 2008). The conversion of palm oil from lipase enzymes produced by three bacterial strains that support transesterification was tested, but it exhibited low conversion efficiency (~20% compared with commercial lipases at 90+%) (Meng and Salihon, 2011).

After the reaction is complete, the FAAEs must be separated from the catalyst, excess alcohol, water, free fatty acids, and the glycerol. The glycerol is relatively simple to isolate from the products, as it precipitates to the bottom of the tank and does not mix with the product. For the others, additional steps, such as distillation, would be needed.

Notice that in Figure 8.14 the alkyl groups $R_1$, $R_2$, and $R_3$ maintain their identities as functional groups in the reactant and the product. Consequently, the fatty acid profile of biodiesel corresponds in large measure to that of its feedstock. As seen in Section 8.4.1, vegetable oils are concentrated at higher carbon number (mostly at C16 to C18), relative to petroleum jet fuel (roughly C8 to C16), as shown in Section 8.3.1.

For sustainability, the overall process must be streamlined to reduce the number of energy-intensive steps and replace any toxic or hazardous compounds with more environmentally friendly ones. Consequently, the development of effective catalysts that are nontoxic, inflammable, and recyclable are important areas of research. A feasibility study of lipid extraction from cyanobacteria was able to extract 97% of the lipids with liquefied dimethyl ether, a nontoxic compound. Furthermore, this operation was performed on wet biomass, eliminating the need for the drying and cell-wall disruption steps (Kanda and Li, 2011).

The acidic and alkaline catalytic steps require a neutralization step to end the reaction. To avoid this extra step, much work has been devoted to extraction processes using solid catalysts that are more selective, safe, and environmentally friendly. Zeolites and mesoporous compounds can meet these requirements and also have a high concentration of active sites, high thermal stability, and better shape selectivity (Carrero *et al.*, 2011; Cordeiro *et al.*, 2011; Perego and Bosetti, 2010). Conversion efficiencies using these unique materials have improved sufficiently to consider them for commercial biodiesel production (Perego and Bosetti, 2010; Verma *et al.*, 2011).

Other extraction techniques for transesterification that avoid using toxic solvents include supercritical gas extraction (Edwards, 2006; Levine *et al.*, 2010; Li *et al.*, 2010; Soh and Zimmerman, 2011) For further reading on the effects of process variables on FAAE yield,

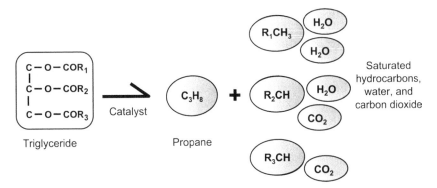

Figure 8.20.   Example of hydroprocessing: deoxygenation of a triglyceride into saturated hydrocarbons, water, and carbon dioxide.

see Alptekin and Canakci (2011) and Rashid, U., F. Anwar, and G. Knothe (2009), and for production-scale transesterification see Van Gerpen and Knothe (2005).

### 8.5.2.2   *Hydroprocessing*

Hydroprocessing is another biorefining technique, using catalysts in the presence of hydrogen to convert a variety of free fatty acids, triglycerides, alkyl esters, and other compounds into paraffinic hydrocarbons by removing oxygen and saturating them with hydrogen. This process can also be used to drive contaminants like sulfur, nitrogen, and trace metals from the hydrocarbons. Hydrotreatment occurs at relatively low temperatures and pressures, which provide sufficient driving force to break the molecular bonds with S, N, or O and replace it with a hydrogen molecule. The residual S, N, and O atoms can combine with hydrogen to form stable compounds. This process works most efficiently on unsaturated oils.

   The deoxygenation reaction is shown in Figure 8.20, which is carried out at low temperature (around 300°C, depending on the specifics of the process) using a dimetallic catalyst, such as nickel-molybdenum (Ni-Mo) or cobalt-molybdenum (Co-Mo). (For better readability, note that the hydrogen atoms have been removed from the display of the triglyceride molecule.) The $R_1$, $R_2$, and $R_3$ still denote alkyl groups (the paraffinic subunits) of a fatty acid. At the completion of the reaction, the terminal carbon in the alkyl group is saturated with hydrogen, rather than bonded to oxygen. One of the alkyl groups has gained an additional link in the hydrocarbon chain ($R_1$-$CH_3$), so that its carbon number increases by one, but the other alkyl groups have become alkanes (paraffins) that maintain the same carbon number. Another product of the reaction is propane ($C_3H_8$), formed from the glycerol backbone, which can be recovered by fractional distillation. The remaining carbon and oxygen atoms in the triglyceride have been converted to $CO_2$ (or CO, depending on the reactant) and $H_2O$. If any carbon atom possesses double bonds in the reactant, these components will become saturated with hydrogen, so that the product consists of long-chain *n*-paraffins only.

   The benefit of converting the triglyceride to paraffins rather than methyl esters is that the stability, specific energy, and cold-temperature and blending properties of deoxygenated hydrocarbons are better suited for jet fuel.

   Since most of the alkyl groups in the reactant vegetable oil are in the range of C16 to C18, there are two strategies for reducing the freeze point: (1) convert the dense, straight-chained paraffins into more highly branched hydrocarbons; or (2) crack the dense hydrocarbons into shorter-chained molecules (~C12 to C14). Recall that Figure 8.6(f) showed that the freeze temperature decreases as carbon number decreases. The freeze temperature also decreased for the more complex, branched isoparaffins compared to that of straight-chain *n*-paraffins. Higher carbon numbers can be better tolerated if the hydrocarbons are isomerized into branched molecules.

There are trade-offs to be made here. The isomerization process is similar to the deoxygenation reaction described above in that it operates at moderate temperatures (perhaps 250 to 350°C) and pressures (typically less than 5 MPa). It uses excess $H_2$ and catalysts as reactants, and it will produce the most jet fuel when the feedstock molecules are in the range of C10 to C14. If the feedstock has carbon chains that are either shorter or longer, the resulting hydrocarbon mix can undergo fractional distillation to separate out the heavier and lighter hydrocarbons. The less desirable components can be used for other purposes, such as green diesel or cooking fuel, but it will supply less jet-grade fuel. Isomerisation is also more effective when the feedstock is not fully saturated.

For the case of vegetable oil that is heavily weighted in the C16 to C18 range and/or is fully saturated, the more extreme hydrocracking option can be considered. At higher temperatures (~350 to 420°C), much higher pressures (7 to 14 MPa), with excess $H_2$ and the right catalyst, the carbon-carbon bonds of long-chained *n*-paraffins are ripped apart to form shorter *n*-paraffins and branched isoparaffins. The resulting product will have a higher proportion of hydrocarbons in the range needed for jet fuel, but because of the large energy input, it will come at a cost, both in terms of fuel and in environmental impact.

For a good reference on hydroprocessing, see Robinson and Dolbear (2006), and for hydrogenization of unsaturated MEs, see Bouriazos *et al.* (2010). Refining vegetable oils removes most of the natural antioxidants (Holser and Harry-O'Kuru, 2006). For a discussion of additives and blending to improve cold-flow properties see Chastek (2011), Coutinho *et al.* (2010), Joshi *et al.* (2011), Kerschbaum and Rinke (2004), Kerschbaum *et al.* (2008), Moser (2009), and Wang *et al.* (2011); and for emissions see Moser (2009).

### 8.5.2.3   *Other strategies*

Another option for creating jet fuel is to start with relatively low-weight alcohols, such as propanol (C3) or butanol (C4), and perform an oligomerization step. In this reaction, the short-chained hydrocarbons undergo a reaction to extend their lengths, thus building up from short-chain hydrocarbons towards jet-fuel-range hydrocarbons.

Some studies have examined the use of microbes to produce jet fuel precursors directly from sunlight (Atsumi *et al.*, 2008 and Tan *et al.*, 2011), from fatty acid feedstocks (Dellomonaco *et al.*, 2010), or through fermentation of sugars from lignocellulosic decomposition (Ha *et al.*, 2010). This technology is still in the early stages and is not a near-term solution.

### 8.5.3   *Coproducts*

One of the most critical aspects of sustainable-process development and economic viability will be the identification of value-added chemicals, energy, and materials from the remnants of the biofuel production process, such as biogas, animal feed, fertilizers, industrial enzymes and chemicals, bioplastics, and surfactants. It will also require finding creative new applications for these "coproducts" as needed. Glycerol, a byproduct of transesterification, is still cited as a valuable coproduct for the cosmetics and chemical industries, but at a practical level, the market has become saturated because of the increasing manufacture of biodiesel (Yazdani and Gonzalez, 2007). Since the creation of glycerol is intrinsic to the process, new uses will have to be found for glycerol, such as microbial fermentation into fuels and marketable chemicals (Yazdani and Gonzalez, 2007).

*Jatropha* press-cake has hemicelluloses, cellulose, and lignin, which can be converted through anaerobic digestion into biogas (Demirbas, 2011) or through pyrolysis into bio-oils, gas, and char. Its biomass can be used as animal feed (with appropriate detoxification) or as fertilizer. For a detailed description of *Jatropha* fruit, shell, husks, and wood that could be used to produce energy, see Jingura *et al.* (2010).

For algae, there are three major components of biomass: lipids, carbohydrates, and proteins. Lipids and carbohydrates can be converted into fuel, and the proteins can become coproducts, such as animal feed. Another option: anaerobic digestion of algal biomass and cellulose can be used as a means of $H_2$ production (Carver *et al.*, 2011). For other sorts of coproducts from algae, see Cardozo *et al.* (2007). The remediation of wastewater (Aresta *et al.*, 2005; Park and Craggs, 2011; Park *et al.*, 2011) or toxic compounds (Petroutsos *et al.*, 2007) or the sequestration of $CO_2$ (Aresta *et al.*, 2005; Sayre, 2010) can be considered coproducts for the purposes of calculating environmental impact in life cycle analysis.

## 8.6   LIFE CYCLE ASSESSMENT

In general terms, sustainability requires that our activities do no harm to the planet or the life that depends upon it. This means that people take priority in the competition for food and water. Balat passionately argues that there is a direct link between using edible oils for biofuel and starvation (Balat, 2011a, 2011b). Thus, the search for energy production should avoid the use of food crops, arable land, or fresh water. The huge preponderance of scientific evidence indicates that we are foolhardy to continue recklessly pumping greenhouse gases (GHGs) into the earth's atmosphere. As a result, energy production techniques are sought that are at least carbon neutral: the net effect of all parts of the process maintains the same level of GHG as currently exists. However, the smarter strategies will result in a net reduction of GHG in the atmosphere. In considering environmental impact, it is not sufficient to consider only the emissions resulting from fuel burn. The environmental cost of materials and energy as well as changes in land use that are required to produce the fuel must also be part of the equation. Particularly for the aviation industry, noise and particulate emissions are of concern for public health and safety and could legitimately be considered criteria for sustainability. To quantify the net impact of any particular process or set of processes, a relatively new field of life cycle assessment (LCA) has emerged to provide solid guidance for determining which processes are in fact sustainable.

The heart of sustainability evaluations resides in the LCA, which is a means of quantitatively evaluating a process system from start to finish for the purpose of objectively comparing different process systems. The goal of an LCA may be to evaluate environmental impact, inform market strategies, and/or analyze socioeconomic impact of a particular set of choices. The following example will consider the outline of an LCA for the environmental impact of conventional versus alternative aviation fuels.

The first task is to decide on the purpose of the LCA and thereby decide on the "functional units" that will be used in the analysis. Conversion factors will be used to relate everything in the analysis to this functional unit, which could be almost anything—but it has to relate to the problem of interest. To analyze how much could be saved over the next 5 years by changing from incandescent light bulbs to compact fluorescent light bulbs in a living room, the functional unit might be related to the amount of light that is needed to read the paper at night. In this case, "lumens per square foot per Euro" may be chosen as a functional unit.

To compare the change in greenhouse gas emissions due to changing from conventional jet fuel to an alternative fuel while flying a fully loaded Boeing 787 from New York to Tokyo, grams of $CO_2$ emissions per kg fuel (g $CO_2$/kg) may be chosen. If the kilograms of fuel burned on such a trip is known, the mass of $CO_2$ emitted during this flight can be calculated. For this case, everything in the problem must be converted to something representing the mass of $CO_2$ that is emitted for every kilogram of fuel that is burned. This becomes a little tricky, because $CO_2$ is not the only greenhouse gas of interest, and the effects of other emissions should be included. Most environmental LCAs incorporate other greenhouse gas emissions by introducing the concept of equivalent $CO_2$ emissions ($CO_{2,eq}$). The equivalence is developed through conversion factors, such as global warming potential or alternatively radiative forcing, which account for the manner in which the pollutant affects the buildup of heat in the atmosphere. When the mass of the greenhouse gas is multiplied by its corresponding conversion factor, the result is in units of grams $CO_{2,eq}$, and the sum is calculated of the effects of all of the greenhouse gases to find a number that represents the mass of $CO_2$ equivalent that was pumped into the atmosphere by the flight.

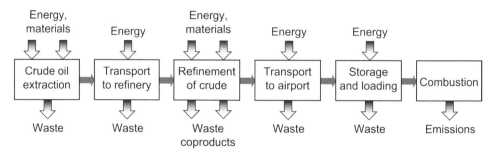

Figure 8.21.   Schematic of life cycle assessment for conventional petroleum fuel.

Knowing something about the chemistry of the fuel combustion process and the aircraft performance and operating conditions, it may be relatively straightforward to estimate how much $CO_2$, $CO$, $NO_x$, $SO_x$, $H_2O$, and particulate matter is created by burning a kilogram of jet fuel. The really tricky part is to include the entire history of $CO_2$-equivalent emissions that were generated by manufacturing that kilogram of fuel. So now the process that created this jet fuel must be reviewed and analyzed.

The life cycle of conventional fossil fuel is depicted in Figure 8.21. The LCA must capture the environmental impact of all of the materials and processes that are required to create aviation fuel, as well as all of the things that are left behind after each step of the process is performed. In the simplest type of analysis, each of these things is expressed in terms of the functional units. It may not at first be immediately obvious how to do that, but it becomes clear by identifying all of the stuff along the life cycle path.

The first step is to extract the crude oil, noting all of the energy and materials needed to perform this step, such as chemicals, lubricants, and water or other coolants used for drilling. Also take care to identify all of the waste that is generated during the extraction. For example, there will be gaseous emissions associated with using fuel to power machinery.

Next, the crude oil is transported to the refinery in a pipeline, in a supertanker, by truck, and/or by rail. Each of these modes of transport requires energy (e.g., fuel to run motors or pumps) and generates waste (e.g., gas emissions). At the refinery, the crude oil is processed to yield aviation-grade fuel. Inputs for this step include electricity and/or fuel to heat reaction chambers and pump fuel through distillation columns, as well as the environmental cost of input gases, chemicals, and catalysts. Crude oil is a mixture of hydrocarbons, only some of which are suitable for aviation fuel. The remaining (non-aviation) fuel can be sold for other purposes and would be considered a coproduct of the refining process. Coproducts are materials, energy, or other benefits that are created during the processing step, and they add value in the overall life cycle assessment. Inputs that are consumed and waste that is generated yield a value that represents their environmental cost. Coproducts thus offset some of the net environmental cost, so if the inputs and waste are expressed as positive values, then the coproducts are given negative values.

After refinement, the aviation fuel is transported to the airport, where it is stored and eventually some proportion of that fuel is loaded onto an aircraft. When the aircraft takes off, the fuel is consumed and the combustion products are exhausted into the atmosphere. The "system boundaries" in this LCA then extend from the oil well where the crude was extracted to the exhaust gases left behind in the wake of the aircraft, leading to the widely used description of this sort of LCA as "well to wake."

A similar LCA for an alternative fuel, such as castor-derived hydrotreated esters and fatty acids fuel, might look quite similar to Figure 8.16, but it requires additional steps to describe it. In Figure 8.22, a couple of steps are added at the front end to represent the production of the castor feedstock. The first box represents the growth of the castor in the fields, harvesting, and the separation of seeds from the rest of the plant material. Inputs might include the environmental costs of seeds, fertilizer, herbicides and pesticides, irrigation, gasoline and electricity for farm

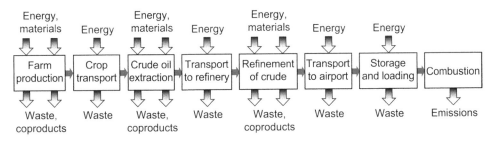

Figure 8.22.   Schematic of life cycle of a first-generation biofuel.

machinery, and packaging materials. Output waste might consist of gaseous emissions, fertilizer runoff, and plant debris. However, perhaps the plant debris left over on the farm could be transformed into a useful coproduct by running it through an anaerobic digester to create methane fuel or turned into compost to improve soil quality. In that case, some of the waste stream might be turned back in toward the crop production process and reduce the amount of input energy or fertilizers needed to produce the crop.

Once the castor seeds are delivered to the processing plant, depicted in the third box of Figure 8.22, they could be crushed to extract vegetable oil. The debris left over from the crushing process could be used for animal feed, so this would be tallied as a coproduct. The vegetable oil must be further processed and blended to yield qualified aviation fuel, generally done in a different facility, so it travels on to the crude oil refinery. Beyond this point, the fuel's journey has the same stages as that of petroleum-based jet fuel.

LCAs are meant to systematically evaluate processes, and since processes change over time, the LCAs must also be capable of evolving over time. These changes might include new farming techniques for feedstocks that can increase yield or reduce fertilizer or irrigation requirements, more efficient equipment, development of markets for coproducts, water utilization, land use, and waste management. Clearly, there is some degree of subjectivity in defining the components of an LCA, and published studies may not always agree because they include different sets of processes or different inputs and outputs, or they may not assign consistent values to each component. To address this issue, the inputs and outputs to the LCA can be considered part of a life cycle inventory (LCI), which is a database that provides the numbers that are assigned to each constituent process/material/energy usage in the functional units of interest. Standardized LCIs have become available; some are free (such as the European Reference Life Cycle Database,[8] whereas others are proprietary (such as ecoinvent[9]). With such databases, one can drill down to the smallest details and quantify the environmental impact of using material A versus material B for a pipeline of length $L$.

LCAs are relatively new methodology, so that a literature survey of aviation fuel LCAs can be somewhat confusing because of conflicting results and methodologies. In 2006, ISO 14040 (2006) and ISO 14044 (2006) were established to provide a uniform approach to environmental LCAs. Since then, the field has matured considerably. GREET (Greenhouse Gases, Regulated Emissions, and Energy Use in Transportation) is a software package developed by Argonne National Laboratories in the United States to perform LCAs for vehicular fuels. This code is also commonly used in LCAs for aviation fuels. Another option is the evolving OpenLCA (GreenDelta), which is freeware developed for generalized LCAs.

Startup costs, or capital expenditures, are often ignored in economics-focused LCAs, but they can make all the difference in predicting whether a specific venture will be profitable or whether a particular fuel will be environmentally friendly. If the extraction of the petroleum fuel is at a

---

[8]See *http://lca.jrc.ec.europa.eu/lcainfohub/datasetArea.vm*
[9]See *http://www.ecoinvent.ch*

remote location and requires a road to be built to get there, this can be a significant cost. Not only is there the direct material cost due to asphalt and other building materials and the energy cost associated with road-building equipment, there may also be indirect costs associated with land use change, such as deforestation if that is needed to build the road. Slashing and burning forest to create farmland on which to grow biofuel feedstocks comes with significant economic cost, and the environmental impact can be even worse.

In the process of photosynthesis, plants turn $CO_2$, a greenhouse gas, into gaseous $O_2$ and stored carbon. As such, forests are living, breathing systems that store, or sequester, carbon in their limbs and roots. When trees die, fall to the forest floor, and decompose, the sequestered carbon in the wood becomes nutritious soil for the next generation of plants. When forestland is cleared to create farmland, the carbon that was stored in the plant matter may be released into the atmosphere, typically by burning. The upfront greenhouse gas emissions due to clearing a hectare (1 ha = 10,000 $m^2$) of forest can result in the release of 604 to 1146 MT (megatons) of greenhouse gases into the atmosphere (Searchinger *et al.*, 2008). Clearing a savannah or grass-land can release 74 to 305 MT/ha into the atmosphere. Since the net environmental benefit of growing corn for ethanol is measured as a couple of megatons per hectare per year (MT/ha/yr) at best, it can take decades to recover the environmental cost of preparing untouched land for farming (Searchinger *et al.*, 2008). In fact, some have argued that land-use issues end up being *the* dominant issue in sustainability (Melillo *et al.*, 2009; Righelato and Spracklen, 2007).

More subtly, the effect of land use change can be indirect. Suppose that a farmer in Iowa decides to sell the corn harvest to an ethanol refinery rather than a grocery store chain in order to benefit from agricultural subsidies provided in the United States. This corn will no longer be available to feed human beings or animals, and the price of corn rises. The globally rising price of corn encourages another farmer in Mexico to clear virgin land to produce corn. The net effect on the environment is to increase the amount of greenhouse gas pumped into the atmosphere because of land use change. Some LCAs do not attempt to quantify indirect land use change, since it is difficult to formulate a meaningful model of the process that adequately describes the socioeconomic factors that are involved. Nevertheless, this is a real phenomenon, and this is an active area of research, particularly for LCAs geared towards analysis of global issues in sustainability.

Other land-use concerns revolve around the use of nutrients to more efficiently grow crops. Nitrogen fertilizers tend to outgas nitrogen oxide, which can lead to smog formation when com-bined with hydrocarbons and sunlight. Nitrogen and phosphorous fertilizer are also responsible for eutrophication as nutrient-rich runoff from farmland escapes into a lake or ocean and creates algal blooms. As the algae die off, their decomposition robs the water of dissolved oxygen that fish and other marine life depend on for survival. Not only are we creating unnecessary problems for ourselves by our cavalier use of fertilizers, the global phosphate supply should be considered a precious resource that may not be readily available in a couple of decades (U.S. Department of Energy, 2010; Vaccari, 2009).

Many of the early sustainability analyses for corn-based ethanol ignored the issue of land use change altogether, which was a grievous omission. They often did not include the entire system boundaries, nor did they have a complete inventory (von Blottnitz and Curran, 2007). When properly accounting for land use change, system boundaries, and a more thoughtful assessment of the inventory required, corn-based ethanol production shifts from being a good thing for the environment to a strategy that is ultimately harmful to the environment.

Arguably, the most essential feature of any LCA is to thoroughly document the approach taken, including purpose of the LCA, system boundaries, critical assumptions, key parameters, variability in fuel components, and uncertainties (Stratton *et al.*, 2011). In many historical LCAs, the values for a given input or output is given as a single fixed number (such as the yield of *Jatropha* oil/ha/yr). However, in fact, there are uncertainties in that value associated with real demographic differences. More uncertainties creep into the analysis from other variable quanti-ties, outdated data, or incomplete problem description. To compare different strategies for alter-native fuels, feedstocks need to be compared within an equivalent framework. If the approach is fully documented, then future analyses can harmonize the assessment by updating or supplementing the data (e.g., Sun *et al.*, 2011; von Blottnitz and Curran, 2007; Warner *et al.*, 2010).

Defining the appropriate criteria by which to measure energy sustainability is still a work in progress (McBride *et al.*, 2011). Moreover, the hard data on biofuel production that are needed for such analyses are full of uncertainty (Murphy *et al.*, 2011; Slade *et al.*, 2011), because it is not available, still emerging, or preliminary. In order to accurately estimate the amount of land that would be needed globally for producing some specified amount of biofuel, the model must account for local and regional variations for many variables such as climate, water availability, crop production techniques, land suitability, and refining capabilities. Wigmosta and coworkers have developed a comprehensive framework that can account for most of these variables in great detail for the United States down to a resolution of 30 miles (Wigmosta *et al.*, 2000; also see Chapter 11). As such comprehensive tools mature, they will provide even better insight into modeling various scenarios for biofuel production.

For further reading on LCAs for aviation fuels, see Dray *et al.* (2010), Jorquera *et al.* (2010), Kinsel (2010), Kreutz *et al.* (2008), Stratton *et al.* (2010), and Yang *et al.* (2011).

## 8.7   CONCLUSIONS

Aviation fuel requirements are demanding because of the extreme environment in which aircraft must operate. Without an alternative to the internal combustion engine in the near term, the industry must rely on drop-in fuels that are compatible with existing engines. Although biodiesel based on methyl esters are adequate for automotive applications, their poor cold-weather properties and reduced specific energy make them unsuitable for aviation fuel. Both synthetic paraffinic kerosene and hydrotreated esters and fatty acids fuels can meet aviation's needs, but it is necessary to find sustainable ways of manufacturing these fuels. Some of the environmental and economic improvements that will bring closure to the problem will include new efficiencies in strain selection, crop management and production, and more creative approaches to finding uses for coproducts.

## ACKNOWLEDGEMENTS

The author would like to gratefully acknowledge the support of NASA Glenn Research Center and the alternative fuels component of the Subsonic Fixed Wing Program, as well as the wisdom and guidance of Bob Hendricks and Arnon Chait, the heroic efforts of the NASA Glenn Technical Library, especially Marcia Stegenga, and the expert editing by Laura Becker and Lorie Passe.

## REFERENCES

Achten, W. M. J., E. Mathijs, L. Verchot, *et al.* 2007. Jatropha biodiesel fueling sustainability? *Biofuels Bioprod. Biorefin.* 1:283–291.

Achten, W. M. J., L. Verchot, Y. J. Franken, *et al.* 2008. Jatropha bio-diesel production and use. *Biomass Bioenergy* 32:1063–1084.

Agency for Toxic Substances and Disease Registry. 1995. Toxicological profile for jet fuels JP–4 and JP–8. Atlanta, GA: U.S. Department of Health and Human Services. *http://www.atsdr.cdc.gov/ToxProfiles/TP.asp?id=773&tid=150* (accessed September 1, 2011).

Akbar, E., Z. Yaakob, S. K. Kamarudin, *et al.* 2009. Characteristic and composition of *Jatropha Curcas* oil seed from Malaysia and its potential as biodiesel feedstock. *European J. Sci. Res.* 29:396–403.

Alptekin, E., and M. Canakci. 2008. Determination of the density and the viscosities of biodiesel-diesel fuel blends. *Renew. Energy* 33:2623–2630.

Alptekin, E., and M. Canakci. 2009. Characterization of the key fuel properties of methyl ester-diesel fuel blends. *Fuel* 88:75–80.

Alptekin, E., and M. Canakci. 2011. Optimization of transesterification for methyl ester production from chicken fat. *Fuel* 90:2630–2638.

Anand, K., Y. Ra, R. D. Reitz, et al. 2011. Surrogate model development for fuels for advanced combustion engines. *Energy Fuels* 25:1474–1484.

Araujo, G. S., L. J. Matos, L. R. Goncalves, et al. 2011. Bioprospecting for oil producing microalgal strains: evaluation of oil and biomass production for ten microalgal strains. *Bioresour. Technol.* 102:5248–5250.

Aresta, M., A. Dibenedetto, and G. Barberio. 2005. Utilization of macro-algae for enhanced $CO_2$ fixation and biofuels production: development of a computing software for an LCA study. *Fuel Process. Technol.* 86:1679–1693.

ASTM D1319. 2013. Standard test method for hydrocarbon types in liquid petroleum products by fluorescent indicator adsorption—e-learning course. ASTM International, West Conshohocken, PA.

ASTM D1655. 2016. Standard specification for aviation turbine fuels. ASTM International, West Conshohocken, PA.

ASTM D6751-15ce1. 2015. Standard specification for biodiesel fuel blend stock (B100) for middle distillate fuels. ASTM International, West Conshohocken, PA.

ASTM D7566. 2016. Standard specification for aviation turbine fuel containing synthesized hydrocarbons. ASTM International, West Conshohocken, PA.

ASTM DS4B. 1991. Physical constants of hydrocarbon and non-hydrocarbon compounds. ASTM International, West Conshohocken, PA.

Atsumi, S., T. Hanai, and J. C. Liao. 2008. Non-fermentative pathways for synthesis of branched-chain higher alcohols as biofuels. *Nature* 451:86–89.

Balat, M. 2011a. Challenges and opportunities for large-scale production of biodiesel. *Energy Educ. Sci. Technol. A Energy Sci. Res.* 27:427–434.

Balat, M. 2011b. Potential alternatives to edible oils for biodiesel production—a review of current work. *Energy Convers. Manage.* 52:1479–1492.

Bartis, J. T., and L. Van Bibber. 2011. Alternative fuels for military applications. Santa Monica, CA: Rand National Defense Research Institute.

Ben-Amotz, A. 1995. New mode of Dunaliella biotechnology: two-phase growth for β-carotene production. *J. Appl. Phycol.* 7:65–68.

Blakey, S., L. Rye, and C. W. Wilson. 2011. Aviation gas turbine alternative fuels: a review. *Proc. Comb. Inst.* 33:2863–2885.

Boeing. 2011. Long-term market: current market outlook 2011–2030. *http://www.boeing.com/commercial/cmo/index.html* (accessed September 1, 2011).

Bouriazos, A., S. Sotiriou, C. Vangelis, et al. 2010. Catalytic conversions in green aqueous media: Part 4. Selective hydrogenation of polyunsaturated methyl esters of vegetable oils for upgrading biodiesel. *J. Organomet. Chem.* 695:327–337.

Brennan, L., and P. Owende. 2010. Biofuels from microalgae—a review of technologies for production, processing, and extractions of biofuels and co-products. *Renew. Sust. Energy Rev.* 14:557–577.

Bruno, T. J., and E. Baibourine. 2011. Comparison of biomass-derived turbine fuels with the composition-explicit distillation curve method. *Energy Fuels* 25:1847–1858.

Bruno, T. J., and M. L. Huber. 2010. Evaluation of the physicochemical authenticity of aviation kerosene surrogate mixtures. Part 2: Analysis and prediction of thermophysical properties. *Energy Fuels* 24:4277–4284.

Byron, D. 2011. Air Force officials tackle current, future energy needs. In *Air Force Print News Today,* Washington, DC: U.S. Air Force. *http://www.af.mil/News/Article-Display/Article/112403/air-force-officials-tackle-current-future-energy-needs/* (accessed April 25, 2017).

Cardozo, K. H. M., T. Guaratini, M. P. Barros, et al. 2007. Metabolites from algae with economical impact. *Com. Biochem. Phys. C Toxicol. Pharmacol.* 146:60–78.

Carrero, A., G. Vicente, R. Rodriguez, et al. 2011. Hierarchical zeolites as catalysts for biodiesel production from nannochloropsis microalga oil. *Catal. Today* 167:148–153.

Carvalho, A. P., L. A. Meireles, and F. X. Malcata. 2006. Microalgal reactors: a review of enclosed system designs and performances. *Biotechnol. Progr.* 22:1490–1506.

Carver, S. M., C. J. Hulatt, D. N. Thomas, et al. 2011. Thermophilic, anaerobic co-digestion of microalgal biomass and cellulose for $H_2$ production. *Biodegradation* 22:805–814.

CEN EN 14214. 2014. Liquid petroleum products—fatty acid methyl esters (FAME) for use in diesel engines and heating applications—requirements and test methods. Brussels: European Committee for Standardization.

CGSB–3.22. 2012. Wide-cut type aviation turbine fuel (grade jet b). Canadian General Standards Board, Gatineau, Canada.

Chastek, T. Q. 2011. Improving cold flow properties of canola-based biodiesel. *Biomass Bioenergy* 35:600–607.

Chen, C.-Y., K.-L. Yeh, R. Aisyah, *et al.* 2011. Cultivation, photobioreactor design and harvesting of microalgae for biodiesel production: a critical review. *Bioresource Technol.* 102:71–81.

Cheng, K. C., and K. Ogden. 2011. Algal biofuels: the research. *Chem. Eng. Progr.* 107:42–47.

Chevron. 2006. Aviation fuels technical review. San Ramon, CA: Chevron Corporation. *http://www.cgabusinessdesk.com/document/aviation_tech_review.pdf* (accessed September 1, 2011).

Chèzea, B., P. Gastineauc, and J. Chevallier. 2011. Forecasting world and regional aviation jet fuel demands to the mid-term (2025). *Energy Policy* 39:5147–5158. *http://www.sciencedirect.com/science/article/pii/S0301421511004496* (accessed April 17, 2017).

Chisti, Y. 2007. Biodiesel from microalgae. *Biotechnol. Adv.* 25:294–306.

Colket, M., T. Edwards, S. Williams, *et al.* 2007. Development of an experimental database and kinetic models for surrogate jet fuels. AIAA–2007–770.

Convery, F. J. 2009. Reflections—the emerging literature on emissions trading in Europe. *Rev. Environ. Econ. Policy* 3:121–137.

Cooney, M. J., G. Young, and R. Pate. 2011. Bio-oil from photosynthetic microalgae: case study. *Bioresource Technol.* 102:166–177.

Cordeiro, C. S., F. R. da Silva, F. Wypych, *et al.* 2011. Heterogeneous catalysts for biodiesel production. *Quim. Nova* 34:477–486.

Corporan, E., T. Edwards, L. Shafer, *et al.* 2011. Chemical, thermal stability, seal swell, and emissions studies of alternative jet fuels. *Energy Fuels* 25:955–966.

Coutinho, J. A. P., M. Goncalves, M. J. Pratas, *et al.* 2010. Measurement and modeling of biodiesel cold-flow properties. *Energy Fuels* 24:2667–2674.

Dahlquist, E., ed. 2013. *Biomass Utilization—Combustion, Gasification, Torrification and Fermentation.* Boca Raton, FL: CRC Press.

Day, J. G., and A. J. Tsavalos. 1996. An investigation of the heterotrophic culture of the green alga Tetraselmis. *J. Appl. Phycol.* 8:73–77.

DEF STAN 91–87. 2002. Turbine Fuel, Aviation Kerosine Type, Containing Fuel System Icing Inhibitor. Ministry of Defence, Westminster, London.

DEF STAN 91–91. 2016. Turbine Fuel, Aviation Kerosine Type, Jet A-1. Ministry of Defence, Westminster, London.

Dellomonaco, C., C. Rivera, P. Campbell, *et al.* 2010. Engineered respiro-fermentative metabolism for the production of biofuels and biochemicals from fatty acid-rich feedstocks. *Appl. Environ. Microbiol.* 76:5067–5078.

Demirbas, A. 2011. Competitive liquid biofuels from biomass. *Appl. Energy* 88:17–28.

Doan, T. T. Y., B. Sivaloganathan, and J. P. Obbard. 2011. Screening of marine microalgae for biodiesel feedstock. *Biomass Bioenergy* 35:2534–2544.

Dooley, S., S. H. Won, M. Chaos, *et al.* 2010. A jet fuel surrogate formulated by real fuel properties. *Combust. Flame* 157:2333–2339.

Dray, L., A. Evans, T. Reynolds, and A. Schäfer. 2010. Mitigation of aviation emissions of carbon dioxide: analysis for Europe. *Transport. Res. Rec.* 2177:17–26.

Edwards, T. 2006. Cracking and deposition behavior of supercritical hydrocarbon aviation fuels. *Combust. Sci. Technol.* 178:307–334.

Edwards, T. 2007. Advancements in gas turbine fuels from 1943 to 2005. *J. Eng. Gas Turb. Power* 129:13–20.

Edwards, T. 2010. Jet fuel composition. In *Jet Fuel Toxicology,* ed., M. L. Witten, E. Zeiger, and G. D. Ritchie, Boca Raton, FL: CRC Press, ch. 2, pp. 21–26.

ExxonMobil Aviation. 2005. World jet fuel specifications. *http://www.exxonmobil.com/AviationGlobal/Files/WorldJetFuelSpecifications2005.pdf* (accessed July 2011).

Fazal, M. A., A. S. M. A. Haseeb, and H. H. Masjuki. 2010. Biodiesel feasibility study: an evaluation of material compatibility; performance; emission and engine durability. *Renew. Sust. Energy Rev.* 15:1314–1324.

230  *E.S. Nelson*

Foidl, N., G. Foidl, M. Sanchez, *et al.* 1996. Jatropha curcas L. as a source for the production of biofuel in Nicaragua. *Bioresource Technol.* 58:77–82.

Freitas, S. V. D., M. J. Pratas, R. Ceriani, *et al.* 2011. Evaluation of predictive models for the viscosity of biodiesel. *Energy Fuels* 25:352–358.

Gong, Y., and M. Jiang. 2011. Biodiesel production with microalgae as feedstock: from strains to biodiesel. *Biotechnol. Lett.* 33:1269–1284.

Goodrum, J. W., and D. P. Geller. 2005. Influence of fatty acid methyl esters from hydroxylated vegetable oils on diesel fuel lubricity. *Bioresource Technol.* 96:851–855.

Gordon, D. R., K. J. Tancig, D. A. Onderdonk, *et al.* 2011. Assessing the invasive potential of biofuel species proposed for Florida and the United States using the Australian Weed Risk Assessment. *Biomass Bioenergy* 35:74–79.

Gourlay, P., J. Leak, and T. L. Wright. 2011. Managing jet fuel and carbon in Europes's new emissions trading system: an OPIS primer. Gaithersburg, MD: Oil Price Information Service.

Gouveia, L., and A. C. Oliveira. 2009. Microalgae as a raw material for biofuels production. *J. Ind. Microbiol. Biotechnol.* 36:269–274.

Griffiths, M. J., and S. T. L. Harrison. 2009. Lipid productivity as a key characteristic for choosing algal species for biodiesel production. *J. Appl. Phycol.* 21:493–507.

Grima, E. M., E.-H. Belarbi, F. G. A. Fernandez, *et al.* 2003. Recovery of microalgal biomass and metabolites: process options and economics. *Biotechnol. Adv.* 20:491–515.

Ha, S.-J., J. M., Galazka, S. R., Kim, *et al.* 2010. Engineered Saccharomyces cerevisiae capable of simultaneous cellobiose and xylose fermentation. *Proc. Natl. Acad. Sci. U.S.A.* 108:504–509.

Hadaller, O. J., and J. M. Johnson. 2006. World fuel sampling program. Coordinating Research Council (CRC) Report No. 647.

Hendricks, R. C. 2008. Potential carbon negative commercial aviation through land management. Paper presented at the 12th International Symposium on Transport Phenomena and Dynamics of Rotating Machinery, Honolulu, HI, p. 1422.

Hendricks, R. C., D. M. Bushnell, and D. T. Shouse. 2011. Aviation fueling: a cleaner, greener approach. *International Journal of Rotating Machinery* Article ID 782969.

Herbinet, O., W. J. Pitz, and C. K. Westbrook. 2008. Detailed chemical kinetic oxidation mechanism for a biodiesel surrogate. *Combust. Flame* 154:507–528.

Herbinet, O., W. J. Pitz, and C.K. Westbrook. 2010. Detailed chemical kinetic mechanism for the oxidation of biodiesel fuels blend surrogate. *Combust. Flame* 157:893–908.

Hileman, J. I., Stratton, R. W., and P. E. Donohoo. 2010. Energy content and alternative jet fuel viability. *J. Propul. Power* 26:1184–1195.

Hirsch, R. L. 2005. The inevitable peaking of world oil production. *Bulletin of the Atlantic Council of the United States* 16:1–9.

Ho, S.-H., C.-Y. Chen, D.-J. Lee, and J.-S. Chang. 2011. Perspectives on microalgal $CO_2$-emission mitigation systems—a review. *Biotechnol. Adv.* 29:189–198.

Holser, R. A., and R. Harry-O'Kuru. 2006. Transesterified milkweed (Asclepias) seed oil as a biodiesel fuel. *Fuel* 85:2106–2110.

Huber, M. L., E. W. Lemmon, and T. J. Bruno. 2010. Surrogate mixture models for the thermophysical properties of aviation fuel Jet-A. *Energy Fuels* 24:3565–3571.

International Energy Agency. 2010. World energy outlook. Paris, France: OECD/IEA.

IRENA (International Renewable Energy Agency). 2017. Biofuels for aviation: technology brief. *http://www.irena.org/DocumentDownloads/Publications/IRENA_Biofuels_for_Aviation_2017.pdf* (accessed February 16, 2017).

ISO 14040:2006. 2006a. Environmental management—life cycle assessment—principles and framework. International Organization for Standardization, Geneva, Switzerland.

ISO 14044:2006. 2006b. Environmental management—life cycle assessment—requirements and guidelines. International Organization for Standardization, Geneva, Switzerland.

Jain, S., and M. P. Sharma. 2010. Stability of biodiesel and its blends: a review. *Renew. Sust. Energy Rev.* 14:667–678.

Jingura, R. M., D. Musademba, and R. Matengaifa. 2010. An evaluation of utility of Jatropha curcas L. as a source of multiple energy carriers. *Int. J. Eng. Sci. Technol.* 2:115–122.

Jorquera, O., A. Kiperstok, E. A. Sales, *et al.* 2010. Comparative energy life-cycle analyses of microalgal biomass production in open ponds and photobioreactors. *Bioresource Technol.* 101:1406–1413.

Joshi, H., B. R. Moser, J. Toler, *et al.* 2011. Ethyl levulinate: a potential bio-based diluent for biodiesel which improves cold flow properties. *Biomass Bioenergy* 35:3262–3266.

Kanda, H., and P. Li. 2011. Simple extraction method of green crude from natural blue-green microalgae by dimethyl ether. *Fuel* 90:1264–1266.

Karpovitch, E. A. 2011. The green road to energy independence. In *Surface Warfare, Web exclusive,* Arlington, VA: U.S. Navy. *http://surfwarmag.ahf.nmci.navy.mil/green_road.html* (accessed September 2011).

Kerschbaum, S., and G. Rinke. 2004. Measurement of the temperature dependent viscosity of biodiesel fuels. *Fuel* 83:287–291.

Kerschbaum, S., G. Rinke, and K. Schubert. 2008. Winterization of biodiesel by micro process engineering. *Fuel* 87:2590–2597.

Kinder, J. D., and T. Rahmes. 2009. Evaluation of bio-derived synthetic paraffinic kerosenes (Bio-SPK). White paper, Sustainable Biofuels Research & Technology Program, The Boeing Company.

Kinsel, W. C. 2010. Environmental life cycle assessment of coal-biomass to liquid jet fuel compared to petroleum-derived JP–8 jet fuel. Master of Science in Engineering Management thesis, Air Force Institute of Technology, Wright-Patterson Air Force Base.

Knothe, G. 2005. Dependence of biodiesel fuel properties on the structure of fatty acid alkyl esters. *Fuel Process. Technol.* 86:1059–1070.

Knothe, G. 2007. Some aspects of biodiesel oxidative stability. *Fuel Process. Technol.* 88:669–677.

Knothe, G. 2008a. "Designer" biodiesel: optimizing fatty ester composition to improve fuel properties. *Energy Fuels* 22:1358–1364.

Knothe, G. 2008b. Evaluation of ball and disc wear scar data in the HFRR lubricity test. *Lubr. Sci.* 20:35–45.

Knothe, G. 2010a. Biodiesel and renewable diesel: a comparison. *Prog. Energy Combust. Sci.* 36:364–373.

Knothe, G. 2010b. Biodiesel: current trends and properties. *Top. Catal.* 53:714–720.

Knothe, G., J. Van Gerpen, and J. Krahl. 2005. *The biodiesel handbook.* Champaign, IL: AOCS Press.

Knothe, G., and K. R. Steidley. 2005. Lubricity of components of biodiesel and petrodiesel: the origin of biodiesel lubricity. *Energy Fuels* 19:1192–2000.

Kohse-Höinghaus, K., P. Osswald, T. A. Cool, *et al.* 2010. Biofuel combustion chemistry: from ethanol to biodiesel. *Angew. Chem., Int. Ed.* 49:3572–3597.

Krahl, J., A. Munack, O. Schroder, *et al.* 2005. Influence of biodiesel and different petrodiesel fuels on exhaust emissions and health effects. In *The Biodiesel Handbook,* eds. G. Knothe, J. Van Gerpen, and J. Krahl, Champaign, IL: AOCS Press, pp. 173–180.

Krahl, J., G. Knothe, A. Munack, *et al.* 2009. Comparison of exhaust emissions and their mutagenicity from the combustion of biodiesel, vegetable oil, gas-to-liquid and petrodiesel fuels. *Fuel* 88:1064–1069.

Kreutz, T. G., E. Larson, G. J. Liu, *et al.* 2008. Fischer-Tropf fuels from coal and biomass. Presented at the 25th Annual International Pittsburgh Coal Conference, Pittsburgh, PA. *http://web.mit.edu/mitei/docs/reports/kreutz-fischer-tropsch.pdf* (accessed August 2011).

Kumar, P., A. Robins, S. Vardoulakis, *et al.* 2010. A review of the characteristics of nanoparticles in the urban atmosphere and the prospects for developing regulatory controls. *Atmos. Environ.* 44:5035–5052.

Lang, W., S. Sokhansanj, and F. W. Sosulski. 1992. Modelling the temperature dependence of kinematic viscosity for refined canola oil. *J. Amer. Oil Chem. Soc.* 69:1054–1055.

LeClercq, P., and M. Aigner. 2009. Impact of alternative fuels physical properties on combustor performance. Presented at 11th Triennial International Conference on Liquid Atomization and Spray Systems, Vail, CO.

Lee, D. S., G. Pitari, V. Grewe, *et al.* 2010. Transport impacts on atmosphere and climate: aviation. *Atmos. Environ.* 44:4678–4734.

Lee, S. J., S. Go, G. T. Jeong, *et al.* 2011. Oil production from five marine microalgae for the production of biodiesel. *Biotechnol. Bioprocess Eng.* 16:561–566.

Levine, R. B., T. Pinnarat, and P. E. Savage. 2010. Biodiesel production from wet algal biomass through in situ lipid hydrolysis and supercritical transesterification. *Energy Fuels* 24:5235–5243.

Li L., E. Coppola, J. Rine, *et al.* 2010. Catalytic hydrothermal conversion of triglycerides to non-ester biofuels. *Energy Fuels* 24:1305–1315.

Li, Y., M. Horsman, N. Wu, C. Q. Lan, *et al.* 2008. Biofuels from microalgae. *Biotechnol. Progress* 24:815–820.

Li, Y. G., L. Xu, Y.-M. Huang, *et al.* 2011. Microalgal biodiesel in China: opportunities and challenges. *Appl. Energy* 88:3432–3437.

Mahmoud, A., P. Arlabosse, and A. Fernandez. 2011. Application of a thermally assisted mechanical dewatering process to biomass. *Biomass Bioenergy* 35:288–297.

McBride, A. C., V. H. Dale, L. M. Baskaran, *et al.* 2011. Indicators to support environmental sustainability of bioenergy systems. *Ecol. Indicators* 11:1277–1289.

McDowell Bomani, B. M., D. L. Bulzan, D. I. Centeno-Gomez, et al. 2009. Biofuels as an alternative energy source for aviation—A survey. NASA/TM—2009-215587.

Melillo, J. M., J. M. Reilly, D. W. Kicklighter, et al. 2009. Indirect emissions from biofuels: How important? Science 326:1397–1399.

Meng, L., and J. Salihon. 2011. Conversion of palm oil to methyl and ethyl ester using crude enzymes. J. Biotechnol. Biomaterials 1.

Mensch, A., R. J. Santoro, T. A. Litzinger, et al. 2010. Sooting characteristics of surrogates for jet fuels. Combust. Flame 157:1097–1105.

Mercer, P., and R. E. Armenta. 2011. Developments in oil extraction from microalgae. Eur. J. Lipid Sci. Tech. 113:539–547.

MIL–DTL–5624U. 2004. Detail specification: turbine fuel, aviation, grades JP–4 and JP–5. U.S. Department of Defense, Washington, DC.

MIL–DTL–83133E. 2004. Detail specification: turbine fuel, aviation, kerosene type, JP–8. U.S. Department of Defense, Washington, DC.

Moavenzadeh, J., M. Torres-Montoya, and T. Gange. 2011. Repowering transport. World Economic Forum, Geneva, Switzerland, 2011.

Moser, B. R. 2009. Biodiesel production, properties, and feedstocks. In Vitro Cell. Dev. Biol. Plant 45:229–266.

Moses, C. A. 2008. Comparative evaluation of semi-synthetic jet fuels: final report. CRC Project No. AV-2-04a.

Moses, C. A. 2009. Comparative evaluation of semi-synthetic jet fuels. Addendum: Further analysis of hydrocarbons and trace materials to support Dxxxx. CRC Project No. AV-2-04a.

Moses, C. A., and P. N. J. Roets. 2009. Properties, characteristics, and combustion performance of Sasol fully synthetic jet fuel. J. Eng. Gas Turb. Power 131: Article 041502.

Mtui, G. Y. S. 2009. Recent advances in pretreatment of lignocellulosic wastes and production of value added products. Afr. J. Biotechnol. 8:1398–1415.

Murphy, R., J. Woods, M. Black, et al. 2011. Global developments in the competition for land from biofuels. Food Policy 36:S52–S61.

Naik, C. V., C. K. Westbrook, O. Herbinet, et al. 2011. Detailed chemical kinetic reaction mechanism for biodiesel components methyl stearate and methyl oleate. Proc. Combust. Inst. 33:383–389.

Nelson, N., and C. F. Yocum. 2006. Structure and function of photosystems I and II. Annu. Rev. Plant Biol. 57:521–565.

Nogueira, Jr., C. A., F. X. Feitosa, F. A. N. Fernandes, et al. 2010. Densities and viscosities of binary mixtures of babassu biodiesel + cotton seed or soybean biodiesel at different temperatures. J. Chem. Eng. Data 55:5305–5310.

Novillo, E., M. Pardo, and A. Garcia-Luis. 2011. Novel approaches for the integration of high temperature PEM fuel cells into aircrafts. J. Fuel Cell Sci. Technol. 8:011014.

Park, J. B., and R. J. Craggs. 2011. Algal production in wastewater treatment high rate algal ponds for potential biofuel use. Water Sci. Technol. 63:2403–2410.

Park, J. B. K., R. J. Craggs, and A. N. Shilton. 2011. Wastewater treatment high rate algal ponds for biofuel production. Bioresource Technol. 102:35–42.

Perego, C., and A. Bosetti. 2010. Biomass to fuels: the role of zeolite and mesoporous materials. Micropor. Mesopor. Mat. 144:28–39.

Petroutsos, D., P. Katapodis, P. Christakopoulos, et al. 2007. Removal of p-chlorophenol by the marine microalga Tetraselmis marina. J. Appl. Phycol. 19:485–490.

Pitz, W. J., and C. J. Mueller. 2011. Recent progress in the development of diesel surrogate fuels. Prog. Energy Combust. Sci. 37:330–350.

Pratas, M. J., S. Freitas, M. B. Oliveira, et al. 2011a. Densities and viscosities of minority fatty acid methyl and ethyl esters present in biodiesel. J. Chem. Eng. Data 56:2175–2180.

Pratas, M. J., S. V. D. Freitas, M. B. Oliveira, et al. 2011b. Biodiesel density: experimental measurements and prediction models. Energy Fuels 25:2333–2340.

Quintana, N., F. Van der Kooy, M. D. Van de Rhee, et al. 2011. Renewable energy from cyanobacteria: energy production optimization by metabolic pathway engineering. Appl. Microbiol. Biotechnol. 91:471–490.

Raikos, V., S. S. Vamvakas, J. Kapolos, et al. 2011. Identification and characterization of microbial contaminants isolated from stored aviation fuels by DNA sequencing and restriction fragment length analysis of a PCR-amplified region of the 16S rRNA gene. Fuel 90:695–700.

Ramos, M. J., C. M. Fernandez, A. Casas, *et al.* 2009. Influence of fatty acid composition of raw materials on biodiesel properties. *Bioresource Technol.* 100: 261–268.

Rashid, U., F. Anwar, and G. Knothe. 2009. Evaluation of biodiesel obtained from cottonseed oil. *Fuel Process. Technol.* 90:1157–1163.

Razon, L. F. 2009. Alternative crops for biodiesel feedstock. *CAB Reviews: Perspectives in Agriculture, Veterinary Science, Nutrition and Natural Resources* 4:1–15.

Refaat, A. A. 2009. Correlation between the chemical structure of biodiesel and its physical properties. *Int. J. Environ. Sci. Technol.* 6:677–694.

Renouard-Vallet, G., M. Saballus, P. Schumann, *et al.* 2012. Fuel cells for civil aircraft application: on-board production of power, water and inert gas. *Chem. Eng. Res. Des.* 90:3–10.

Righelato, R., and D. V. Spracklen. 2007. Environment. Carbon mitigation by biofuels or by saving and restoring forests? *Science* 317:902.

Robinson, P. R., and G. E. Dolbear. 2006. Hydrotreating and hydrocracking: fundamentals. In *Practical Advances in Petroleum Engineering*, eds. C. S. Hsu and P. R. Robinson, New York, NY: Springer, pp. 177–218.

Rodolfi, L., G. Chini Zittelli, N. Bassi, *et al.* 2009. Microalgae for oil: strain selection, induction of lipid synthesis and outdoor mass cultivation in a low-cost photobioreactor. *Biotechnol. Bioeng.* 102:100–112.

Rubin, E. M. 2008. Genomics of cellulosic biofuels. *Nature* 454:841–845.

Santos, C. A., M. E. Ferreira, T. L. da Silva, *et al.* 2011. A symbiotic gas exchange between bioreactors enhances microalgal biomass and lipid productivities: taking advantage of complementary nutritional modes. *J. Ind. Microbiol. Biotechnol.* 38:909–917.

Saravanan, S., and G. Nagarajan. 2011. Effect of single double bond in the fatty acid profile of biodiesel on its properties as a CI engine fuel. *Int. J. Energy Environ.* 2:1141–1146.

Sayre, R. 2010. Microalgae: the potential for carbon capture. *Bioscience* 60:722–727.

Searchinger, T., R. Heimlich, R. A., Houghton, *et al.* 2008. Use of U.S. croplands for biofuels increases greenhouse gases through emissions from land-use change. *Science* 319:1238–1240.

Sheehan, J., T. Dunahay, J. Benemann, *et al.* 1998. A look back at the U.S. Department of Energy's Aquatic Species Program: biodiesel from algae. NREL/TP–580–24190.

Sims, R. E. H., W. Mabee, J. N. Saddler, *et al.* 2010. An overview of second generation biofuel technologies. *Bioresource Technol.* 101:1570–1580.

Singh, D., T. Nishiie, and L. Qiao. 2011. Experimental and kinetic modeling study of the combustion of n-decane, Jet-A, and S–8 in laminar premixed flames. *Combust. Sci. Technol.* 183:1002–1026.

Singh, J., and S. Gu. 2010. Commercialization potential of microalgae for biofuels production. *Renew. Sust. Energy Rev.* 14:2596–2610.

Singh, S. P., and D. Singh. 2010. Biodiesel production through the use of different sources and characterization of oils and their esters as the substitute of diesel: a review. *Renew. Sust. Energy Rev.* 14: 200–216.

Sivakumar, G., D. R. Vail, J. Xu, *et al.* 2010. Bioethanol and biodiesel: alternative liquid fuels for future generations. *Eng. Life Sci.* 10:8–18.

Slade, R., R. Gross, and A. Bauen. 2011. Estimating bio-energy resource potentials to 2050: learning from experience. *Energy Environ. Sci.* 4:2645–2657.

Soh, L., and J. Zimmerman. 2011. Biodiesel production: the potential of algal lipids extracted with supercritical carbon dioxide. *Green Chem.* 13:1422–1429.

Stockenreiter, M., A.-K. Graber, F. Haupt, *et al.* 2012. The effect of species diversity on lipid production by micro-algal communities. *J. Appl. Phycol.* 24:45–54.

Stratton, R. W., H. M. Wong, and J. I. Hileman. 2010. Life cycle greenhouse gas emissions from alternative jet fuels. PARTNER Project 28 report, Version 1.2.

Stratton, R.W., H. M. Wong, and J. I. Hileman. 2011. Quantifying variability in life cycle greenhouse gas inventories of alternative middle distillate transportation fuel. *Environ. Sci. Technol.* 45:4637–4644.

Su, C.-H., L.-J. Chien, J. Gomes, *et al.* 2011. Factors affecting lipid accumulation by Nannochloropsis oculata in a two-stage cultivation process. *J. Appl. Phycol.* 23:903–908.

Sun, A., R. Davis, M. Starbuck, *et al.* 2011. Comparative cost analysis of algal oil production for biofuels. *Energy* 36:5169–5179.

Sydney, E. B., T. E. da Silva, A. Tokarski, *et al.* 2011. Screening of microalgae with potential for biodiesel production and nutrient removal from treated domestic sewage. *Appl. Energy* 88:3291–3294.

Tan, X., L. Yao, Q. Gao, W. Wang, *et al.* 2011. Photosynthesis driven conversion of carbon dioxide to fatty alcohols and hydrocarbons in cyanobacteria. *Metabolic Eng.* 13:169–176.

Tang, H., N. Abunasser, M. E. D. Garcia, *et al.* 2011. Potential of microalgae oil from Dunaliella tertiolecta as a feedstock for biodiesel. *Appl. Energy* 88:3324–3330.

Timko, M. T., T. B. Onasch, M. J. Northway, *et al.* 2010. Gas turbine engine emissions—Part II: chemical properties of particulate matter. *J. Eng. Gas Turb. Power* 132:061505.

Timko, M. T., S. C. Herndon, E. de la Rosa Blanco, *et al.* 2011. Combustion products of petroleum jet fuel, a Fischer-Tropsch synthetic fuel, and a biomass fatty acid methyl ester fuel for a gas turbine engine. *Combust. Sci. Technol.* 183:1039–1068.

Touloukian, Y. S., S. C. Saxena, and P. Hestermans. 1975. Viscosity. New York, NY: IFI/Plenum Data Company.

TRC. 2011. TRC thermodynamic tables. College Station, TX: Thermodynamic Research Center, Texas A&M University.

Uduman, N., Y. Qi, M. K. Danquah, G. M. Forde, *et al.* 2010. Dewatering of microalgal cultures: a major bottleneck to algae-based fuels. *J. Renew. Sustain. Energy* 2: 012701.

U.S. Department of Energy. 2010. National algal biofuels technology roadmap. DOE/EE–0332. U.S. Department of Energy, Office of Energy Efficiency and Renewable Energy, Biomass Program, *http://www1.eere.energy.gov/biomass/pdfs/algal_biofuels_roadmap.pdf* (accessed September 2011).

U.S. Energy Information Administration. 2011. Monthly U.S. refinery yield of kerosene-type jet fuel. *http://www.eia.gov/dnav/pet/hist/LeafHandler.ashx?n=PET&s=MKJRYUS3&f=M* (accessed July 2011).

Vaccari, D. A. 2009. Phosphorus: a looming crisis. *Sci. Amer.* 300:54–59.

van der Westhuizen, R., M. Ajam, P. De Coning, *et al.* 2011. Comprehensive two-dimensional gas chromatography for the analysis of synthetic and crude-derived jet fuels. *J. Chromatogr. A* 1218: 4478–4486.

Van Gerpen, J., and G. Knothe. 2005. Basics of the transesterification reaction. In *The Biodiesel Handbook*, eds. G. Knothe, J. Van Gerpen, and J. Krahl, Champaign, IL: AOCS Press.

Vaughan, N. E., T. M. Lenton, and J. G. Shepherd. 2009. Climate change mitigation: trade-offs between delay and strength of action required. *Clim. Chang.* 96:29–43.

Verma, D., R. Kumar, B. S. Rana, *et al.* 2011. Aviation fuel production from lipids by a single-step route using hierarchical mesoporous zeolites. *Energy Environ. Sci.* 4: 1667–1671.

von Blottnitz, H., and M. A. Curran. 2007. A review of assessments conducted on bio-ethanol as a transportation fuel from a net energy, greenhouse gas, and environmental life cycle perspective. *J. Clean. Product.* 15:607–619.

Wang, Y., S. Ma, M. Zhao, *et al.* 2011. Improving the cold flow properties of biodiesel from waste cooking oil by surfactants and detergent fractionation. *Fuel* 90:1036–1040.

Warner, E., Heath, G., and P. O'Donoughue. 2010. Harmonization of energy generation Life Cycle Assessments. (LCA) FY2010 LCA milestone report November 2010. Golden, CO: National Renewable Energy Laboratory, pp. 1–17.

Westbrook, C. K., W. J. Pitz, H. J. Curran, *et al.* 2009. Recent advances in detailed chemical kinetic models for large hydrocarbon and biodiesel transportation fuels. LLNL–CONF–411689, 7th Asia-Pacific Conference on Combustion, Taipei, Taiwan.

Westbrook, C. K., C. V. Naik, O. Herbinet, *et al.* 2011. Detailed chemical kinetic reaction mechanisms for soy and rapeseed biodiesel fuels. *Combust. Flame* 158:742–755.

Weyer, K. M., D. R. Bush, A. Darzins, *et al.* 2010. Theoretical maximum algal oil production. *Bioenergy Res.* 3:204–213.

Wigmosta, M. S., A. M. Coleman, R. J. Skaggs, *et al.* 2011. National microalgae biofuel production potential and resource demand. *Water Resour. Res.* 47.

Xiao, J., H. Zhang, L. Niu, X. Wang, *et al.* 2011. Evaluation of detoxification methods on toxic and antinutritional composition and nutritional quality of proteins in Jatropha curcas meal. *J. Agr. Food Chem.* 59:4040–4044.

Xiong, W., C. Gao, D. Yan, C. Wu, *et al.* 2010. Double $CO_2$ fixation in photosynthesis-fermentation model enhances algal lipid synthesis for biodiesel production. *Bioresource Technol.* 101:2287–2293.

Yang, C.-y., X. Deng, Z. Fang, *et al.* 2010. Selection of high-oil-yield seed sources of Jatropha curcas L. for biodiesel production. *Biofuels* 1:705–717.

Yang, J., M. Xu, X. Zhang, *et al.* 2011. Life-cycle analysis on biodiesel production from microalgae: water footprint and nutrients balance. *Bioresource Technol.* 102:159–165.

Yazdani, S. S., and R. Gonzalez. 2007. Anaerobic fermentation of glycerol: a path to economic viability for the biofuels industry. *Curr. Opin. Biotechnol.* 18:213–219.

Yuan, W., A. C. Hansen, and H. Zhang. 2003. Predicting the physical properties of biodiesel for combustion modeling. *Trans. ASAE* 46:1487–1493.

Yuan, W., A. C. Hansen, Q. Zhang, and Z. Tan. 2005. Temperature-dependent kinematic viscosity of selected biodiesel fuels and blends with diesel fuel. *J. Amer. Oil Chem. Soc.* 82:195–199.

Yuan, W., A. C. Hansen, and Q. Zhang. 2009. Predicting the temperature dependent viscosity of biodiesel fuels. *Fuel* 88:1120–1126.

Zhang, X., Q. Hu, M. Sommerfeld, *et al.* 2010. Harvesting algal biomass for biofuels using ultrafiltration membranes. *Bioresource Technol.* 101:5297–5304.

APPENDIX.  BASIC TERMINOLOGY AND CONCEPTS IN HYDROCARBON
          CHEMISTRY

For our purposes, we can think of molecular structure and chemical reactions as being all about putting a stable number of electrons in atomic orbitals. Each orbital can accommodate a discrete number of electrons in successive rings; those numbers are 2, 6, 10, and 14. Hydrogen has a single proton in its nucleus and one electron in the first orbital, so that hydrogen can share one electron and one vacancy with those of another atom to form a covalent bond of two electrons. Carbon has vacancies for four electrons in its outer shell, so it can share electrons and form covalent bonds with up to four hydrogen atoms. With a series of single bonds (i.e., each set of atoms share only two electrons and two vacancies total), chains of hydrogen and carbon can also be built up to form stable hydrocarbon molecules. The chemical formula for these molecules can be written as $C_nH_{n+2}$ where $n$ is the number of carbon atoms present in the molecule. The extra hydrogen atoms are needed to stabilize the carbon atoms at the end of a chain, as shown in Figures 8.23(a) and (b). As the number of carbons in the molecule increase, the straight-chain hydrocarbons have higher viscosities, boiling points, and lubricating indices. However, the molecule does not have to follow this straight-chained structure. The molecule in Figure 8.23(c) is an isomer of butane; that is, it has the same chemical formula, but it has a different configuration—in this case a branched structure. Hexane has a ladder-like structure with 6 carbon and 14 hydrogen atoms. Cyclohexane exhibits a stable ring configuration, shown in Figure 8.23(d), and is an isomer of hexane. All of the molecules from Figure 8.23 are called saturated hydrocarbons, meaning that each carbon atom still shares a single electron with neighboring carbon and hydrogen atoms. Such molecules consisting of single bonds are called alkanes (also known as paraffins, particularly in petroleum engineering). Alkanes with a cyclic structure (Fig. 8.23(d)) are also called cycloalkanes, or naphthenes.

Alkanes react with oxygen in a combustion process to produce carbon dioxide ($CO_2$), water ($H_2O$), and energy in a chemical reaction of the form:

$$C_nH_{2n+2} + (3n+1)/2 O_2 \rightarrow nCO_2 + (n+1)H_2O + energy$$

Note that this is the net reaction; the actual chemical reaction will almost always take a series of steps in which intermediate species are created and consumed. Hydrocarbons with double or triple covalent bonds between adjacent carbon atoms are termed "unsaturated." An alkene (also called olefin) is an unsaturated hydrocarbon with one or more carbon-to-carbon double bonds. The simplest noncyclic alkenes have only one double bond and can be described by the chemical formula $C_nH_{2n}$, such as ethylene, shown in Figure 8.24(a). There are also ring-shaped alkenes, such as benzene (Fig. 8.24(b)). Aromatics are stable molecules that contain benzene or similar ring structures; some of the earliest identified aromatics smell good, hence the name. Their properties are sufficiently different from alkenes so that they are considered as a separate class of chemical compounds.

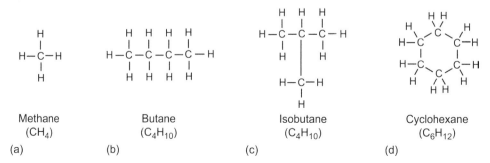

Figure 8.23.   Molecular structure of alkanes. (a) Methane. (b) Butane. (c) Isobutane. (d) Cyclohexane.

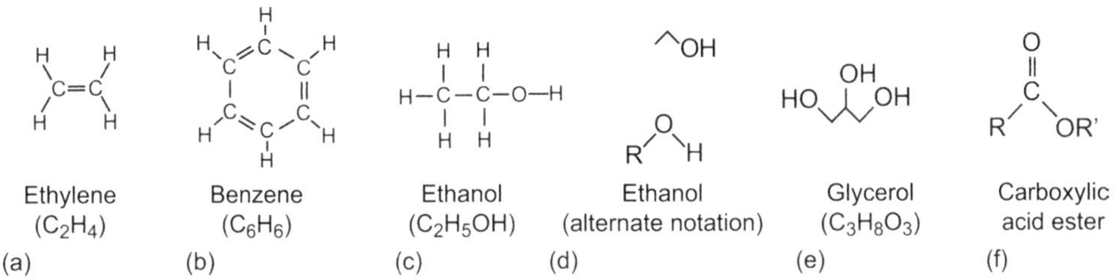

Figure 8.24.    Molecular structure of more complex hydrocarbons.(a) Ethylene, an alkene. (b) Benzene, an aromatic compound. (c) Ethanol. (d) Alternate notations of ethanol. (e) Glycerol. (f) Carboxylic acid ester.

Alcohols, such as ethanol in Figure 8.24(c), have a similar structure to alkanes, but in this case one or more hydrogens are replaced by a hydroxyl ($OH^-$) group. The negative charge of $OH^-$ results in a net neutral charge for the alcohol molecule. Functional groups are atomic sub-units within a molecule where chemical reactions usually occur. Higher carbon alcohols include ethanol, a straight-chain molecule with a terminal hydroxyl functional unit (Figs. 8.24(c) and (d)). As the complexity of the molecule increases, notation becomes more challenging and a sort of shorthand is used. Some simplifications are shown that represent ethanol in Figure 8.24(d). Carbon atoms and their supporting hydrogens are represented by kinks and the empty terminus in the version on top. On the bottom, R is used to designate a representative alkyl, in this case the ethyl group $C_3H_5$, while the other atoms are explicitly shown. Glycerol (Fig. 8.24(e)) is a reaction product from the refining process and mimics half of a sugar molecule. Esters are present in vegetable oils and animal fats, and they may also appear as reaction products. They come from the reaction of an alcohol with an acid. Carboxylic acid esters (Fig. 8.24(f)) are based on a carbon atom and have the general form RCOOR', where R is an alkyl group such as $C_3H_5$ and R' is an aryl (aromatic) group. A moiety is a subunit within a molecule that may itself include functional groups within it. There may be several ways of parsing a molecule into its constituent moieties.

Other molecules that use hydrogen, carbon, and oxygen atoms—but in more complicated configurations—include lipids, fatty acids, sugars, and carbohydrates (i.e., saccharides) with the formula $C_m(H_2O)_n$. Smaller carbohydrates (mono- and disaccharides) are sugars. Polysaccharides are polymeric carbohydrates;, that is, they comprise simpler units, such as mono- or disaccharides, in a repeating chain. Fatty acid methyl esters (MEs) are key components of vegetable oils and animal fats. Free fatty acids are carboxylic acids with a long unbranched hydrocarbon chain attached to it. The hydrocarbons may be saturated without any carbon-carbon double bonds, monounsaturated with a single double bond, or polyunsaturated with multiple double bonds in the hydrocarbon chain. Glycerol, shown in Figure 8.24(e), has three ends at which it may couple with fatty acids. When the coupling occurs, the molecules are termed "glycerides." A glycerol with one fatty acid is a monoglyceride; likewise, a glycerol with two fatty acids is a diglyceride. A triglyceride (also called triacylglycerol or triacylglyceride, or TAG) comprises three fatty acids joined to the glycerol subunit. TAGs are lipids that some algae use as its main carbon storage molecule. These compounds are important for the biorefining discussed in Section 8.5.

As discussed in Section 8.3.1 the double bonds of the MEs involved in biofuel production (Table 8.3) have chemical structures with *cis* and *trans* double bonds. The *cis* configuration is such that like functional groups are located on the same side of the double carbon-to-carbon bond (Fig. 8.25(a)). In the *trans* configuration, like functional groups are located on opposite sides of the double bond (Fig. 8.25(b)).

For fatty acids, Cx:y is the lipid number, in which x is the number of carbon atoms and y is the number of double bonds. Cx:0 is a saturated compound. C18:3 denotes an acid with 18 carbon atoms and 3 double bonds, but there are several molecules that fit this description. To be precise, it is appended with notation that indicates where the unsaturated bond is located relative to the

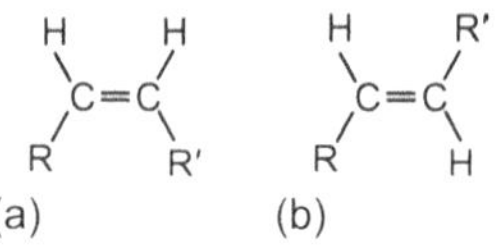

Figure 8.25.   Molecular structures showing configuration of functional groups about a double bond. (a) *cis*. (b) *trans*.

carboxylic acid end, n–$x$, $\Delta$–$x$ or $\Delta^x$ (where the $x$th carbon molecule has a double bond with the $x+1$th carbon molecule), depending on the field of study. Palmitic acid is C16:0; stearitic acid is C18:0; oleic acid is C18:1; linoleic acid (ω-6 fatty acid) is C18:2; α-linoleic acid (ω-3 fatty acid) is C18:3. Other examples and discussion may be found in Section 8.3.1.

# CHAPTER 9

## Overview of alternative fuel drivers, technology options, and demand fulfillment

Kirsten Van Fossen, Kristin C. Lewis, Robert Malina, Hakan Olcay and James I. Hileman

### 9.1   INTRODUCTION

Drop-in alternative aviation fuels (i.e., fuels that are chemically similar to petroleum-based jet fuel and compatible with existing equipment/facilities) may provide a near-term opportunity to improve aviation sustainability without requiring changes in current equipment and without hindering current approaches to improving operational practices. Strong aviation sector interest in drop-in alternative aviation fuels is driven by the potential benefits for environment, economy, and energy security (Air Transport Action Group (ATAG), 2011; Bauen *et al.*, 2009; Federal Aviation Administration, 2011; Sustainable Aviation Fuels Northwest, 2011; Sustainable Way for Alternative Fuels and Energy for Aviation, 2011). Large increases in jet fuel price and growing concerns about an uncertain supply have led the commercial aviation sector and the military to embrace the potential for alternative jet fuels as a countermeasure against supply shortages and price volatility. Concerns about aviation's contributions to climate change as well as the long-term potential for greenhouse gas regulation have also led to greenhouse gas reduction commitments for which alternative fuels are a critical component. This introduction provides background on the three elements motivating alternative aviation fuels—environment, economy, and energy security—and introduces the currently available technology options and the requirements to fulfill demand.

### 9.2   ALTERNATIVE FUEL DRIVERS

#### 9.2.1   *Environment and human health*

Alternative aviation fuels provide the potential opportunity to reduce carbon dioxide ($CO_2$) and other greenhouse gas (GHG) emissions, consequently reducing the aviation sector's contribution to climate change. Global aviation was contributing 2% of global $CO_2$ emissions as of 2012 (ATAG, 2012). The aviation industry has experienced rapid growth and, hence, increased fuel consumption in recent decades. To enable continued growth to meet anticipated increased demand without increasing emissions that further alter the global climate, the global aviation community has committed to the pursuit of carbon-neutral growth starting in 2020, formalized in a Joint Statement between the International Civil Aviation Organization (ICAO) and the Air Transport Action Group (ATAG) in May 2013 (ICAO and ATAG, 2013). Within the United States, the Federal Aviation Administration (FAA) has further articulated a goal of carbon-neutral growth at 2005 emissions levels starting in 2020 (FAA, 2011). Based on ICAO forecasts, sector technological and operational improvements alone will not enable aviation to achieve these goals (ICAO, 2010). Alternative aviation fuels may offer the aviation industry an additional opportunity to achieve carbon-neutral growth goals.

To evaluate the GHG emissions reduction achieved by a fuel, scientists perform a life-cycle analysis that measures the emissions released and absorbed along the entirety of the fuel supply chain through to combustion. For conventional jet fuel, that supply chain starts at petroleum extraction, and for alternative jet fuel, it starts at feedstock production. As explained further in Chapter 13, life-cycle GHG analyses of alternative jet fuels reveal that there are production pathways that lower GHG emissions relative to those calculated for conventionally extracted petroleum-based fuel. By introducing GHG-reducing alternative fuels to commercial flight, the aviation industry has the opportunity to continue along its current growth trajectory while reducing its impact on the environment.

The motivation to reduce jet-fuel-related emissions is related not only to environmental concerns but to human health concerns. Air pollutants are responsible for a range of health problems, which include aggravating or inducing asthma (or other respiratory disease), cancer, birth defects, and permanent lung or brain damage (U.S. Environmental Protection Agency (EPA), 2007). Much research has been done on the impact that jet fuel combustion has on local air quality, leading to the establishment of air quality standards that U.S. airports must meet under the National Environmental Policy Act in conjunction with the Clean Air Act (Kim *et al.*, 2012). In a recent comparison of conventional and alternative jet fuel combustion emissions, the alternative aviation fuels showed reduced combustion-related particulate matter (PM) emissions, with certain alternative fuel blends reducing the emitted number and mass of PM by an average of 54% when considering the fuel consumed during a standard landing and takeoff cycle (Carter *et al.*, 2011; Lobo *et al.*, 2011). A wide range of gas turbine engines have shown reduction in PM emissions when operating on 50–50 blends of alternative jet fuel with conventional jet fuel, with the exact reduction being engine technology dependent (Carter *et al.*, 2011). Part of this shift may be due to the low levels of sulfur found in certain alternative jet fuels (Bulzan *et al.*, 2010), as sulfur oxides generated during combustion can form sulfate particulates. A shift to alternative aviation fuels could provide significant health benefits to exposed populations near airports.

In addition to the climate and human health opportunities related to alternative jet fuels, advanced drop-in fuels also provide the opportunity to address societal needs for the management and environmentally friendly disposal of municipal solid waste and sewage, as well as the utilization of the biomass available from the growth and spread of invasive species. Other research has demonstrated that some feedstock crops have the ability to render marginal land productive, protect agricultural lands during fallow periods, and enrich soil quality (discussed in Chapter 10). Finally, some alternative fuel feedstocks consume very little water and/or require only salt water, addressing fresh water consumption and quality concerns related to fuel production, particularly those linked to irrigation. As the petroleum industry shifts toward more water intensive crude oil extraction methods and the global supply of high-quality freshwater continues to diminish, alternative aviation feedstocks that minimize the consumption and degradation of fresh water will be desirable. Considering the positive environmental results that the industry aims to achieve with alternative fuels, it is important that all the impacts that a particular fuel has with respect to land use, water consumption, water quality, air quality, soil quality, and biodiversity are fully investigated. The consideration of environmental impacts and avoidance of environmental harm are addressed in detail in Chapter 13.

### 9.2.2 *Economy*

The most important economic driver behind alternative aviation fuels has been the price of petroleum (a.k.a., crude oil). The difference in price between jet fuel and crude oil, termed the "crack spread," has averaged 19% of the total jet fuel price over the past 28 years, with a standard deviation of just 5.8% (Hileman and Stratton, 2014; U.S. Energy Information Administration, 2014); thus, the price of petroleum-based jet fuel is closely tied to the change in crude oil price. As one of the most high profile commodities, petroleum is subject to price volatility due to supply and demand, as well as to the added impacts of market speculation (Alquist and Kilian, 2010; Juvenal and Petrella, 2012; Kilian, 2009; Tang and Xiong, 2012). A review of historical crude oil prices since 1986 shows large and unpredictable price spikes. Not only is the volatile nature of the petroleum price a concern, but also the relatively steady increase of its price over the years; both of these affect the price behavior of other commodities (Tang and Xiong, 2012).

The average price of petroleum in 2013 was greater than five times its average price in 1993. Part of the increase in price can be attributed to the increasing trend of global petroleum consumption. Another factor is that the oil and gas industry is increasingly taking advantage of new extraction sources and methods such as shale gas, oil sands, and hydraulic fracturing (a.k.a., fracking) to fulfill demand, but these extraction methods are more costly than traditional extraction and may have greater environmental impacts. Understanding that crude oil is a finite resource and that the global population is growing, the U.S. Energy Information Administration expects the crude oil price to increase further if the current level of dependency on petroleum persists (U.S. Energy Information Administration, 2013).

The use of market-based measures (MBMs) to reduce the $CO_2$ emissions from aviation provides an additional economic motivation for alternative jet fuel use. Carbon-emissions-reducing MBMs incentivize carbon-emitting entities to lower their carbon emissions by introducing a financial penalty for high emissions and a financial reward for low emissions. MBMs may take the form of a fuel tax, a carbon emissions tax, or carbon trading (a.k.a., cap-and-trade). The concept of cap-and-trade can be traced back to Canadian economics professor John H. Dales, who visualized the system in his 1968 book *Pollution, Property & Prices* (Dales, 1968). Early implementation of the scheme demonstrated its value, as it was successfully employed to drive lead removal from gasoline and decrease sulfur pollution in the air during the Ronald Reagan and George H. W. Bush administrations (Fialka, 2011a, 2011b). Finally, cap-and-trade entered onto the international stage in 1997 in the meeting sessions of the Kyoto Protocol (Fialka, 2011a, 2011b).

As of 2014, many carbon-trading schemes had already been implemented across the world, including Alberta's Specified Gas Emitters Regulation, the European Union Emissions Trading System (EU ETS), the New Zealand Emissions Trading System, the Northeastern United States' Regional Greenhouse Gas Initiative, the Tokyo Emissions Trading System, and the California-Quebec Cap-and-Trade Program (International Emissions Trading Association, 2013). The one aviation-specific trading scheme in existence is implemented through the EU ETS. Although the EU ETS was originally intended to apply to all flight to and from Europe starting in 2013, the EU delayed implementation on international flights until at least 2016, such that currently the EU ETS applies only to flight within European airspace[1] (Buyck, 2014). The EU intends to include aviation in the EU ETS until the introduction of the ICAO emission-reducing MBM (European Parliament and Council of the European Union, 2008). ICAO is currently developing a global MBM for aviation based on the resolution of the 38th ICAO Assembly (Oct. 2013), which will be proposed in time for the 39th ICAO Assembly in 2016, for implementation beginning in 2020 (ICAO 38th Assembly, 2013). In the meantime, ICAO encourages States to submit voluntary actions plans to manage $CO_2$ emissions from international aviation (ICAO 38th Assembly, 2013).

A third economic benefit of alternative jet fuels is their potential effect on the global economy. In 2012, aviation directly supported 8.36 million jobs and enabled roughly another 47 million through its close relationship to tourism (ATAG, 2012). ATAG also calculated that in that same year, aviation's $539 billion contribution to gross domestic product (GDP) would have ranked aviation 19th in the GDP ranking by country if the aviation sector were a country. The contribution of aviation to the global economy is expected to further increase; the industry projects that aviation will contribute $1 trillion to the global GDP within 13 years (ATAG, 2012; FAA, 2013).

A scaled-up alternative aviation fuels industry will add to economic opportunities associated with aviation and will enable aviation industry growth. Alternative fuel production, particularly for bio-based fuels, is seen as a valuable opportunity for rural development and jobs expansion within the agricultural sector. The U.S. Department of Agriculture (USDA) Secretary Tom Vilsack has said that, "By continuing to work together to produce American made 'drop-in' aviation fuels from renewable feed-stocks, we will create jobs and economic opportunity in rural America, lessen America's reliance on foreign oil and develop a thriving biofuels industry

---

[1] EU ETS makes an exception for operators that do not exceed 243 flights in or out of Europe within a 4-month period, as long as that operator flies for three consecutive 4-month periods (European Parliament and Council of the European Union, 2008).

that will benefit commercial and military enterprises" (USDA, 2013). To take full advantage of these opportunities, bioeconomy innovators can use existing tools to aid in supply-chain development to ensure a resilient industry. Chapter 10 explores the use of ecological models to help guide industry toward optimal diversity of feedstocks to ensure availability and reliability, and to explore system interactions. Such models can assist industry and governments in planning robust, equitable supply chains. By enabling aviation industry growth and creating an entirely new industry, alternative aviation fuels may provide a dual benefit to the global economy.

### 9.2.3    *Energy security*

The current lack of diversity within the aviation fuel supply places the aviation industry in a vulnerable position. The demand for aviation fuel is fulfilled almost completely by petroleum-derived fuel. International dependence on a subset of countries for such a critical resource has important geopolitical consequences beyond just petroleum industry commerce. The Organization of Petroleum Exporting Countries (OPEC) is empowered—both economically and politically—by their countries' vast supplies of oil and has exercised this power in the past (as observed in the 1973 Oil Embargo).[2] Although new extraction methods introduced in the U.S. petroleum production sector have ramped up domestically sourced petroleum and at least temporarily stabilized supply, petroleum is still a finite resource.

Even if U.S. petroleum production could meet U.S. needs indefinitely, energy security concerns from a military perspective still would not be eradicated. The geographic limitations of petroleum production require the use of precarious convoys to deliver fuel to troops in forward-operating bases. These military operations are targets for opposing forces, and there have been numerous occasions where they have been intercepted. Secretary of the Navy Ray Mabus expanded on this danger while promoting alternative aviation fuels in a National Public Radio interview in 2011, "We look at it tactically because every convoy of fuel that we take into Afghanistan is costly in a lot of ways. The fuel is very costly. But also for every 50 convoys we lose a Marine, either killed or wounded" (National Public Radio, 2011). Some alternative fuels offer the flexibility of little or no geographic constraint on fuel production, such as the fuel coming from the Navy's solar-powered algae generators or locally generated biomass or waste materials, which could help ensure that the military can acquire the fuel that enables them to carry out their missions (National Public Radio, 2011). By increasing the mix of available aviation fuels, a more robust fuel supply is achieved. The knowledge gained through the advancement of alternative aviation fuels would also be applicable to the broader energy sector. The resulting increase in petroleum independence translates to improved energy security overall.

### 9.3    TECHNOLOGY OPTIONS

As of 2014, two alternative fuel production pathways were available and approved for use as jet fuel. A fuel conversion pathway must undergo a thorough approval process before it is considered acceptable for use in aircraft. This certification process has already been completed for three different pathways—Fischer-Tropsch (FT) processing of syngas for synthetic paraffinic kerosene (FT–SPK) in 2009, hydroprocessing esters and fatty acids for synthetic paraffinic kerosene (HEFA–SPK) in 2011, and synthesized isoparaffins (SIP) fuel in 2014—and it is expected to be completed for roughly 10 more over the next several years. The certified and nearest-term pathways are:

- *FT fuels*: The FT process involves the steam reforming or gasification of any carbon-containing feedstock—such as natural gas, coal, or biomass—into syngas, which is then

---

[2] The 1973 Oil Embargo was the Arab OPEC response to U.S. and other foreign support for Israel during the 1973 Arab-Israeli War. The embargo was placed on the U.S., South Africa, the Netherlands, and Portugal in the last few months of 1973 and was not lifted until March 1974, a long enough period to render the embargoed countries extremely fuel-strained (U.S. Department of State, 2013).

purified (e.g., from sulfur) and converted to paraffinic hydrocarbons by the FT synthesis. A cracker unit can then be utilized to shorten long hydrocarbon chains to maximize fuel cuts like diesel or jet fuel.

- *HEFA fuels*: Triglycerides, which are found in animal fats, and vegetable and algal oils, have one glycerol molecule and three fatty acid molecules in their structures. Triglycerides can be hydrotreated into straight-chain alkanes, known as HEFA fuels. Based on the type of feedstock, the fatty acids can vary in length, which will have a direct effect on the final product distribution.
- *SIP fuels*: The SIP process uses a yeast-based fermentation process to transform sugar feedstocks into farnesene, which is then converted to the hydrocarbon farnesene. This process is different from that for FT and HEFA fuels in that the product is a single molecule rather than a range of hydrocarbons. SIP fuel has been certified for use up to a 10% blend with standard petroleum-based jet fuels.
- *Other sugar-based jet fuel processes*: There are other sugar-based jet fuel pathways under development that rely on technologies that break down the building blocks of sugary, starchy, or lignocellulosic biomass feedstocks into five-carbon and/or six-carbon simple sugars, and then increase their carbon length to the jet fuel range through carbon-carbon coupling reactions (Huber *et al.*, 2006). Simple sugars can be used to produce intermediate products (such as ethanol or butanol), known as platform molecules, which are then converted into jet fuel, or the sugars may be converted to alkanes directly via "advanced fermentation" by engineered microbes or algae.

See Chapter 8 for additional discussion of the biochemical properties of alternative fuels and the associated growth and manufacturing processes. Further details on the certified technologies and the approval process that were required to get those technologies certified are provided in Chapter 12.

## 9.4   MEETING DEMAND FOR ALTERNATIVE JET FUEL

Each alternative fuel production process requires a feedstock to provide the carbon backbone of the alternative fuel hydrocarbons. These feedstocks may be coal, natural gas, or biological materials such as sugars, starches, lignocellulosic materials, or plant or animal oils. Feedstock selection plays an important role in overall environmental impacts, system resilience, and greenhouse gas benefits. The choice of feedstock and process efficiency also determine the land area required to fulfill the need for alternative jet fuels. Areal productivity in terms of fuel produced per unit area of feedstock production can impose severe restrictions on the viability of a certain biofuel for substituting large volumes of jet fuel. Alternative jet fuel pathways generally produce a product slate of different fuels which can be optimized to maximize certain individual products or total output. From an aviation perspective, the desire would be to maximize jet fuel production, but this optimization comes at an additional production cost which would have to be carefully weighed against the additional jet fuel production volume that can be obtained. Ultimately, the fraction of the total product slate that would be jet fuel depends on market conditions (Pearlson *et al.*, 2013; Tijm, 1994). For example, hydroprocessing esters and fatty acids fuel production from crops such as *Camelina sativa*, soy, *Jatropha curcas*, and *Salicornia* species results in a product slate that is only 15% to 57% HEFA jet fuel, with the remaining portion being diesel fuel and naphtha. Fischer-Tropsch synthesis from biomass, natural gas, or coal also results in a product slate with the majority of the output being diesel fuel with jet fuel and naphtha coproducts.

Biofuel yield will determine the amount of land required and the ability to meet articulated GHG emissions reduction targets. (See Chapter 11 for a discussion of microalgae feedstocks and land use.) Both commercial and military aviation have committed to alternative fuel goals. The Federal Aviation Administration has set an aspirational goal for domestic alternative jet fuel consumption of 1 billion gallons per year (65,200 barrels per day) by 2018 (FAA, 2011), which corresponds to approximately 5% of the total jet fuel consumption in the United States. The U.S. Navy and Air Force have also put in place goals or mandates in terms of alternative fuel sources (U.S. Air Force, 2010; U.S. Navy, 2010). The FAA aspirational goal provides a useful means to compare crop yields and their impact on acreage requirements. If the goal were met with an exclusive reliance on a low areal-productivity HEFA feedstock such as *Camelina sativa*, then approximately 8 million to 35 million hectares would be needed (Hileman *et al.*, 2013), which is a land area equivalent in size to 5% to 21% of all currently used cropland in the United States. High-fuel-yield feedstock such as lignocellulosic biomass using FT synthesis or microalgae-based HEFA fuels, on the other hand, could satisfy the FAA goal with a substantially smaller land footprint, amounting to less than 1 million hectares (Hileman *et al.*, 2013), which is less than 1% of all cropland available in the United States.

## 9.5   CONCLUSIONS

A number of governmental research programs and public-private partnerships within the United States and globally are working to overcome the barriers to alternative jet fuel industry commercialization and scale up. Organizations such as the Commercial Aviation Alternative Fuels Initiative® and similar stakeholder partnerships around the world (e.g., the Australian Initiative for Sustainable Aviation Fuels, and Germany's Aviation Initiative for Renewable Energy in Germany, or Aireg, among others) are working to identify complementary opportunities to facilitate alternative jet fuel technical testing and certification and sustainability evaluations. U.S. agencies such as the Federal Aviation Administration, Department of Agriculture, Department of Energy, National Aeronautics and Space Administration, National Institute of Standards and Technology, Environmental Protection Agency, and Department of Defense are investing resources into fuel research, development, testing, and certification as well. These concerted efforts are focused on developing a sustainable alternative jet fuel supply chain.

While the aviation community has clearly stated its intent to move toward alternative aviation fuel, the scaleup and commercialization of alternative aviation fuels still present obstacles. The burgeoning interest in bioproducts has already triggered a competitive bioeconomy. Although the industry and government goals strongly encourage a market for alternative jet fuels, feedstock and fuel producers may find it more beneficial to divert their resources and/or products to other sectors. The following chapters address the key drivers for alternative aviation fuel use in more detail, and discuss some of the critical challenges for achieving the full potential benefits of alternative jet fuels—quantification of environmental benefit, avoidance of environmental harm, testing and certification of fuels, and development of economically viable, equitable, and environmentally responsible supply chains when the industry is scaled up, as well as some of the potential solutions for addressing these challenges.

## REFERENCES

Air Transport Action Group. 2011. Powering the future of flight: the six easy steps to growing a viable aviation biofuels industry.

Air Transport Action Group. 2012. Facts & figures. *http://www.atag.org/facts-and-figures.html* (accessed September 1, 2016).

Alquist, R., and L. Kilian. 2010. What do we learn from the price of crude oil futures? *J. Appl. Econom.* 25: 539–573.

Bauen, A., J. Howes, L. Bertuccioli, *et al.* 2009. Review of the potential for biofuels in aviation. In *Review of the Potential for Biofuels in Aviation,* London, UK: E4Tech.

Bulzan, D., B. Anderson, C. Wey, *et al.* 2010. Gaseous and particulate emissions results of the NASA Alternative Aviation Fuel Experiment (AAFEX). ASME GT2010–23524.

Buyck, C. 2014. Talks yield provisional ETS "stop-the-clock" extension. *Aviation Week Aviation Daily* 6:1.01.

Carter, N. A., R. W. Stratton, M. K. Bredehoeft, *et al.* 2011. Energy and environmental viability of select alternative jet fuel pathways. AIAA 2011–5968.

Dales, J. H. 1968. *Pollution, property and prices: an essay in policy-making and economics*. Glasgow, U.K.: Edward Elgar Publishing.

European Parliament and Council of the European Union. 2008. Directive 2008/101/EC of the European Parliament and of the Council of 19 November 2008 amending Directive 2003/87/EC so as to include aviation activities in the scheme for greenhouse gas emission allowance trading within the community. *Official Journal of the European Union* 13.1.2009:3–21.

Federal Aviation Administration. 2011. FAA destination 2025. *https://www.faa.gov/about/plans_reports/media/Destination2025.pdf* (accessed September 6, 2016).

Federal Aviation Administration. 2013. FAA aerospace forecast: fiscal years 2013–2033. *https://www.faa.gov/data_research/aviation/aerospace_forecasts/media/2013_Forecast.pdf* (accessed September 6, 2016).

Fialka, J. J. 2011a. Policy: Cap and trade—the meandering path of an idea to curb pollution—Part 1. *ClimateWire*.

Fialka, J. J. 2011b. Policy: How a Republican anti-pollution measure, expanded by Democrats, got rooted in Europe and China—Part 2. *ClimateWire*.

Hileman, J. I., and R. W. Stratton. 2014. Alternative jet fuel feasibility. *Transport Policy* 34:52–62.

Hileman, J. I., E. De la Rosa Blanco, P. A. Bonnefoy, *et al.* 2013. The carbon dioxide challenge facing aviation. *Progr. Aerosp. Sci.* 63:84–95.

Huber, G. W., S. Iborra, and A. Corma. 2006. Synthesis of transportation fuels from biomass: chemistry, catalysts, and engineering. *Chem. Rev.* 106:4044–4098.

International Civil Aviation Organization. 2010. ICAO 2010 Environmental Report. Montreal, Canada: ICAO.

International Civil Aviation Organization. 2013. Assembly—38th session: resolutions adopted by the Assembly: Provisional Edition, *http://www.icao.int/Meetings/a38/Documents/Resolutions/a38_res_prov_en.pdf* (accessed September 6, 2016).

International Civil Aviation Organization and Air Transport Action Group. 2013. Joint statement to cooperate on the promotion of sustainable approaches to global aviation emissions reduction. *http://www.icao.int/Newsroom/News%20Doc%202013/Joint%20Statement_ICAO-ATAG_US-Letter_2013-05-13_signed.pdf* (accessed September 6, 2016).

International Emissions Trading Association. 2013. The world's carbon markets: a case study guide to emissions trading. *http://www.ieta.org/The-Worlds-Carbon-Markets* (accessed September 6, 2016).

Juvenal, L., and I. Petrella. 2012. Speculation in the oil market. Working Paper 2011–027E, St. Louis, MO: Federal Reserve Bank of St. Louis. *http://research.stlouisfed.org/wp/2011/2011-027.pdf* (accessed September 6, 2016).

Kilian, L. 2009. Not all oil price shocks are alike: disentangling demand and supply shocks in the crude oil market. *Am. Econ. Rev.* 99:1053–1069.

Kim, B., J. Rachami, D. Robinson, *et al.* 2012. Guidance for quantifying the contribution of airport emissions to local air quality. Airport Cooperative Research Program Report 71, Washington, DC: Transportation Research Board.

Lobo, P., D. E. Hagen, and P. D. Whitefield. 2011. Comparison of PM emissions from a commercial jet engine burning conventional, biomass, and Fischer-Tropsch fuels. *Environ. Sci. Technol.* 45:10744–10749.

National Public Radio. 2011. Environmental outlook: the military and alternative energy. Interview with Secretary of the Navy Ray Mabus transcript. *http://www.navy.mil/navydata/people/secnav/Mabus/Interview/NPR%20Transcript%20-%20Operational%20Energy%20(5%20Jul%2011).pdf* (accessed September 6, 2016).

Pearlson, M., C. Wollersheim, and J. Hileman. 2013. A techno-economic review of hydroprocessed renewable esters and fatty acids for jet fuel production. *Biofuels, Bioprod. Biorefin.* 7:89–96.

Sustainable Aviation Fuels Northwest. 2011. Sustainable Aviation Fuels Northwest: powering the next generation of flight. *https://www.portseattle.org/Environmental/Air/Airport-Air-Quality/Documents/SAFN_ExecSummary.pdf* (accessed January 20, 2017).

Sustainable Way for Alternative Fuels and Energy for Aviation. 2011. Final Report to the European Commission SWAFEA formal report D.9.2 v1.1. Brussels, Belgium: European Commission's Directorate General for Mobility and Transport.

Tang, K., and W. Xiong. 2012. Index investment and financialization of commodities. *Financ. Anal. J.* 68:54–74.

Tijm, P. J. A. 1994. Shell middle distillate synthesis: the process, the plant, the products. *https://web.anl.gov/PCS/acsfuel/preprint%20archive/Files/39_4_WASHINGTON%20DC_08-94_1146.pdf* (accessed September 6, 2016).

U.S. Air Force. 2010. Air Force energy plan 2010.

U.S. Department of Agriculture. 2013. Agriculture Secretary Vilsack and Transportation Secretary LaHood renew agreement to promote renewable fuels in the aviation industry: resolution with the FAA and private-sector partners will continue the work to develop a robust aviation biofuels industry. Release No. 0070.13, Washington, DC: USDA Office of Communications.

U.S. Department of State Office of the Historian. 2013. Oil embargo, 1973–1974. In *Milestones: 1969–1976. http://history.state.gov/milestones/1969-1976/oil-embargo* (accessed September 1, 2016).

U.S. Energy Information Administration. 2013. International Energy Outlook with projections to 2040. DOE/EIA–0484(2013).

U.S. Energy Information Administration. 2014. Petroleum Navigator. Spot prices (crude oil in dollars per barrel, products in cents per gallon). Independent Statistics & Analysis. *http://www.eia.gov/dnav/pet/pet_pri_spt_s1_d.htm* (accessed September 6, 2016).

U.S. Environmental Protection Agency. 2007. The plain English guide to the Clean Air Act: why should you be concerned about air pollution? Publication No. EPA–456/K–07–001.

U.S. Navy. 2010. A Navy energy vision for the 21st century.

# CHAPTER 10

## Biofuel feedstocks and supply chains: how ecological models can assist with design and scaleup

Kristin C. Lewis, Dan F.B. Flynn and Jeffrey J. Steiner

## 10.1   INTRODUCTION

The ability of plants to fix atmospheric carbon provided the original ancient sources for today's fuel, such as coal, oil, and natural gas, which drives modern society. Plants also have provided food, feed, and fiber throughout human history, and can now be the building blocks for the production of new renewable fuels and other bio-based products. However, large-scale replacement of fossil-based energy sources with new plant-based ones will require the production of large amounts of dedicated bioenergy crops. The cultivation of agricultural crops and their conversion into energy products present numerous technological, economic, environmental, and social challenges. However, with the careful design of renewable fuel supply chains and their associated sectors, a myriad of benefits could result for feedstock producers, biorefineries, and consumers alike. Economically viable pathways will emerge from the diverse array of feedstock and processing combinations, resulting from optimal matches with the available resources within different regions. In this way, truly sustainable, regionally appropriate production systems can evolve that will provide dependable amounts of feedstocks to produce reliable supplies of sustainable biofuels. However, production of dedicated bioenergy crops will need to be facilitated in a coordinated way by both end users and the agricultural sector. Bioenergy crops competing for land and processing facilities competing for feedstock will affect the optimal structure of a scaled-up biofuel industry within a given region.

The many possible bioenergy crops can be categorized by the kinds of feedstock material they produce: cellulose from herbaceous or woody plants, sugars produced directly by plants or derived from cellulosic plant material, and lipids extracted from seed crops or plant material such as algae. Renewable fuels can also be made from waste fats and greases or from biomass or gases derived from sewage, manure, or municipal solid waste. As of August 2016, there are five approved specifications for alternative jet fuel production from sugars, biomass-based feedstocks, and agricultural lipids: "synthetic isoparaffinic kerosene" and alcohol-to-jet, Fischer-Tropsch (FT), with and without synthetic aromatics, and hydroprocessing technologies, respectively. The aviation community is actively pursuing specifications for additional fuel production pathways (Fig. 10.1). As with most bio-based fuel production processes, FT and hydroprocessing processes produce a palette of energy products, including diesel and naphtha as well as jet fuels, and may also produce electricity and other bio-based coproducts (e.g., bioplastics or synthetic chemicals). The synthetic isoparaffinic kerosene process produces a single-platform molecule, farnesene, that can be converted into a jet fuel blendstock or other bio-based chemicals. Alcohol-to-jet processes can be tuned to make other hydrocarbons or just jet fuel. As more biorefineries are deployed, the demand for biomass will increase, as will competition among biorefineries and biomass users for feedstocks (e.g., for power generation) from associated growing areas. Optimal feedstock production must therefore develop within a context of complete supply chains defined by specific product targets and must take into account the multiple economic, environmental, and social services that rural landscapes provide, along with the needed transportation and distribution infrastructure, and the associated values of communities and allied businesses (Boody *et al.*, 2005).

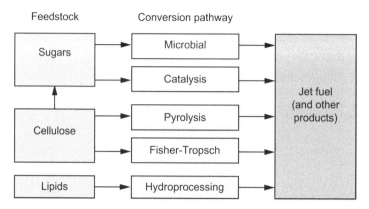

Figure 10.1.   Generalized relationships between biomass feedstocks and conversion pathways that lead to renewable jet fuel.

This chapter highlights the challenges and opportunities for commercial-scale feedstock production from agricultural crops, and presents a novel approach for discovering and designing optimally linked bioenergy crop and conversion process deployment strategies for application at regional scales. Specifically, we describe the potential use of competition-based ecological conceptual frameworks to model the scaling-up process for biofuel production and to identify beneficial levels of diversity for bioenergy crops and conversion processes. We discuss a set of contrasting scenarios that can identify the potential impacts of different land use structures, number and distribution of bioenergy crops, and number and distribution of conversion processes within a region. Data required for this modeling approach would include agricultural yields, conversion efficiencies, and process breakeven capacities, among other factors that may come into play as multiple biorefineries are deployed and the demand for feedstocks within a region increases. Adapting ecological models for this use may facilitate the development of supply chains that address the challenges and enhance the potential benefits of alternative jet fuels.

## 10.2   CHALLENGES OF DEVELOPING AN AGRICULTURALLY BASED ADVANCED BIOFUEL INDUSTRY

The expansion of an advanced biofuels industry will require creation of an entirely new biomass production sector in addition to what already exists for biodiesel, ethanol, and other bioproducts. Furthermore, other industries (such as electric power generation) are also likely to increase demand for biomass. As demand expands, key challenges to scaling up dedicated biomass crop production include:

- Identifying production within suitable growing areas
- Determining ways to incorporate feedstock production into existing agricultural systems without disrupting present markets or degrading the natural resources base
- Developing effective cultivation techniques to reduce production and transaction costs among supply chain participants

All of these challenges must be overcome while complying with environmental and other regulatory requirements, particularly for crops that have not previously been produced at commercial scales. Successful supply chain development will require the establishment of dependable advanced biofuel markets that ensure long-term offtake agreements to provide stability and minimize risks for all supply-chain participants. All of these considerations will affect whether bioenergy crops can be grown successfully in a given region. Full-scale production of dedicated biomass crops is just beginning, so careful monitoring of indicators that measure the effects of expanded production on natural resource quality, the welfare of all supply chain participants, and existing markets will be necessary to anticipate possible adverse long-term consequences and to take corrective actions as needed.

There is little suitable land in developed countries that has not already been converted to existing agricultural purposes (Food and Agriculture Organization of the United Nations, 2003). As an example in North America, biomass and other dedicated feedstocks needed for biofuels, biopower, and bio-based products will be primarily produced on existing agricultural and forestry lands that are already used to produce food, feed, and fiber for established markets. The land area required to support dedicated bioenergy crops necessary to meet the 21 billion gal (79.5 billion L) demand for advanced transportation fuels under the Renewable Fuel Standard (RFS2) could exceed 24 million acres (about 10 million ha), not including areas that can provide crop and timber harvest residues (U.S. Department of Agriculture, 2010). This is an area approximately equal to present corn and soybean production in Iowa (National Agricultural Statistics Service, 2010). Previous estimates suggest that large amounts of feedstock could be available in the continental United States without significantly altering current food production patterns—600 million to over a billion tons of feedstocks harvested per year (Milbrandt, 2005; Perlack and Stokes, 2011; Perlack *et al.*, 2005).

However, it is not just the amount of available feedstock that matters, but also the amount of fuel that can be produced through a specific feedstock-conversion technology pathway. Available bioenergy crops and their yields within a specific region combined with the conversion efficiency of a conversion technology pathway will determine the land area required to meet specific kinds of biofuel demand. Previous work by Stratton *et al.* (2010) provides perspective on the amount of land that would be required to fulfill 50% to 100% of jet fuel demand based on theoretical average yields of fuel per hectare of biomass crop and the first two certified bio-based jet fuel production pathways, Fischer-Tropsch and hydroprocessing (Fig. 10.2). This graphic shows how crucial high feedstock yields will be to minimize land area requirements for scaled-up alternative

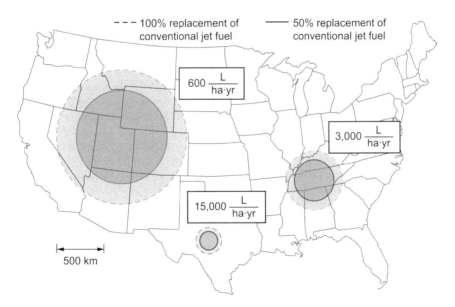

Figure 10.2.   Land area requirements to replace 100% or 50% of conventional jet fuel use in the United States with alternative jet fuels, given a range of theoretical average yields of jet fuel per hectare (ha). The circles representing land area are shown overlain on the continental United States for scale and are not specific to any region. (From Stratton, R. W., H. M. Wong, *et al.*, 2010. Life cycle greenhouse gas emissions from alternative jet fuel, Version 1.2. Cambridge, MA: PARTNER/MIT. With permission.)

fuel production. The amount of biofuel that can be produced from any region based on realistic feedstock estimates are needed to assess whether the requirements of different conversion technology pathways can be fulfilled, particularly for technologies that require certain scales of production to feasibly operate.

As with expanded food production, energy production should also consider how changing populations, socio-economic trends, and water availability affect the allocation of the Earth's finite land mass (Steiner and Griffith, 2010). Competition among different land uses will shift land allocation toward the most profitable enterprises and products. The extent of other demands on feedstocks, e.g., for power or bio-based materials, will also determine the land area available to fulfill various end-use goals and needs, including the desire for bio-based jet fuels, although some of these products can be produced in parallel from the same processes or in series (e.g., use of biomass to make bioplastics, which, after use, are disposed of by "recycling" into fuels). However, it is clear that a scaled-up aviation biofuel industry would place demands not only on crop lands, but also on existing agricultural pasturelands and rangelands, as well as lands currently set aside for conservation (such as the Conservation Reserve Program administered by the U.S. Department of Agriculture (USDA) Farm Service Agency). Shifts in land use to intensive crop production have already occurred in response to increased corn and other crop demand. Continued future demand for advanced biofuels may result in further direct use or indirect changes in land allocation. Concurrently, there is also the potential that biomass crops (as with any other more intensified use of land) may affect biodiversity and greenhouse gas emissions resulting from land use change as result of shifts in the kinds of crops grown, the arrangement of them across landscapes, and the practices used to produce them (Faber *et al.*, 2012). Biodiversity and greenhouse gas emissions impacts are discussed in greater detail in Chapter 13.

Beyond the biophysical limitations that determine where feedstocks will be produced, socio-economic factors will also come into play. Significant effort will be required to build confidence in the grower community to begin growing new biofuel crops. Timely information on costs, contracting, and the risks and benefits of attempting new business enterprises will be required to demonstrate the feasibility of new crops (Steiner *et al.*, 2012). Also, appropriate new relationships will need to be built among the biofuel supply chain participants to establish the needed business transactions and to secure financing, not only to construct biorefineries, but also to produce the required annual supplies of feedstocks. Many of the participants needed to form new biofuel supply chains will likely have not worked together before, and therefore the industry will require proactive considerations as the supply chain network develops.

There are obvious economic considerations that will affect the development of the bioenergy industry. Expanded production of biomass and other dedicated bioenergy crops and purpose-grown wood will likely affect domestic demand and exports, prices, and relative land use of existing crops and timber (USDA, Economic Research Service, 2011). Local economies will be directly affected by the construction and operation of biorefineries within their effective range of operation, as would regional- and national-scale economies that respond to the expansion of this new sector. The required investment for biorefinery construction is significant—the USDA reports that more than 500 new biorefineries would be needed to produce the 21 billion gallons of advanced biofuels, that will require approximately $160 billion in capital investments (USDA, 2010). Funding these up-front capital costs will pose challenges and necessitate that the facilities have assurances of long-term, cost-effective feedstock supplies.

Development of reliable, low-risk supply chains also needs to take into account some level of diversity for bioenergy crop supplies to provide the most dependable aggregated production of feedstocks over time, taking into account inter-annual variation in water availability, temperature and growing season length, and other factors that may influence agricultural productivity such as disease resistance (Smith *et al.*, 2015). Furthermore, the ability of those feedstocks to reliably supply biofuel conversion pathways will depend on the flexibility and resilience of specific conversion pathways, their minimum feedstock requirements to break even and maintain fuel production, and competing alternative uses for feedstocks. These aspects of biofuel supply chain development have scarcely been looked at in a holistic manner. Modeling approaches that can take into account resource availability, infrastructure, competition among feedstocks for land and among conversion processes for feedstocks, and variability in these components over time could assist greatly in targeting viable feedstock and process diversity levels for a given region.

## 10.3   POTENTIAL BENEFITS OF SCALED-UP BIOFUEL FEEDSTOCK PRODUCTION

In spite of the challenges, American farms and forests have the capacity to support expanded biofuels production from biomass, but care will be needed to produce bioenergy crops in ways that simultaneously consider the multiple economic, environmental, and social risks and benefits (Hansen *et al.*, 2007). Potential benefits of biofuel feedstock production may include opportunities for enhanced rural development, proactive land management to improve productivity and land use efficiency, and selection of feedstock sources and production strategies that further other environmental goals such as land restoration, maintenance of biodiversity, and invasive species management.

The Food and Agriculture Organization contends that "[bioenergy] can ... be an important tool for improving the well-being of rural people if governments take into account environmental and food security concerns" (Matthews, 2007). Bioenergy production can result in synergy with food production, depending on the implementation of bioenergy crop production as well as current food production conditions in the region. In the developing world, food production may be limited by access to agricultural inputs, irrigation, electricity and transportation infrastructure, or markets, so bioenergy crop production could be expanded in a way that actually enhances opportunities for food production by improving access to these resources (Diaz-Chavez *et al.*, 2010).

The U.S. Department of Agriculture Rural Development Agency has promoted rural bioenergy production through Farm Bill programs and the Business and Industry Guaranteed Loan Program that can be used to finance first-of-a-kind commercial projects (USDA Rural Development, 2012). Similar commercial development programs are available through the U.S. Department of Energy (DOE) Office of Energy Efficiency and Renewable Energy (DOE, 2007). USDA's analyses suggest that a new biorefinery will result directly in local job creation as well as spur demand for locally grown bioenergy crops, but these also must be weighed against changes in local conditions such as air quality (Pender *et al.*, 2012).

Multiple regional strategies will have to be employed to acquire adequate land areas to produce bioenergy crops. Bioenergy crops will either have to be produced on lands that have been abandoned or are less productive than those being used for other purposes, be produced at times of the year when other crops are not grown, or be incorporated into existing agricultural systems where the benefits of including of the energy crop in the overall system exceed the loss of production from the existing crop. Bioenergy crops could also be useful for rehabilitating degraded lands if the selected species have the ability to remediate contamination (Garbisu and Alkorta, 2001; Gordon *et al.*, 1998) or help restore nutrients and improve local hydrological function (e.g., removal of native trees that have invaded once productive rangelands).

Marginally productive lands have often been promoted as widely available areas where bioenergy crops could be produced without affecting existing food production (McKendry, 2002). Utilization of abandoned or other nonproduction lands provides one option for expanded bioenergy crop production, but the low productivity and environmental sensitivity of these lands generally provide limited potential for energy crop production at a large scale. The term "marginal" is probably a poor characterization of these kinds of lands, because this term refers to a relative comparison to intensively managed lands used for food crop production that often require large amounts of purchased inputs such as agricultural chemicals and irrigation to achieve potential yields. Also, marginal areas may not be as productive as more intensively managed agricultural lands, but have other important functions supporting livestock grazing and providing a number of ecosystem services such as sources of clean water for downstream use, carbon sequestration, maintenance of biodiversity, and wildlife habitat.

In spite of these deterrents, there are examples in which marginal lands may provide opportunities for producing bioenergy crops. Hundreds of thousands of acres on the western side of the highly productive San Joaquin Valley of California have lost irrigation water rights or are degraded because of highly saline soil conditions. Winter-grown cruciferous oil seed crops (e.g., *Sinapis alba*, white mustard) that require little or no supplemental irrigation could produce lipids on such lands (Bañuelos *et al.*, 2010). Similarly, hundreds of thousands of acres in the trans-Pecos region of Texas are available in which fresh water supplies for irrigation during summer are limited due to saline and high-temperature conditions. Castor bean (*Ricinus communis*), a summer perennial shrub that produces seeds high in lipids, is adapted to these otherwise poor growing conditions and could be grown on these lands without competing with other crop options that would require good quality irrigation water (Severino *et al.*, 2012). Depending on the productivity of the lipid crops, more than 30 million gal (114 million L) of oil could be produced annually in each region—enough feedstock to possibly support one alternative jet fuel biorefinery at each location (USDA, Agricultural Research Service, 2012). Intriguingly, noncrop native successional vegetation alone on marginal land also has been shown to potentially yield up to 25% of the cellulosic biofuel target for 2022 (Gelfand *et al.*, 2013).

An alternative strategy could be to produce energy crops on productive lands in seasons when summer crops are not grown (Heggenstaller *et al.*, 2008). More than 20 million acres (8 million ha) of land are left fallow during the winter period in the upper Midwestern United States. The winter annual, field pennycress (*Thlaspi arvense*), could be inserted as a cover crop between summer-grown corn in one summer season and harvested for seed before soybean is planted in June the following summer season (Isbell, 2009). Pennycress grown on winter fallow in soybean production areas has the potential to produce 1.2 billion gal (4.5 billion L) of vegetable oil annually with little disruption to existing summer crop production (USDA, Agricultural Research Service, 2012).

A third strategy for expanding the land area needed for the production of bioenergy crops is to incorporate energy crops into existing production systems such that the benefits to overall system performance from growing the energy crop grown exceed the loss of production from one or more existing crops. For example, annually rotating an oil seed crop such as industrial rapeseed (*Brassica napus*) into 10% of the 50 million acres (20 million ha) of wheat grown in the Western United States would provide 426 million gal (1613 million L) of vegetable oil[1]—the equivalent of 65 to 239 million gal (246 to 905 million L) of renewable jet fuel and 239 to 115 million gal (905 to 435 million L) of diesel in maximum distillate or maximum jet fuel production scenarios, respectively (Pearlson *et al.*, 2013). One in ten years of an oil seed crop grown in sequence with wheat would provide complementary benefits including reduction in weeds, improved soil structure from the deep-rooted nature of the oil seed crop, and diversified income streams that reduce economic risks (Johnston *et al.*, 2002). Contracting arrangements between the feedstock supplier and biorefinery could also provide a dependable income stream that is uncoupled from otherwise volatile commodity markets. Thus, the loss of a portion of the cereal grain production due to the rotation with oilseeds could result in increased overall net benefits to the predominant regional wheat-based system.

The production of alternative jet fuels can also provide an additional value-added market outlet for biomass to relieve other environmental challenges. Tens of millions of acres of western rangelands have been invaded over the past century by invasive tree species such as Eastern red cedar (*Juniperus virginiana*), Western juniper (*Juniperus occidentalis*), and Pinyon pine (*Pinus edulis*). Harvesting these native but invasive woody species for use in biofuel production could provide a relatively low-cost source of biomass, while at the same time provide an opportunity to restore rangeland to productive condition. Conservation program payments to remove invasive trees could help reduce the cost of biomass for biorefineries and could provide incentives to land managers for re-establishing productive rangelands with greater diversity of

---

[1]Based on baseline rapeseed yield value of 3.35 mg rapeseed per hectare per year and 44% oil fraction described in Stratton *et al.* (2010).

native species and restored ecosystem functions such as water yield, improved wildlife habitat, and reduced wildfires risks (Starks *et al.*, 2011). Harvesting at a rate that stops further expansion of these trees could provide enough biomass to supply three or more 40 million gal/yr (150 million L/yr) biorefineries in Oklahoma, Nevada, Oregon, and Idaho (USDA 2012).

The competition for land use on farm and forest lands may also be relieved by the development of biomass production systems based on purpose-grown wood (Hinchee *et al.*, 2010), algae production (Wigmosta *et al.*, 2011; also see Chapter 11), animal manure (Cantrell *et al.*, 2008), and municipal solid waste—all of which could be utilized as feedstocks for biofuel production. This will be important because as discussed above, there are few productive lands available that are presently unused, and less productive lands are often ecologically vulnerable and economically marginal for production of high-value crops with established markets, likely including dedicated bioenergy crops.

If a multifunctional landscape approach is utilized to design future feedstock production systems (Boody *et al.*, 2005), bioenergy crop cultivation could be used to simultaneously contribute to renewable energy production while enhancing biodiversity and ecosystem functions. Strategies have been suggested to manage feedstock production to minimize environmental impacts and maximize ecosystem functions, including biodiversity management approaches in conjunction with the development of major bioenergy projects to optimize biological diversity and wildlife benefits (Fargione *et al.*, 2009; Tilman *et al.*, 2006). See Chapter 13 on environmental impacts for further discussion. It is not clear whether such approaches would be economically viable compared to more intensively managed grassland systems, which modeled and field-scale-validated studies have shown to produce 700% (Farrell *et al.*, 2006) and 540% (Schmer *et al.*, 2008) more output than nonrenewable energy consumed, respectively. Minimal research has been conducted that directly compares the ecological benefits of monocultures and mixtures (Mitchell *et al.*, 2010). Some incentive programs restrict benefits to projects that use native species or enhance local environmental conditions (e.g., USDA's Biomass Crop Assistance Program). Optimizing feedstock diversity as well as the kinds of technology pathways used by conversion facility may provide more opportunities for maintaining and enhancing ecosystem functions and processes. Careful consideration will be required to manage landscapes producing bioenergy feedstocks, particularly as more biorefineries are built and greater amounts of biofuels are produced.

## 10.4   REGIONALIZED BIOMASS PRODUCTION AND LINKAGE TO CONVERSION TECHNOLOGY

As described previously, biomass crops and their production systems will be inherently regional, with producers selecting bioenergy crops that are suited to a given location depending on combinations of climate and soil compatibility, production costs and competitiveness with existing land uses and business operations, and availability of market outlets. Alternative jet fuel production within a given region will require a certified conversion process that can be supplied by regionally sourced biomass feedstock crops, or economically supplemented by feedstocks imported from elsewhere to ensure continuous and dependable operation of the conversion facilities. Since we assume no single region will be able to provide all the required biomass feedstocks needed for the Nation's commercial and military aviation fuels, multiple regions producing best-adapted dedicated bioenergy crops (along with purpose-grown wood species) will participate in biofuel production. Therefore, a diversity of bioenergy crop options will be needed that meet the feedstock specifications of the alternative jet fuel biorefineries based on the conversion technology, delivery schedule required to meet facility operational specifications, and feedstock storage capacity and storability.

For optimal aviation biofuel supply chains to emerge that benefit all system participants, upstream biomass supply components cannot be developed without downstream biorefining and biofuel purchase market signals, including cost points and biorefinery feedstock quality specifications. Feedstocks may vary in their chemical constituents, and these may influence the costs of production, costs of conversion, and quality of the resulting fuel. Similarly, given options for choosing among downstream technologies, upstream signals from farms and rural communities should be included in technologic-economic decision frameworks to provide the most robust opportunities for farm and rural community economic development (Steiner, 2010). Downstream alternative fuel processing technology options will therefore be competing with each other and with other agricultural-based industries (e.g., bio-based chemicals and power generation) for feedstocks and transport capacity. The impacts of expanded advanced biofuel production on land use and associated markets will not be as great for the first 30, 40, or 50 new biorefineries that are initially built, as they will be for the last 300, 400, or 500 facilities that are built (Steiner *et al.*, 2011). Thus, biofuel supply chain development and industry scaleup will have to account for both microlevel factors that can directly influence feedstock supply to individual biorefineries and macrosupply considerations at larger regional and national scales.

In addition, as the industry deployment scales up, novel conversion technologies are likely to emerge that will affect supply chain structure and focus within regions where first-of-a-kind conversion facilities and their linked feedstock supplies have already been established because of early technology readiness (which can be measured by the industry's Fuel Readiness Level, or FRL, used to evaluate new fuel production technology (Commercial Aviation Alternative Fuels Initiative, 2009)). Early established technology pathways could eventually be less efficient than later maturing conversion processes. Preference and previous investment decisions that may favor less-efficient conversion technology pathways, including their choice of feedstock supplies, should give way to better performing approaches, but scenarios and analyses incorporating this kind of industry flexibility and pragmatism have not been reported. The ability to produce the feedstocks needed to meet conversion facility specifications will likely be more flexible than the ability of biorefiners to make significant changes in their highly capitalized technology platforms.

First-generation biofuels, for example, were generally produced with existing corn and soybean commodity crops. Based on concerns regarding conflicts between food and fuel production and the perception that such conflicts caused increased food prices, the commercial air transportation industry and U.S. Department of Defense agencies were intent on using nonfood feedstocks to produce renewable jet fuels (Blakeley, 2012; IATA, 2013). Retrospective analyses have shown that food price increases in 2007 and 2008 were not predominantly caused by increased ethanol production from corn, but were affected by a number of factors including global growth in population, rising per capita incomes, increasing world per capita consumption of animal products, rising energy prices, growing global biofuel production, depreciation of the U.S. dollar, and slower growth in agricultural productivity (Hamelinck, 2013; Trostle *et al.*, 2011). The selected modeling approach also strongly influences estimates of biofuel impacts on food prices, such that modeling of the individual sectors of food and agriculture find much greater impacts on food prices than models that address sectoral interactions and the larger economy (Timilsina and Shrestha, 2010). In the long term, it is the price of crude oil that appears to have been the greatest driver of food prices (Baffes and Dennis, 2013). Nevertheless, as a result of concerns about biofuels and food prices, investment has been made in new, nonfood crops such as *Jatropha curcas* and *Camelina sativa* (Li *et al.*, 2010). These bioenergy crops have been popularly reported as being highly productive on marginal lands and required minimal agrochemical inputs. However, little is known about how to produce these crops at commercial scale in North America, what their actual performance at large scales of production would be across a wide range of regional environments, and very importantly, what their relative performance would be compared to other potentially suitable regionally adapted feedstock crops.

Examples of a range of feedstocks adapted to different regions and their potential for conversion to alternative jet fuel production by example conversion technologies are shown in Table 10.1. Because little agronomic information is available for many of these and other proposed bioenergy crop species, there is a great deal of uncertainty about their potential economic viability, and this has resulted in resistance to growing dedicated energy crops in some grower communities.

Table 10.1. Relative amounts of alternative jet fuel produced per land area within the region of adaptation of potential feedstocks for Fischer-Tropsch (FT), hydroprocessed esters and fatty acids (HEFA), or alcohol-to-jet (ATJ), and pyrolysis conversion technologies, showing Estimated Feedstock Readiness Level (FSRL) and Fuel Readiness Level (FRL) of the feedstock crop and conversion technology.

| Energy crop Common name | Scientific name | Regional adaptation | Yield lignocellulosic material or raw vegetable oil, Ton/acre | Jet fuel equivalent[a] gal/acre | L/ha | FSRL | Jet fuel conversion technology | FRL |
|---|---|---|---|---|---|---|---|---|
| Energy cane | Saccharum hybrids/cultivars | Semitropical, tropical | 27.14 | 686–813 | 6,421–7,607 | 5 | ATJ | 6 |
| Energy beet | Beta vulgaris cultivars | Semitropical | 17.86 | 452–535 | 4,224–5,005 | 2–3 | ATJ | 6 |
| Oil palm | Elaeis guineensis | Tropical | 2.14 | 92–342 | 861–3,199 | 8 | HEFA | 8 |
| Napier grass | Pennisetum purpureum | Semitropical | 10.00 | 253–300 | 2,365–2,803 | 5 | FT | 9 |
| | | | | | | | ATJ | 6 |
| | | | | | | | Pyrolysis | 6 |
| Switchgrass | Panicum virgatum | Temperate | 6.00 | 152–180 | 1,419–1,682 | 5–6 | FT | 9 |
| | | | | | | | ATJ | 6 |
| | | | | | | | Pyrolysis | 6 |
| Corn grain starch | Zea mays | Temperate | 6.14 | 155–184 | 1,453–1,722 | 9 | ATJ | 6 |
| Jatropha | Jatropha curcas | Tropical | 0.68 | 29–109 | 271–1,020 | 4–5 | HEFA | 8 |
| Ethiopian mustard | Brassica carinata | Temperate | 0.21 | 9–33 | 84–309 | 4–7 | HEFA | 8 |
| Castor bean | Ricinus communis | Temperate, semitropical | 0.51 | 22–81 | 206–758 | 2–3 | HEFA | 8 |
| Rapeseed | Brassica rapa/napus | Temperate | 0.43 | 18–68 | 168–636 | 4–7 | HEFA | 8 |
| Sunflower | Helianthus annuus | Temperate | 0.34 | 15–55 | 140–514 | 9 | HEFA | 8 |
| Safflower | Carthamus tinctorius | Temperate | 0.28 | 12–45 | 112–421 | 9 | HEFA | 8 |
| Camelina | Camelina sativa | Temperate | 0.21 | 9–34 | 84–318 | 4–7 | HEFA | 8 |
| Corn oil | Zea mays | Temperate | 0.06 | 3–10 | 28–94 | 9 | HEFA | 8 |
| Industrial hemp | Cannabis sativa | Temperate, semitropical | 0.13 | 6–21 | 21–78 | 2–3 | HEFA | 8 |

[a] Jet fuel equivalent is only an estimate and is meant to give relative comparisons between feedstocks. Yields could vary greatly by regional and local production conditions. FT and HEFA are the likely to be used in the near term to produce certified jet fuel, but many species are also suitable for ATJ and pyrolysis conversion processes. Jet fuel equivalents are developed using crop yield information adapted from Kurki et al. (2010); USDA (2010); Agrisoma (2012); and Knoll et al. (2012) and assuming 70.9 (Davis, 2013) to 84 (Dutta, 2013) gal of ethanol per dry ton of cellulosic biomass, a yield of 0.357 gal jet fuel per gallon of ethanol via the ATJ process (Staples et al., 2014), and 0.15 to 0.56 gal of jet fuel produced per gal vegetable oil (Pearlson et al., 2013). Note that in addition to jet fuel, each of these processes will also produce diesel and other productions from the same feedstock, the amount of which depends on the process and product slate tuning.

These and other factors, such as a lack of available financing for building biorefineries, remain primary barriers to aviation biofuels being competitively priced compared to conventional fuels, even with certified conversion technology pathways available to produce the fuels and markets willing to purchase them. In the future, it will be critical to systematically evaluate the potential productivity of any proposed feedstock so that optimal feedstock supply strategies evolve that meet the specifications of biorefiners and are competitive with existing regional land uses.

These issues highlight the need for the application of standardized assessment approaches such as the Feedstock Readiness Level (FSRL) tool (Steiner *et al.*, 2012) and the FRL tool (CAAFI, 2009) that can help identify potential gaps in aviation biofuel supply chains due to disconnects between the capacity to supply feedstocks for a particular conversion process and markets for feedstock producers, including the availability of technologies and bioconversion facilities to create feedstock markets. Using an integrated feedstock and conversion technology approach allows for the identification of all feedstock production costs and activities to scale up production and allocation of resources to effectively develop the needed supply chains to create a viable aviation biofuels industry (Steiner *et al.*, 2012).

As the aviation biofuel industry scales up, certain feedstocks and processing technologies may become more prominent. To develop high-performance supply chains, efforts should concentrate on the best use of available regional resources, using insights into how the present best-use strategies may change over time. The bioenergy crops that are deployed first will depend on the readiness of commercial-scale aviation biofuel refiners, with subsequent feedstocks depending on the performance of the next generation of commercially ready conversion technologies.

## 10.5    APPLYING ECOLOGICAL MODELS TO BIOFUEL PRODUCTION

Even though the lack of commercial-scale biorefineries is a chief limitation to the immediate creation of widely distributed dedicated biomass and other bioenergy crop markets, the eventual success of advanced biofuels production will depend on the development of efficient complete supply chains that optimize regional productivity. Significant computational power and modeling capacity are now available to describe how our past actions have affected the world around us, and these can also be used to help identify opportunities to build more sustainable agricultural systems that also include biofuel production (Steiner *et al.*, 2011). Modeling approaches could be developed and validated with emerging region-based information to assess newly established biofuel production supply chain performance, particularly if combined with key indicators of system conditions such as those internationally agreed upon by the Global Bioenergy Partnership (Global Bioenergy Partnership, 2011). Such directed approaches can also identify corrective options to help alleviate conflicts and smooth the transition from a petroleum-based economy to a new bioeconomy.

Macro-econometric modeling approaches have been used to develop supply curves and inventories for the amounts of biomass that could be available (Perlack and Stokes, 2011). However, a documented history of dedicated biomass production at large scales does not exist because a commercial sector is just beginning to emerge. Mechanistic modeling approaches are available to predict the impacts of agricultural production on natural resources and economic indicators of sustainability, and when linked, biophysical and economic models allow integrated analyses to be made of management impacts on environmental and economic factors across geographic scales (Whittaker *et al.*, 2007). Such approaches provide opportunities to optimize multiple objectives and policy alternatives that would be acceptable to (or ideally optimal for) the greatest number of stakeholders, and then design renewable energy systems that meet those requirements. As with the intent of the Global Bioenergy Partnership sustainability framework, quantitative integrated analyses are important because maximized desired outcomes for any individual supply chain component may not capture important feedbacks and interactions (Antle and Capalbo, 2002). Integrated analyses could be used to identify the most sustainable options to manage agricultural and forest landscapes from which biofuel feedstocks are produced as well as provide details about the potential tradeoffs that should be considered across regions and competing sectors (U.S. Department of Energy Office and U.S. Department of Agriculture, 2009).

Even with these quantitative approaches for gaining insights into how an emergent biofuels sector may impact economic, biophysical, and social aspects of complex agricultural systems, these modeling approaches do not address the development of the bioenergy industry structure in terms of mutual interactions between regional feedstock supplies, their use by biofuel conversion facilities, or the expansion of complete supply chains over time, nor do they address the potential competitive relationships among conversion processes and facilities across a regional landscape. Therefore, an approach is needed that complements econometric and mechanistic modeling to determine how biomass supplies and biorefineries for an emerging aviation biofuel industry will change as the industry is scaled up.

With relatively few advanced biofuel bioconversion facilities presently coming into operation compared to the number needed to achieve industry and government goals, there will be little initial competition for biomass supplies within the biofuel industry, although there may be competition with existing power generation facilities or other biomass users. This will change as increasing numbers of facilities are built that compete for similar feedstocks within a region. To approach this challenge, it is important to consider the possible development pathways for biofuel production under different feedstock availability scenarios, including the advantages or disadvantages of concentrating efforts on one or a few bioenergy crops or conversion processes versus distribution among a greater number of bioenergy crops or processes. Such work requires defining the key elements of the biofuel production system, describing the linkages between them, and exploring the consequences of changing key system parameters.

The suggested modeling approach shares essential features with community ecology analyses, allowing the use of existing theory and models. The key questions in this chapter are centered on the number, abundance, and distribution of regionalized bioenergy crops used in biofuel production, and their associated region-based processing options. With the modeling approach described here, community ecology theory can be applied to relationships between conversion facilities and feedstocks in ways similar to those for biological patterns and processes for species coexistence. Emphasis is placed on interactions between different members of a given part of the supply chain. This approach also allows evaluation of the consequences of external environmental changes such as extinctions (loss of economic viability) and effects of environmental disturbance on feedstock availability to conversion facilities.

Ecological models often describe relationships between predators and prey in food chains, which we correspond to biorefineries and feedstocks, respectively, as parts of an aviation biofuel supply chain. Final stable conditions may result with the dominance of a single predator species (a single biorefinery technology pathway), or the coexistence of two or more species (a diversity of biorefining technology pathways). In this chapter we propose that an ecological modeling approach may be translated into tools for optimizing the level of diversity of bioenergy crops (prey) and conversion processes (consumers, predators) within a region. The level of diversity to optimize feedstock supply is likely to be dependent on priority (the order in which bioenergy crops or biorefinery processes that utilize them enter a regional system), competition among crops and among conversion processes, and crop resilience to environmental perturbation such as drought that may alter crop yields. We propose ecological modeling approaches to elucidate those relationships and their impacts on scaled-up biofuel supply chains. Below we describe four types of ecological models that are relevant to biofuel supply chain modeling:

1. *Patch-occupancy models.* Patch-occupancy models (Caswell and Cohen, 1993; Hanski, 1994) were developed for small isolated populations that are not likely to persist. These models have been used extensively to predict the long-term dynamics of rare species (Hanski *et al.*, 1995), and have well-understood properties. The relevance of these models for biofuel feedstock production is that a bioenergy crop field or conversion facility can be considered a population, and the success of that crop or facility may depend in part on the size of the operation (patch size)

and the distance and distribution of other similar fields or facilities (connectivity) for economies of scale in preprocessing and transportation.

2. *Lotka-Volterra competition.* This classic model (Wangersky, 1978) is based on predator-prey dynamics, most well-known for application to lynx and hare population dynamics, where boom and bust cycles arise because of geometric growth in populations and the time lag in a predator's responses to change in the prey's population size. These dynamics can be modeled in a simple two-species system where the population size of each depends on the intrinsic growth rate, initial population size, and interaction coefficient with the prey (Chesson, 2000). Lotka-Volterra models work as well for competing species of the same category (e.g., between predators). Such models would be applicable to bioenergy crops if the focus was on how several bioenergy crops compete with each other in one location. The competition coefficient between bioenergy crops might incorporate factors such as yield differences, required agricultural inputs (e.g., fertilizer or irrigation water), and resilience to disturbance (e.g., drought). Table 10.2 shows how the parameters of the Lotka-Volterra model parallel phenomena in the proposed biofuel scaleup supply-chain modeling. This type of model may also be used to understand how bioenergy crops will compete with existing agricultural crops in the region.

3. *Food webs.* Food web models focus on system stability as a function of the number of links between members of the community, as well as the strength of interactions (Paine, 1980). Communities with many weak links (such as community members that are not highly dependent on any one prey or resource) tend to be more stable. By analogy, this would suggest that having a diversity of bioenergy crop options within a region may stabilize biofuel supply chains, although competition with other land uses and existing crops may also influence the viability of maintaining diverse bioenergy feedstocks. A high diversity of predators has been found to make systems more stable in both experiments (O'Gorman *et al.*, 2008) and mathematical simulations (Brose, 2008). With biofuel conversion facilities considered as "predators" consuming a bioenergy crop (Fig. 10.3), the implication is that more stable biofuel production over the long term many be achieved by promoting several conversion technologies within a region, rather than the dominance of one kind of technology, despite economies of scale achieved with a single technology. Ecological experiments designed to test food-web models have also found substantial priority effects, meaning outcomes of competition between species depends on the order of community assembly (Fukami *et al.*, 2010). In the current application, the order of deployment of a crop or conversion technology introduced into a region would be parallel to the ecological example.

The three-trophic-level food web shown in Figure 10.3 has bold lines showing focal species linkages between the resources for which they compete and a predator that affects both of them. High density of a focal species increases predator density, leading to greater predation on both focal species (apparent competition). Feedback loops through resources create resource competition. For modeling biofuel feedstocks, food webs provide a logical starting point if the goal is to assess the stability of a system over time. The three trophic levels can be analogous to components of the biofuel supply chain, as shown in Figure 10.3. The bidirectionality of relationships also reflects the interplay between bioenergy crop and process selection, since suitability of bioenergy crops for a region will determine appropriate processes, but available processing capacity will determine whether a grower will produce the bioenergy crops. Table 10.2 shows in greater detail how food-web modeling parameters can be transferred to the biofuel scaleup applications. The table provides examples of two ecological model approaches that can be transferred to aspects of biofuel industry interactions, with descriptions of the key parameters in the common formulation of these models and their analogies in biofuel production. Parallels have been shown between food-web approaches to predator-prey relationships and cost relationships among complex supply chains by translating biomass flow (in ecological models) to economic flow (Nagurney and Nagurney, 2011). Multiple predators with different efficiencies for capturing and processing prey would be analogous to different biomass uses such as ethanol, biodiesel, or bioplastics. This existing work will be helpful in modeling the biofuel industry scaleup.

4. *Biodiversity-ecosystem functioning.* For the last 25 years, ecologists have been investigating how biological species diversity itself drives ecosystem functioning, rather than simply

Table 10.2.   Examples of two well-established ecological models adapted to address aspects of biofuel supply chain components and their interactions.

| Model | Parameter | Biological meaning | Possible feedstock analogies |
|---|---|---|---|
| Food webs | $\rho$ | Niche overlap between two species | Extent of overlap in land occupied; overlap in type of resources needed for production |
| | $\kappa$ | Relative fitness of a particular species | Economic value per unit produced; greenhouse gas emission level |
| | $\alpha_{ij}$ | Interaction coefficient of species $i$ on species $j$ | Sum of all competitive effects between two bioenergy crops |
| | $\alpha_{jj}$ | Interaction coefficient of species $j$ on itself; intra-specific competition | |
| Lotka-Volterra | $\alpha$ | Interaction coefficient Total effect of species $i$ on species $j$ | As for food webs |
| | $K$ | Carrying capacity | Total available land; total available resources |
| | $R$ | Per capita growth rate | Rate of expansion of a bioenergy crop over the landscape |
| | $N$ | Population size Usually number of individuals, but can also use total biomass or area | Area of cropland; tons of feedstock produced |

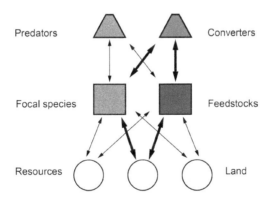

Figure 10.3.   Schematic of a food web model used in ecology, with analogies in biofuel feedstock production. (Adapted from and reprinted by permission from Macmillan Publishers Ltd: Chesson, P., and J. J. Kuang. 2008. The interaction between predation and competition. *Nature* 456:235–238. Copyright 2008.)

being a byproduct of particular abiotic conditions. In other words, do diverse systems function differently than nondiverse systems? To evaluate this, researchers often evaluate what the likely general consequences of extinctions are for the way ecosystems function. This work has led to the consensus that within a particular environment, species loss leads to deterioration of eco-system function (Cardinale *et al.*, 2006; Hooper *et al.*, 2005). Experimental work has focused on manipulations of the producer trophic level (often grassland plants) and measurement of net primary productivity (peak annual aboveground biomass) as a key ecosystem function. The clear analogy with biofuel feedstock production makes this approach an interesting one: the focus is on how to maximize total production with an appropriate mixture of feedstock sources, given the advantages and disadvantages for each bioenergy crop (Dickson and Gross, 2015). A mechanistic model initially developed by Loreau (1998), focused on resource depletion and complementary resource use (i.e., partitioning of natural resource base among species), provides a starting point for this type of modeling with regard to biofuel production supply chains.

### 10.5.1   *Description of on ecological model for biofuel production*

Defining elements of the biofuel supply-chain modeling requires limiting the complexity of the system to the essential elements. In the case of biofuel production, the system is reduced to three core elements: (1) production of bioenergy crops, (2) logistics of pre-processing and transportation, and (3) conversion of raw feedstocks into fuel (Fig. 10.4). Each of these elements has several core components and processes, within which analogies to ecological models can be found:

1. *Production.* Bioenergy crop production depends at the most basic level on crop type and available land. Land availability for production depends regionally on the distribution and size of patches. In the scenarios described below, the difference between many large and few small patches of available land can have large implications for the rest of the supply chain. The analogous model in ecology is a patch-occupancy modeling approach, where the success of a species depends on the rate at which it can disperse to available patches and the degree to which it can compete effectively with any existing species already established in that patch.

Bioenergy crop type certainly will also play a large role in production volume and patterns. The competition between bioenergy crops for available land will depend on the success of a particular feedstock, given the proximity to and type of conversion facilities in the region, the agronomic suitability of the bioenergy crop for the region, and the competitive economic value of the feedstock in relationship to existing land uses considering the end product. Such competition between bioenergy crops can be modeled following the Lotka-Volterra approach. Lotka-Volterra modeling minimally requires knowing the population size (acreage) and growth rate (expansion of a bioenergy crop) of each species, and a general interaction term that describes the extent to which one species decreases the growth of another. Such an interaction term could be used to separately describe agronomic suitability, energy yield per area of production, or economic value given the rest of the supply chain.

2. *Logistics/Transportation.* The logistical elements of the supply chain involve the transportation network and preprocessing such as densification that may occur before raw feedstocks are transported. These transportation and processing elements connect bioenergy crops to conversion facilities and vary in response to the type of crop and the distribution of available land.

3. *Conversion*: Biofuel refineries can vary in both number and type at the regional scale. Reducing each technology to the essential elements, we consider only the conversion efficiency of a facility for a particular feedstock and the minimum quantity of feedstock required for economical operation. Modeling the relationship between conversion facilities and feedstocks, given the logistics, can be accomplished following a food web approach. Here the number and identity of "consumers" (conversion facilities or other users of biomass such as power generation facilities) can be manipulated, given a set of bioenergy crops, and the total biofuel output can be modeled in response.

Underlying each element is the natural resource base and the man-made infrastructure. Figure 10.4 illustrates the simplified framework allowing individual components to be modeled by using ecological models, as well as modeling the whole supply chain by linking together the three components with the natural-resource base (namely land-based resources) and the man-made infrastructure.

4. *Caveats.* The use of ecological models for biofuel feedstock production has clear benefits. Such ecological models have been developed to consider how species compete with one another, as bioenergy crops will for available land, and how linkages between different trophic levels (conversion facilities and feedstock producers) determine the stability of the system and its total productivity over the long term. However, differences clearly exist between such models and the reality of biofuel production. As a profitability-driven system, external influences such as regulations, interest-rate changes, or subsidies can strongly affect the success of a particular technology or bioenergy crop. While such external influences could be thought of as disturbances like hurricanes or wildfires, which promote the growth of some species while reducing others, these influences are difficult to capture in simple models of species interactions.

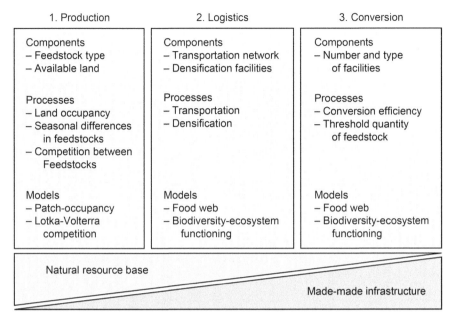

Figure 10.4.   Schematic for modeling biofuel production.

In addition, predation in this system has a much different implication than in an entirely biological system, in that in a biological system the producer/prey species are not determined by the predator/consumer species that enter the system, whereas the selection of a given bioenergy crop by producers is likely to depend on the availability of suitable conversion facilities, and dedicated bioenergy crop production cannot occur in the absence of conversion facilities. Furthermore, feedstock producers want their product to be consumed by the processing facilities, so contractual agreements must be established to pass feedstocks from producers to conversion facility users. These caveats will pose additional challenges to the use of ecological models for assessing biofuel supply chain development scenarios.

10.5.2   *Scenarios to test biofuel production*

Predicting exactly how biofuel production will scale up at the continental level is beyond the ability of any one model. However, we can test the consequences of our assumptions about how such production will play out at a regional level by considering specific scenarios where only one key element is manipulated. Of the many possible scenarios, here we consider the following seven contrasts, and their potential implications for a regional biofuel industry. These contrasts are also outlined with simple graphical summaries in Table 10.3.

1. *Many small versus few large patches of available land for bioenergy crop production.* The distribution of the size of available land areas for cultivating bioenergy crops (patches) could have a dramatic influence on all the downstream parts of the supply chain. As an example, in a scenario where many small patches are available, such as from smaller farms independently choosing to produce bioenergy crops, the location of densification facilities and the logistics of transporting prepared materials to biorefineries would likely be more complicated. Such a scenario could lead to greater diversity of feedstocks and technologies, but could present challenges for conversion technologies that have a high minimum threshold of raw materials to maintain profitable operation.

Table 10.3. Summary of scenarios to model the outcomes of different assumptions in the scaling-up of biofuel feedstock production.

| Schematic | Dichotomy | Key factor/Gatekeeper, modeling methodology, and data needed | Implications of first option relative to second option |
|---|---|---|---|
| | 1. Many small patches versus few large patches | *Key factor:* Land ownership distribution<br>*Model:* Patch occupancy<br>*Data:* Distribution and spatial arrangement of farm sizes in given area | More variability in production, greater logistical challenges (transportation and densification), more complex contracting Example: British Petroleum, 2012, Highlands County, Florida |
| | 2. Many conversion processes versus few conversion processes | *Key factor:* Fuel Readiness Level (FRL)<br>*Model:* Food web<br>*Data:* Conversion efficiency of given land area and feedstock, breakeven feedstock volumes, and current fuel price | Greater competition among consuming biorefineries for feedstocks, more complex logistics |
| | 3. Many bioenergy crops versus few bioenergy crops | *Key factor:* Feedstock Readiness Level (FSRL)<br>*Model:* Lotka–Volterra<br>*Data:* Yield, conversion efficiencies for processing, cost, and resilience to environmental change | Competition among bioenergy crops, increased system complexity if bioenergy crops are also from different classes, potentially greater production due to resilience in face of environmental variability |
| | 4. Long-term stability versus interannual variation (e.g., long-term contracts or environmental stability versus lack of contracts/stability) | *Key factor:* Regional climate variability<br>*Model:* Biodiversity-ecosystem functioning<br>*Data:* Bioenergy crop yield under varying precipitation and temperature | Diversity of bioenergy crops less important, less robust transportation network required |
| | 5. Use of easily converted land versus landscape redesign | *Key factor:* Current patterns of land use and bioenergy supply chain requirements<br>*Model:* Food web, patch occupancy<br>*Data:* Current land use "compatibility" with switch to a bioenergy crop type for easily converted land versus all land available for landscape redesign, effect of design on efficiency of bioenergy supply chains, as required by each configuration | Fewer contiguous patches of bioenergy crop, more complex logistics |
| | 6. Certified process priority and technology maturation over time versus equal opportunity for all conversion processes | *Key factor:* ASTM certification<br>*Model:* Food web<br>*Data:* Current status of certification for different technologies and anticipated timeframe for future certifications | Dominance of some bioenergy crops, even if they are potentially less competitive (ease of logistical switching between bioenergy crops and conversion compatibility determine length of dominance) |
| | 7. Priority for certain bioenergy crops versus equal opportunity for all bioenergy crops (i.e., one feedstock followed by several others versus everything arriving at same time) | *Key factor:* FRL or FSRL<br>*Model:* Food web<br>*Data:* Ease of crop switching among feedstock types | Largest patches will be occupied first by earliest bioenergy crops (features shared with scenario 1; ease of switching in terms of both logistics and conversion determine priority effects) |

A scenario with few large patches of available land for cultivation could arise either by the historical contingency or due to contractual agreements. For example, in central Florida, a large tract of land was initially developed for bioenergy crop production by British Petroleum, but this scenario was abandoned (Biofuel Digest, 2012). With a single large landholding available and the biorefiner the first in the region, there would be an advantage to the consumer of a minimal number of contracts and reduced logistical complexity for coordinating delivery of feedstocks to the conversion facility.

Patch-occupancy modeling would be an appropriate model to adapt to test the consequences of these contrasting scenarios. In ecological research, it has been found that there is a tradeoff between having many available patches for distinct populations to colonize (promoting the survival of a species overall) and having sufficient size of patches to support a viable population (promoting the survival of each particular population).

2. *Many versus few conversion processes.* The number of conversion technologies employed within a region, independent of the land available, could likewise have systemic consequences for biofuel production. The simplified logistics and demand for only a particular set of bioenergy crops under the scenario of fewer conversion processes could ensure that each facility would more easily meet its minimum tonnage requirements of raw material. However, such a scenario could also be more susceptible to external disturbances, as from economic or policy changes or major environmental perturbations such as drought.

Food-web models would be appropriate for testing the consequences of such scenarios, where the number of consuming biofuel refineries varies. Both the total output of the production system and the stability of this system in response to disturbances would be appropriate response variables. The number of different processes used to produce non-fuel products—such as bioplastics—in addition to biofuels, rather than the number of processes just to produce biofuels, could also be modeled in this framework.

3. *Many versus few bioenergy crops.* The number of bioenergy crops in a region may be determined by many factors, including feedstock readiness, technology readiness, the availability of suitable logistics and transportation networks, and the history of cultivation in an area. For modeling purposes, we can consider two extreme scenarios by contrasting monocultures with multiple bioenergy crop species on the land available for bioenergy crops. With multiple bioenergy crops, competition will arise between bioenergy crops for land resources, and logistics will become increasingly complex.

Such competition could be modeled by Lotka-Volterra models, in which a species' population size is determined by its own growth rate (rate of spread by technology transfer or other popularization), its carrying capacity in the region (the maximum acreage it could reach), and the interactions with other species. These interactions could be modeled in terms of economic advantage, energy yield per acre, or relative environmental benefits.

In the case of many bioenergy crops over the long term, net fuel production may be enhanced relative to monocultures due to resilience in the face of environmental variability. Such resilience can be modeled as well in the Lotka-Volterra framework.

4. *Long-term stability versus interannual environmental variation.* Isolating the effect of climate or other environmental variability would allow the testing of how resilient a region may be in biofuel production. Holding the number of bioenergy crops, number of technologies, and patch size distribution constant, we can examine the consequences of future climate scenarios for biofuel production. Under stable environmental conditions with predictable growing season precipitation and few temperature extremes outside of a normal anticipated range, it is likely that optimal production would be achieved with only a few most suitable bioenergy crops and technologies that would involve simpler transportation logistics. Conversely, the more variable the climate, the more important resilience is likely to be, which may translate into greater desired bioenergy crop diversity that ensures consistent biofuel production over time, even if any year would be slightly lower than the absolute optimal achievable by focusing on one or a few bioenergy crops. Biodiversity-ecosystem functioning models could be employed to estimate the

optimal diversity level for a bioenergy crop production system in a region as well as the composition of bioenergy crop types.

5. *Use of available land versus landscape redesign.* Beyond the size of land patches available for bioenergy crop production, the spatial arrangement of these patches can have a substantial effect on the whole system. If advanced bioenergy crops are cultivated simply in a patchwork way, it is possible that a less efficient supply chain with overly complex infrastructure could result.

The degree of compatibility between current land uses and potential bioenergy crops will determine the ease of changes in land use, and therefore the chances of such landscape redesign being achieved. For example, switching from softwood production for paper in the Southeastern United States to hardwood biomass production likely presents fewer challenges than switching from corn to agricultural lipid production. Switching between annual crops may be simpler than switching from one perennial to another. Data on the compatibility of current land uses and bioenergy crops are thus the most important type of information for investigating these scenarios. Both food-web modeling and patch-occupancy modeling would be useful in investigating this scenario.

6. *Priority of technologies: Certified processes—first versus all.* Since conversion technologies have different levels of readiness for commercial production, it is reasonable to assume that only one or a few technologies will initially be dominant in any particular region. As other technologies mature towards certification and commercial readiness, the optimal solution for spatial distribution, number, and type of bioenergy crops being produced would also likely change. This complex scenario can be modeled by holding other variables constant and contrasting a situation where several technologies are all certified from the start versus when only one is certified initially, gradually followed by others. As for the landscape redesign scenario, the ease of switching between bioenergy crops will be a crucial variable.

This scenario is clearly analogous to food-web models, with substantial evidence for "priority effects" leading to the dominance of a consumer (i.e., a conversion technology) arising simply from the earlier arrival of that species at a location. Interestingly, such priority effects could lead to dominance even by a less efficient conversion technology, if the barriers to switching bioenergy crops and technologies are high.

This type of scenario would also be useful to investigate the effects of existing processes that use biomass resources (such as power generation or existing ethanol or biodiesel processes) that may compete with advanced biofuel production.

7. *Priority of bioenergy crops.* Similar to the previous scenario, disparities in the maturity of advanced biofuel bioenergy crops measured by FSRL can also have large consequences for the production supply chain. As above, a simple comparison can be made between situations where all of several bioenergy crops are equally ready for commercial production from the start versus where one crop initially dominates due to greater readiness or greater community interest.

With priority effects, it is reasonable to assume that the largest patches will be occupied first by the most-ready bioenergy crop, leading to scenarios similar to scenario 1 contrasting many small versus few large patches. The ease of switching production systems, both in terms of transportation and densification logistics and in conversion technologies, will determine how strong and long lasting such priority effects will be. Food-web models again provide a possible tool for exploring the consequences of such priority effects arising from disparities in feedstock readiness levels.

These contrasting scenarios provide insight into the potential use of existing or adapted ecological models to describe biofuel supply-chain development and optimization. Development of a composite model that takes into account the different strengths of the aforementioned models and can be more specifically customized to details of the biofuel supply chain would allow for the inclusion of factors such as costs, product pricing, and other parameters that may have no simple parallel in existing ecological models. By introducing these conceptual opportunities, we hope to stimulate further application of such modeling approaches to facilitate the development of sustainable, efficient biofuel supply chains that maximize productivity of bioenergy crops and fuels.

## 10.6 SUMMARY

Bioenergy crop and renewable aviation biofuel production can provide potential economic and possibly environmental benefits with careful planning. A key challenge for the nascent bioenergy feedstock production industry is how to produce sufficient feedstocks to supply one or more biorefining processes in a given region based on the constraints of available land, bioenergy crop compatibility with the local conditions, feedstock compatibility with processing options, and minimization of adverse impacts on existing land uses and the environment. The selection and cultivation of different sources of biomass (e.g., use of planted dedicated bioenergy crops versus agricultural or forestry residues versus invasive species that could be used for biomass) will strongly affect the regional yield and the processes that are viable for the region. The establishment of facilities for a given process or biomass user will also affect bioenergy crop selection, such that early arrivers are likely to strongly influence the development of the regional feedstock industry.

We propose that existing ecological modeling approaches can be used to identify optimal levels of bioenergy crop and producer diversity under various conditions for different regions. Envisioning bioenergy crops as the "producers" and conversion processes as the "consumers" in ecologically oriented models can highlight key aspects of the relationships among bioenergy crops, among processes, and between these two levels, as well as clarify their dependence on the natural resource base and man-made infrastructure such as transportation and other logistical options. Adapting ecological models for use in the biofuel industry may allow policymakers and industry stakeholders to better understand the dynamics that will affect the optimal number and types of bioenergy crops and conversion processes for a given region. These traditional ecological models also provide the opportunity to evaluate the potential importance of priority (being first to establish in a given region), and the optimal level of bioenergy crop diversity that would maintain regional output on an inter-annual basis in spite of potential extreme weather variations such as drought. Other aspects of land use, such as agricultural land holding size, and improvements or changes in technology efficiencies over time, may also affect these outcomes. Given the complexity of supply chains and feedstock process interactions, it is likely that for any given region there will be multiple "optimal" solutions, and these would likely change as the industry scales up and deploys greater numbers of facilities. Developing models to explore potential options may provide insights that would facilitate effective scaleup of the biofuel industry at the regional level and help to ensure that sustainable outcomes are achieved that ensure productivity and profitability to all supply chain participants, protect the natural resources base, and benefit rural communities.

## REFERENCES

Agrisoma. 2012. Company Web site. *http://agrisoma.com/#pageID=109* (accessed August 29, 2016).

Antle, J. M., and S. M. Capalbo. 2002. Agriculture as a managed ecosystem: policy implications. *J. Agr. Resour. Econ.* 27:1–15.

Baffes, J., and A. Dennis. 2013. Long-term drivers of food prices. Policy Research Working Paper 6455, The World Bank Development Prospects Group & Poverty Reduction and Economic Management Network, Trade Department. *http://www-wds.worldbank.org/servlet/WDSContentServer/WDSP/IB/2013/05/21/000 158349_20130521131725/Rendered/PDF/WPS6455.pdf* (accessed August 29, 2016).

Bañuelos, G. S., J. D. Roche, and J. Robinson. 2010. Developing selenium-enriched animal feed and biofuel from canola planted for managing Se-laden drainage waters in the westside of central California. *Int. J. Phytorem.* 12:243–254.

Biofuel Digest. 2012. The October surprise: BP cancels plans for US cellulosic ethanol plant. *http://www.biofuelsdigest.com/bdigest/2012/10/26/the-october-surprise-bp-cancels-plans-for-us-cellulosic-ethanol-plant/* (accessed August 29, 2016).

Blakeley, K. 2012. DOD alternative fuels: policy, initiatives and legislative activity. *C.R. Report for Congress*. Washington, DC: Congressional Research Service.

Boody, G., B. Vondracek, D. A. Andow, *et al.* 2005. Multifunctional agriculture in the United States. *BioScience* 55:27–38.

Brose, U. 2008. Complex food webs prevent competitive exclusion among producer species. *Proc. R. Soc. B.* 275:2507–2514.

Cantrell, K. B., T. Ducey, K. S. Ro, *et al.* 2008. Livestock waste-to-bioenergy generation opportunities. *Bioresource Technol.* 99:7941–7953.

Cardinale, B. J., D. S. Srivastava, J. Emmett Duffy, *et al.* 2006. Effects of biodiversity on the functioning of trophic groups and ecosystems. *Nature* 443:989–992.

Caswell, H., and J. E. Cohen. 1993. Local and regional regulation of species-area relations: a patch-occupancy model. In *Species diversity in ecological communities*. Chicago: The University of Chicago, 99–107.

Chesson, P. 2000. Mechanisms of maintenance of species diversity. *Annu. Rev. Ecol. Evol. Syst.* 31:343–366.

Chesson, P., and J. J. Kuang. 2008. The interaction between predation and competition. *Nature* 456:235–238.

Commercial Aviation Alternative Fuels Initiative. 2009. Fuel readiness level. *http://caafi.org/information/fuelreadinesstools.html* (accessed August 29, 2016).

Davis, R. 2013. Biochemical Platform Analysis: biochemical platform review. *http://energy.gov/sites/prod/files/2016/05/f31/biochem_davis_2611.pdf* (accessed August 29, 2016).

Diaz-Chavez, R., S. Mutimba, H. Watson, *et al.* 2010. Mapping food and bioenergy in Africa. A report prepared for FARA. Ghana: Forum for Agricultural Research in Africa.

Dickson, T. L., and K. L. Gross. 2015. Can the results of biodiversity-productivity studies be translated to bioenergy production? *PLoS ONE* 10:e0135253.

Dutta, A. 2013. DOE Bioenergy Technologies Office: Project Peer Review—syngas mixed alcohol cost validation. *http://www.energy.gov/sites/prod/files/2015/06/f24/thermochemical_conversion_dutta_3611.pdf* (accessed August 29, 2016).

Faber, S., S. Rundquist, and T. Male. 2012. Plowed under—how crop subsidies contribute to massive habitat losses. Washington, DC: Environmental Working Group.

Fargione, J. E., T. R. Cooper, D. J. Flaspohler, *et al.* 2009. Bioenergy and wildlife: threats and opportunities for grassland conservation. *BioScience* 59:767–777.

Farrell, A. E., R. J. Plevin, B. T. Turner, *et al.* 2006. Ethanol can contribute to energy and environmental goals. *Science* 311:506–508.

Food and Agriculture Organization of the United Nations. 2003. World agriculture towards 2015/2030, an FAO perspective. Ed. J. Bruinsma, Rome: FAO, 432.

Fukami, T., I. A. Dickie, J. P. Wilkie, *et al.* 2010. Assembly history dictates ecosystem functioning: evidence from wood decomposer communities. *Ecol. Lett.* 13:675–684.

Garbisu, C., and I. Alkorta. 2001. Phytoextraction: a cost-effective plant-based technology for the removal of metals from the environment. *Bioresource Technol.* 77:229–236.

Gelfand, I., R. Sahajpal, X. Zhang, *et al.* 2013. Sustainable bioenergy production from marginal lands in the US Midwest. *Nature* 493:514–517.

Global Bioenergy Partnership. 2011. The global bioenergy partnership sustainability indicators for bioenergy. Rome, Italy: Food and Agriculture Organization of the United Nations.

Gordon, M., N. Choe, J. Duffy, *et al.* 1998. Phytoremediation of trichloroethylene with hybrid poplars. *Environ. Health Perspect.* 106:1001.

Hamelinck, C. 2013. Biofuels and food security: risks and opportunities. Utrecht, Netherlands: Ecofys. *http://www.ecofys.com/files/files/ecofys-2013-biofuels-and-food-security.pdf* (accessed August 29, 2016).

Hansen, T. M., C. Francis, J. Dixon Esseks, *et al.* 2007. Multifunctional rural landscapes: economic, environmental, policy, and social impacts of land use changes in Nebraska. In *Theses, Dissertations, and Student Research in Agronomy and Horticulture,* Lincoln, NE: DigitalCommons@University of Nebraska. *http://digitalcommons.unl.edu/agronhortdiss/45* (accessed August 29, 2016).

Hanski, I. 1994. Patch-occupancy dynamics in fragmented landscapes. *Trends Ecol. Evol.* 9:131–135.

Hanski, I., T. Pakkala, M. Kuussaari, *et al.* 1995. Metapopulation persistence of an endangered butterfly in a fragmented landscape. *Oikos* 72:21–28.

Heggenstaller, A. H., R. P. Anex, M. Liebman, *et al.* 2008. Productivity and nutrient dynamics in bioenergy double-cropping systems. *Agron. J.* 100:1740–1748.

Hinchee, M. A. W., L. N. Mullinax, and W. H. Rottmann. 2010. Woody biomass and purpose-grown trees as feedstocks for renewable energy. In *Plant Biotechnology for Sustainable Production of Energy and Co-products*, eds. P. N. Mascia, J. Scheffran, and J. M. Widholm, 66:155–208, Heidelberg: Springer Berlin Heidelberg.

Hooper, D. U., F. S. Chapin, J. J. Ewel, *et al*. 2005. Effects of biodiversity on ecosystem functioning: a consensus of current knowledge. *Ecol. Monogr.* 75:3–35.

IATA. 2013. Fact sheet: alternative fuels (updated June 2013). *http://www.iata.org/publications/pages/alternative-fuels.aspx/* (accessed August 29, 2016).

Isbell, T. A., 2009. U.S. effort in the development of new crops (Lesquerella, Pennycress, Coriander and Cuphea). *OCL* 16:205–210.

Johnston, A. M., D. Tanaka, P. R. Miller, *et al*. 2002. Oilseed crops for semiarid cropping systems in the northern Great Plains. *Agron. J.* 94:231–240.

Knoll, J. E., W. F. Anderson, T. C. Strickland, *et al*. 2012. Low-input production of biomass from perennial grasses in the coastal plain of Georgia, USA. *Bioenergy Res.* 5:206–214.

Kurki, A., A. Hill, and M. Morris. 2010. Biodiesel: the sustainability dimensions. *ATTRA* 34.

Li, L., E. Coppola, J. Rine, *et al*. 2010. Catalytic hydrothermal conversion of triglycerides to non-ester biofuels. *Energy Fuels* 24:1305–1315.

Loreau, M. 1998. Biodiversity and ecosystem functioning: a mechanistic model. *Proc. Natl. Acad. Sci.* 95:5632–5636.

Matthews, C. 2007. Bioenergy could drive rural development: experts weigh bio-power impact. In *FAO Newsroom,* Rome, Italy: Food and Agriculture Organization of the United Nations.

McKendry, P. 2002. Energy production from biomass (part 1): overview of biomass. *Bioresource Technol.* 83:37–46.

Milbrandt, A. 2005. A geographic perspective on the current biomass resource availability in the United States. Golden, CO: National Renewable Energy Laboratory.

Mitchell, R., L. Wallace, W. Wilhelm, *et al*. 2010. Grasslands, rangelands, and agricultural systems. In *The Ecological Society of America's History and Records,* Washington, DC: ESA Historical Records Committee, *http://www.esa.org/esa/sustainable-biofuels-from-forests-grasslands-and-rangelands/* (accessed August 29, 2016).

Nagurney, A., and L. S. Nagurney. 2011. Spatial price equilibrium and food webs: the economics of predator-prey networks. In *2011 IEEE International Conference on Supernetworks and System Management*, ed. F.-Y. Xu, and J. Dong, Beijing, China: IEEE Press, 1–6.

National Agricultural Statistics Service. 2010. Quick stats U.S. and all states data—crops. Washington, DC: U.S. Department of Agriculture.

O'Gorman, E., R. Enright, and M. Emmerson. 2008. Predator diversity enhances secondary production and decreases the likelihood of trophic cascades. *Oecologia* 158:557–567.

Paine, R. T. 1980. Food webs: linkage, interaction strength and community infrastructure. *J. Anim. Ecol.* 49:667–685.

Pearlson, M., C. Wollersheim, and J. Hileman. 2013. A techno-economic review of hydroprocessed renewable esters and fatty acids for jet fuel production. *Biofuels, Bioprod. Biorefin.* 7:89–96.

Pender, J., A. Marré, and R. Reeder. 2012. Rural wealth creation: concepts, strategies and measures. In *ERS Report Summary*, Vol. ERR–131, Washington, DC: U.S. Department of Agriculture, Economic Research Service.

Perlack, R. D., and B. J. Stokes. 2011. U.S. billion-ton update: biomass supply for a bioenergy and bioproducts industry. ORNL/TM–2011/224.

Perlack, R. D., L. L. Wright, A. F. Turhollow, *et al*. 2005. Biomass as feedstock for a bioenergy and bioproducts industry: the technical feasibility of a billion-ton annual supply. DOE/GO–102995–2135 and ORNL/TM–2005/66, 78.

Schmer, M. R., K. P. Vogel, R. B. Mitchell, *et al*. 2008. Net energy of cellulosic ethanol from switchgrass. *Proc. Natl. Acad. Sci.* 105:464–469.

Severino, L., D. Auld, M. Baldanzi, *et al*. 2012. A review on the challenges for increased production of Castor. *Agron. J.* 104:853–880.

Smith, V. H., R. C. McBride, J. B. Shurin, *et al*. 2015. Crop diversification can contribute to disease risk control in sustainable biofuels production. *Front. Ecol. Environ.* 13:561–567.

Staples, M. D., R. Malina, H. Olcay, *et al*. 2014. Lifecycle greenhouse gas footprint and minimum selling price of renewable diesel and jet fuel from fermentation and advanced fermentation production technologies. *Energy Environ. Sci.* 7:1545–1554.

Starks, P. J., B. C. Venuto, J. A. Eckroat, *et al*. 2011. Measuring Eastern Redcedar (Juniperus virginiana L.) mass with the use of satellite imagery. *Rangeland Ecol. Manage.* 64:178–186.

Steiner, J. 2010. Unpublished notes from research coordination meeting. U.S. Department of Energy, Energy Efficiency and Renewable Energy/ U.S. Department of Agriculture, Agricultural Research Service Headquarters Staff meeting.

Steiner, J., and T. Griffith. 2010. World food availability and the natural land resources base. In *Perspectives on political and social regional stability impacted by global crises—a social science context,* Washington, DC: The Strategic Multi-Layer Assessment (SMA) and U.S. Army Corps of Engineers Research and Development Directorate, 170–177.

Steiner, J., K. Lewis, H. Baumes, *et al*. 2012. A feedstock readiness level tool to complement the aviation industry fuel readiness level tool. *Bioenergy Res.* 1–12.

Steiner, J. J., M. O'Neill, and W. Goldern. 2011. The national biofuels strategy – Importance of sustainable feedstock production systems in region-based supply chains. In *Sustainable Feedstocks for Advanced Biofuels—Sustainable Alternative Fuel Feedstock Opportunities, Challenges, and Roadmaps for Six U.S. Regions,* ed. R. Braun, D. Karlen, and D. Johnson, Ankeny, IA: Soil and Water Conservation Society, 361–375.

Stratton, R. W., H. M. Wong, *et al*. 2010. Life cycle greenhouse gas emissions from alternative jet fuel, Version 1.2. Cambridge, MA: PARTNER/MIT.

Tilman, D., J. Hill, and C. Lehman. 2006. Carbon-negative biofuels from low-input high-diversity grassland biomass. *Science* 314:1598–1600.

Timilsina, G. R., and A. Shrestha. 2010. Biofuels: markets, targets and impacts. Washington, DC: The World Bank Development Research Group, Environment and Energy Team. *http://papers.ssrn.com/sol3/papers.cfm?abstract_id=1645735* (accessed August 29, 2016).

Trostle, R., D. Marti, S. Rosen, *et al*. 2011. Why have food prices risen again? A report of the Economic Research Service, Washington, DC: U.S. Department of Agriculture, Economic Research Service, *http://www.ers.usda.gov/media/126752/wrs1103.pdf* (accessed August 29, 2016).

U.S. Department of Agriculture. 2010. A USDA regional roadmap to meet the biofuels goals of the Renewable Fuel Standard by 2022. *http://www.usda.gov/documents/USDA_Biofuels_Report_6232010.pdf* (accessed August 29, 2016).

U.S. Department of Agriculture. 2012. *Unpublished data.*

U.S. Department of Agriculture, Agricultural Research Service. 2012. *Unpublished data.*

U.S. Department of Agriculture, Economic Research Service. 2011. Measuring the indirect land-use change associated with increased biofuel feedstock production—a review of modeling efforts. In *Report to Congress,* Washington, DC: U.S. Department of Agriculture, Economic Research Service.

U.S. Department of Agriculture Rural Development. 2012. Rural development energy programs. *http://www.rurdev.usda.gov/energy.html* (accessed August 25, 2016).

U.S. Department of Energy. 2007. EERE Project Management Center: business opportunities. *https://www.eere-pmc.energy.gov/Business.aspx* (accessed August 29, 2016).

U.S. Department of Energy Office and U.S. Department of Agriculture. 2009. Sustainability of biofuels: future research opportunities; report from the October 2008 workshop, DOE/SC–0114.

Wangersky, P. J. 1978. Lotka-Volterra population models. *Annu. Rev. Ecol. Evol. Syst.* 9:189–218.

Whittaker, G. W., J. R. Confesor, S. Griffith, *et al*. 2007. A hybrid genetic algorithm for multiobjective problems with activity analysis-based local search. *Eur. J. Oper. Res.* 193:195–203.

Wigmosta, M. S., A. M. Coleman, R. J. Skaggs, *et al*. 2011. National microalgae biofuel production potential and resource demand. *Water Resour. Res.* 47:W00H04.

# CHAPTER 11

## Microalgae feedstocks for aviation fuels

Mark S. Wigmosta, Andre M. Coleman, Erik R. Venteris and Richard L. Skaggs

## 11.1 INTRODUCTION

In both the commercial and military sectors, there is significant global interest in developing, testing, and using alternative jet fuels to create a sustainable and stable fuel supply while reducing greenhouse gas emissions. Currently, the aviation industry is entirely dependent on a finite supply of petroleum-based fuel sourced in part from politically and economically unstable regions of the world. In 2009, 17.8 billion gal (67.4 billion L) of commercial jet fuel was used in the contiguous United States, and in 2010, 1.5 (5.7), 0.6 (2.3), and 0.8 (3.0) billion gal (L) of jet fuel were used by the U.S. Air Force, Navy, and Army respectively (Carter *et al.*, 2011). U.S. commercial and military aviation sectors have set ambitious near-term alternative fuel and environmental performance targets. This includes a tentative Federal Aviation Administration goal of 1 billion gal/yr (BGY; 3.8 billion L/yr, BLY) of alternative fuel use by commercial aircraft by 2018. The U.S. Air Force has set a usage target of 50% alternative fuels for U.S. Air Force domestic aviation by 2016 (0.73 BGY; 2.8 BLY), and the Navy has set a total energy consumption (0.3 BGY; 1.1 BLY) of 50% alternative fuels by 2020 (Carter *et al.*, 2011). If these targets become policy, at least 2 BGY (7.6 BLY) of domestically produced alternative jet fuel will be required by 2020.

The Energy Independence and Security Act (EISA) of 2007 established production requirements for domestic alternative fuels under the Renewable Fuel Standard (RFS). For example, 36 BGY (136 BLY) of renewable fuel must be produced by 2022, of which 21 BGY (79 BLY) must be advanced biofuels. EISA defines advanced biofuels as non-cornstarch-derived biofuels with lifecycle greenhouse gas emissions that are 50% lower than those of gasoline. There are a number of potential fuel pathways for meeting the RFS. One of these is biomass-based diesel, including jet fuel (Schnepf and Yacobucci, 2013). The U.S. Department of Energy (DOE) Bioenergy Technologies Office has a stated goal in its 2013 Multi-Year Program Plan (U.S. DOE, 2013) to support the RFS through the development of "commercially viable biomass utilization technologies to encourage the creation of a new domestic bioenergy industry...." The Bioenergy Technologies Office also recognized the potential for aviation biofuels to support the bioenergy industry, seeing drop-in bio-based jet fuels as one of the viable alternatives for the aviation industry and the military to meet their ambitious near-term greenhouse gas reduction targets (U.S. DOE, 2014). One of the important Multi-Year Program Plan Targets (U.S. DOE, 2013) is to establish feedstock resource assessment models to evaluate the geographic, economic, quality, and environmental criteria for which 20 million metric tonnes of ash-free dry weight algal biomass can be produced by 2022.

Algal biofuels may offer a number of advantages toward meeting the EISA requirements. They can produce a range of biofuel feedstocks suitable for diesel and aviation fuels. Microalgae, on a strain-specific basis, can be cultivated using impaired water, including saline and/or brackish pumped groundwater or seawater, treated industrial wastewater, municipal sewage effluent, and water produced during oil- and gas-drilling operations. Microalgae require nitrogen and phosphates as essential nutrients and could provide water treatment cobenefits for municipalities, industry, and the environment.

Table 11.1.   Water footprint, land use, energy, and biofuel yield of various energy crops.

| Feedstock | Water footprint, m³/GJ | Land use, m²/GJ | Energy, GJ/ha | Biofuel yield, L/ha |
|---|---|---|---|---|
| Soybean | 383 | 689 | 15 | 446 |
| *Jatropha* | 396 | 162 | 62 | 1,896 |
| Rapeseed (canola) | 383 | 258 | 39 | 1,190 |
| Sunflower | 61 | 323 | 31 | 951 |
| Oil palm | 75 | 52 | 192 | 5,906 |
| Microalgae | <379 | 2 to 13 | 793 to 4,457 | 24,355 to 136,886 |

*Source:* Adapted from Singh, A., P. S. Nigam, and J. D. Murphy. 2011. Renewable fuels from algae: an answer to debatable land based fuels. *Bioresour. Technol.* 102:10–16.

A potentially significant advantage of microalgae as an alternative feedstock source for fuels is that, relative to land-based feedstocks, microalgae are a high-density, rapidly growing feedstock resource. For example, the biomass yields and energy content of algae in gigajoules per hectare are on the order of 12 to 144 times greater than those of *Jatropha*, rapeseed (canola), or sunflower (Table 11.1). This productivity is driven by the low-order, high efficiency of algae. These single-cell microorganisms are controlled primarily by water temperature and available light from the photosynthetically active radiation (PAR) portion of the solar spectrum, which is about 45% of the incident solar radiation. In the context of these two variables, each strain of algae has its own preferential growing characteristics: minimum, optimum, and maximum light intensity thresholds and water temperatures, and optimal growth range as a function of water temperature. When growing conditions are outside a given algal strains' optimum light intensity or water temperature, there is a decline in growth efficiency. Stable and productive microalgae growth favors locations with ample sunlight and warm year-round temperatures. Colder or cloudy climates reduce biochemical processes and associated growth rates, while warmer climates may result in excessive water temperatures that slow growth or eventually compromise the viability of the algae culture.

The capture of essential solar radiation and the operating temperature range can be controlled to some extent by the type of cultivation system used. Currently, most feedstock production occurs in open raceway ponds. These relatively shallow ponds (15- to 30-cm deep) have a large surface-to-depth ratio (Fig. 11.1) to maximize the capture of sunlight and to minimize "dark zones" that may result in loss of biomass through dark respiration. Cooling is generally achieved through evaporation. The thermal mass of water and the surrounding soil provide some buffering against rapid changes in air temperature, although an obvious, but a damped, diurnal signal in water temperature is still observed in most ponds.

In theory, closed photobioreactor (PBR) systems (Fig. 11.1) provide a more controlled environment than open ponds, and culture-crashing pathogens and predators are less likely to be introduced into the growth medium. However, PBR systems have historically been known to overheat because of the lack of natural evaporative cooling, though these systems can employ various engineering solutions to control water temperature (e.g., spray cooling, water bath, cool water injections, and heat exchangers, etc.). The potential for overheating, along with additional operational difficulties and a higher capital investment, generally make PBR systems significantly more expensive to initiate and operate than open ponds (Amer et al., 2011; Davis et al., 2011; Lundquist et al., 2010; Sun et al., 2011; Pate, 2013). For these reasons, this chapter focuses on open pond feedstock production.

Algae growth characteristics are described in Section 11.2, including fundamental biophysical processes and the influence of spatial and temporal variations in climate on algae growth. Large-scale feedstock production potential along with land, water, and nutrient resource demand is considered in Section 11.3, and Section 11.4 provides a detailed examination of the 2 BGY (7.6 BLY) renewable aviation fuel target.

Figure 11.1.   Example algae cultivation systems. (a) Open raceway ponds. (b) Tubular glass photobioreactor (PBR) system. (c) Plastic flat-plate PBR. ((b) and (c) From Wikipedia. 2017. Photobioreactors. *https://commons.wikimedia.org/wiki/Category:Photobioreactors*. IGV Biotech *https://creativecommons.org/licenses/ by-sa/3.0/deed.en* (accessed March 29, 2017).

## 11.2   ALGAE GROWTH CHARACTERISTICS

### 11.2.1   *Biophysics*

Microalgae grow by converting solar energy to chemical storage in the form of biomass via photosynthesis. With adequate nutrients, the growth rate of microalgae is primarily influenced by the intensity of the incident solar radiation and the corresponding water temperature of the growth media. In particular, solar radiation in the form of photosynthetically active radiation, which operates at the 0.4- to 0.7-$\mu$m portion of the electromagnetic spectrum, provides available light for photosynthesis, whereas shortwave radiation, 0.285 to 2.8 μm, has a dominant influence on heating water within open ponds. For any photosynthesizing plant, available light intensities below the optimum range will cause a resource-limited environment in the photosynthetic process and, thus, the conversion of photons to carbon (biomass) will not function at the full potential. Similarly, if the light intensity exceeds the optimum range, or extends past the maximum light intensity threshold, proteins within the photosynthetic units will begin to turn off in an attempt to prevent cell damage and will avoid photon absorption and/or release absorbed photons as heat, resulting in a decline in biomass productivity (Béchet *et al.*, 2013; Rubio *et al.*, 2003; Weyer *et al.*, 2010). PAR is limited by normal diurnal and seasonal fluctuations as a function of the sun's changing zenith angle throughout the year. Consequently, algae cultivation sites at lower latitudes experience less change in solar insolation and generally have a more consistent daily availability of PAR because of the limited change in solar insolation. Cloud cover and storms have a significant impact on available PAR; however, photosynthesis still occurs at a reduced rate using available diffuse radiation (Churkina and Running, 1998). Although some areas within the United States, such as the Southwest, receive high percentages of available and uninhibited PAR, the lack of cloud cover and low relative humidity can also present issues with thermal energy loss from open ponds at night due to the resulting low nighttime temperatures. Thus, from a climate-resource perspective, areas where strain-specific optimal temperature ranges exist and have limited variability within the diurnal and seasonal air temperature regimes, tend to be more suitable locations for growth.

The water temperature within shallow microalgae cultivation ponds is bounded by the principle of the conservation of energy within a fluid volume and is thus influenced by pond water depth, water density (which varies by level of salinity), the specific heat of water, and net surface heat-flux, including net solar shortwave radiation, downward atmospheric longwave radiation, longwave back radiation, and heat flux due to evaporation and conduction, that are driven by meteorological variables including air temperature, wind, and relative humidity. Open pond systems are subject to dominant control from environmental conditions barring engineered solutions, such as industrial waste heat via heat exchangers during cool temperature months or the introduction of cool makeup water into the ponds during warm months to help keep pond temperatures at the optimum for the algae strain. As noted previously, the water temperature in an open pond will be impacted by large diurnal swings in air temperature but also by the degree to which evaporative cooling takes place. For example, in humid regions, evaporative cooling is much less effective due to the reduced ability of the air to absorb additional water vapor and thus creates a reduced rate of heat exchange through the liquid-to-gas phase change. Because of the thermal properties of water, the water temperature will respond to air temperatures with varying degrees of latency and dampening.

Optimal water temperatures vary between types and strains of microalgae (Chisti, 2007; Pate, 2013; Sheehan *et al.*, 1998). Many microalgae can tolerate temperatures down to 15°C below their optimal, but exceeding the optimal temperature range by 2 to 4°C can cause total culture loss (Mata *et al.*, 2010). Photosynthetic reactions become limiting outside the optimal temperature range, and exceeding the minimum or maximum temperature will more than likely lead to reduced cell viability. Understanding the basic growth characteristics of specific strains of microalgae is fundamental to determining what and where to grow to maximize biomass production potential. Figure 11.2 shows the minimum, optimal minimum, optimal maximum, and maximum temperature ranges for four strains of microalgae: the freshwater strains *Chlorella* and Sphaeropleales, the seawater strain *Nannochloropsis salina* (*N. salina*), and the brackish water strain *Arthrospira*.

The growth model of Wigmosta *et al.* (2011) is used to describe the key components in the conversion of solar energy to algal biomass, with the rate of biomass production ($P_{mass}$ in mass per unit area per unit time) given by Equation (11.1):

$$P_{mass} = \left(\tau_p C_{PAR} E_s\right)\left[\frac{E_c \varepsilon_b}{E_a Q_r E_p}\right]\left(\varepsilon_s \varepsilon_t\right) \qquad (11.1)$$

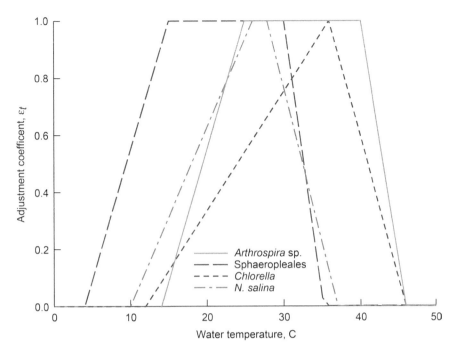

Figure 11.2. Minimum, optimal minimum, optimal maximum, and maximum water temperatures for four strains of microalgae: freshwater strains *Chlorella* and Sphaeropleales, seawater strain *N. salina*, and brackish water strain *Arthrospira*. The adjustment coefficient of 1.0 indicates the optimal temperature range.

The first term on the right-hand side of Equation (11.1) represents the amount of PAR available, where $E_s$ is the full spectrum solar energy at the land surface in megajoules per meter squared (MJ/m$^2$), $C_{PAR}$ is the fraction of PAR, and $\tau_p$ is the transmission efficiency of incident solar radiation to the pond microalgae. The middle term on the right-hand side is a strain-specific term representing the conversion of PAR to biomass under optimal light and water temperature, where $E_a$ is the energy content per unit biomass in megajoules per kilogram (MJ/kg), the photon energy $E_p$ in megajoules per mole (MJ/mol) converts PAR as energy to the number of photons, and $\varepsilon_b$ is a function of species, water temperature, and other growing conditions accounting for the energy required for cell functions, such as respiration, that do not produce biomass. The quantum requirement $Q_r$ is the number of photons required to liberate one mole of $O_2$ and, together with the carbohydrate energy content $E_c$, represents the conversion of light energy to chemical energy through photosynthesis (Weyer et al., 2010). The biomass accumulation efficiency $\varepsilon_b$ is a poorly understood function of species, water temperature, and other growing conditions accounting for the energy required for cell functions that do not produce biomass (e.g., respiration). The final term in Equation (11.1) represents a reduction in photon absorption from suboptimal light $\varepsilon_s$ and/or water temperature $\varepsilon_t$. The light utilization efficiency curve for two groups of algae strains with known light saturation constants of 150 and 250 µmol/m$^2$·s were determined using the Bush equation and are represented in Figure 11.3 (Chisti, 2007; Huesemann et al., 2009). The strain-specific correction for water temperature varies between 0 and 1, as depicted in Figure 11.3.

## 11.2.2 *Climate variability*

The dominant controls over microalgae production are influenced by a scale of processes that range from larger global and regional drivers down to site-specific conditions, all of which vary over space and time. Short-range meteorology shapes longer-term statistical trends of climate and is observed to have distinct characteristics that are revealed through spatial patterns over both horizontal and vertical

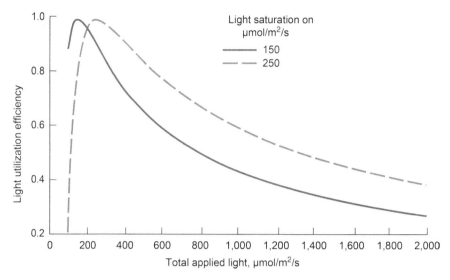

Figure 11.3.   Light utilization efficiency curve for two groups of algae with light saturation constants of 150 and 250 μmol/m²·s.

space. At a broader scale, macroclimates are driven by major processes around solar geometry/time-of-year, latitude, and long-term ocean and atmospheric circulation patterns. These broader scale processes interface with the mesoclimate scale where finer scale weather and climate patterns are influenced by land surface characteristics such as elevation, aspect, slope, soils, land cover, and water bodies. Beyond the mesoclimate scale, fine-resolution microclimatic conditions that represent an individual site are critically important to realize the potential production and variability around that potential. There are, of course, additional factors that influence growth rates, including bioavailable nutrients, type and purity of $CO_2$, pond mixing, culture density, pond ecology, as well as other managed operations and engineering of open pond or photobioreactor systems.

As with spatial variability in climate, there is also temporal variability that is witnessed at multiple temporal scales from diurnal, monthly, seasonal, inter-annual, through to longer-term climate signals that can statistically denote nonstationary trends (i.e., earlier spring warmup) along with changes in the timing, frequency, and magnitude of meteorological elements, for example, more frequent occurrence of high temperatures or shorter duration, higher intensity rainfall. Adding to the complexity of temporal variability, individual meteorological variables can trend differently and will be influenced by their particular location and scaled meteorological conditions (i.e., the macroclimate to the microclimate), for example, humid environments exhibit different behaviors of variability than do arid environments, and local meteorological conditions (solar radiation, air temperature, wind, humidity, and cloud cover) within a microclimate will affect the annual, seasonal, and diurnal water temperature regimes in open ponds.

### 11.2.3   *Productivity*

As noted previously, individual strains of algae have differing absolute bounds and optimal environmental conditions that dominantly influence biomass growth. Using the Wigmosta *et al.* (2011) combined pond water temperature and microalgae growth model (Eq. (11.1)) driven by site-scale hourly meteorology for 30 years, the spatial and temporal patterns of four different strains of algae are represented for the conterminous United States in both their long-term annual and seasonal rates of biomass production. Biomass growth was modeled using ash-free dry weight in units of g/m²·day, which could then be scaled based on the available pond area and then further evaluated to total volumes of biomass, lipid content, and/or fuel per unit time, depending on the downstream processing pathways, for example, lipid extraction, fast pyrolysis, or hydrothermal liquefaction.

The point locations shown on Figures 11.4 and 11.5 represent large, contiguous potentially usable land locations for large-scale algal production facilities. Each location has a minimum unit area of 485 ha (1200 ac), which assumes 100 4-ha (~10-ac) ponds with a total 405 ha (1,000 ac) of pond area and 80 ha (200 ac) for supporting infrastructure such as access roads, site pipelines, water treatment systems, dewatering systems, nutrient recycling, and office space (Benemann *et al.*, 1982; Maxwell *et al.*, 1985; Sheehan *et al.*, 1998). These land areas were determined through spatial suitability modeling using several criteria: lands with ≤1% slope that were not protected, environmentally sensitive, or developed and that did not have existing agriculture on them (Wigmosta *et al.*, 2011). The pond water temperature model for the freshwater strain *Chlorella* uses a 30-cm deep pond for cultivation, whereas the brackish water strain *Arthrospira* has shown better growth performance using a 15-cm-deep pond because it is more heat tolerant. The two types of algae presented here represent only two of thousands of possible strains.

From Figure 11.4, it is evident that *Chlorella* has a dominant production period in the summer months, primarily in the Salton Sea region of southern California and southern Arizona through the Texas plains, Gulf Coast areas, Florida panhandle, and mid-southern U.S. Atlantic coastal area. The spring and fall bring moderate production for these regions with the remainder of the country at low production levels. For the winter months, the very southern tip of Texas and the middle to southern Florida panhandle are the only areas with appreciable production. The overall annual production of *Chlorella* is, at best, a moderate performer in southern Texas and the Florida panhandle. The spatial patterns in Figure 11.5 clearly show that *Arthrospira* is a high-performing strain, with its dominant area of productive growth in the Texas plains, Gulf Coast region, Florida panhandle, mid-southern U.S. Atlantic coastal area, and a pocket in the Salton Sea region of southern California. Its performance dominates in the summer months and tapers to very low production throughout most of the U.S. in the winter months and a mid-level of production in the spring months for the same referenced southerly locations.

As noted earlier, the unique growth characteristics of each strain cause microalgae to thrive under different environmental conditions. Figure 11.6 provides a spatial and temporal perspective for four strains of microalgae including *Chlorella*, *Arthrospira*, the order Sphaeropleales, and *N. salina* at four favorable growing locations around the United States. At all four sites, the long-term mean monthly growth of *Chlorella* and *Arthrospira* follow a predictable pattern that follows seasonal warming and longer days represented by high production rates in the summer months. The magnitude and duration of high-growth varies between sites, with some exception to the Davis, California, site, where the two strains follow a similar growth pattern. Sphaeropleales has superior growth at Davis, California, with a sustained production rate over the late April to August time period, whereas the other sites appear to exceed the optimal temperature thresholds in the summer months and thus show a reduced level of production. *N. salina* shows a similar pattern of reduced growth in the summer months, with the major production decreases being seen at Houston, Texas, and Miami, Florida.

Strain production is presented in Figures 11.4, 11.5 and 11.6, which illustrate the influence of site-specific climate on algal growth, assuming that water and nutrients are unconstrained. Section 11.3 describes the impacts of land, water, and nutrient demand and constraints.

## 11.3   LARGE–SCALE PRODUCTION POTENTIAL AND RESOURCE CONSTRAINTS

A key consideration for agriculture, including microalgal aquaculture, is the limitations presented by natural resource supplies including land, water, and nutrients. Algae cultivation facilities require large areas of land. In our studies, each unit farm of our studies is 485 ha, and meeting Energy Independence and Security Act targets would require thousands of such sites. These facilities have 405 ha of pond surface area and, therefore, considerable potential

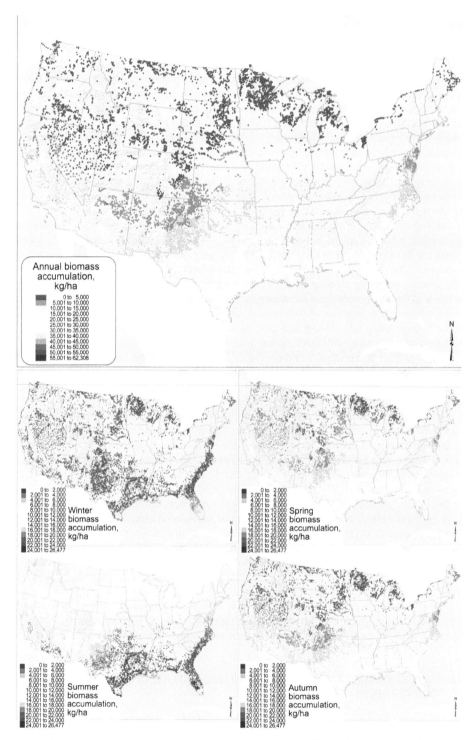

Figure 11.4.   Site-specific long-term mean annual (top) and seasonal (lower four panels from left-to-right: winter, spring, summer, fall) total biomass accumulation for *Chlorella*. Note: the scale range shown for cumulative annual production is 0 to 62,308 kg/ha·year and that the scale range for the seasonal production is 0 to 26,477 kg/ha·season. The annual production is the sum of the production during the four seasons.

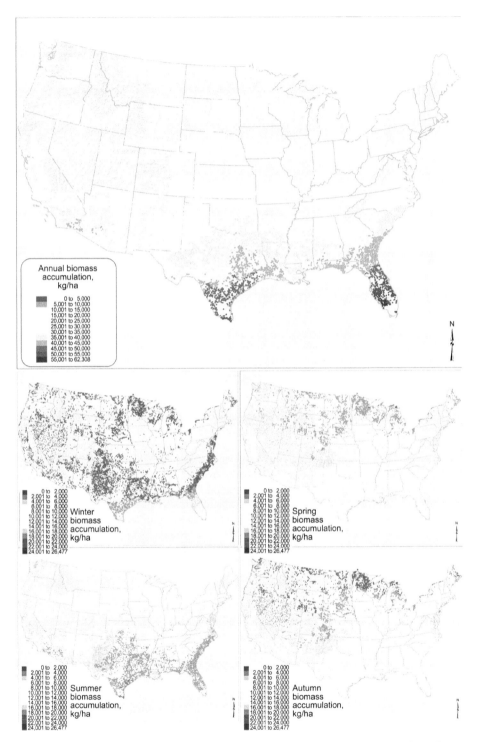

Figure 11.5.   Site-specific long-term mean annual and seasonal total biomass accumulation for *Arthrospira*. Note that the range shown for annual production is 0 to 62,308 kg/ha·year and that the range for seasonal production is 0 to 26,477 kg/ha·season. The annual production is the sum of the production for the four seasons.

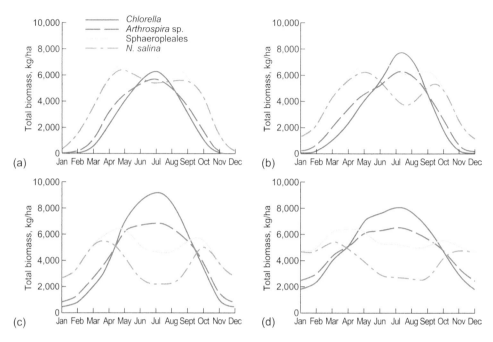

Figure 11.6.   Long-term mean monthly cumulative biomass production for four locations around the United States representing four different strains of algae—*Chlorella, Arthrospira* sp., the order Sphaeropleales, and *N. salina.* (a) Davis, California. (b) Tucson, Arizona. (c) Houston, Texas. (d) Miami, Florida.

for evaporation. The amount of makeup water needed to replace evaporation and the discharge salts needed to maintain salinity can be large, depending on the climate. Rapid and efficient algae cultivation also requires supplemental nitrogen and phosphorus fertilizers, similar to those for traditional field crops. In addition, microalgae also require supplemental $CO_2$, which can be provided from industrial sources or can be drawn from combustion waste streams—for example, from available fossil-fuel-based electricity generation. While theoretical resource assessments can be posed as questions of mass balance, as a practical matter they are inextricably linked to economics and government regulations generally intended to attain broader environmental and/ or socioeconomic goals. Assessment of natural resource limitations in the face of environmental, societal, and economic impacts is crucial to assessing the sustainability of any feedstock including algae-based biofuels. It is within this framework that we discuss the supplies of natural resources relative to the demands of algae cultivation and biofuel manufacture at scales sufficient to be meaningful to the budgets for aviation and U.S. energy.

Resource modeling and sustainability assessment is highly dependent on the selection of biomass-to-biofuel conversion process. Here we introduce two technologies of interest. The traditional fuel production approach is to exploit the chemical similarity between algal lipids and diesel fuels. In lipid extraction (LE) (Davis *et al.*, 2012) the algae are harvested and lipids extracted through one of many technology pathways (e.g., hexane solvents, cell lysis, etc.). The extracted lipids are upgraded to one of many possible fuel precursors, and the remaining biomass can either be recycled for nutrients and power generation (i.e., anaerobic digestion) or sold as byproducts such as fertilizers or animal feed. The second, more recent technology is hydrothermal liquefaction (HTL) (Elliott *et al.*, 2013; Frank *et al.*, 2013; Jena and Das, 2011), where the moist algal biomass is fed into a reactor vessel where it is subjected to heat and pressure to produce a product similar to fossil petroleum. This fuel precursor is then hydrotreated to remove excess oxygen and nitrogen and is refined (distilled) into fuel products including aviation kerosene. These early studies of HTL have shown much promise for the production of large quantities of biofuels, because the amount of biofuel produced per unit biomass is large, and dewatering requirements (e.g., costs, energy) are generally less than they are for LE.

While models have been developed to inventory and evaluate the mass balances, costs, and sustainability issues associated with individual natural resource components, sustainability ultimately can only be evaluated with full consideration of the interaction between biomass productivity, efficiency of biomass to biofuel and nutrient recycling processes, and key resource components. In general, we evaluate sustainability issues by specifying a target amount of biofuel to be produced. Typical targets (discussed in Sec. 11.1) include those set by the EISA, such as 36 billion gal/yr (136 billion L/yr) of renewable fuel or 21 BGY (79.5 BLY) of advanced biofuels (non-corn-based ethanol) by the year 2022. A common production goal in recent DOE-funded research is 5 BGY (18.9 BLY), or roughly one quad of energy (Davis *et al.*, 2012, 2014; Venteris *et al.*, 2014b). The aviation sector has a renewable fuel consumption goal of 2 BGY (7.6 BLY) to be attained by 2020.

The size of the biofuel production target is the key to framing discussions of industry viability because sustainability challenges generally increase with the size of the target. While there is value in estimating maximum production capacities for the contiguous United States (Pate *et al.*, 2011; Quinn *et al.*, 2012; Venteris *et al.*, 2013; Wigmosta *et al.*, 2011), in general, such models arrive at biofuel amounts well in excess of EISA goals and involve thousands of cultivation installations consuming hundreds of thousands of square kilometers of land and volumes of water well in excess of current irrigation demands for food production (Pate, 2013; Wigmosta *et al.*, 2011). Achievement of such change is unlikely within the 10- to 50-year planning cycles for energy systems, and realistic modeling of the broad economic and sustainability impacts at this scale is nearly impossible. Getting a realistic picture of the overall viability for relatively modest 5-BGY (18.9-BLY) systems has proved sufficiently challenging (Davis *et al.*, 2012, 2014; Venteris *et al.*, 2014b). With a set production goal, the relationships between biofuel production and resource consumption become clear. Overall, increases in strain growth rate, biomass-to-biofuel efficiency, and nutrient utilization and recycling efficiency all lead to reduced resource demands and a higher probability of sustainable biofuel production.

A recent report by the National Research Council (2012) presents a useful inventory of the key natural resources required for algal cultivation, their sustainability issues, and a review of results from the literature (with a heavy emphasis on studies prior to the renewed DOE effort funded by American Reinvestment and Recovery Act). In this report, land, water, and nutrient supplies were termed "Concerns of High Importance," and therefore their assessment has been given priority. At the time of the writing of that report, these issues had not yet been evaluated in a combined, comprehensive manner. Here, we summarize the latest findings surrounding these resources components, with emphasis on recent integrated studies of resource demand that were not available at the time of the National Research Council review.

## 11.3.1   *Land*

One of the oft-stated advantages of algae cultivation is that fertile soils are not required, providing a potential opportunity for employing underutilized lands such as deserts, steppe, and marginal crop lands. Land-screening studies have generally shown that large amounts of appropriate land may be available. A land-screening model based on the contiguous United States (Coleman *et al.*, 2014) identified $4.0 \times 10^5$ km$^2$ of non-crop, non-protected land with a slope of less than 1% to minimize the costs of leveling required for pond construction. Similarly, an economic analysis of the land value of low-slope lands (<1%) and produced agricultural products (Venteris *et al.*, 2012) identified over $5.0 \times 10^4$ km$^2$ of cropland with poor production value (unlikely to be profitable without government subsidy) and over $1.0 \times 10^6$ km$^2$ of land expected to have low owner resistance to sale. Especially promising, at least at its surface, is the identification of $2.5 \times 10^5$ km$^2$ of desert, scrub, and barren land, utilization of which could minimize the food security and environmental impacts of land use change.

The studies discussed in the preceding paragraph established that land supply alone is unlikely to be a significant limitation. The second set of more advanced questions are more difficult to answer: How much land is required, where is that land best located, and what are the potential impacts of converting the land? Pate (2013) showed land requirements of $1.0 \times 10^4$ to $2.0 \times 10^4$ km$^2$ and $3.9 \times 10^4$ to $8.1 \times 10^4$ km$^2$ to meet targets of 5 and 20 BGY (18.9 and 75.7 BLY), respectively, based on biomass-to-biofuel conversion through LE and on a modest range of biomass growth rates in open ponds (the first two production scenarios in Table 3 of that work). A more site- and strain-specific model based on *Chlorella* production in open ponds (Venteris *et al.*, 2014c) supplied with fresh or brackish water (also based on LE) required $9.1 \times 10^4$ km$^2$ to meet a 21-BGY (79.5-BLY) target. The agreement between these independent estimates of the land requirements for open ponds is remarkable. However, this same strain only required $3.3 \times 10^4$ km$^2$ to achieve 21 BGY (79.6 BLY) when paired with biomass-to-biofuel conversion by HTL. The result demonstrates that whatever technologies are ultimately adopted, maximizing the fuel output per unit area through increasing productivity and the selection of efficient biomass-to-biofuel conversion methods are both promising avenues to reducing resource impacts.

The results presented in the previous paragraph do not consider the impact of resource colocation on land requirements. When locating for appropriate water and flue gas resources, *Chlorella* and HTL were found to require $3.6 \times 10^4$ km$^2$ of land (Venteris *et al.*, 2014c) to meet the 21-BGY (79.5-BLY) target, a modest increase from the base scenario. However, statistical characterization of the uncertainty in these estimates is currently difficult, if not impossible. Currently we are limited to comparing scenarios with the modest goal of identifying the range of potential outcomes. A more conservative colocation scenario was presented in Venteris *et al.* (2014b) with a 0.23 lower biomass-to-biofuel HTL efficiency, with *Chlorella* production penalized for brackish salinity (by 0.13 for freshwater and by 0.25 for brackish water (Venteris and Wigmosta, 2013), and with a broader range of colocated resources. They found that an area of $6.7 \times 10^4$ km$^2$ was required to produce 21 BGY (79.5 BLY). Such results (Fig. 11.7) demonstrate that seemingly subtle changes in biomass production and conversion efficiency can have dramatic impacts on resource consumption.

The final consideration is the location and type of land to be converted. As discussed in Section 11.2, the Wigmosta *et al.* (2011) growth model consistently demonstrates higher algal growth in the southeastern U.S. rather than the southwestern U.S. indicated in other open pond (Pate *et al.*, 2011) and PBR (Quinn *et al.*, 2012) models. This is mainly due to the combined effects of photo-inhibition and warmer nights on the growth model, which serve to extend the length of the productive growing season (Venteris *et al.*, 2014c; Wigmosta *et al.*, 2011). However, studies based on Pacific Northwest National Laboratory's (PNNL's) Biomass Assessment Tool (BAT) model (Wigmosta *et al.*, 2011) have also found that resource colocation also strongly favors the Southeast. For example, the transportation and utility infrastructure is more dense (Venteris *et al.*, 2014a), the water costs and availability are generally more favorable—freshwater availability is greater, evaporative water demand less, and access to marine water far less expensive (Venteris *et al.*, 2013), and there are a larger number of $CO_2$ flue gas sources (Pate, 2013; Venteris *et al.*, 2014b, 2014c) in the Southeast relative to the Southwest. However, the land use per land cover targeted in the region is roughly split between pasture and forest lands (Venteris *et al.*, 2014c), so their conversion could have a potentially significant sustainability impacts. For the optimistic HTL and *Chlorella* scenario, production based solely on barren lands (located primarily in the Southwest) would require $5.0 \times 10^4$ km$^2$ of land, with a penalty to economic efficiency of at least 50% (Venteris *et al.*, 2014c). Therefore, the question of converting forest and pasture lands in the Southeast remains a critical issue for the large-scale production of algae biofuels.

## 11.3.2   Water

Evaluating the sustainability of water supplies is a challenging task because water consumption rates can be very large, especially for arid regions such as the southwestern United States.

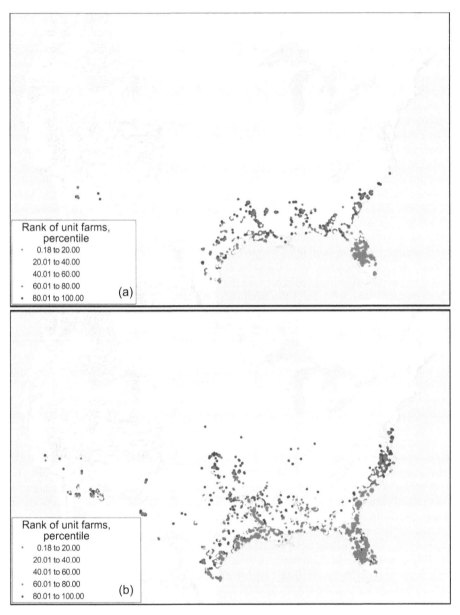

Figure 11.7.   Maps of the contiguous United States showing unit farm (485-ha) selections to produce 21 billion gallons per year of renewable diesel based on *Chlorella* and hydrothermal liquefaction (HTL). (a) Model for 7,478 unit farms using production not penalized by salinity, as well as a higher biomass-to-biofuel conversion efficiency. (b) Similar model for 13,833 unit farms with growth reduced as a function of operating salinity, a more conservative biomass-to-biofuel conversion efficiency (Frank *et al.*, 2013), and a broader set of resources (with minor effects). ((a) Adapted from Venteris, E. R., R. C. McBride, A. M. Coleman, *et al.*, 2014a. Siting algae cultivation facilities for biofuel production in the United States: trade-offs between growth rate, site constructability, water availability, and infrastructure. *Environ. Sci. Technol.* 48:3,559–3,566. With permission. (b) Adapted from Venteris, E. R., R. L. Skaggs, M. S. Wigmosta, *et al.*, 2014c. Regional algal biofuel production potential in the coterminous United States as affected by resource availability trade-offs. *Algal Research* 5:215–225. With permission.)

The process of open-pond evaporation results in both water loss and the buildup of salts. To reduce salt content to an acceptable range (dictated by the requirement of the algal strain), saline water in the pond is removed and replaced by water with a lower salinity. The water removed is termed blowdown. The amount of makeup water required is driven by both evaporation and the amount of blowdown. Geographic constraints in water demand are dramatic. For freshwater, open ponds on the Florida peninsula consume on average $7 \times 10^5$ m³/yr per 405-ha total pond area (100 10-ac ponds), whereas those in the Southwest average $4 \times 10^6$ m³/yr (Venteris et al., 2013). The amount of water needed each year for a single unit farm (in our studies, 405 ha of pond with a total of 485 ha of occupied land) would be sufficient to supply thousands of single-family homes. The contrast is even greater when accounting for saline water use and the attendant blowdown stream. The amount of water required to attain around 20 BGY (75.7 BLY) is also very large. The modest production rates of Pate (2013)—$9.36 \times 10^3$ and $19.7 \times 10^3$ L/ha·yr oil production—show evaporative freshwater demands of approximately $1.0 \times 10^5$ BLY. For *Chlorella* and LE, Venteris et al. (2014b) showed a mixed demand (fresh and brackish water, accounting for blowdown) of $1.34 \times 10^5$ BLY to produce 21 BGY (79.4 BLY), but this was reduced to $2.5 \times 10^4$ BLY with the application of HTL. The dramatic reduction in water use is due to the higher fuel output per water consumed, a greater portion of the production being based on sites with low evaporation, and fewer sites requiring saline groundwater with its larger blowdown stream. Accordingly, production of ~20 BGY of biofuel could increase the U.S. irrigation water consumption by 156% to 114% of current levels. With the well-documented stresses to land-based (surface and ground) water supplies due to unsustainable groundwater withdrawal (Konikow, 2013) and drought (Seager et al., 2009), any increase in freshwater consumption requires careful analysis.

Evaluation of the amount of water available for microalgae production is difficult. The use of seawater is an attractive option because it avoids the complex supply issues inherent to terrestrial-based waters. Another positive aspect is that disposing of saline concentrate by returning it to the ocean has an established analog in the desalinization literature (Voutchkov, 2011); inland disposal of saline concentration is less established and more uncertain (Mickley, 2001). The main challenges to seawater use are the cost of water-delivery infrastructure (Venteris et al., 2013; Vigon et al., 1982), coastal land-use challenges (e.g., right-of-way access, permitting, and environmental sensitivities), and the added risks of extreme weather events such as hurricanes. Analysis of water costs (Venteris et al., 2013) provided an initial estimate of 4 BGY (15.1 BLY) of production potential based on a generic algae strain and LE, with seawater costs limited to 20 percent of the value of the produced biofuel. We expect that seawater can support a significant amount of fuel production, and colocation studies involving broader resources and consideration of issues specific to sensitive coastal areas are in preparation.

Evaluating the supply of terrestrial waters is an even more complex task than for seawater, there are a range of issues to address, coupled with the considerable uncertainty that surrounds the availability of water resources, particularly those in the subsurface. Beginning with Venteris et al. (2013), freshwater utilization for BAT-based studies was limited to 5% of the mean annual flow of each hydrologic unit code 6 (HUC–6) (U.S. Geological Survey, 2009) watershed to minimize sustainability impacts, loosely based on the amount of non-consumptive withdrawal that the U.S. Environmental Protection Agency allows for electricity generation. Current cost models in BAT are based on freshwater extraction from shallow wells without distinguishing between surface and groundwater withdrawals (a non-trivial task, considering that it is common to locate wells near surface water sources). The 5% limitation is designed to ensure minimal impact to surface flows and avoidance of groundwater mining. However, in some locations, water supplies could be further limited, especially due to water rights in the southwestern United States and supply disruptions due to drought. Supply issues are further complicated by the geochemical diversity of groundwater. Water cost models described in Venteris and Wigmosta (2013), Venteris et al. (2014b, 2014c), and Davis et al. (2014) account for the salinity of the water supply, estimated from national water chemistry databases. As illustrated in Venteris et al. (2014a), local groundwater resources may be too saline to support operating salinities compatible with freshwater strains. In addition, these studies have not constrained the amount of saline groundwater. Estimates of saline aquifer volumes

(Androwski *et al.*, 2011; National Energy Technology Laboratory, 2012; Venteris *et al.*, 2009) illustrate that the simple dimensions of deep water-containing geologic layers are highly uncertain, let alone petrophysical properties such as porosity. Nevertheless, these studies suggest very large volumes. For example, Androwski *et al.* (2011) estimate the total volume at $3.1 \times 10^8$ billion L, enough to supply even a conservative 21-BGY (79.5-BLY) LE-based scenario for over 2,000 years without accounting for any recharge.

Costs for supplying terrestrial water are a function of demand, the salinity of the source, and the operating salinity. Supply costs go up with evaporative demand, but they go down as the difference between the source water and pond operating salinity is increased. Within BAT, blow-down is calculated explicitly and sensitivity analyses are conducted to select operating salinities that approximate the economic optimum (Venteris and Wigmosta, 2013). However, the criteria stated in Venteris *et al.* (2014a) of an operating salinity of 5 times that of the source water is a reasonable estimate of the difference required to maintain a small blowdown stream. Terrestrial water costs can be less than $100,000/yr·(unit farm) for shallow freshwater to over $500,000/yr·(unit farm) for highly saline and deep resources (Venteris *et al.*, 2013). Additional challenges and costs are incurred for inland disposal of saline concentrates; current analyses suggest modest additional costs similar in magnitude to those of supply. While not insignificant, water costs alone do not present a financial barrier to algae biofuel production.

At question is the amount of inexpensive fresh and alternative water resources available to support algae biofuel production. The amount of municipal waste water (MWW) available has been estimated to be a maximum of $1.2 \times 10^4$ BLY in the productive southern half of the United States (Chiu and Wu, 2013), with a rough estimate of ~$6.0 \times 10^3$ BLY with proximal appropriate land. At best, this resource is sufficient to produce roughly 5 GGY (18.9 BLY) based on the parameters used for the optimistic version (Venteris *et al.*, 2014c) of the *Chlorella* and HTL scenario. However, there is an accounting issue that needs to be considered if MWW is used. Because most MWW is returned to the surface water system (rivers and streams), its evapora-tion in algae cultivation ponds removes the contribution, making it questionable that MWW is really an alternative to freshwater. However, utilization of MWW as a nutrient source (Lundquist *et al.*, 2010) for algae cultivation likely has economic and sustainability benefits at energy scales (Venteris *et al.*, 2014b). Similarly, produced water from oil and gas may be utilized as cultivation water and perhaps could be provided to ponds free of charge as a form of disposal. However, this resource has at most $3.24 \times 10^3$ BLY available (Clark and Veil, 2009). These alternative waters may be important locally, but the most prevalent and cost-effective water resource is freshwater, so the supply of freshwater and the legal and regulatory constraints require further investigation.

### 11.3.3 *Nutrients*

Nutrients represent the final natural resource "Concern of High Importance" from the National Research Council (2012) report. Algae cultivation at high production rates requires supplemen-tal carbon (in the form of $CO_2$ or bicarbonate) and nitrogen and phosphorus fertilizers. Carbon dioxide may be limiting because of its potentially high delivery costs (roughly $40/t commer-cially (National Energy Technology Laboratory, 2010), and excessive consumption of nitrogen and phosphorus may compete with that needed for food production. Currently, nutrient demands are estimated from the stoichiometry of algal species, either estimated from the Redfield ratio (Pate, 2013; Pate *et al.*, 2011; Redfield, 1934) or from nutrient profiles taken from algae-specific analyses (Davis *et al.*, 2012, 2014; Venteris *et al.*, 2014b; Williams and Laurens, 2010). Such assessments are only rough estimates, and do not account for the complexities of biochemical availability, denitrification, and other factors. It is also important to note that nutrient consump-tion is a function of algae growth rate, so unlike land and water resources, increasing the growth rate alone does not increase nutrient use efficiency. Instead, nutrient consumption relative to fuel production depends on the efficiency of biomass-to-biofuel conversion and on the fate of nutrients in the remaining byproducts (there is an inherent tradeoff between recycling versus coproducts) (Venteris *et al.*, 2014b) and removed during fuel upgrading and refining.

Currently, algae resource studies emphasize the reuse of $CO_2$ from point combustion sources such as fossil-fuel-burning power plants, but purchase on the open market is also a viable option. Assuming zero recycling (unlike with nitrogen and phosphorus recycling, the benefits of returning $CO_2$ to the pond are not clear (Frank *et al.*, 2013). For example, in two studies, demand has been estimated for *Chlorella* at $6.97 \times 10^8$ t/yr producing 20 BGY (75.7 BLY) (Pate, 2013) and $9.42 \times 10^8$ t/yr producing 21 BGY (79.5 BLY) (Venteris *et al.*, 2014b). Both estimates are based on the less efficient LE pathway. This can be compared to the $3.05 \times 10^9$ t/yr that are emitted from a point combustion source, so that $CO_2$ supplies alone are not likely to be limiting. The consumption per unit biofuel can also be reduced by nearly half through HTL technology ($5.24 \times 10^8$ t/yr (Venteris *et al.*, 2014b). However, flue gas delivery pipelines are potentially expensive, and current analyses show them to be the most expensive resource infrastructure for many locations (Orfield *et al.*, 2014; Quinn *et al.*, 2012; Venteris *et al.*, 2014b, 2014c). It should be noted, however, that the current market price for $CO_2$ is typically stated at $40/t (National Energy Technology Laboratory, 2010), and more detailed studies suggest capture and transport costs ranging from $38/t to $55/t (National Enhanced Oil Recovery Initiative, 2012). Average annual $CO_2$ costs per unit farm for the 21 BGY (79.5 BLY) *Chlorella* and HTL scenario (Venteris *et al.*, 2014c) are therefore capped at ~$2 million/yr·(unit farm) relative to fuel values ranging from $8 million to $10 million/yr·(unit farm). Carbon dioxide capture is also of high interest to the fossil fuel industry for carbon capture and storage (Bachu and Adams, 2003) and for enhanced oil recovery (National Enhanced Oil Recovery Initiative, 2012), so there is considerable political motivation to incentivize $CO_2$ capture to increase supplies and reduce prices. Where proximal colocation is available, it may be possible to obtain $CO_2$ below market prices. Even at market prices, the purchase of $CO_2$ does not appear to be cost prohibitive in and of itself.

The potential for a significant increase in nitrogen and phosphorus consumption and the subsequent impacts on fertilizer and food prices is a serious concern. Here, choices in the biomass-to-biofuel processing and recycling pathways can have a dramatic impact on demand. As a baseline, LE requirements at the 20- to 21-BGY level—assuming full consumption of the nitrogen and phosphorus contained in the biomass (through coproducts (Richardson *et al.*, 2010) or perhaps burial for carbon sequestration (Sayre, 2010)—is around $3.0 \times 10^7$ t/yr for nitrogen and $3.0 \times 10^6$ to $4.0 \times 10^6$ t/yr for phosphorus. The contrasting phosphorus consumption is due to differences in lipid content and stoichiometric specifications (Pate, 2013; Venteris *et al.*, 2014b). Such consumption would result in roughly a 250% increase in nitrogen and 200% increase in phosphorus consumption relative to that of current U.S. agriculture, raising serious sustainability concerns. In general, consumption of nutrients in coproducts or other forms of non-reuse does not appear to be feasible at energy scales. Fortunately, there are technologies in development that can greatly reduce the amount of nitrogen and phosphorus consumed per unit of produced biofuel. For example, for *Chlorella*, biofuel-to-biomass conversion by HTL and nutrient recovery from the non-oil phase (nitrogen and phosphorus in the oil are currently specified as lost during the upgrading process) through catalytic hydrothermal gasification can reduce nitrogen and phosphorus demand to $5.0 \times 10^6$ t/yr and $6.0 \times 10^5$ t/yr, respectively (Venteris *et al.*, 2014b). This reduction in demand not only eases sustainability concerns, but also increases the contribution and viability of alternative nutrient supplies (Venteris *et al.*, 2014b) such as from municipal waste water (Lundquist *et al.*, 2010)—although these advanced biomass-to-biofuel and recycling processes (Elliott *et al.*, 2013) have not yet been incorporated into collocation studies of alternative nutrient resources (Fortier and Sturm, 2012; Orfield *et al.*, 2014).

## 11.4   TWO-BILLION GALLON PER YEAR CASE STUDY

At question are the land, water, and nutrient resources required to produce 2 billion gal/yr (7.6 billion L/yr) of aviation kerosene to meet renewable fuel targets, based on the cultivation of algae in open ponds. We estimated the amount of competitive freshwater (water salinity <2,000 mg/L and a depth <300 m), waters generally targeted for municipal and agricultural use

(Venteris *et al.*, 2013), and seawater required to meet the production target. Freshwater allows greater flexibility in the choice of operating chemistry and is generally inexpensive to procure. However, the water demand for a single typical 405-ha (pond area) algae facility is comparable to that of thousands of single family homes, raising serious sustainability issues. In contrast, seawater presents a nearly inexhaustible water supply with potentially consistent ion chemistry (when collected away from coastal areas with substantial freshwater influx). The main disadvantages of seawater utilization are supply infrastructure costs and the requirement that the selected microalgae species perform at salinities in excess of seawater (to permit salinity maintenance with a small amount of blowdown).

We calculated the resource requirements to provide 2 BGY (7.6 BLY) of jet fuel (aviation kerosene) from the cultivation of *Chlorella* in freshwater and *Arthrospira* in seawater. *Chlorella* was selected because it is a productive freshwater species and there are available lab-derived growth model parameters for Department of Energy strain number 1412 (Huesemann, 2014). *Arthrospira* was selected because it is a productive saltwater strain that is already cultivated commercially in open ponds (Ahsan *et al.*, 2008; Earthrise Nutritional, LLC, 2013).

For both scenarios, biomass-to-biofuel conversion was based on hydrothermal liquefaction (Elliott, 2010; Frank *et al.*, 2013). As discussed in Section 11.3, previous studies (Venteris *et al.*, 2014b) have shown significant resource and economic advantages for HTL technology when compared to lipid extraction, so this pathway was selected for analysis. We calculate biomass production based on the climate and growth models of Wigmosta *et al.* (2011) and on estimates of the amount of HTL oil and fuel products produced on results from recent laboratory experiments (Jones, 2014). Biomass and biofuel calculation parameters are presented in Table 11.2, and the calculation methods are presented in Venteris *et al.* (2014a, 2014b). For these scenarios, the HTL oil is upgraded (hydrotreated) and distilled to three forms of biofuel, the proportions selected to be realistic considering distillation and economic factors (Table 11.2). Carbon dioxide, nitrogen, and phosphorus demand were based on a nutrient profile (Table 11.2) and full consumption as either byproducts (animal feed, etc.) or waste (zero recycle) to provide a conservative estimate of the potential nutrient impacts of meeting the 2-BGY (7.6 BLY) target.

Production sites are prioritized and selected from a nationwide dataset of 88,692 potential locations (Venteris *et al.*, 2014c) on the basis of produced fuel value relative to resource availability and costs. For simplicity, we limit our considered resources to $CO_2$ from flue gas and water. *Chlorella* grows best in fresh, low-salinity water (see Venteris *et al.*, 2014c) and *Arthrospira* can be grown in seawater (Tredici *et al.*, 1986). For *Chlorella*, the freshwater availability model of Venteris *et al.* (2013) was applied, limiting consumptive withdrawal to 5% of the mean annual flow of each hydrologic unit code 6 watershed, as discussed in Section 11.3.2 (U.S. Geological Survey, 2009). The previous freshwater cost model was refined by inclusion of salinity estimates and blowdown demand (an operating salinity of 4,000 mg/L was specified based on sensitivity analyses (Venteris and Wigmosta, 2013). For *Arthrospira*, the seawater demand was based on an operating salinity of 45,000 mg/L and an enhanced version of the geographic information systems (GIS) based pipeline cost model presented in Venteris *et al.* (2013). The most rather than a single pipeline for each unit farm. Cost models have also been developed for seawater intake structures and vertical turbine pump systems. The $CO_2$ model is unchanged from previous presentations (Venteris *et al.*, 2014a, 2014b). Carbon dioxide is sourced from flue gas, with availability (supply based on the size of each flue gas source) and costs determined from a GIS-based cost-distance model that estimates the capital and operating costs to transport $CO_2$ and $N_2$ mixtures to the ponds (without phase change) through pipelines.

## 11.4.1   *Results*

Not all of the produced HTL oil can be realistically converted to aviation kerosene. Therefore, the manufacture of 2 BGY (7.6 BLY) of jet fuel requires the coproduction of renewable diesel and naphtha (gasoline). For our model scenario, the production of 2 BGY (7.6 BLY) of

Table 11.2.   Parameters and values used in the growth and biofuel production models.

| Parameter | Value | Units | Source |
|---|---|---|---|
| Growth model light saturation constant, $s_o$, *Spirulina/Chlorella* | 150/250 | $\mu mol/m^6 \cdot s$ | Sapphire Energy (2013); Huesemann (2014) |
| Growth model biomass accumulation efficiency, $\varepsilon_b$, *Spirulina/Chlorella* | 0.5/0.61 | – | Sapphire Energy (2013); Huesemann (2014) |
| Growth model minimum water temperature for zero productivity, $T_{min}$, *Spirulina/Chlorella* | 5.0/12.8 | °C | Sapphire Energy (2013); Huesemann (2014) |
| Growth model lower water temperatures for optimal productivity, $T_{opt\_low}$, *Spirulina/Chlorella* | 15.0/36.0 | °C | Sapphire Energy (2013); Huesemann (2014) |
| Growth model upper water temperatures for optimal productivity, $T_{opt\_high}$, *Spirulina/Chlorella* | 30.0/36.2 | °C | Sapphire Energy (2013); Huesemann (2014) |
| Growth model maximum water temperature for zero productivity, $T_{max}$, *Spirulina/Chlorella* | 35.0/45.0 | °C | Sapphire Energy (2013); Huesemann (2014) |
| Carbon utilization efficiency, $E_{JET_2}$ | 0.82 | – | Davis *et al.* (2012) |
| Biomass to hydrothermal liquefaction (HTL) oil efficiency, $E_{HTL}$, *Spirulina/Chlorella* | 0.38/0.606 | – | Toor *et al.* (2013); Jones (2014) |
| Renewable diesel upgrading efficiency, $E_{RD}$ | 0.60 | – | Jones (2014) |
| Aviation kerosene upgrading efficiency, $E_{JET}$ | 0.30 | – | Jones (2014) |
| Naphtha upgrading efficiency, $E_N$ | 0.10 | – | Jones (2014) |
| Renewable diesel density, $\rho_{RD}$ | 0.793 | kg/L | Jones (2014) |
| Aviation kerosene density, $\rho_{JET}$ | 0.810 | kg/L | ASTM International (2014) |
| HTL naphtha density, $\rho_N$ | 0.780 | kg/L | Jones (2014) |
| Renewable diesel wholesale price, $P_{RD}$ | 3.05 | US$ | U.S. Energy Information Administration (2013) |
| Aviation kerosene wholesale price, $P_{JET}$ | 3.06 | US$ | U.S. Energy Information Administration (2013) |
| Naphtha (gasoline) wholesale price, $P_N$ | 2.82 | US$ | U.S. Energy Information Administration (2013) |
| C content, dry biomass, $W_{CB}$ | 0.550 | fraction | Williams and Laurens (2010) |
| N content, dry biomass, $W_{NB}$ | 0.078 | fraction | Williams and Laurens (2010) |
| P content, dry biomass, $W_{PB}$ | 0.0081 | fraction | Williams and Laurens (2010) |

kerosene requires 4.05 BGY (15.3 BLY) of renewable diesel and 0.69 BGY (2.6 BLY) of gasoline. Therefore, we calculate the resources realistically required to meet aviation renewable fuel targets, with the understanding that the majority of fuel products (70%) and resource demands are supplying other transportation fuels with similar economic value (Table 11.3). The entire system contributes 6.74 BGY (25.5 BLY) to general renewable fuel targets such as the Energy Independence and Security Act.

We estimate that 1,488 *Chlorella* or 2,559 *Arthrospira* 485-ha sites would be required to meet the 2-BGY (7.6-BLY) target. First, the average modeled productivity and biomass-to-HTL conversion efficiencies for *Chlorella* are significantly higher than for *Arthrospira*, so more land is required to meet the production target for seawater-based production. The relative species and strain performance over the range of potential water sources (fresh to hypersaline) and operating salinities (0 to ~200,000 mg/L) is unknown, so the result does not indicate that freshwater strains will outperform seawater strains in general. Regardless of strain, the use of freshwater in a manner that reduces local sustainability impacts requires geographic dispersal (Fig. 11.8). So even

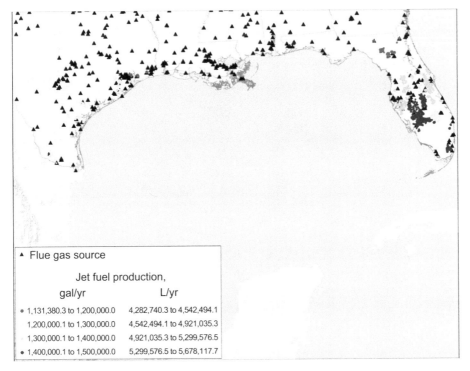

| **▲ Flue gas source** | |
|---|---|
| Jet fuel production, | |
| gal/yr | L/yr |
| • 1,131,380.3 to 1,200,000.0 | 4,282,740.3 to 4,542,494.1 |
| 1,200,000.1 to 1,300,000.0 | 4,542,494.1 to 4,921,035.3 |
| 1,300,000.1 to 1,400,000.0 | 4,921,035.3 to 5,299,576.5 |
| • 1,400,000.1 to 1,500,000.0 | 5,299,576.5 to 5,678,117.7 |

Figure 11.8. Highest performing locations (1,488 unit farms) for *Chlorella* cultivation that meet the 2-billion-gallons-per-year (7.6-billion-liters-per-year) aviation kerosene target.

though the maximum production potential for *Chlorella* is in the southern Florida peninsula, less productive sites in northern Florida and along the Gulf of Mexico coast in Texas are required and add to the land requirement. In contrast, the *Arthrospira* selection is located on the Florida peninsula and is far more concentrated geographically (Fig. 11.9), but it requires an additional 518,000 ha of land.

## 11.4.2   *Discussion and conclusions*

A feasibility assessment of any biofuel scenario requires a careful consideration of the tradeoffs between economics and environmental sustainability. In this example, we do not conduct a full technoeconomic analysis, so the ultimate economic viability remains an open question. Nonetheless, it is clear that even with cost reduction through shared resources, supplying seawater is far more expensive (nearly 10 times, Table 11.3) than supplying freshwater. However, several unaccounted-for factors could narrow this gap. First, current research is working to assess the potential impacts of freshwater supply variability on production, for example, interruptions in supply due to drought (Seager *et al.*, 2009) and unsustainable groundwater usage (Konikow, 2013) have economic consequences that need quantification. Second, the ability to spatially concentrate cultivation facilities introduces potential efficiencies such as infrastructure utilization. For example, the increased siting flexibility of seawater ponds relative to freshwater ponds can permit greater colocation with flue gas sources otherwise limited by the freshwater capacity of each watershed. The final selection of water sources also depends on the performance of the organisms under consideration. Here, the seawater-based scenario is only $0.25/gal more than the freshwater-based scenario, which may be a small price to pay for decreased sustainability issues and increased water supply reliability.

This and previously published analyses (Davis *et al.*, 2012; Venteris *et al.*, 2014a, 2014b, 2014c; Wigmosta *et al.*, 2011) have yet to identify key resource limitations (National Research

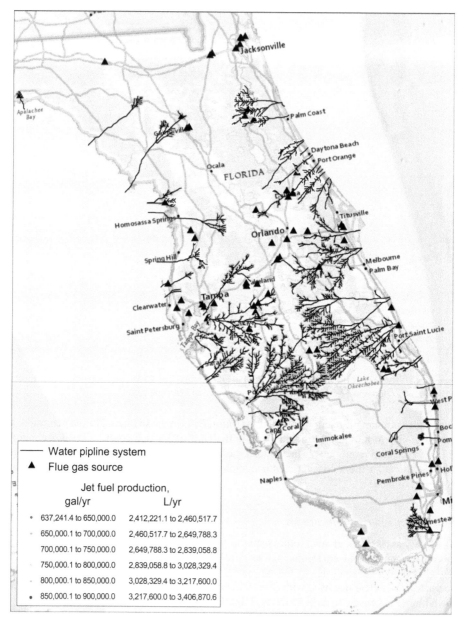

Figure 11.9.  Seawater pipeline routes and highest performing sites (2,559 unit farms) for *Arthrospira* cultivation to meet the 2-billion-gallons-per-year (7.6-billion-liters-per-year) aviation kerosene target.

Council, 2012) that would prevent the production of algal biofuel to meet modest production targets (~5 BGY; 18.9 BLY). While water and $CO_2$ costs are substantial, for many sites they are not economically limiting. Flue gas supply is not limiting at this level, and careful use of freshwater (i.e., choosing sites with low net evaporation near major rivers) and incorporation of sites using alternative waters can alleviate concerns presented by water supply. The major challenges currently center on the site economics of growing and harvesting and on the technology pathway to process the microalgae into "drop-in" hydrocarbon fuels, which analyses show are currently not cost-competitive with petroleum.

Table 11.3.   Results for *Chlorella* and *Arthrospira* biofuel production and resource assessment models.

| Scenario | 485-ha sites | Aviation kerosene, gal/yr (L/yr) | Renewable diesel, gal/yr (L/yr) | Gasoline, gal/yr (L/yr) | Total biofuel, gal/yr (L/yr) |
|---|---|---|---|---|---|
| *Chlorella* | 1,488 | $2.00\times10^9$ ($7.57\times10^9$) | $4.05\times10^9$ ($1.53\times10^{10}$) | $6.92\times10^8$ ($2.62\times10^9$) | $6.74\times10^9$ ($2.55\times10^{10}$) |
| *Arthrospira* | 2,559 | $2.00\times10^9$ ($7.57\times10^9$) | $4.05\times10^9$ ($1.53\times10^{10}$) | $6.92\times10^8$ ($2.62\times10^9$) | $6.74\times10^9$ ($2.55\times10^{10}$) |

| Scenario | Land, ha | Make-up water, m³/yr | N, percent current U.S. agriculture (no recycle) | P, percent current U.S. agriculture (no recycle) |
|---|---|---|---|---|
| *Chlorella* | $7.22\times10^5$ | $2.13\times10^9$ | 23.5 | 16.8 |
| *Arthrospira* | $1.24\times10^6$ | $9.25\times10^9$ | 23.5 | 16.8 |

| Scenario | Water cost, $/gal biofuel | $CO_2$ cost, $/gal biofuel | Total water and flue gas cost, $/gal biofuel | Fuel value— Resource cost, $/gal biofuel |
|---|---|---|---|---|
| *Chlorella* | $0.02 | $0.28 | $0.30 | $2.70 |
| *Arthrospira* | $0.21 | $0.34 | $0.55 | $2.45 |

The scenario presented here assumes that these economic limitations are overcome through the sale of high-value coproducts such as animal feed and cosmetics. A consequence of such an approach is the full consumption of the nitrogen and phosphorus used to support biomass growth. This would result in a 24% increase in nitrogen and a 17% increase in phosphorus consumption relative to current U.S. agricultural consumption (Table 11.3). The current increase in natural gas production (Newell, 2011) makes enhanced nitrogen production feasible (through the Haber-Bosch process), but phosphorus limitations (Vaccari, 2009) are a serious concern. Fortunately, there is considerable potential for offsets from nitrogen and phosphorus contained in municipal sewerage and animal manure (Venteris *et al.*, 2014b), and recycling of leftover biomass (through anaerobic digestion or catalytic hydrothermal gasification) rather than sale through coproducts could reduce demands by at least 50% (Venteris *et al.*, 2014b). Although considerable technological challenges remain to improve the economic production of microalgae-based biofuel, algal biofuel remains a promising avenue for the U.S. Department of Defense and the airline industry to meet greenhouse gas emission reduction targets and renewable fuel requirements.

## 11.5   SUMMARY AND CONCLUSIONS

The land, water, and nutrient demands of algal biofuel production at energy scales are a serious sustainability concern. This review shows that careful planning is needed and that spatiotemporal geographic-information-systems-based models such as the Biomass Assessment Tool have a critical role to play in decision support because they bring holistic analysis to the drive toward sustainable renewable fuel production. For land, use change is a critical issue for further exploration. The production potential of steppe and desert lands is limited by water supply (Venteris *et al.*, 2013, 2014a) and infrastructure concerns (Venteris *et al.*, 2014b). Pasture, forest, and marginal croplands in the southeastern United States currently have the most promise for economically viable algae cultivation facilities, but careful consideration of the tradeoffs between fuel, food, and fiber production are required. The impact can be reduced through maximizing biomass productivity and fuel conversion efficiency, and by targeting lands within these categories that are the least productive under their current land use (Venteris *et al.*, 2012). Water use can also be made more efficient by maximizing fuel production, but the central issue is the tradeoffs between sustainability and costs.

Deep saline groundwater resources are potentially large, and seawater is effectively inexhaustible; but delivery costs are high, and the disposal of large amounts of saline concentrate adds costs and regulatory concerns. Alternative water sources such as municipal waste water and produced water can provide local solutions, but these have limited potential at energy scales. In contrast, freshwater supplies are relatively inexpensive and hold the potential for supporting large production targets. However, a serious knowledge gap currently exists in how much freshwater is available in the face of groundwater sustainability, current and future streamflow regulations, water rights, climate change, and subsequent spatial and temporal variability.

Finally, nutrient demands do not appear to be an insurmountable sustainability concern so long as efficient biomass-to-biofuel conversion and nutrient recycling prove economically viable. Carbon dioxide flue gas supplies are plentiful, and while expensive, commercial options are not prohibitively so. The production of coproducts requires careful consideration of nitrogen and phosphorus consumption, but recycling through anaerobic digestion or gasification can dramatically reduce demands. In this review, it has been demonstrated that reducing sustainability impacts requires careful planning and design, but the resource limitations alone do not preclude the production of biofuels at magnitudes that are well in excess of the goals set for the U.S. Department of Defense and the U.S. airline industry.

## ACKNOWLEDGMENTS

Support for this research was provided by the Office of the Biomass Program of the U.S. Department of Energy. The Pacific Northwest National Laboratory is operated by Battelle Memorial Institute for the U.S. Department of Energy under contract DE–AC06–76RLO 1830.

## REFERENCES

Amer, L., B. Adhikari, and J. Pellegrino. 2011. Technoeconomic analysis of five microalgae-to-biofuels processes of varying complexity. *Bioresour. Technol.* 102:9350–9359.
Androwski, J., A. Springer, T. Acker, *et al.* 2011. Wind-powered desalination: an estimate of saline groundwater in the United States. *J. Am. Water Resour. Assoc.* 47:93–102.
ASTM International. 2014. Standard Specification for Aviation Turbine Fuels, ASTM D1655-14a. ASTM International, West Conshohocken, Pennsylvania, United States. doi: 10.1520/D1655-14A.
Bachu, S., and J. J. Adams. 2003. Sequestration of $CO_2$ in geological media in response to climate change: capacity of deep saline aquifers to sequester $CO_2$ in solution. *Energy Convers. Manage.* 44:3151–3175.
Béchet, Q., A. Shilton, and B. Guieysse. 2013. Modeling the effects of light and temperature on algae growth: state of the art and critical assessment for productivity prediction during outdoor cultivation. *Biotechnol. Adv.* 31:1648–1663.
Benemann, J. R., R. P. Goebel, J. C. Weissman, *et al.* 1982. Microalgae as a source of liquid fuels. Final Technical Report DOE/ER/30014–T1.
Carter, N. A., R. W. Stratton, M. K. Bredehoeft, *et al.* 2011. Energy and environmental viability of select alternative jet fuel pathways. AIAA 2011–5968.
Chisti, Y. 2007. Biodiesel from microalgae. *Biotechnol. Adv.* 25:294–306.
Chiu, Y.-W., and M. Wu. 2013. Considering water availability and wastewater resources in the development of algal bio-oil. *Biofuels, Bioprod. Biorrefin.* 7:406–415.
Churkina, G., and S. W. Running, 1998. Contrasting climatic controls on the estimated productivity of global terrestrial biomes. *Ecosystems* 1:206–215.
Clark, C. E., and J. A. Veil. 2009. Produced water volumes and management practices in the United States. Argonne National Laboratory Technical Report ANL/EVS/R–09/1.

Coleman, A. M., J. M. Abodeely, R. L. Skaggs, *et al.* 2014. An integrated assessment of location-dependent scaling of microalgae biofuel production facilities. *Algal Res.* 5:79–94.

Davis, R., A. Aden, and P. T. Pienkos. 2011. Techno-economic analysis of autotrophic microalgae for fuel production. *Appl. Energy* 88:3524–3531.

Davis, R., D. Fishman, E. D. Frank, *et al.* 2012. Renewable diesel from algal lipids: an integrated baseline for cost, emissions, and resource potential from a harmonized model. Argonne National Laboratory Technical Report ANL/ESD/12–4 (NREL/TP–5100–55431 and PNNL–21437).

Davis, R. E., D. B. Fishman, E. D. Frank, *et al.* 2014. Integrated evaluation of cost, emissions, and resource potential for algal biofuels at the national scale. *Environ. Sci. Technol.* 48:6035–6042.

Earthrise Nutritional, LLC. 2013. Welcome to Earthrise.com—#1 Spirulina on Earth. *http://www.earthrise. com/* (accessed September 25, 2013).

Elliott, D. C. 2010. Hydrothermal liquefaction of biomass. *IEA Bioenergy Task 34 Pyrolysis (PNNL– SA–76509)*, 28:21–23.

Elliott, D. C., T. R. Hart, A. J. Schmidt, *et al.* 2013. Process development for hydrothermal liquefaction of algae feedstocks in a continuous-flow reactor. *Algal Res.* 2:445–454.

Fortier, M-O. P., and B. S. M. Sturm. 2012. Geographic analysis of the feasibility of collocating algal biomass production with wastewater treatment plants. *Environ. Sci. Technol.* 46:11426–11434.

Frank, E., A. Elgowainy, J. Han, *et al.* 2013. Life-cycle comparison of hydrothermal liquefaction and lipid extraction pathways to renewable diesel from algae. *Mitig. Adapt. Strat. Glob. Change* 18:137–158.

Habib, M. A. B., M. Parvin, T. C. Huntington, *et al.* 2008. A review on culture, production and use of spirulina as food for humans and feeds for domestic animals and fish. FAO Fisheries and Aquaculture Circular No. 1034 (FIMA/C1034 (En), Rome: Food and Agriculture Organization of the United Nations.

Huesemann, M. March 2014. Pacific Northwest National Laboratory. *Personal communication.*

Huesemann, M. H., T. S. Hausmann, R. Bartha, *et al.* 2009. Biomass productivities in wild type and a pigment mutant of Cyclotella sp. (Diatom). *Appl. Biochem. Biotechnol.* 157:507–526.

Jena, U., and K. C. Das. 2011. Comparative evaluation of thermochemical liquefaction and pyrolysis for bio-oil production from microalgae. *Energy Fuels* 25:5472–5482.

Jones, S. March 2014. Pacific Northwest National Laboratory. *Personal communication.*

Konikow, L. F. 2013. Groundwater depletion in the United States (1900–2008). U.S. Geological Survey Scientific Investigations Report 2013–5079. *http://pubs.usgs.govir/2013/5079/* (accessed November 18, 2014).

Lundquist, T. J., I. C. Woertz, N. W. T. Quinn, *et al.* 2010. A realistic technology and engineering assessment of algae biofuel production. Berkeley, CA: University of California.

Mata, T. M., A. A. Martins, and N. S. Caetano. 2010. Microalgae for biodiesel production and other applications: a review. *Renew. Sustain. Energy Rev.* 14:217–232.

Maxwell, E. L., A. G. Folger, and S. E. Hogg. 1985. Resource evaluation and site selection for microalgae production systems. Solar Energy Research Institute Report SERI/TR–215–2484.

Mickley, M. C. 2001. Membrane concentrate disposal: practices and regulation. Desalination and Water Purification Research and Development Program Report No. 69.

National Energy Technology Laboratory. 2010. Carbon dioxide enhanced oil recovery—untapped domestic energy supply and long term carbon storage solution. *http://www.netl.doe.gov/file library/research/oil-gas/CO2_EOR_Primer.pdf* (accessed August 28, 2014).

National Energy Technology Laboratory. 2012. United States 2012 carbon utilization and storage atlas. Fourth ed. *http://www.netl.doe.gov/File Library/Research/Coal/carbon-storage/atlasiv/Atlas-IV-2012. pdf* (accessed August 28, 2014).

National Enhanced Oil Recovery Initiative. 2012. Carbon dioxide enhanced oil recovery: a critical domestic energy, economic, and environmental opportunity. *http://www.c2es.org/docUploads/EOR-Report.pdf* (accessed November 25, 2014).

National Research Council. 2012. Sustainable development of algal biofuels in the United States. Washington, DC: The National Academies Press.

Newell, R. G. 2011. The long-term outlook for natural gas. Washington, DC: U.S. Energy Information Administration.

Orfield, N., G. A. Keoleian, and N. G. Love. 2014. A GIS based national assessment of algal bio-oil production potential through flue gas and wastewater co-utilization. *Biomass & Bioenergy* 63:76–85. *http:// dx.doi.org/10.1016/j.biombioe.2014.01.047* (accessed December 1, 2014).

Pate, R. C. 2013. Resource requirements for the large-scale production of algal biofuels. *Biofuels* 4:409–435.

Pate, R., G. Klise, and B. Wu. 2011. Resource demand implications for US algae biofuels production scale-up. *Appl. Energy* 88:3377–3388.

Quinn, J. C., K. Catton, N. Wagner, *et al*. 2012. Current large-scale US biofuel potential from microalgae cultivated in photobioreactors. *Bioenerg. Res.* 5:49–60.

Redfield, A. C. 1934. On the proportions of organic derivatives in sea water and their relation to the composition of plankton. 176–192. Liverpool: University Press of Liverpool.

Richardson, J. W., J. L. Outlaw, and M. Allison. 2010. The economics of microalgae oil. *AgBioForum* 13:119–130.

Rubio, F. C., F. G. Camacho, J. M. F. Sevilla, *et al*. 2003. A mechanistic model of photosynthesis in microalgae. *Biotechnol. Bioeng.* 81:459–473.

Sapphire Energy, February 2013. Sapphire Energy, Inc. *Personal communication*.

Sayre, R. 2010. Microalgae: the potential for carbon capture. *BioScience* 60:722–727.

Sayre, R., P. Comer, H. Warner, *et al*. 2009. A new map of standardized terrestrial ecosystems of the conterminous United States. U.S. Geological Survey Professional Paper 1768.

Schnepf, R., and B. D. Yacobucci. 2013. Renewable fuel standard (RFS): overview and issues. Congressional Research Service 7–5700, R40155.

Seager, R., A. Tzanova, and J. Nakamura. 2009. Drought in the southeastern United States: causes, variability over the last millennium, and the potential for future hydroclimate change. *J. Climate* 22:5021–5045.

Sheehan, J., T. Dunahay, J. Benemann, *et al*. 1998. A look back at the U.S. Department of Energy's Aquatic Species Program: biodiesel from algae. National Renewable Energy Laboratory Report NREL/TP–580–24190.

Singh, A., P. S. Nigam, and J. D. Murphy. 2011. Renewable fuels from algae: an answer to debatable land based fuels. *Bioresour. Technol.* 102:10–16.

Sun, A., R. Davis, M. Starbuck, A. Ben-Amotz, *et al*. 2011. Comparative cost analysis of algal oil production for biofuels. *Energy* 36:5169–5179.

Toor, S. S., H. Reddy, S. Deng, *et al*. 2013. Hydrothermal liquefaction of Spirulina and Nannochloropsis salina under subcritical and supercritical water conditions. *Bioresour. Technol.* 131:413–419.

Tredici, M. M., T. Papuzzo, and L. Tomaselli. 1986. Outdoor mass culture of Spirulina maxima in sea-water. *Appl. Microbiol. Biotechnol.* 24:47–50.

U.S. Department of Energy. 2013. Bioenergy Technologies Office Multi-Year Program Plan. U.S. Department of Energy & Renewable Energy DOE/EE–0915. *http://www1.eere.energy.gov/bioenergy/pdfs/mypp_may_2013.pdf* (accessed November 24, 2014).

U.S. Department of Energy. 2014. Aviation fuels. Bioenergy Technologies Office. *http://www.energy.gov/eere/bioenergy/aviation-fuels* (accessed September 28, 2014).

U.S. Energy Information Administration. 2013. Petroleum & other liquids—spot prices. *http://www.eia.gov/dnav/pet/pet_pri_spt_s1_d.htm* (accessed June 6, 2013).

U.S. Geological Survey. 2009. The National Map. Hydrography—What is the WBD? *http://nhd.usgs.gov/wbd.html* (accessed August 24, 2011).

Vaccari, D. A. 2009. Phosphorus famine: the threat to our food supply. New York: Scientific American.

Venteris, E. R., R. R. Riley, J. McDonald, *et al*. 2009. Establishing a regional geologic framework for carbon dioxide sequestration planning: a case study. In *Carbon dioxide sequestration in geological media—state of the science*. ed., M. Grobe, J. C. Pashin, and R. L. Dodge, American Association of Petroleum Geologists Studies in Geology, 59:191–225.

Venteris, E. R., R. L. Skaggs, A. M. Coleman, *et al*. 2012. An assessment of land availability and price in the coterminous United States for conversion to algal biofuel production. *Biomass Bioenergy* 47:483–497.

Venteris, E. R., R. L. Skaggs, A. M. Coleman, *et al*. 2013. A GIS cost model to assess the availability of freshwater, seawater, and saline groundwater for algal biofuel production in the United States. *Environ. Sci. Technol.* 47:4840–4849.

Venteris, E. R., R. C. McBride, A. M. Coleman, *et al*. 2014a. Siting algae cultivation facilities for biofuel production in the United States: trade-offs between growth rate, site constructability, water availability, and infrastructure. *Environ. Sci. Technol.* 48:3559–3566.

Venteris, E. R., R. L. Skaggs, M. S. Wigmosta, *et al*. 2014b. A national-scale comparison of resource and nutrient demands for algae-based biofuel production by lipid extraction and hydrothermal liquefaction. *Biomass Bioenergy* 64:276–290.

Venteris, E. R., R. L. Skaggs, M. S. Wigmosta, *et al*. 2014c. Regional algal biofuel production potential in the coterminous United States as affected by resource availability trade-offs. *Algal Res.* 5:215–225.

Venteris, E. R., and M. S. Wigmosta. 2013. Water cost and availabilty for algae cultivation—salinity issues. *ABO Webinars*, ed. M. Rosenthal, Washington, DC: Algae Biomass Organization.

Vigon, B. W., M. F. Arthur, L. G. Taft, *et al.* 1982. Final report on resource assessment for microalgal/ emergent aquatic biomass systems in the arid southwest to Solar Energy Research Institute. Columbus, OH: Battelle.

Voutchkov, N. 2011. Overview of seawater concentrate disposal alternatives. *Desalination* 273:205–219.

Weyer, K. M., D. R. Bush, A. Darzins, *et al.* 2010. Theoretical maximum algal oil production. *Bioeng. Res.* 3:204–213.

Wigmosta, M. S., A. M. Coleman, R. J. Skaggs, *et al.* 2011. National microalgae biofuel production potential and resource demand. *Water Resour. Res.* 47.

Wikipedia. 2017. Photobioreactors. *https://commons.wikimedia.org/wiki/Category:Photobioreactors* (accessed March 29, 2017).

Williams, P. J. le B., and L. M. L. Laurens. 2010. Microalgae as biodiesel & biomass feedstocks: review & analysis of the biochemistry, energetics & economics. *Energy Environ. Sci.* 3:554–590.

# CHAPTER 12

## Certification of alternative fuels

Mark Rumizen and Tim Edwards

## 12.1  INTRODUCTION

Within the last several years, the technical feasibility of combining biofuel blending components with aviation fuel has been proven by a variety of demonstration flights that used a mixture of aviation fuel and alternative fuel, as well as by extensive testing to develop the first
alternative jet fuel specifications for bio-based fuels through American Society for Testing and
Materials (ASTM) International (West Conshohocken, PA). The first alternative jet fuels to be
certified were from Fischer-Tropsch (FT) processing, which has had a long history of use of
fossil-based synthetic fuels; the new specification also allowed the use of biomass feedstocks in
the fuel production. In the summer of 2011, an aviation biofuel pathway called hydroprocessed
esters and fatty acids (HEFA) was approved by ASTM International as the second annex to the
ASTM D7566 (2014) "Standard Specification for Aviation Turbine Fuel Containing Synthesized
Hydrocarbons" for fuels derived from oil feedstocks such as *Camelina*, canola, animal fats, and
algae. This was followed by the approval in June 2014 of a third aviation biofuel pathway called
synthesized isoparaffins (SIPs). These approvals were a necessary first step toward commercializing aviation biofuels and the eventual scale-up of the feedstock production and fuel conversion
processes. Task forces have been working to complete research reports that will form the basis
for the certification of additional bio-based fuel production pathways. The biofuels that result
from these efforts will permit even more widespread field testing, which is eagerly anticipated
by many airlines. This chapter describes precertification testing, the certification process, the key
biofuel properties, and blending issues.

## 12.2  BACKGROUND

### 12.2.1  *Aviation fuel specifications*

All current aviation equipment has been developed, tested, and certified to use petroleum-based
jet fuel, which meets stringent specifications to ensure the performance and, more importantly, the safety of operations. Operating with alternative jet fuels was approved after it was
first demonstrated that the new alternative fuels were essentially identical to the existing petroleum-derived fuel, and then those fuels were incorporated into existing military and commercial
fuel specifications. Aviation fuel specifications are used to define the operating limitations for
virtually all gas-turbine-powered aircraft in operation today. The most commonly used aviation
fuel types and their corresponding specifications are given by:

- Jet A/Jet A–1: ASTM D1655 (2014, United States) and ASTM D7566 (2014, United States)
- Jet A–1: DEFSTAN 91–91 (2011, United Kingdom)
- Jet propellant (JP–8): MIL–DTL–83133 (2013, U.S. Department of Defense)
- JP–5: MIL–DTL–5624 (2013, U.S. Department of Defense)

See Exxon (2005) for additional information on jet fuel specifications from the United States,
United Kingdom, and other countries.

Table 12.1.    Principal specification requirements.

| Property | Limits | ASTM test method(s)[a] |
|---|---|---|
| Aromatics, % v/v, max | 25 | D1319 |
| Sulfur, total, wt%, max | 0.30 | D1266, D2622, D4294, or D5453 |
| 10% distillation temperature, °C:, max | 205 | D86, D2887 |
| Final boiling point, temperature, °C:, max | 300 | D86, D2887 |
| Flash point, °C, min | 38 | D56 or D3828 |
| Density at 15°C, kg/m$^3$ | 775 to 840 | D1298 or D4052 |
| Freezing point, °C max | −40 (Jet A) −47 (Jet A-1) | D5972, D7153, D7154, or D2386 |
| Viscosity −20°C, mm$^2$/s, max | 8.0 | D445 |
| Net heat of combustion, MJ/kg, min | 42.8 | D4529, D3338, or D4809 |
| Smoke point, mm, min | 18 | D1322 |
| Napthalenes, % v/v, max | 3.0 | D1840 |

[a]Full ASTM specification names are provided in the reference list.
*Adapted from:* ASTM D1655, Table 1. Republished with permission of STM International, from ASTM D1655. 2014. Standard Specification for Aviation Turbine Fuels. ASTM International, West Conshohocken, PA; permission conveyed through Copyright Clearance Center, Inc.

Incorporation of the new jet fuel pathway into the existing jet fuel specifications is effectively the only means for entering the new fuel into service. However, jet fuel specifications only control the fuel properties necessary to support quality control and purchasing, so other properties must be evaluated for new, alternative fuel pathways during the qualification process. The principal specification requirements are excerpted from ASTM D1655 (2014), as shown in Table 12.1. Note that only coarse compositional limits are listed and that the specification is fairly broad: for example, allowable fuel densities cover the broad range from 775 to 840 kg/m$^3$. In addition to the properties listed, ASTM D1655 defines the fuel processing generically through the following: "Aviation turbine fuel, except as otherwise specified in this specification, shall consist of refined hydrocarbons … derived from conventional sources including crude oil, natural gas liquid condensates, heavy oil, shale oil, and oil sands."

### 12.2.2    Foundational elements of the approval process

When the military and commercial aviation fuel communities became seriously involved in alternative fuels in 2006 through a joint effort, the first question to be addressed was, "What kind of alternative fuel do we want?" After relatively little discussion, the aviation community decided that a "drop-in" alternative fuel was needed, in which any changes in fuel due to feedstock or processing would be transparent to the fuel handlers and fuel users and no changes to aircraft, pipelines, or infrastructure would be required. In essence, what was being sought was not an alternative fuel composition, but rather a similar, interchangeable composition produced from alternative feedstocks. In other words, what was desired was an alternative feedstock, not an alternative fuel *per se*. If security of supply and reduction of petroleum imports were the goals, the alternative feedstock could be coal or natural gas. If a smaller greenhouse gas emissions "footprint" was an additional goal, the feedstock could be some type of biomass. Note that a measure of the lifecycle greenhouse gas footprint is an accounting method that considers all aspects of feedstock generation and conversion to a fuel product. It cannot be assessed by simply testing the fuel, so it is not considered a "fuel performance property," and therefore is not included in the jet fuel specifications. Instead, the greenhouse gas footprint will be handled separately by regulatory compliance or taxation.

In essence, the "drop-in" performance requirement constrains the alternative aviation fuel(s) to be composed solely of hydrocarbons. This is in contrast to biodiesel and ethanol, which contain heteroatoms such as oxygen. Another way of looking at this is that avoiding the need for a "flex fuel" aircraft became the key aviation fuel goal. The logistical infrastructure for aviation fuel is much different than that for ground vehicles. Most airports are fed by a single pipeline and have one set of storage tanks, so it would not be feasible to separately handle a fuel such as ethanol or biodiesel—even if these fuels could meet the performance demands of aircraft jet engines.

It was clear from the early stages of the alternative aviation fuels effort that all the stakeholders needed to be involved—fuel buyers (airlines and the Defense Logistics Agency), fuel producers, fuel users, equipment manufacturers, pipeline companies, airports, and the research community. The U.S. Federal Aviation Administration (FAA) recognized this interdependency and created the Commercial Aviation Alternative Fuels Initiative (CAAFI®) to enable the aviation fuel community to coordinate efforts. The military also participated in CAAFI, although additional military requirements for fuel had to be addressed separately (such as the use of jet fuel in afterburners, augmentors, and diesel engines, which are common practice in the military but not in commercial aviation).

As the aviation fuel community began to address alternative fuels, the wide variety of feedstocks and production processes led to the decision to group the new fuels by production process or pathway for the purposes of evaluation and ultimate incorporation into the jet fuel specification. The main driving force behind this decision was the need to group fuels in such a way that control could be kept over the process and composition. The precedent for using an alternative process for aviation fuel was set in 1999 when the Fischer-Tropsch process that was used to make Sasol isoparaffinic kerosene (IPK), was approved for blending into petroleum-derived jet fuel at a 50% blend concentration. However, this was not a generic process approval because only the IPK made in Sasol's refinery was approved (Moses *et al.*, 1997).

As alternative fuel efforts began to gear up around 2006, the approval of a generic FT fuel blending component seemed to be a logical initial target, given the industry's desire to avoid approving individual manufacturer's products. The discussion that follows describes the certification path taken by the three approved alternative aviation fuels and potential follow-on certification targets. The point of view is that of the ASTM certification process—the military process is similar (by design) because harmonization of the specifications is in the interest of the entire aviation community. This harmonization has been enhanced with the recent U.S. military's conversion from JP–8 to Jet A fuel in the continental United States.

## 12.3   ASTM CERTIFICATION PROCESS

### 12.3.1   *Basis of the approval process*

The original Sasol isoparaffinic kerosene approval in 1999 was used by the aviation fuel community as the basis to develop an alternative aviation fuel qualification process. The ASTM International Subcommittee on Aviation Fuels required a comprehensive data report to support the incorporation of Sasol's IPK into the existing ASTM D1655 (2014) aviation turbine fuel specification. The data that Sasol provided included the basic specification properties as well as an expanded set of fuel properties that were based on experience with petroleum-derived jet fuel. This expanded set of properties, called "fit-for-purpose" (FFP) properties, were intended to evaluate all aspects of fuel performance over the intended operational range of the fuel. The FFP properties were then compared to the same properties for petroleum-derived jet fuel to determine if the new fuel properties were within the "experience base" of the current fuel. For example, viscosity, surface tension, density, and specific heat were evaluated over broad temperature ranges. The complete composition of the fuel, including trace materials, was analyzed. Other key properties investigated included dielectric constant, thermal stability breakpoint, and compatibility with fuel system materials.

### 12.3.2   *Guidebook for the approval process (ASTM D4054)*

The ASTM International Subcommittee on Aviation Fuels determined that this fundamental approach should be applied to the approval of generic classes of alternative fuels. This decision

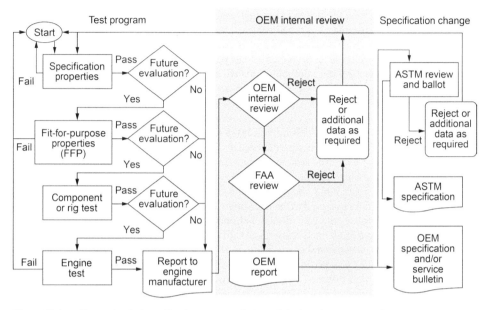

Figure 12.1.    Alternative fuel certification process. OEM, original equipment manufacturer; FAA, Federal Aviation Administration. (Republished with permission of ASTM International, from ASTM D4054. 2009. Standard Practice for Qualification and Approval of New Aviation Turbine Fuels and Fuel Additives. ASTM International, West Conshohocken, PA; permission conveyed through Copyright Clearance Center, Inc.)

led to the issuance of ASTM D4054 (2009), "Standard Practice for Qualification and Approval of New Aviation Turbine Fuels and Fuel Additives," in December 2009. ASTM D4054 describes the technical data and process needed to develop and issue a specification for alternative aviation fuel. In a parallel effort, the U.S. Air Force issued MIL–HDBK–510 (2010), "Aerospace Fuels Certification," initially in October 2007 and later updated in 2010 and 2014 (MIL–HDBK–510A, 2014). This specification provided similar guidance for the military qualification of these fuels.

Both of these processes describe the data required to demonstrate that a candidate alternative fuel performs essentially identical to conventional petroleum-derived jet fuel in terms of performance and composition. In the case of ASTM, the data generated in accordance with ASTM D4054 (2009) is compiled in an ASTM Research Report, which is used to support the incorporation of the new fuel pathway or process in the aviation fuel specification. The ASTM International Subcommittee on Aviation fuels utilizes a balloting process to review and approve the research report and any proposed changes to the fuel specification.

Perhaps the most unique part of ASTM D4054 (2009) is the requirement that fuel producers work with the rest of the aviation industry to evaluate the data as they become available. Figure 12.1 shows the overall outline of the process (ASTM D4054, 2009). As data from groups of fuels going through the process are collected, they are presented to the aviation community as one or more ASTM Research Reports and are used to support the case for modifying the specification (or not).

Figure 12.2 is a more detailed breakout of the various tests in ASTM D4054 (2009). Typically, the properties shown in Figure 12.2 are assessed for the alternative fuel or fuel blend, and then compared to the "experience base" for current fuels. Ideally, a given fuel property falls within the experience base of current fuels, thus ensuring "drop-in" performance for that property. The experience base can come from various compilations of properties, such as the World Fuel Sampling Program (Hadaller and Johnson, 2006), the Aviation Fuel Properties Handbook

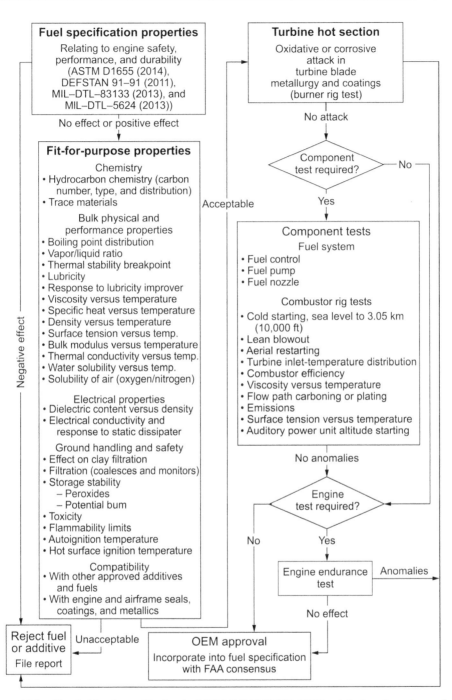

Figure 12.2.    Detailed breakout of testing in ASTM D4054 (2009). OEM, original equipment manufacturer; FAA, Federal Aviation Administration. Republished with permission of ASTM International, from ASTM D4054. 2009. Standard Practice for Qualification and Approval of New Aviation Turbine Fuels and Fuel Additives. ASTM International, West Conshohocken, PA; permission conveyed through Copyright Clearance Center, Inc.)

(CRC, 2014), the Petroleum Quality Information System (PQIS) (Gross, 2011), or other sources. The recent alternative fuel development and qualification activity has highlighted the need to obtain additional data on conventional aviation fuels to update and refine the experience base.

### 12.3.3    *Drop-in fuel specification (ASTM D7566)*

The ASTM International Subcommittee on Aviation Fuels also determined that alternative fuels would be defined in a separate specification rather than including these fuels in ASTM D1655 (2014). This was intended to ease the adoption of the new specification and to allow the subcommittee to specify additional controls for the alternative fuels. The new specification was issued in September 2009 as ASTM D7566 (2014).

ASTM D7566 "Standard Specification for Aviation Turbine Fuel Containing Synthesized Hydrocarbons" is the "drop-in fuel specification." It is structured with annexes that define property and compositional requirements for synthetic blending components that can be mixed with conventional, petroleum-derived jet fuel at specified percentage volumes. The finished, blended fuel must then meet the requirements of Table 1 of the specification, which is very similar to the highlights above Table 1 in ASTM D1655 (2014). However, a few additional requirements were added to ensure that the properties of the blend were bounded. A minimum aromatic level of 8% v/v (by ASTM D1319, 2013) and a lubricity requirement were added. In addition, two measures of distillation slope, specified as the difference between key ASTM D86 distillation temperature points, were added to avoid "narrow-boiling" fuels, which were outside the experience base. These distillation slope criteria are T50–T10 > 15 and T90–T10 > 22°C.

The original version of ASTM D7566 (2014) includes only one annex that defined the criteria for a Fischer-Tropsch blending component denoted as Synthesized Paraffinic Kerosene. The requirements for the SPK fuel, which are collected in Annex A, include a production pathway description in Section A1.4.1:

> *FT–SPK synthetic blending components shall be comprised of hydroprocessed synthesized paraffinic kerosene wholly derived from synthesis gas via the Fischer-Tropsch (FT) process using Iron or Cobalt catalyst. Subsequent processing of the product shall include hydrotreating, hydr[o]cracking, or hydroisomerization and is expected to include, but not be limited to, a combination of other conventional refinery processes such as polymerization, isomerization, and fractionation.*

Only fuels produced from production pathways that meet this description can be called FT–SPK fuels.

Several of the usual specification tests are required for the synthetic blend component in Annex A (freeze point, flash point, and distillation), but two major differences can be seen. First, the thermal stability test (ASTM D3241, 2014, jet fuel thermal oxidation test) requires a much higher temperature limit (325 versus 260°C for conventional Jet A). This makes the test much more stringent and helps to ensure an absence of contaminants in SPK. In addition, compositional requirements are imposed in their Table A1–2, including cycloparaffin and aromatic limits within a hydrocarbon composition requirement. Non-hydrocarbons are also specifically limited, with stringent limitations on nitrogen, sulfur, and halogens, and a 100-ppb limit on a long list of "metals": Al, Ca, Co, Cr, Cu, Fe, K, Li, Mg, Mn, Mo, Na, Ni, P, Pb, Pd, Pt, Sn, Sr, Ti, V, and Zn. These impurity limits were made very stringent (much too stringent for a conventional jet fuel to pass) to ensure the absence of any process-related materials in the final SPK.

In July 2011, a second annex was added for hydroprocessed esters and fatty acids. This annex included many of the same requirements as the FT annex. This was followed by the recent addition of a third annex for synthesized isoparaffins. The requirements in the SIP annex differ from FT and HEFA because of the unique compositional nature of this fuel. An overview of each of these synthetic fuel blending components and the associated ASTM qualification process is provided in Sections 12.3.4 and 12.3.6.

### 12.3.4 *Fischer-Tropsch–synthesized paraffinic kerosene, 2006 to 2009*

The ASTM International Subcommittee on Aviation Fuels conducted its evaluation of the first generic synthetic fuel in parallel with the development of ASTM D4054 (2009). As data became available from the U.S. Air Force's FT jet fuel evaluation program and from Sasol's fuel development program, five fuels appeared to be available in sufficient quantities to present the data required for the research reports. These five fuels were thought to encompass a representative range of FT aviation fuels that could be produced, and all were termed collectively as SPKs. The U.S. Air Force purchased (through the Defense Logistics Agency) 100,000 gal (378,500 L) of Syntroleum S–8 in 2006, 315,000 gal (1,190,000 L) of Shell FT kerosene in 2007, and 395,000 gal (about 1,500,000 L) of Sasol IPK in 2008. The U.S. Air Force tested essentially all aircraft and engines with these SPK fuels and generated a significant amount of data and confidence in the fuels.

Because of the low density of SPK fuels and the absence of aromatic hydrocarbons (which promote compatibility with some fuel system elastomers), it became apparent early on that SPK fuels (such as IPK specifically) could only be used as blends with conventional jet fuel, not as 100% synthetic fuels. With the availability of FT fuels produced by Sasol, Shell, and Syntroleum, covering both the low-temperature and high-temperature FT processes, researchers concluded that there were sufficient data and experience to publish the SPK Research Report in September 2008 (Moses, 2008). The report (naturally) followed the precedent of including the types of data collected for the successful IPK ballot in 1999, and it covered blends of up to 50% SPK in petroleum-derived jet fuel. Although the report passed the ballot, the number of comments requesting more information on potential contaminants and other measures to ensure that the composition was based predominantly on hydrocarbons led to the collection of more data. This additional data was published in April 2009 in a Research Report Addendum entitled "Further Analysis of Hydrocarbons and Trace Materials To Support Dxxxx" (Moses, 2009). The information collected was used to modify the required testing for the SPK component in the fuel specification.

### 12.3.5 *Hydroprocessed esters and fatty acids*

The next target for fuel certification was a type of fuel that went through various names— hydrotreated renewable jet and bio-SPK—before finally being called HEFA. This fuel uses the same types of feedstocks as might be used by biodiesel (plant oils, animal fats, and waste greases), except that the feedstocks undergo conversion into a pure hydrocarbon fuel blending component. The HEFA fuel production process, which evolved from the Defense Advanced Research Projects Agency BioJet program that began in 2007, received significant support from Boeing-led flight tests in late 2008 to early 2009 (Kinder, 2010; Rahmes, *et al.*, 2009) and from subsequent large U.S. Air Force and Navy development efforts. The HEFA process was successfully balloted and added to ASTM D7566 (2014) as Annex A2 in July 2011. The chemical composition of HEFA fuels (predominantly n-paraffins and isoparaffins, similar to SPK) limited their use to up to 50% blends with conventional jet fuel. HEFA is described in Section A2.1.1 of ASTM D7566 as follows:

> *Synthetic blend components shall be comprised of hydroprocessed synthesized paraffinic kerosene wholly derived from hydrogenation and deoxygenation of fatty acids esters and free fatty acids. Subsequent processing of the product shall include...*

This is very similar to the structure of the SPK description in Annex A1. The requirements in Annex A2 for HEFA are very similar to those imposed on SPK, including the stringent limits on metals and other non-hydrocarbons. The Research Report for HEFA was much more extensive than that for SPK (Kinder, 2010) and included varying amounts of data from 10 manufacturers.

### 12.3.6    Synthesized isoparaffin

The annex for the SIP fuel was developed under the ASTM's Direct Sugar to Hydrocarbons task force. The name of the task force was derived from a key characteristic of the SIP fuel production process, specifically the fermentation of sugar directly into a hydrocarbon molecule. This is accomplished with genetically modified yeast that redirects the fermentation from producing ethanol to directly producing an unsaturated C15 hydrocarbon molecule called farnesene, which is then saturated into farnesane. SIP fuel is renewable because sugar is used as the feedstock. SIP fuel differs from the FT and HEFA blending components by having only one type of molecule, rather than a broad range of molecules covering the kerosene carbon number range. Consequently, SIP is limited to 10% in blends with petroleum-derived jet fuel, rather than the 50% blends permitted with the initial two pathways discussed above. However, a very conservative approach was taken during the ASTM D4054 (2009) test program to require that blends of 20% SIP undergo testing. Much of the rig and component test data utilized in the ASTM process was generated under the U.S. Federal Aviation Administration's Continuous Lower Emissions, Energy, and Noise (CLEEN) research and development program. Test results, along with an extensive amount of fuel property and composition data were compiled in an ASTM Research Report that was balloted in February 2014. This was followed by the balloting of the proposed specification annex in May and the final approval of Annex A3 on June 15, 2014. Annex A3 does include some different property requirements than Annex A1 and A2 because of SIP's unique single-molecule composition. For example, the distillation slope (T90–T10) is specified as a maximum value of 5°C, rather than as a minimum value as for FT and HEFA. The maximum value ensures a very narrow carbon number range (i.e., one molecule), whereas the minimum value would ensure a broad carbon number range as required by both HEFA and FT. Similarly, a much narrower density range is specified for SIP that reflects the single molecule composition.

### 12.4    U.S. FEDERAL AVIATION ADMINISTRATION CERTIFICATION

Historically, the commercial aviation industry has relied on a limited number of well-proven, conventional fuels for the certification and operation of aircraft and engines. The vast majority of today's engines and aircraft were designed and certified to operate on one of two basic fuels— kerosene-based fuel for turbine-powered aircraft and leaded aviation gasoline (avgas) for spark ignition reciprocating engine-powered aircraft. These fuels are produced and handled as bulk commodities, characterized by multiple producers sending fuel through the distribution system to airports and aircraft. These fuels are defined and controlled by industry consensus-based fuel specifications ASTM D1655 (2014) for jet fuel and ASTM D910 (2013) for aviation gasoline. These specifications, along with the oversight of the ASTM International Subcommittee on Aviation Fuels, accommodate the need to move the fuel as a commodity.

The FAA regulations pertaining to aircraft, engines, and aviation fuel were structured to complement the use of industry specifications similar to those issued by ASTM. They require that type certificate applicants identify the fuel specifications that are used in their products during the certification process. Once compliance with the airworthiness regulations has been demonstrated, the grade designation or specification becomes part of the airplane, rotorcraft, and engine operating limitations. These operating limitations are specified in the type certificate data sheet in the Airplane Flight Manual (AFM) or the Rotorcraft Flight Manual (RFM). Aircraft operators are required by 14 CFR §91.9 (U.S. Government, 2011) to only use fuels and oils listed in the AFM or RFM (see Fig. 12.3).

ASTM D7566 (2014) includes a provision to allow fuels that meet this specification to be re-identified as conventional "D1655" fuels when they enter the distribution infrastructure. The re-identification provision allows the ASTM D7566 drop-in fuels to be seamlessly integrated into the infrastructure and onto the aircraft without the need for separate tracking or regulatory approval. This is because the infrastructure is already designed to support ASTM D1655 jet

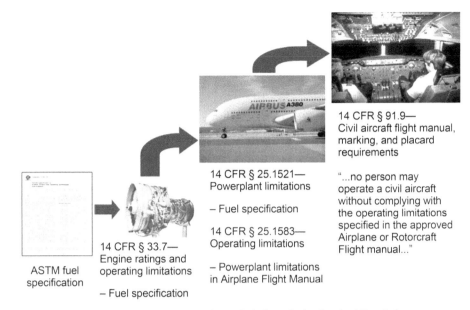

14 CFR § 91.9—
Civil aircraft flight manual, marking, and placard requirements

14 CFR § 25.1521—
Powerplant limitations

– Fuel specification

14 CFR § 25.1583—
Operating limitations

– Powerplant limitations in Airplane Flight Manual

"...no person may operate a civil aircraft without complying with the operating limitations specified in the approved Airplane or Rotorcraft Flight manual..."

14 CFR § 33.7—
Engine ratings and operating limitations

– Fuel specification

ASTM fuel specification

Figure 12.3.   Airworthiness regulations for aviation fuel. CFR, Code of Federal Regulations.

fuel, and virtually all civil aircraft are certified to operate with jet fuel that meets ASTM D1655. Consequently, once a new, alternative jet fuel is added as an annex to ASTM D7566, it is ready to fly on commercial airliners. The FAA issued an informational bulletin in September 2011 to explain this concept and to provide information on the use of hydroprocessed esters and fatty acids or Fischer-Tropsch fuels in aircraft and engines that have been certificated by the FAA (Federal Aviation Administration Aviation Safety, 2011).

## 12.5   FUTURE PATHWAYS

The aviation fuel community has accomplished a great deal with the approvals of hydroprocessed esters and fatty acids, synthesized isoparaffins, and Fischer-Tropsch fuels, but the production capacity still needs to be scaled up significantly to realize the environmental and economic benefits of these new fuels. Although industry is working hard to facilitate the commercialization of fuels that have already been approved, other prospective alternative fuel producers are working with the ASTM International Subcommittee on Aviation Fuels to approve new types of bio-derived fuels, as shown in Figure 12.4.

Several companies are working on pathways that convert sugars or other oxygenates to hydrocarbon fuel components. In addition to SIPs, researchers are evaluating pathways that rely on the conversion of sugar to intermediate products, such as alcohols, which are then converted to pure hydrocarbons via the use of catalysts or other processing methods. Eventually, lignocellulosic material such as plant stalks or wood chips will be used as the feedstock in place of sugar. Pyrolysis and technologies based on pyrolysis are also being developed to produce jet fuel blending components

In the future, we hope to include several more annexes into ASTM D7566 (2014) to help advance the burgeoning alternative jet fuel industry. We also hope that the addition of these new pathways and associated production volumes will have a more immediate impact on the jet biofuel supply.

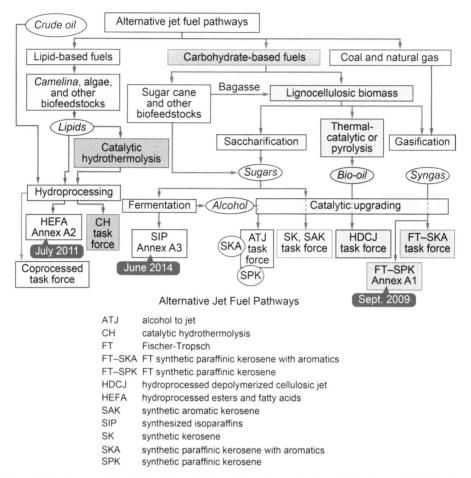

Alternative Jet Fuel Pathways

| | |
|---|---|
| ATJ | alcohol to jet |
| CH | catalytic hydrothermolysis |
| FT | Fischer-Tropsch |
| FT–SKA | FT synthetic paraffinic kerosene with aromatics |
| FT–SPK | FT synthetic paraffinic kerosene |
| HDCJ | hydroprocessed depolymerized cellulosic jet |
| HEFA | hydroprocessed esters and fatty acids |
| SAK | synthetic aromatic kerosene |
| SIP | synthesized isoparaffins |
| SK | synthetic kerosene |
| SKA | synthetic paraffinic kerosene with aromatics |
| SPK | synthetic paraffinic kerosene |

Figure 12.4.    Future aviation fuel pathways (*Adapted from* Brown, R. C., and T. R. Brown, 2012. Why Are We Producing Biofuels? Brownia LLC, Ames, IA. With permission).

REFERENCES

ASTM D56. 2005. Standard Test Method for Flash Point by Tag Closed Cup Tester. ASTM International, West Conshohocken, PA.

ASTM D86. 2012. Standard Test Method for Distillation of Petroleum Products at Atmospheric Pressure. ASTM International, West Conshohocken, PA.

ASTM D445. 2014. Standard Test Method for Kinematic Viscosity of Transparent and Opaque Liquids (and Calculation of Dynamic Viscosity). ASTM International, West Conshohocken, PA.

ASTM D910. 2013. Standard Specification for Aviation Gasolines. ASTM International, West Conshohocken, PA.

ASTM D1266. 2002. Standard Test Method for Flow Characteristics of Preformed Tape Sealants. ASTM International, West Conshohocken, PA.

ASTM D1298. 2012. Standard Test Method for Density, Relative Density, or API Gravity of Crude Petroleum and Liquid Petroleum Products by Hydrometer Method. ASTM International, West Conshohocken, PA.

ASTM D1319. 2013. Standard Test Method for Hydrocarbon Types in Liquid Petroleum Products by Fluorescent Indicator Adsorption. ASTM International, West Conshohocken, PA.

ASTM D1322. 2012. Standard Test Method for Smoke Point of Kerosine and Aviation Turbine Fuel. ASTM International, West Conshohocken, PA.

ASTM D1655. 2014. Standard Specification for Aviation Turbine Fuels. ASTM International, West Consho-hocken, PA.

ASTM D1840. 2007. Standard Test Method for Naphthalene Hydrocarbons in Aviation Turbine Fuels by Ultraviolet Spectrophotometry. ASTM International, West Conshohocken, PA.

ASTM D2386. 2012. Standard Test Method for Freezing Point of Aviation Fuels. ASTM International, West Conshohocken, PA.

ASTM D2622. 2010. Standard Test Method for Sulfur in Petroleum Products by Wavelength Dispersive X-Ray Fluorescence Spectrometry. ASTM International, West Conshohocken, PA.

ASTM D2887. 2013. Standard Test Method for Boiling Range Distribution of Petroleum Fractions by Gas Chromatography. ASTM International, West Conshohocken, PA.

ASTM D3241. 2014. Standard Test Method for Thermal Oxidation Stability of Aviation Turbine Fuels. ASTM International, West Conshohocken, PA.

ASTM D3338. 2009. Standard Test Method for Estimation of Net Heat of Combustion of Aviation Fuels. ASTM International, West Conshohocken, PA.

ASTM D3828. 2012. Standard Test Method for Flash Point by Small Scale Closed Cup Tester. ASTM International, West Conshohocken, PA.

ASTM D4052. 2011. Standard Test Method for Density, Relative Density, and API Gravity of Liquids by Digital Density Meter. ASTM International, West Conshohocken, PA.

ASTM D4054. 2009. Standard Practice for Qualification and Approval of New Aviation Turbine Fuels and Fuel Additives. ASTM International, West Conshohocken, PA.

ASTM D4294. 2010. Standard Test Method for Sulfur in Petroleum and Petroleum Products by Energy Dispersive X-Ray Fluorescence Spectrometry. ASTM International, West Conshohocken, PA.

ASTM D4529. 2001. Standard Test Method for Estimation of Net Heat of Combustion of Aviation Fuels. ASTM International, West Conshohocken, PA.

ASTM D4809. 2013. Standard Test Method for Heat of Combustion of Liquid Hydrocarbon Fuels by Bomb Calorimeter (Precision Method). ASTM International, West Conshohocken, PA.

ASTM D5453. 2012. Standard Test Method for Determination of Total Sulfur in Light Hydrocarbons, Spark Ignition Engine Fuel, Diesel Engine Fuel, and Engine Oil by Ultraviolet Fluorescence. ASTM International, West Conshohocken, PA.

ASTM D5972. 2005. Standard Test Method for Freezing Point of Aviation Fuels (Automatic Phase Transition Method). ASTM International, West Conshohocken, PA.

ASTM D7153. 2010. Standard Test Method for Freezing Point of Aviation Fuels (Automatic Laser Method). ASTM International, West Conshohocken, PA.

ASTM D7154. 2005. Standard Test Method for Freezing Point of Aviation Fuels (Automatic Fiber Optical Method). ASTM International, West Conshohocken, PA.

ASTM D7566. 2014. Standard Specification for Aviation Turbine Fuel Containing Synthesized Hydrocarbons. ASTM International, West Conshohocken, PA.

ASTM International. 2011. Evaluation of Bio-Derived Synthetic Paraffinic Kerosenes (Bio-SPKs), Research Report D02–1739, Committee D02 on Petroleum Products and Lubricants, Subcommittee D02.J0.06 on Emerging Turbine Fuels. ASTM International, West Conshohocken PA.

Brown, R. C., and T. R. Brown, 2012. Why Are We Producing Biofuels? Brownia LLC, Ames, IA.

Coordinating Research Council. 2014. Aviation Fuel Properties Handbook. CRC Report 663.

DEFSTAN 91–91. 2011. Turbine Fuel, Aviation Kerosine Type, Jet A–1, NATO Code: F–35, Joint Service Designation: AVTUR, Issue 7. *http://www.seta-analytics.com/documents/Defstan_91-91_R7.pdf* (accessed March 14, 2017).

Exxon Mobil Aviation. 2005. World Jet Fuel Specifications—With AvGas Supplement. *http://www.exxon-mobil.com/AviationGlobal/Files/WorldJetFuelSpecifications2005.pdf* (accessed March 14, 2017).

Federal Aviation Administration Aviation Safety. 2011. Semi-synthetic jet fuel. SAIB NE–11–56.

Gross, M. 2011. PQIS 2010 annual report. LMI Report No. DES10C1.

Hadaller, O. J., and J. M. Johnson, 2006. World fuel sampling program. CRC Report 647.

MIL–DTL–5624 (JP–5). 2013. Turbine Fuel, Aviation, Grades JP–4 and JP–5. Department of Defense.

MIL–DTL–83133 (JP–8). 2013. Turbine Fuel, Aviation, Kerosene Type, JP–8 (NATO) F–35, and JP8+100 (NATO F–37). Department of Defense.

MIL–HDBK–510. 2010. Aerospace Fuels Certification. Department of Defense.

MIL–HDBK–510A. 2014. Aerospace Fuels Certification. Department of Defense.

Moses, C. A. 2008. Comparative evaluation of semi-synthetic jet fuels—final report. CRC Project No. AV–2–04a.

Moses, C. A. 2009. Comparative evaluation of semi-synthetic jet fuels—addendum: further analysis of hydrocarbons and trace materials to support Dxxxx. CRC Project No. AV–2–04a.

Moses, C. A., L. L. Stavinoha, and P. Roets. 1997. Qualification of Sasol semi-synthetic Jet A–1 as commercial jet fuel. SwRI Report No. 8531.

Rahmes, T. F., J. D. Kinder, and T. M. Henry, *et al.* 2009. Sustainable bio-derived synthetic paraffinic kerosene (bio-SPK) jet fuel flights and engine tests program results. AIAA–2009–7002.

U.S. Government. 2011. Code of Federal Regulations, Title 14, Aeronautics and Space, General Operating and Flight Rules, Civil aircraft flight manual, marking, and placard requirements. 14 CFR § 91.9.

# CHAPTER 13

# Environmental performance of alternative jet fuels

Hakan Olcay, Robert Malina, Kristin Lewis, Jennifer Papazian, Kirsten van Fossen, Warren Gillette, Mark Staples, Steven R.H. Barrett, Russell W. Stratton and James I. Hileman

## 13.1   INTRODUCTION

As described in the preceding chapters, a variety of feedstocks, feedstock types, fuel production processes, and supply chain structures are being considered as the advanced alternative jet fuel industry is developing. How feedstocks and fuels are produced and how supply chains are structured affect the ability of the resulting fuels to not only meet the technical requirements of the fuel, but also address economic drivers (e.g., price) and environmental performance. This chapter focuses on understanding how feedstock selection and processing decisions and the entire life cycle of the fuel production process affect the environmental performance of the resulting alternative jet fuels.

A life cycle approach considers the complete supply chain, including feedstock extraction or growth, transportation of raw materials to production facilities, refining to finished products, blending and distribution of finished products, and finally end use. Each step in this process has the potential to contribute to the environmental impacts associated with the fuel. These impacts are evaluated separately for different resource categories, or "indicators." This chapter focuses on three key environmental indicators:[1] life cycle greenhouse gas (GHG) emissions in Section 13.2, water use and quality in Section 13.3, and biodiversity in Section 13.4.

Alternative jet fuels may lead to improvements in environmental performance for these indicators if production and processing choices are made carefully. However, there is also the risk of environmental damage from scaling up a large new industry. Other aspects of environmental performance may be worse with alternative fuels than with conventional fuels, and there may be tradeoffs such that improved performance in one area is accompanied by worsened performance in another area. Such tradeoffs will require societal decisions prioritizing different environmental considerations. Alternative jet fuel producers and users should choose to produce and use fuels that minimize environmental harm and maximize benefit. The measurement and analysis of impacts for each of these indicators is discussed further in this chapter.

An "impact matrix," such as that shown in Table 13.1, is useful for showing the potential environmental risks associated with different parts of the economic supply chain for alternative fuels. The impact matrix is the result of work done by the Commercial Aviation Alternative Fuels Initiative (CAAFI®), Life Cycle Associates, and FuturePast. The impact matrix shown here is a subset of the full matrix, found in the CAAFI Environmental Sustainability Overview (CAAFI, 2016a). The impact matrix qualitatively covers the indicators that are considered in this chapter. As shown in the impact matrix, feedstock production may be associated with a number of potential environmental impacts. The overarching reasons for feedstock production-related impacts

---

[1] These are a subset of the many sustainability indicators that could be considered; organizations such as the Roundtable on Sustainable Biomaterials (Roundtable on Sustainable Biofuels, 2010) and the Global Bioenergy Partnership (Global Bioenergy Partnership, 2011) have defined many additional principles, indicators, and criteria that should be addressed to cover full sustainability considerations, including economic, social, and environmental concerns.

Table 13.1.   Impact matrix depicting the qualitative potential for direct impacts on selected indicators along the alternative fuel production supply chain.

| | Supply chain | | | | |
|---|---|---|---|---|---|
| | Feedstock producer | Fuel producer | Fuel blender/ distributor | Fuel end user | LoSU[a] |
| Impact indicator | Impact severity or LoSU[a] | | | | |
| Life cycle GHG[b] emissions | | | | | |
|    Direct $CO_2$ emissions | High | High | Low | High | High |
|    Indirect $CO_2$ emissions | High | Low | Low | Low | Low |
|    Non-$CO_2$ emissions | High | Medium | Low | High | Low |
| Biodiversity | High | Medium | Low | Low | Medium |
| Water quality (pollutants, eutrophication) | High | Medium | Low | Low | Medium |
| Freshwater use (withdrawal, consumption) | High | High | Low | Low | High |

[a]LoSU, level of scientific understanding.
[b]GHG, greenhouse gas.
*Source:* Adapted from (subset of) Commercial Aviation Alternative Fuels Initiative. 2014. Alternative jet fuel environmental sustainability overview. Version 4.1. With permission.

due to biofuels include land use/habitat conversion, the potential need for irrigation, the potential to introduce invasive species, impacts on water quality due to fertilizer, pesticides, or soil runoff, and the emissions of GHGs associated with each of these activities. For non-bio-based alternative fuels (e.g., coal or natural gas-based fuels made through processes such as Fischer-Tropsch (FT) gasification and synthesis), potential environmental issues arise from land-use change (LUC, e.g., mountaintop removal for coal extraction), water and energy use, potential ground water contamination, pollutant and sediment emissions, and the release of GHG. For conversion facilities or refineries, the activities that may contribute most to potential environmental impacts include energy and water usage for conversion, chemical emissions and accidental releases, and some potential for LUC and impacts on biodiversity. Depending on how the feedstock is handled in the production stages, the fuel producer could have an impact on biodiversity due to the potential for inadvertent dispersal of nonnative species propagules that might facilitate the spread of potentially invasive species. The use of the fuel results in GHG emissions. The impact matrix indicates that explicit planning and management activities should be implemented by feedstock producers and fuel producers to limit the potential direct environmental impacts related to these parts of the supply chain. CAAFI has also developed an "Environmental Progression" (CAAFI, 2016b) that assists a feedstock or fuel producer in identifying when in development particular environmental analyses should be performed. Most should be repeated at every stage of feedstock or fuel process development and scale up.

The matrix also indicates the current level of scientific understanding of the indicator, which in this instance refers to the ability to measure the indicator but does not address the understanding of how to determine actual impacts in comparison to particular thresholds, as those are societal decisions rather than scientific ones. For example, we know to measure the amount of water that is withdrawn and consumed by a given activity, so the level of scientific understanding entry on the impact matrix is "high," but how we determine whether that amount of withdrawal or consumption is an acceptable level of impact depends on conservation decisions regarding the scale of impact considered (e.g., stream-level versus watershed level) and societal decisions regarding acceptability of the effects on that resource at the selected scale, particularly in the context of tradeoffs with other environmental priorities (e.g., water use versus GHG reduction).

Depending on the environmental indicator being considered and the question that is being answered, the environmental impacts of each part of the supply chain can be considered separately or the impact can be considered on a life cycle basis, wherein the impacts for a given indicator are summed across the different components of the supply chain. Because the finished fuel could have varied energy content per unit volume and mass, life-cycle-based evaluations are best expressed on an energy basis to allow for comparisons among fuels. A life cycle approach is useful for examining environmental impacts that have similar consequences regardless of the geographical location where the activity occurs. Long-lived GHG emissions meet this criterion and, as such, they are often considered on a life cycle basis. In addition, life cycle evaluations of other indicators, such as water use, provide a measure of efficiency for a given fuel production pathway (i.e., how much fuel is produced per unit of water withdrawn or consumed). However, the environmental impacts associated with air quality, water use, and water quality depend on the geographical region wherein the activity occurs. For example, if the production of two different fuels requires the same number of gallons of water to be extracted per unit of fuel, the actual impacts of that water use will differ depending on the corresponding environment from which the water is extracted (e.g., groundwater in a relatively arid area versus surface water in a mesic area), as well as the competing uses for that water.

One of the challenges for evaluating the sustainability of alternative fuels is that there are many sustainability frameworks, each with its own list of indicators, its own desired or provided methodology or approach, and its own specific focal level for the analysis. This is largely due to the varying needs of stakeholders, including standards developers, bioenergy industry participants, governmental policy makers, regulators, local communities, bioenergy industry workers, fuel users, and the general public (Futurepast, 2012). For example, the Global Bioenergy Partnership is an international coalition brought together under the auspices of the Food and Agriculture Organization of the United Nations to identify key indicators of bioenergy sustainability at the national level (Global Bioenergy Partnership, 2011). The focus is the evaluation of sustainability of an entire bioenergy industry in a country. Others, like the Roundtable on Sustainable Biomaterials (formerly Roundtable on Sustainable Biofuels) focus on a specific set of criteria, categorized into 12 "principles" that are used to quantify biofuel performance and measurement, but the unit of evaluation is at the economic operator level, covering "Feedstock Producers, Feedstock Processors, Biofuel Producers and Biofuel Blenders" (Roundtable on Sustainable Biofuels, 2010). There are also product labeling approaches that focus on the final product as the unit of evaluation, such as Environmental Product Declarations based on Product Category Rules, which are defined by the International Organization for Standardization (ISO) and require third-party verification of environmental information that can be compared across products (ISO, 2006). The Publicly Available Specification 2050 (PAS 2050:2011; see British Standards Institution, 2011) is another example of a sustainability methodology for a given product, but this is specifically focused on GHG accounting and does not cover other environmental indicators. Therefore, it is critical to select an appropriate sustainability evaluation tool based on the objectives of the analysis.

## 13.2   EVALUATING GREENHOUSE GAS EMISSIONS AND IMPACTS OF ALTERNATIVE FUELS ON GLOBAL CLIMATE CHANGE

### 13.2.1   *Greenhouse gas life cycle analysis background*

Carbon dioxide ($CO_2$), methane ($CH_4$), nitrous oxide ($N_2O$), and other gases that are released by human activities, such as combustion of fossil fuels and other agricultural and industrial processes, are tied to changes in global climate. The use of alternative fuels may offer an opportunity to reduce these emissions in comparison with petroleum-based fuels when considered on a life cycle basis. Life cycle analysis (LCA) for greenhouse gases takes into account the emissions associated with producing the final fuel from initial feedstock production (e.g., an oil well, planting of seeds, etc.) through to combustion emissions. Currently the aviation sector is focusing on "drop-in" alternative fuels that consist of hydrocarbons and are compatible with existing equipment and infrastructure. Biomass-derived alternative jet fuels have the potential to reduce life cycle GHG emissions in comparison with conventional jet fuel, since biomass feedstocks absorb $CO_2$ from the atmosphere when they grow and the $CO_2$ emitted during fuel combustion is equal to that absorbed during biomass cultivation. The $CO_2$ biomass "credit" offsets the combustion $CO_2$ in the LCA. This biomass credit is the primary difference between biomass and fossil fuels in terms of their GHG emissions. However, a biofuel does not necessarily have lower GHG emissions than a fossil fuel, since there can be GHG emissions associated with cultivating and harvesting the feedstock, with fuel production, with feedstock and fuel transportation, as well as with land-use change attributable to the cultivation of biomass.

Non-biomass-based feedstocks, such as municipal solid waste or other wastes, may also offer GHG emissions benefits because of the re-use of the carbon, although whether this is accounted to offset combustion emissions may depend on which party (producer of the waste or converter of the waste) takes credit for the re-use (Seber *et al.*, 2014).

A variety of guidances, methodologies, and tools exist to perform GHG LCAs. While all of these tools attempt to address the same challenge of accounting for life cycle GHG emissions, and there is a high level of scientific understanding of what elements need to be addressed to calculate direct $CO_2$ emissions, there are still areas of uncertainty for indirect $CO_2$ and non-$CO_2$ emissions, and the analyst must make decisions about how to proceed. Thus, there is no single standard methodology for performing a full life cycle GHG analysis. Some of the key points at which processes, guidances, and tools diverge are (Allen *et al.*, 2009):

1. *System boundaries.* A GHG LCA is supposed to capture all relevant steps in the life cycle. Consequently, an LCA should investigate not only the direct but also all the indirect material and energy flows and activities associated with the primary production chain that lead to GHG emissions. In practice, because resources are limited in terms of funds and time, cutoff criteria have to be defined to constrain the size of the system in the LCA, while still capturing the impacts as well as the scope. Even though these cutoff criteria are easy to set from a normative perspective, in reality they will always be subjective. Consequently, it is of paramount importance to be transparent about where the system boundaries are drawn so that the analysis can be compared, criticized, and modified (if necessary) by other researchers.

2. *Coproduct allocation.* A second key issue with LCA is the allocation of GHG emissions among various coproducts—such as animal feeds, electricity, other fuels, and other commodity chemicals—that are produced at different stages of the life cycle. Using different allocation rules can significantly impact the GHG results, particularly for renewable jet fuel production, during which larger quantities of different coproducts are often being produced. Choosing a method entails subjective judgment, and the appropriate method may depend to a large extent on the question that the particular analysis is aiming to answer. The choice of coproduct allocation could also be used to manipulate the life cycle GHG emissions for the fuel being examined; thus, it is critical that the allocation rules being employed are clearly stated. In general, emissions can be allocated among the coproducts based on their most representative physical characteristics, such as their mass or energy content, or on their economic value, as represented by their market price. Some coproducts produced during the fuel life cycle can be used to substitute for other products. Hence, another approach to deal with the coproducts is to expand the system boundaries to take into account the consequences of fuel production on the greater economic system and thus account for these changes with substitution credits; this is known as the displacement method. However, this method can be arbitrary and can result in very low or very high emission values based on the choice of market products to be substituted. It has also been shown that if the product of interest is not the primary product, displacing the primary product can lead to erroneous results. The reader is referred to Stratton *et al.* (2010) for a more detailed discussion on the properties and consequences of using various coproduct allocation rules.

3. *Data quality and uncertainty.* The data used to carry out the LCA need to be of sufficient quality to accurately represent all the processing steps and phases of the operation under

consideration. The data quality and uncertainty highly depend on the time frame and scale. Most of the technologies under consideration for producing advanced alternative fuels are in the process of being commercialized or have not yet been commercialized. As such, the data quality is poorer than one would expect for many of the pathways considered than it would be for well-established processes that have higher quality data from a variety of operating facilities.

4. *Approach for accounting for land-use change.* One of the most discussed issues concerning fuel production from renewable sources are climate impacts due to changes in the use of land which are brought about by the cultivation of biomass feedstocks. Biomass cultivation can change the biomass, soil, and organic waste contained on and within the land, which, in turn, changes the $CO_2$ flux because of additional $CO_2$ emissions or sequestration. This impact might be referred to as the "biogeochemical" impact of biomass growth. In addition to this effect, growing biomass might also have a "biogeophysical" effect on climate by changing surface characteristics such as reflectivity (albedo) and the green-up and die-down of vegetation (phenology), leading to impacts on transport and heat into the atmosphere and changes to the radiative balance of the Earth (Caiazzo *et al.*, 2014; Georgescu *et al.*, 2011; Hallgren *et al.*, 2012). Changes in land use can be divided into direct and indirect change (Gawel and Ludwig, 2011; Plevin *et al.*, 2010; Wang *et al.*, 2011). Direct LUC occurs when land is converted to grow biomass for fuel production. Indirect LUC occurs when this displacement induces changes in market prices of crop(s) or land value, leading to conversion of land somewhere else to satisfy the demand for the displaced crop.

LCA is not only a valuable tool for scientists interested in the mitigation potential of alternative fuels in terms of GHG emissions but also as a *de facto* mandatory tool for fuel producers and regulatory agencies that need to assess the compliance of fuel options with Federal regulations, such as the Energy Independence and Security Act (EISA), the Renewable Fuels Standard, or the California Low Carbon Fuel Standard (LCFS), which incorporate GHG LCA into their requirements. Although these three regulations are similar in that they require LCA for GHG estimation, they differ in key ways in terms of the details required for the analysis.

Although it is not the first to require LCA, Section 526 of the EISA of 2007 provides a relatively straightforward use of LCA, with language stating that "no Federal agency shall enter into a contract for procurement of an alternative or synthetic fuel, including a fuel produced from nonconventional petroleum sources, for any mobility-related use, other than for research or testing, unless the contract specifies that the life cycle GHG emissions associated with the production and combustion of the fuel supplied under the contract must, on an ongoing basis, be less than or equal to such emissions from the equivalent conventional fuel produced from conventional petroleum sources." This requirement is relatively straightforward in that it stipulates that the life cycle GHG emissions must simply be below a baseline value for petroleum-based fuels. To meet this requirement, one needs to show that the life cycle GHG emissions from their fuel pathway are below the baseline value, with the margin relative to the baseline being irrelevant. This requirement is important to the jet fuel industry because it can limit purchases by Defense Logistics Agency Energy for the U.S. Air Force and U.S. Navy, which are all parts of the U.S. Government.

The first renewable fuel mandate in the United States was implemented by the first Renewable Fuels Standard (RFS) program, created under the Energy Policy Act of 2005. This use of LCA is more complex in that it requires a minimum margin relative to the baseline. The program originally set the minimum amount of renewable fuel that would be required to blend into gasoline by 2012. This program was then expanded, under the EISA of 2007, to include diesel, and it revised the blend volumes while establishing new categories for renewable fuels and setting GHG reduction thresholds for each category. According to this new program, termed the "RFS2 program," the renewable fuel categories and the minimum GHG reductions for each category from the 2005 baselines of gasoline or diesel, which the renewable fuel replaces, are as follows:

1. Renewable fuel (e.g., corn ethanol or corn butanol), 20% GHG emissions reduction
2. Advanced biofuel (e.g., sugarcane ethanol), 50%

3. Biomass-based diesel (e.g., biodiesel from soy, or renewable diesel from waste oils, fats, greases, or algal oil), 50%
4. Cellulosic biofuels (e.g., cellulosic ethanol or diesel), 60%

In relation to jet fuel, it is important to note that although RFS2 does not set a specific goal for renewable jet fuel, the renewable jet fuel could still qualify under the biomass-based diesel, undifferentiated advanced biofuel category or, depending on the pathway involved, under cellulosic biofuel (U.S. Environmental Protection Agency (EPA), 2010). This program is further differentiated from the requirements under Section 526 of EISA in that it is the U.S. EPA that calculates the life cycle GHG emission values, as opposed to the fuel producer.

In 2006, the state of California adopted the LCFS program with the goal of reducing by 10% the state's transportation carbon intensity, as measured on a life cycle basis, by 2020. The California LCFS adds a further level of fidelity to that of the RFS2 program because it is based on a carbon intensity instead of a threshold requirement; thus, whereas RFS2 and Section 526 are essentially binary, with the fuel being either below or above a predetermined GHG threshold requirement, the LCFS is a continuous requirement. The LCFS does not apply to jet fuel, but it does point to another way in which life cycle GHG emissions are used by regulators to reduce GHG emissions from transportation.

Thus, LCA is being used to determine which alternative fuel pathways can be sold to the U.S. Air Force, the U.S. Navy, and other departments of the U.S. Government, and which can qualify as renewable fuel under the RFS2 program, as well as to track progress toward meeting California's carbon intensity target.

Section 13.2.2 assesses the climate impacts of alternative drop-in aviation fuels through the LCA of associated GHG emissions. The alternative jet fuel pathways considered herein include jet fuel production from petroleum (including oil sands as a variation in petroleum source); Fischer-Tropsch technology; hydroprocessed esters and fatty acids (HEFA) from vegetable, waste, and algal oils as well as from animal fats; and biochemical conversion of simple sugars through fermentation and advanced fermentation technologies. These pathways have been evaluated and compared with conventional jet fuel production as the reference case.

### 13.2.2  Overview of greenhouse gas life cycle analysis results for drop-in jet fuels

In this section, we summarize existing work on alternative jet fuel GHG LCA that has been performed under the Partnership for Air Transport Noise and Emissions Reduction (PARTNER), a U.S. Federal Aviation Administration (FAA) Center of Excellence, over the course of 5 years of research. It rests upon a report published in 2010 (Stratton et al., 2010) and subsequent papers and theses (Carter, 2012; Carter et al., 2011; Pearlson et al., 2013; Seber et al., 2014; Staples et al., 2014). The research was conducted using a current benchmark for LCA of transportation fuels called Greenhouse Gases, Regulated Emissions, and Energy Use in Transportation (GREET). The GREET model was originally developed by Argonne National Laboratory for ground transportation. It was then updated and expanded, based primarily on Stratton et al. (2010), to include a jet fuel module. When a particular fuel technology was not available in GREET, the analysis involved modifying the GREET framework and its database to incorporate the new jet-fuel-producing pathway into its structure. The necessary process inputs and characteristics for the pathway were obtained from the open literature. The simulations were carried out for the year 2015 using GREET v1.8a and v1.8b, except for Camelina sativa- and algae-derived HEFA fuel and the biochemical conversion of simple sugars. Camelina and algae analyses were carried out using GREET 2011 for the years 2017 and 2012, respectively. Results for biochemical conversion of simple sugars were conducted using GREET NET v1.0.0.8377 for the year 2012. All of the values given here include $CO_2$, $N_2O$, and $CH_4$, and are in terms of GHG emissions per unit of energy of jet fuel (based on the lower heating value)—more specifically, in grams of $CO_2$-equivalent[2] per megajoules of jet fuel (g $CO_2$e/MJ).

---

[2] Global warming potentials (relative to $CO_2$) employed in the analyses: $CH_4$, 25; $N_2O$, 298.

The GHG LCAs of the feedstock-to-fuel pathways under consideration were carried out at a screening level, using a consistent methodology and capturing all steps of the life cycle, thus allowing for comparison among multiple processes, feedstocks, and LUC scenarios. The analysis started from the field, well, or mine where the feedstock was grown and/or extracted and was extended to the wake behind the aircraft ("well to wake," WTW) while employing consistent coproduct allocation techniques. The screening-level analyses provided high-level assessments of various alternatives in order to inform decisionmakers on the overall viability of these alternatives, outlining crucial factors that define the GHG emissions profile for each feedstock and life cycle step. Because of space restrictions, the information provided here has been limited to brief process descriptions, the primary factors that define the variability among the scenarios created, and primary factors driving the life cycle GHG emissions. The reader is encouraged to refer to the PARTNER Project 28 report (Stratton *et al.*, 2010) and the other references cited in each subsection for additional information about the analyses. A comparison of all the pathways in terms of WTW GHG emissions results is given in Figure 13.1.

With regard to data quality and uncertainty, the ranges associated with these pathways have been established using optimistic, nominal, and pessimistic scenarios and have been embedded into the screening-level analysis, where the variability in key parameters has been explored. Hence, the GHG emissions are presented as low (optimistic assumptions), baseline (nominal assumptions), and high (pessimistic assumptions) for each feedstock-to-fuel pathway.

In the sections that follow, we first present the LCA for the conventional crude-oil-derived jet fuel production that forms the reference against which the alternative jet fuel technologies are compared, as well as the GHG effects of using oil sands as an unconventional resource. Alternative jet fuel production using the FT technology and the production of HEFA jet fuel are presented next, since these were the first two approved pathways for production of certified alternative jet fuel. Finally, we summarize the results for biochemical technologies that are currently being developed to convert biomass-derived sugar feedstocks into jet fuel. The GHG LCA summaries do not include any LUC in the evaluation, but Section 13.2.3 explores how direct LUC scenarios would affect the GHG LCA for each fuel.

### 13.2.2.1   *Conventional jet fuel from crude oil*

The LCA of the petroleum-derived jet fuel production (Stratton *et al.*, 2010) serves as the reference case for the rest of the work presented here. The primary parameters in the variation of the life cycle values associated with this pathway are the crude oil source and the fuel refining efficiency. Two methods were used in the calculation of this efficiency. The baseline-emissions scenario employed a top-down approach that considered the overall U.S. refining energy efficiency. The two extreme cases, i.e., low- and high-emissions scenarios used a bottom-up approach that summed the energy requirements for the individual refining processes. To capture the importance of a crude oil source, Stratton *et al.* (2010) examined the GHG emissions from the extraction of different petroleum sources. The low-emissions case was based on domestic U.S. petroleum production, which has low transportation emissions, whereas the high-emissions case used Nigerian crude, which in 2005 (the year upon which the data were based) had considerable methane venting with petroleum production and higher emissions from transportation to the United States. To capture the variation in refining operations, the low-emissions case assumed that the jet fuel being produced was "straight-run" and only required crude desalting and atmospheric distillation. On the other extreme, the high-emissions case assumed that the jet fuel was produced using a combination of vacuum distillation, hydrotreating, and/or hydrocracking. The low- and high-emissions cases therefore capture the potential breadth of jet fuel lifecycle GHG emissions. The nominal value was based on the mix of petroleum arriving at U.S. refineries in 2005 and refinery data that are representative of domestic U.S. operations.

The WTW life cycle GHG emissions for the baseline case were calculated to be 87.5 grams of $CO_2$-equivalent per megajoule (g $CO_2$e/MJ). The low- and high-emissions cases yield values of 80.7 and 109.3 g $CO_2$e/MJ, respectively. The scenarios employed the energy allocation method.

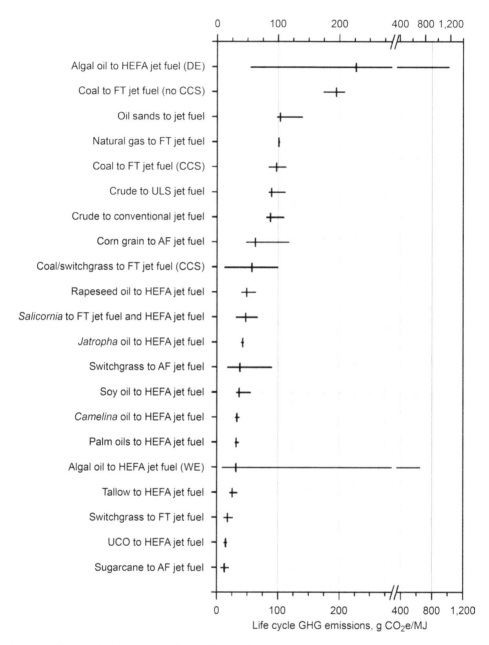

Figure 13.1.   Results in terms of well-to-wake greenhouse gas (GHG) emissions under assumption of no land-use change (LUC) attributable to the fuels under consideration. Non-$CO_2$ combustion emissions are not included. AF, advanced fermentation; CCS, carbon capture and sequestration; DE, dry extraction; FT, Fischer-Tropsch; HEFA, hydroprocessed esters and fatty acids; UCO, used cooking oil; WE, wet extraction.

It should be noted, however, that these results assumed the same average crude oil properties. Hence, they did not include any variation in processing emissions from oil quality yet employed the different calculated refining efficiencies as the only factor for variation. Parameters other than the process efficiency that increased the emissions in the high-emissions case relative to the baseline were the methane venting during oil extraction and the transportation emissions associated with the Nigerian crude oil. Although Stratton *et al.* (2010) used the life cycle GHG emission from aviation fuel produced from conventionally extracted petroleum as their reference case, unconventional methods are increasingly used to extract petroleum from deposits previously viewed as economically nonviable. Petroleum from oil sands emits greater life cycle GHG emissions because unconventional petroleum extraction involves more energy-intensive processes. For jet fuel made from oil sands, the life cycle GHG emissions for low, baseline, and high scenarios were calculated to be 97.9, 103.4, and 139.0 g $CO_2$e/MJ, respectively. Note that the work of Stratton *et al.* (2010) did not consider the life cycle GHG emissions from the use of petroleum derived via hydraulic fracturing.

### 13.2.2.2    *Fischer-Tropsch jet fuel*

The analyses of FT jet fuel production from coal (coal to liquid, CTL) and natural gas (gas to liquid, GTL) were conducted using GREET v1.8a. GREET v1.8b was used to analyze FT jet fuel derived from biomass (biomass to liquid, BTL) and from a combination of coal and biomass (coal and biomass-to-liquid, CBTL). In the following LCAs (Stratton *et al.*, 2010), it was assumed that the energy requirements to produce an FT diesel fuel were directly applicable to FT jet fuel production. It was further assumed that 25% of the product slate would be jet fuel and that the plant would be able to produce enough electricity to support its needs but not enough to be exported. Emissions among coproducts were allocated on the basis of energy.

The analysis for GTL FT considered a scenario in which a non-North American natural gas was supplied to a stand-alone FT plant located internationally that was designed to maximize liquid fuels production by recycling the tail gases. Because of process efficiency variability (which can vary from 60 to 65%), the low-, baseline-, and high-emissions scenarios used efficiencies of 65%, 63%, and 60%, respectively. On the basis of these assumptions, the overall GHG emissions came out to be 100.1, 101.0, and 102.4 g $CO_2$e/MJ. When a carbon capture and sequestration (CCS) process with an efficiency of 85% was assumed for the baseline, calculated emissions were reduced to 86.2 g $CO_2$e/MJ.

For the CTL FT technology LCA, process efficiency and coal type determined the variability in the overall GHG emissions. Only bituminous and subbituminous coal types were considered in this analysis. The former is recovered via both underground and surface mining techniques, whereas the latter is mainly via surface mining. Underground mining results in higher methane emissions. The low-emissions case assumed a 53% process efficiency with surface-mined bituminous coal, the baseline-emissions case assumed a 50% process efficiency with the 2007 average U.S. coal mix of both bituminous and subbituminous types, and the high-emissions case assumed a 47% process efficiency with underground-mined bituminous coal. The relative GHG emissions were calculated as 174.0, 194.9, and 208.0 g $CO_2$e/MJ, for the low-, baseline-, and high-emissions scenarios, respectively. When CCS was applied (with efficiencies of 90%, 85%, and 80%, respectively), these values were lowered to 84.9, 97.2, and 112.6 g $CO_2$e/MJ. This shows that the use of CCS in the CTL processes can allow for approximately 50% reduction in the GHG emissions.

The analysis for the BRL FT process assumed the use of a nonfood energy crop, such as switchgrass (*Panicum virgatum*), for the biomass feedstock. Switchgrass is a perennial warm-season grass native to North America. The assumptions regarding the yield, energy, and emissions associated with switchgrass cultivation were based on a survey of existing cultivation data from the literature. The low-emissions case scenario utilized an optimistic yield of 13.0 t/ha based on large field plots in southern Wisconsin, the baseline-emissions case used a yield of 10.3 t/ha based on a projected national average, and the high-emissions case used a yield of 7.2 t/ha based on average mid-continental U.S. farm-scale observations. Assumptions surrounding process fuels (i.e., gasoline and diesel) and electricity used in farming, fertilizer inputs (i.e., nitrogen fertilizer, $P_2O_5$, $K_2O$, and limestone), herbicide use, and FT process efficiency impact the results of GHG emissions LCAs. Compared with CTL, there is roughly a 5% efficiency drop for BTL plants because of the additional processing energy required for biomass grinding and drying; for this reason, the efficiencies for low-, baseline-, and high-emission scenarios were assumed to be

52%, 45%, and 42%, respectively. The life cycle GHG emissions for the three scenarios were calculated as 11.9, 17.7, and 26.0 g $CO_2e/MJ$, in which the $N_2O$ emissions arising from the use of nitrogen fertilizers were a primary contributor.

Since biomass and coal can be processed into a similar fuel using FT technology, they could be processed at a single FT plant. This way, biomass could mitigate the high GHG emissions from coal, and the coal could mitigate the low energy density and production limitations of biomass. With CBTL technology, the coal and biomass can be gasified separately and the syngas streams can then be mixed, or they can be gasified in the same unit (cogasification). Since the latter is simply a superposition of the previous two sections, the analysis here examines the case of cogasification.

Process efficiency is affected by biomass pretreatment, including torrefaction and milling, but overall plant efficiency depends highly on the relative weights of biomass and coal to be cogasified. When this ratio exceeds 50%, dimethyl ether production becomes more favorable than the desired hydrocarbon fuels. Hence, this ratio is kept below 50% for all the emission cases. Other parameters that define the overall CBTL efficiency, and hence the variability in the emission scenarios, include the biomass grinding energy and CTL and CCS efficiencies.

The same input and CCS assumptions outlined for CTL were employed in this analysis, but switchgrass was used as the biomass feedstock. The biomass-to-coal ratio was defined as 40%, 25%, and 10%, and the torrefaction efficiency as 97%, 90%, and 85% for the low-, baseline-, and high-emissions scenarios, respectively. On the basis of these assumptions, the life cycle GHG emissions for the three scenarios were calculated to be 12.4, 56.9, and 99.8 g $CO_2e/MJ$.

### 13.2.2.3    Hydroprocessed esters and fatty acids jet fuel

As with FT jet fuel, the life cycle GHG emissions of HEFA jet fuels produced from oils and fats vary depending on the production scenario assumptions. These differ with respect to feedstock cultivation practices and processing efficiency. The PARTNER Project 28 Report (Stratton et al., 2010) details various scenarios that were considered in calculating the life cycle GHG emissions of HEFA jet fuels from the oils of soybean (*Glycine max*), oil palm (*Elaeis guineensis*), rapeseed (*Brassica napus*), *Jatropha curcas*, *Salicornia bigelovii*, and algal cultures for lipid production. In addition to these feedstocks, analyses were conducted on *Camelina sativa* (Carter et al., 2011), waste oils, and animal fats (i.e., used cooking oil and beef tallow) (Seber et al., 2014), as well as more detailed analyses on microalgae (Carter, 2012; Vasudevan et al., 2012).

In large part, the work presented herein builds upon that of Stratton et al. (2010) by incorporating a more accurate model of the hydrotreatment process for the production of HEFA jet fuel that is based on the Pearlson et al. (2013) approximation of the UOP hydrodeoxygenation process using the assumption of a product slate that maximizes the production of middle distillate (jet and diesel) fuels. The emissions scenarios for HEFA fuels are driven by several factors, such as variability in feedstock yield, farming energy, nitrogen fertilizer, processing efficiency, and, in the case of waste fats and oils, variability in rendering practices.

Figure 13.2 provides a breakdown of the life cycle GHG emissions of HEFA fuels from the various feedstocks, and Table 13.2 summarizes the range of results.

For vegetable and algal oils, after the oil extraction step a solid coproduct (meal) remains that has high protein content and offers value as an animal feed. In the case of tallow, HEFA tallow and meat and bone meal (MBM) are produced at the rendering step. MBM is high in protein and can be used as feed for monogastric animals such as poultry and swine. If the coproduct of an oil or fat is suitable for animal feed, emissions between the oil/fat and the coproduct were allocated using market value allocation. If the coproduct is used for other purposes, the energy or displacement method was used. Table 13.2 summarizes the allocation choices made for the oil/fat separation step.

Calculations of GHG LCA for HEFA processing of other feedstocks took into account a range of process and feedstock-specific assumptions, which are detailed in Stratton et al. (2011), Carter et al. (2011), Carter (2012), Pearlson et al. (2013), and Seber et al. (2014). The GHG LCA results are summarized in Table 13.2.

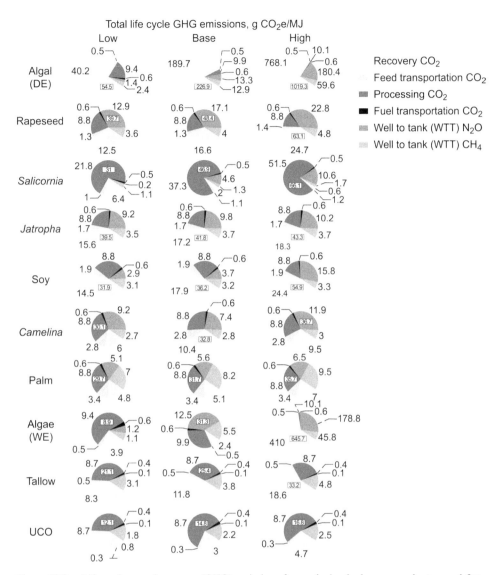

Figure 13.2.   Life cycle greenhouse gas (GHG) emissions for producing hydroprocessed esters and fatty acids fuels from different feedstocks broken down by process step for three emissions scenarios. Results do not include land-use change. DE, dry extraction; UCO, used cooking oil; WE, wet extraction. $CO_2$ emissions associated with $CO_2$ used for algal growth are not included. "Recovery $CO_2$" includes emissions from cultivation and extraction steps as well as the net biomass credit offset by combustion emissions. Because a higher biomass credit is associated with the algal WE low-emissions case, this step has a negative contribution.

### 13.2.2.4   *Renewable jet fuel from sugars*

Various technologies are currently under development to produce sugar-derived renewable jet fuel. Obtaining the simple sugars from sugary, starchy, and lignocellulosic feedstocks requires a pretreatment process followed by a hydrolysis step, which can be carried out either thermochemically or biochemically. The pretreatment step usually involves milling and, for lignocellulosic feedstocks, a thermochemical treatment process such as hot water or dilute acid extraction.

Once extracted from the feedstock, simple sugars can be used to produce intermediate products, known as platform molecules, which are then converted into jet fuel (Bond *et al.*, 2014;

Table 13.2.    Range of lifecycle GHG emissions for HEFA jet fuel by feedstock and employed coproduct method for allocating emissions before the fuel-production step (Carter, 2012; Carter *et al.*, 2011; Seber *et al.*, 2014; Stratton *et al.*, 2010).

| Feedstock | Life cycle greenhouse gas emissions, g $CO_2e/MJ$ | | | |
|---|---|---|---|---|
| | Low-emissions scenario | Baseline-emissions scenario | High-emissions scenario | Upstream coproduct allocation method |
| Algae DE[a] | 54.5 | 226.9 | 1,019.3 | Displacement and market |
| Rapeseed | 39.7 | 48.4 | 63.1 | Market |
| *Salicornia* | 31.0 | 46.9 | 66.1 | Energy and market |
| *Jatropha* | 39.5 | 41.8 | 43.3 | Energy |
| Soybean | 31.9 | 36.2 | 54.9 | Market |
| *Camelina* | 30.1 | 32.8 | 36.7 | Energy |
| Palm | 29.7 | 31.7 | 35.7 | Market |
| Algae WE[b] | 8.9 | 31.3 | 645.7 | Displacement and market |
| Tallow | 21.1 | 25.4 | 33.2 | Market |
| Used cooking oil | 12.1 | 14.8 | 16.8 | N/A |

[a]DE, dry extraction.

[b]WE, wet extraction.

*Sources:* Values from Stratton, R. W., H. M. Wong, and J. I. Hileman, 2010. Life cycle greenhouse gas emissions from alternative jet fuel. Version 1.2, *PARTNER/MIT*, Cambridge, MA, 2010, were augmented with more detailed calculations on the fuel production step from Pearlson, M., C. Wollersheim, and J. Hileman. 2013. A techno-economic review of hydroprocessed renewable esters and fatty acids for jet fuel production. *Biofuels Bioprod. Biorefin.* 7:89–96.

Serrano-Ruiz and Dumesic, 2011). For instance, ethanol or butanol, obtained via fermentation, can serve as a platform molecule that can be oligomerized and hydrotreated into jet fuel compounds (El-Halwagi *et al.*, 2011; Harvey and Meylemans, 2011; Peters and Taylor, 2015). Advanced fermentation techniques have shown that it is possible to obtain platform molecules such as triglycerides, fatty acids, esters, or fatty acid methyl esters, which can be thermochemically hydrodeoxygenated into liquid fuels (Klass, 1998). It is also possible to obtain alkanes directly from simple sugars through other advanced fermentation techniques (e.g., the recently ASTM-approved synthesized isoparaffin (SIP) fuels). Currently, the main challenge of these biochemical pathways is that the microorganisms used in these fermentation steps are generally very sensitive to impurities in the starting sugar solutions, such as furfural, which is a degradation product of xylose. For lignocellulosic pathways, furfural could serve as a platform molecule for the hemicellulose route, and hydroxymethylfurfural (HMF) or gamma-valerolactone (GVL) for the cellulose route (Bond *et al.*, 2014; Serrano-Ruiz and Dumesic, 2011). In such technologies, a series of processes is involved to convert these platform molecules into jet fuel. The GHG emissions of these emerging technologies will be determined by the energy requirements and the conversion efficiencies of the feedstock-to-fuel conversion process.

In an effort under the PARTNER Center of Excellence, Staples *et al.* (2014) assessed life cycle GHG emissions for alternative jet fuel from corn (*Zea mays*), sugar cane (*Saccharum officinale*), and switchgrass using a wide range of biochemical conversion processes that are in various stages of commercialization, and that can be classified as fermentation or advanced fermentation. In these technologies, polymer sugars are extracted from a biomass feedstock and decomposed to monomer sugars using mechanical, chemical, or biological means. The monomer sugars are then metabolized by a microorganism to produce energy-carrying platform molecules, which are then chemically upgraded to a drop-in product slate or to a blendstock specification.

Staples *et al.* (2014) found a significant potential for life cycle GHG reductions through the use of these technologies. Using market-based allocation among upstream coproducts, they calculated nominal emissions of 12.4 g $CO_2e/MJ$ for sugarcane, 37.4 g $CO_2e/MJ$ for switchgrass, and 62.6 g $CO_2e/MJ$ for corn grain. However, significant variability was found in obtained estimates, with values ranging from 6.8 to 19.7 g $CO_2e/MJ$, 7.3 to 89.8 g $CO_2e/MJ$, and 47.5 to 117.5 g $CO_2e/MJ$ for sugar cane, switchgrass, and corn grain jet fuel, respectively. Variability was largely driven by the utility requirements, cogeneration technologies, and conversion efficiencies assumed.

### 13.2.3    *Land-use change*

A research effort was conducted by Stratton *et al.* (2010) to provide an understanding on how emissions from direct LUC impact the total WTW life cycle results presented in the preceding section. Given the large uncertainties involved in estimating changes in emissions due to indirect LUC change (Al-Riffai *et al.*, 2010; Dumortier *et al.*, 2011; Harris *et al.*, 2009; Havlik *et al.*, 2011; Hertel *et al.*, 2009; Laporde, 2011; Plevin *et al.*, 2010; Searchinger *et al.*, 2008; U.S. EPA, 2010; Wang *et al.*, 2011), the scope was limited to emissions directly attributable to the conversion of land for biofuel production. Moreover, only biogeochemical effects were considered. The interested reader is directed to Caiazzo *et al.* (2014) for a recent study on biogeophysical effects and their contributions to the climate impacts of biofuels.

LUC also occurs as a consequence of fossil fuel production when the feedstocks are extracted and fuel processing facilities are placed, but the resulting emissions are negligible in comparison to the rest of the fuel pathway. This is because extraction occurs (mainly) beneath the surface and fuel-processing facilities have large throughputs, which, overall, leads to large volumes of fuel being created per unit area of converted land.

Different scenarios were employed to show the extent to which the life cycle emissions can be influenced by changes in land use. As an important caveat, the reader should note that these scenarios neither capture the full range of potential land-use scenarios nor the full range of feedstocks that might induce changes in land use. The scenarios are rather instructive cases that outline the importance of taking into account emission from changes in land usage. Also, as mentioned in Section 13.2.3, only the biogeochemical effect was captured by the analysis.

Table 13.3 shows the scenarios that were considered for BTL jet fuel from switchgrass and *Salicornia* and HEFA jet fuel from soy, oil palm, rapeseed, and *Salicornia*. Scenarios and subsequent calculations are taken from Stratton *et al.* (2010).

Soybeans, oil palm, and rapeseed are edible food crops requiring arable land for cultivation. Consequently, have a large potential for inducing LUC. However, for emissions stemming from changes in direct land use, the actual magnitude of the effect depends on the actual type of land and usage that is being directly displaced. Scenarios considered for these food crops were the conversion of different types of forest, grassland, and previously set-aside land into cropland.

Because *Salicornia* and switchgrass do not need to grow on high-fertile soils, scenarios were investigated in which carbon-depleted land was used for switchgrass cultivation and desert land was converted into *Salicornia* cultivation fields.

Figure 13.3 shows the relative changes in WTW life cycle GHG emissions of different biomass-to-fuel options under the LUC scenarios investigated compared with WTW emissions without LUC. The figure demonstrates that emissions from LUC can dominate the life cycle, depending on the type of land converted. For example, in the case of peat land rainforest being converted into oil palm plantations for HEFA fuel, the life cycle GHG emissions (in g $CO_2e/MJ$) attributable to this conversion were 22.1 times higher than the emissions from the rest of the life cycle (nominal case). As a consequence, while WTW GHG emissions for HEFA jet from palm oil were calculated to be considerably lower than those of conventional jet fuel if existing plantations were used (i.e., no direct LUC), the GHG emissions were calculated to be 8.0 times higher than those of conventional jet fuel if peatland rainforest was converted into additional palm plantations. The figure also shows that LUC does not necessarily increase WTW GHG emissions of jet fuel. If carbon-depleted land is used for biomass cultivation (as in the scenarios considered for switchgrass and *Salicornia*), more carbon can be sequestered over time and WTW GHG emissions attributable to the particular feedstock-to-jet-fuel pathway decrease substantially. This can

Table 13.3.    Land-use change (LUC) scenarios considered.

| LUC | Scenario 1 | Scenario 2 | Scenario 3 |
|---|---|---|---|
| Switchgrass | Carbon-depleted soils converted to switchgrass cultivation (LUC B1) | N/A | N/A |
| Soy oil | Grassland converted to soybean field (LUC S1) | Tropical rainforest converted to soybean field (LUC S2) | N/A |
| Palm oil | Logged-over forest converted to palm plantation field (LUC P1) | Tropical rainforest converted to palm plantation field (LUC P2) | Peat land rainforest converted to palm plantation field (LUC P3) |
| Rapeseed oil | Set-aside land converted to rapeseed cultivation (LUC R1) | N/A | N/A |
| *Salicornia* | Desert land converted to *Salicornia* cultivation (LUC H1) | N/A | N/A |

*Source:* Adapted from Stratton, R. W., H. M. Wong, and J. I. Hileman, 2010. Life cycle greenhouse gas emissions from alternative jet fuel. Version 1.2, *PARTNER/MIT*, Cambridge, MA, 2010. With permission.

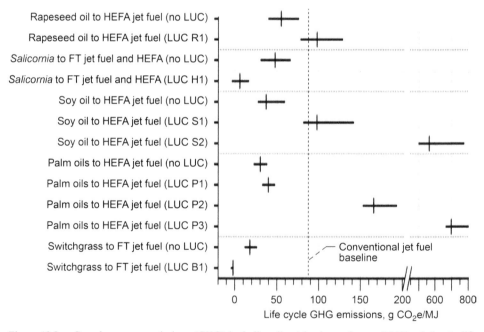

Figure 13.3.    Greenhouse gas emissions (GHG) including direct land-use change (LUC) relative to life cycle emissions without any LUC (Stratton *et al.*, 2010). LUC emissions are assumed to be amortized over 30 years. FT, Fischer-Tropsch; HEFA, hydroprocessed esters and fatty acids. LUC, land-use change, scenarios defined in Table 13.3.

even lead to net carbon sequestration over the full course of the life cycle, as in the case shown for switchgrass FT jet fuel.

Overall, it can be concluded that the assessment of the life cycle GHG emissions from renewable jet fuel would be incomplete, and in some cases misleading, without including the changes in carbon flux that result from the conversion of land.

### 13.2.4   *Discussion of greenhouse gas life-cycle analysis results*

Because of varying crude oil extraction techniques, transportation schemes, and refining effi-ciencies, there exists considerable variability in the life cycle GHG results of petroleum-derived jet fuel. Two of the major factors affecting the variability are methane handling during crude oil extraction, and the extent to which jet fuel produced is straight-run or hydroprocessed. Both factors can change emissions on the order of 10% in comparison with average lifecycle jet fuel emissions from conventional petroleum.

The analyses show that of the fuel options considered herein, conventional petroleum has the lowest GHG emissions of any jet fuel pathway that relies exclusively on fossil fuel resources. Only with extensive CCS and high conversion efficiencies can the FT facilities utilizing fossil fuels reduce overall GHG emissions to levels just below that of conventional jet fuel. Hence, this practice is not likely to be considered as a viable option to reduce life cycle GHG emissions of jet fuel. With regard to fuels made from biomass, all options considered show reductions in lifecycle GHG emissions compared with petroleum-derived jet fuel if adverse emission impacts from LUC can be avoided.

Biofuel FT fuels, for example, have low GHG emissions (though a CBTL facility is largely dependent on the biomass-to-coal ratio in the feedstock). However, FT facilities are capital inten-sive and would operate under significant economies of scale (Carter *et al.*, 2011); as a result, the commercial development of FT facilities that use biomass has been slow.

HEFA fuels present an opportunity to reduce life cycle GHG emissions, assuming adverse LUC emissions are avoided, but many of these feedstocks have relatively small yields (vegetable crops) or limited availability (waste oils, fats, and greases) and, as such, provide limited oppor-tunities to reduce total life cycle GHG emissions of aviation. Algae present an opportunity to achieve large life cycle GHG emissions reductions, but only if technological advances continue. In this regard, the results of Carter (2012) show that life cycle GHG emissions and cost are cor-related with technological improvements, providing reductions in both.

Fermentation and advanced fermentation jet fuels show potential for reducing life cycle GHG emissions in comparison with conventional jet fuel, given appropriate technology selection and implementation. We note, however, that the findings are contingent upon the successful techni-cal development of these technologies and that a number of challenges remain, such as sugar extraction and hydrolysis efficiency, enzyme separation and re-use, and fermentation efficien-cies. As for all other fuels derived from biomass, there might be significant changes in fuel-attributable GHG emissions if biomass growth changes the use of land.

### 13.2.5   *Addition of non-$CO_2$ combustion emissions*

The LCAs presented thus far considered only the $CO_2$ emissions from fuel combustion. In real-ity, the combustion of hydrocarbon fuels, regardless of the source, results in the formation of additional emissions—including water vapor, nitrogen oxides ($NO_x$), sulfur aerosols, and soot. Furthermore, depending on the local atmospheric conditions, aircraft operations can also result in contrails and aviation-induced cloudiness. These emissions and conditions all have a global climate impact.

These non-$CO_2$ combustion emissions and their impacts are generally not included in the fuel life cycle because of the large scientific uncertainties involved in their calculation and the uncertainties associated with how the impacts of these emissions would change with the use of alternative jet fuels. Stratton *et al.* (2011) examined the non-$CO_2$ combustion emissions of alternative jet fuels alongside the other emissions in the fuel life cycle by making several sim-plifying assumptions. The most important of these assumptions was that contrails and avia-tion-induced cloudiness are unchanged with the use of alternative jet fuels. This assumption needs to be revised as emissions measurement data become available from the use of alternative jet fuels. Alternative jet fuels that are free of aromatic- and sulfur-containing compounds (e.g., FT, HEFA, and SIP fuels) have no sulfur aerosols when combusted and have substantially lower soot emissions. However, they result in increased water vapor. The relative importance of these counteracting impacts is as yet unknown; and, as such, the impact of these fuels on contrails and aviation-induced cloudiness is as yet unknown.

The assumptions and methods from Stratton *et al.* (2011) were adopted here to examine life cycle GHG emissions from well to wake for both $CO_2$ and non-$CO_2$ emissions (termed "WTW+"). Mid-range non-$CO_2$ combustion effects of these products have been calculated in terms of $CO_2$ equivalent for both a synthetic paraffinic kerosene fuel and a conventional jet fuel for a time window of 100 years using the climate impacts module of the Aviation Portfolio Management Tool (APMT), a comprehensive tool suite owned by the FAA. Life cycle GHG emissions with the addition of the non-$CO_2$ combustion effects, WTW+, are compared in Figure 13.4. The results show that uncertainty in the non-$CO_2$ combustion impacts is much larger in comparison to the variability in the WTW LCA results. Moreover, the inclusion of these non-$CO_2$ combustion emissions reduces the absolute difference in the emissions of alternative fuels and conventional jet fuel. The interested reader is directed to Stratton *et al.* (2011) for additional details on the analysis.

## 13.3   WATER

Although one of the primary drivers for alternative fuel production has been the reduction of life cycle greenhouse gas emissions and the associated impacts on the global climate, water use and water quality have been identified as critical concerns relating to the production of biofuels and bioenergy. Freshwater use and contamination potential have been identified as critical environmental freshwater concerns relating to the production of biofuels because of the potential demand for water for irrigation of dedicated bioenergy feedstock crops, the potential for nutrient runoff and soil erosion from croplands, and the potential for higher water demands for fuel conversion and processing in comparison with petroleum-based fuels (De La Torre Ugarte *et al.*, 2010; National Research Council of the National Academies, 2008; Tidwell *et al.*, 2011). Globally, many countries are on the cusp of unsustainable water use, which may be exacerbated by biofuel production in the future (Berndes, 2002). One analysis suggests that the United States will exceed 25% withdrawal of available water in the near term, which is considered a water-stress threshold (Berndes, 2002).

Water impacts cover withdrawal (water removal from its source), consumption (water removed but not returned to its source), and/or pollution (via contaminant or thermal emissions or other changes in water quality) (Yeh *et al.*, 2011). Each of these factors deals with different aspects of regional impacts on the availability and quality of water. Section 13.3.1 addresses water use and consumption, and Section 13.3.2 addresses water quality.

### 13.3.1   *Water use and consumption*

For the purposes of considering water withdrawal and consumption, the available water is split into two primary categories: "blue water" from surface water bodies and aquifers, and "green water" held in the soil from precipitation (Yeh *et al.*, 2011). Blue water is split into "renewable" and "nonrenewable" waters. Renewable water reflects the long-term average annual inflow to surface and groundwater bodies in a region (i.e., from precipitation or from sources outside the region) minus natural outflows in surface or groundwater bodies or through evapotranspiration (describes the sum of evaporation and plant transpiration from the Earth's land surface to the atmosphere) from plants and soil (U.S. Geological Survey, 1984). Nonrenewable water resources include resources such as deep aquifers that negligibly recharge on the human time-scale once they are drawn. Even within renewable blue waters, the effects of pumping groundwater are different from withdrawal from a river or irrigation canal; therefore, when the potential impacts of a given activity are evaluated the sources of water need to be identified along with the current status of the resources within potentially affected area (Fingerman *et al.*, 2011). In some cases,

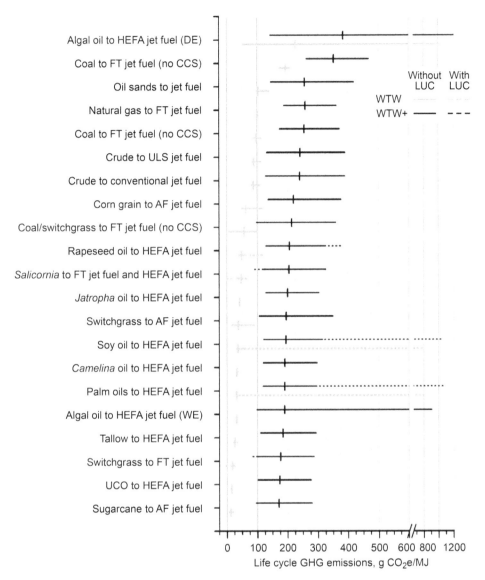

Figure 13.4. Life cycle greenhouse gas emissions for the feedstock-to-fuel pathways under consideration. Uncertainty bars are shown for the low-, baseline-, and high-emissions scenarios for the well-to-well (WTW) and WTW+ analyses, where WTW+ includes the non-$CO_2$ climate impacts of aircraft gas turbine combustion emissions such as soot, sulfate aerosols, $NO_x$, contrails, and contrail cirrus. Data are shown with and without land-use change (LUC) scenarios (full range of scenarios from Table 13.3 were considered). AF, advanced fermentation; CCS, carbon capture and sequestration; DE, dry extraction; FT, Fischer-Tropsch; HEFA, hydroprocessed esters and fatty acids; UCO, used cooking oil; ULS, ultra-low sulfur; WE, wet extraction. (From Stratton, R. W., H. M. Wong, and J. I. Hileman, 2010. Life cycle greenhouse gas emissions from alternative jet fuel. Version 1.2, *PARTNER/MIT*, Cambridge, MA, 2010. With permission.)

water consumption is split into "productive" and "non-productive," with productive use including crop evapotranspiration (i.e., for growing) and water harvested within the crop, and nonproductive including general evapotranspiration from open soil and irrigation pathways (canals, channels, sprinklers, and reservoirs) (Yeh *et al.*, 2011). In addition to blue and green water, some studies use a metric called "gray water" to indicate the amount of water that becomes contaminated or degraded by a process; gray water is the amount of water that would be required to bring pollutant levels to below water quality thresholds such as those required by the Clean Water Act. This is considered part of the "water footprint" of the facility or process (Gerbens-Leenes et al., 2009) that is used to assess the potential impacts on water of a given activity.

The U.S. Department of Energy (DOE) has calculated that freshwater withdrawal in most regions of the United States exceeds the available precipitation (precipitation minus evapotranspiration) and that this problem is more extreme in the western states (U.S. DOE, 2006). The excess water demands in these regions can only be met through withdrawal of groundwater or importation of surface water from other regions (U.S. DOE, 2006). However, groundwater availability is declining as well: in some regions, groundwater levels are 300 to 900 ft (91 to 274 m) lower than they were 50 years ago because of withdrawals greater than the natural recharge rate (U.S. DOE, 2006). The expansion of biofuel production could exacerbate water stresses on the state, regional, and national levels. In the past, the nationwide average for consumption was approximately 25% of withdrawals (Solley et al., 1998). It is important to note, however, that withdrawn water is not all consumed, and therefore some of the withdrawn water may still be available for other uses. In the Eastern United States, for example, freshwater consumption in 1995 was only about 12% of withdrawn water, whereas in the West, about 47% of freshwater withdrawals were actually consumed, which correlates with much greater water use for irrigation (about 90% of all irrigation withdrawals are in the West) (Solley et al., 1998). The remaining water is returned to surface waters, albeit not necessarily in its original condition or location. Therefore, simply quantifying total withdrawals can misrepresent the impacts on a given watershed (Fingerman et al., 2011).

The U.S. DOE projects that water use—both withdrawal and consumption—for energy production is likely to increase greatly if fuel production switches substantially to biofuels (U.S. DOE, 2006). Agriculture, specifically crop irrigation, is the largest water consumer in the United States, accounting for 37% of all freshwater withdrawals in 2005 (Kenny et al., 2009).[3] Irrigation is also the largest consumer of withdrawn water; agricultural irrigation accounted for 85% of total national consumptive use in 1995 (Solley et al., 1998). Agricultural uses dominate water consumption globally as well (Berndes, 2002). A 2012 report from the United Nations Food and Agriculture Organization of the United Nations (Food and Agriculture Organization/Bioenergy and Food Security, 2012) estimates that fresh water use in irrigation in 2000 was 2,384 $km^3$/yr (accounting for 67% of total fresh water use) and expects irrigation water use to fall by 2,050 to 2,049 $m^3$/yr (37% of total fresh water use) mainly on account of fierce competition for fresh water from the industrial and domestic sectors.

The irrigation needed for crop production varies greatly depending on the region. For example, water use for irrigated soy production varies from 600,000 L/ha for Pennsylvania to about 4.3 million L/ha for Colorado, with a national average of 2.4 million L/ha (U.S. DOE, 2006; Stratton et al., 2010). In 2005, 17 conterminous western states accounted for 85% of irrigation water withdrawals in the United States (Kenny et al., 2009). Many areas use no irrigation, while others require very high levels of input. Therefore, the amount of water used to produce dedicated bioenergy feedstocks is likely to be highly dependent on geography, as is the impact of that water use (Fingerman et al., 2011). National-level analyses give some indication of potential water issues, but these do not reflect the great disparities at the regional level with regard to water availability for different uses and therefore may mask more serious water shortages for the production of feedstocks or fuels at a given location (Chapagain et al., 2006; Chiu and Wu, 2012; Fargione et al., 2009; Fingerman et al., 2010; King and Webber, 2008; National Research Council of the National Academies, 2008; Scown et al., 2011; Sheehan, 2009; Staples et al., 2013; Wu et al., 2009; Yang et al., 2009; Zhang et al., 2010).

Large-scale production of energy crops can have regional water availability impacts due to alterations of hydrologic flows and functions such as infiltration, evaporation, transpiration, and

---

[3] Heavy industry, by comparison, accounted for only about 4% of total water withdrawals in 2005.

stream flows (Diaz-Chavez *et al.*, 2011). The potential impacts are related not only to the amount of water withdrawn and consumed, but also to the timing and the context of water use: withdrawal and consumption of a given quantity of renewable water in regions of continuous or seasonal water scarcity are not equivalent to the same withdrawal and consumption in a humid region with high water inflow. In a 2012 study, half of the river basins evaluated experienced severe water scarcity for at least one month per year; that number may be even greater if consideration is given to the amount of water that may be polluted and therefore of limited use (e.g., Hoekstra *et al.*, 2012). Water scarcity metrics (e.g., total water used versus total available water) must therefore be used as the baseline to assess whether a given location can tolerate additional water consumption (e.g., Hoekstra *et al.*, 2012). The placement of large biorefineries and associated water requirements might place additional burdens on water availability in regions where water is already scarce. For example, an ethanol plant proposed for southwestern Minnesota was blocked because the local water system was not able to supply the 350 million gallons of freshwater per year that the plant needed for operation (National Research Council of the National Academies, 2008). In general, it is more efficient in terms of water use (and energy extraction) to use biomass for energy production via combustion than it is to produce liquid fuels (e.g., Gerbens-Leenes *et al.*, 2009).

In addition to the potential for feedstock irrigation to contribute to water withdrawal and consumption for the production of biofuels, conversion of bio-based products into hydrocarbon fuels could require significant water resources. Recent research by Staples *et al.* (2013) funded under the Partnership for Air Transport Noise and Emissions Reduction Center of Excellence estimates life cycle water consumption of conventional jet and diesel fuels to be 4 to 8 liters per liter of fuel from nonirrigated biomass, on the same order of magnitude as conventional jet and diesel fuel. Fischer-Tropsch conversion of natural gas or coal to diesel is 5 to 7 times more water intensive than the refining of conventional petroleum.

A number of approaches are being developed for water footprinting to evaluate both water use and local water availability. Variations among water footprinting approaches are likely to include the system boundaries (e.g., subcatchment, catchment, or region), accounting for gray water, thresholds for "scarcity" designations, and resolution/scale of evaluation. From a broader industry perspective, other challenges will include standardization, comparability, and the ability to associate actual impacts with a given water withdrawal and/or consumption (Ridoutt and Pfister, 2010). Some water footprinting approaches consider it unsustainable to use water at a greater level than necessary, regardless of the local water scarcity context (Hoekstra *et al.*, 2011). Therefore, while it is relatively straightforward to measure the amount of water withdrawn and consumed by a given operator (high level of scientific understanding, Table 13.1), it is less straightforward to determine the actual impacts of that withdrawal and consumption on water availability and scarcity. For additional details, see International Organization on Standardization standard on water footprinting (ISO, 2014).

### 13.3.2   *Water quality*

Biofuel production can both directly and indirectly affect local and regional water quality, a key sustainability consideration. Biofuel processing may affect water quality through the release of pollutants in effluent or accidental spillage and through the release of heated water. As with any agricultural activity, direct effects from feedstock cultivation can include pesticide, fertilizer, and soil runoff and release of stored nutrients from the soil. These effects can be exacerbated by cultivation strategies, particularly tillage techniques; by extraction of residues from cropland or forestry lands; and by other processes (Diaz-Chavez *et al.*, 2011; Lindstrom, 1986). Fertilizer and pesticide residues can affect groundwater as well as surface water (Diaz-Chavez *et al.*, 2011) and can also contaminate municipal water supplies.

High concentrations of nutrients resulting from runoff and nutrient release due to land-use changes cause algae and phytoplankton to proliferate, which in turn leads to elevated biological oxygen demand (BOD) and depletion of dissolved oxygen in the water (eutrophication), which is crucial to the support of aquatic life (Committee on Environment and Natural Resources, 2010). Eutrophication results in the "dead zone" found in the Gulf of Mexico and some 700 other coastal sites around the globe (WRI, 2011) and affects approximately 65% of U.S. coastal and estuarine waters. Approximately of 47% to 58% of coastal waters and estuaries in the United States have reported hypoxia (low dissolved oxygen levels) predominantly resulting from eutrophication due to nutrient influxes (Committee on Environment and Natural Resources, 2010). Hypoxia reduces biodiversity, destroys habitat, and results in altered biogeochemical cycling and release of compounds such as ammonia from sediments (Committee on Environment and Natural Resources, 2010). Although there are microbial processes that can reduce BOD, thus reducing the risk of eutrophication, these processes also release methane, a potent GHG (Diaz-Chavez et al., 2011).

It is possible, however, that careful biofuel production could benefit actually water quality (Berndes, 2002). Oleaginous bacteria that scavenge and store lipids, which can be converted into biofuels, may be able to assist in wastewater treatment (Stephanopoulos, 2007; Subramaniam et al., 2010), and wetland-based water treatment facilities may provide another biomass source for fuel production (Liu et al., 2012), depending on the species included and their tolerance for harvesting (Tanner, 1996). Feedstock selection and cultivation strategies such as shifting toward long-term perennial plant production (e.g., perennial grasses or woody species) from annual row crops may reduce erosion and result in positive impacts on nearby water quality, and woody species cultivation along rivers as a buffer zone may reduce runoff from annual crops (Berndes, 2002; Diaz-Chavez et al., 2011). Some plant species can also perform bioremediation of contaminated sites (e.g., Garbisu and Alkorta, 2001; Gordon et al., 1998), although this may require decontamination prior to use of resulting fuel if the plants accumulate the contaminants, as such compounds may affect fuel performance and conformance with specifications. Transitioning from irrigated crops to crops that do not require water inputs can also reduce regional water use. Algae are most effectively grown in shallow, high-surface-area ponds that will result in high evaporative losses. This constraint will require algae farms to be located near plentiful water supplies, whatever the quality (Vasudevan et al., 2012). However, production of some species of algae, and even some terrestrial biofuel feedstocks such as *Salicornia*, can be accomplished with degraded, saline, or brackish water. In fact, *Salicornia* can be used to clean up the effluent from aquaculture facilities that produce fish (Buhmann and Papenbrock, 2012; Stratton et al., 2010). The use of such water sources may reduce the impact of high water needs for certain water-intensive alternative fuel feedstocks. On the biogeochemical scale, if biofuel production can reduce GHG emissions, it may also reduce ocean acidification, which is associated with increased $CO_2$ uptake as a result of increasing $CO_2$ concentrations in the atmosphere (Doney et al., 2009).

## 13.4    BIODIVERSITY

The impacts of alternative jet fuel and feedstock production on greenhouse gas emissions, water use, and quality can affect local, regional, and, in the case of GHG emissions, global biodiversity. Land conversion and the introduction of novel bioenergy crops can also have direct impacts on local and regional biodiversity. Thus, conservation of biodiversity is a key component in the sustainability of alternative fuel production.

Biodiversity is defined as the variability of living organisms on Earth, encompassing the number of species of plants, animals, and microorganisms; the diversity within those species (genes, populations); the wide variety of habitats and ecosystems of which they are a part, such as grasslands, rainforests, and coral reefs (Millenium Ecosystem Assessment, 2005). Diversity can also include functional diversity of organisms, spatial or temporal diversity, and interaction diversity (the  network of interactions among species that result in transfer of energy and materials) (Naeem et al., 2012). Biological diversity contributes to the survival of all organisms within a system, the stability and resilience of ecosystem functions and services, and the ability to adapt to changing environmental conditions (Millenium Ecosystem Assessment, 2005). Ecosystem functions that are supported or enhanced by biodiversity include carbon storage, water purification and storage, decomposition, nutrient storage and cycling, and ecosystem productivity (i.e., production of biomass) (Cardinale *et al.*, 2012; Maestre *et al.*, 2012; Midgley, 2012; Millenium Ecosystem Assessment, 2005). These ecosystem functions provide humans with essential natural products and services, including oxygen, fresh water, fertile soil, food, and medicines; products and services including protection from environmental hazards such as fires and floods; and climate stability (Millenium Ecosystem Assessment, 2005).

A recent synthesis of two decades of work on "biodiversity and ecosystem function" and "biodiversity and ecosystem services" (Cardinale *et al.*, 2012) indicates that biodiversity in itself is responsible for maintaining biomass productivity and ecosystem resilience to disturbance (e.g., drought, fire, storms, or climate change). This is because diversity tends to promote the presence of species that have a range of functions and tolerances to environmental conditions that are complementary to one another. This allows different species to dominate ecosystem processes in different years or under varying conditions (Cardinale *et al.*, 2012; Isbell *et al.*, 2011). As more complex experimental systems are explored, the effects of biodiversity on healthy ecosystem functioning are revealed to be even greater than originally anticipated (Naeem *et al.*, 2012). Biodiversity is also linked to important social and cultural activities and values, including education, tourism, recreation, science, and cultural identity (e.g., traditional land uses or customs, or traditional uses of species for medicine, food, or religious purposes) (Cardinale *et al.*, 2012). Thus it is important to understand and recognize the value of biodiversity as a whole, as well as the specific components of diversity.

Because of the complex nature of biodiversity, there is no single metric that can be used consistently as an indicator of overall biodiversity, and precise quantification of biodiversity at different levels (ecosystem, community, global, etc.) is difficult (Millenium Ecosystem Assessment, 2005). This also can make standardized evaluations of biodiversity impacts challenging. The most common metrics focus on species-level diversity; of these, the most widely used is species richness, which is the number of species in a given area (Millenium Ecosystem Assessment, 2005). Ecosystems that are species rich are more likely to maintain multiple functions that are beneficial to humans (Maestre *et al.*, 2012).

Species richness, however, only accounts for the number of species present and does not take into account the number of individuals present (abundance), the evenness of species distributions (many species of similar abundance versus domination by one or a few species coexisting with few individuals of many other species), genetic or ecological variability, distribution and role in ecosystem processes, dynamics, trophic position, or functional traits of species (Cardinale *et al.*, 2012; Millenium Ecosystem Assessment, 2005; Naeem *et al.*, 2012). Other studies show that species evenness, species richness, and variability of certain traits among species (e.g., shade tolerance) contribute to the productivity of forest communities (Zhang *et al.*, 2012), whereas biomass production increases with increasing diversity in the low-input mixed grasslands recommended for biofuel production (Tilman *et al.*, 2006). Old field productivity also increases with increasing species evenness of abundance, regardless of the identity of the species present (Wilsey and Potvin, 2000). Furthermore, ecosystems respond nonlinearly, resulting in accelerating change as biodiversity is reduced (Cardinale *et al.*, 2012).

Though the relationship between biodiversity and ecosystem services such as pollination is still somewhat ambiguous, the loss of multiple functional traits or the loss of more than one trophic level in the food web are likely to result in higher impacts as a result of biodiversity loss (Cardinale *et al.*, 2012). Thus conservation of the composition of systems—not merely the maximization of species richness—is critical to maintaining ecosystem services (Millenium Ecosystem Assessment, 2005; Naeem *et al.*, 2012). Other factors lacking in the species richness metric that should be taken into account include resilience or sensitivity to environmental changes, distinction between native and nonnative species, and the scale of the area being measured (Millenium Ecosystem Assessment, 2005).

Two indices commonly used to measure biodiversity are the Simpson Index (Simpson, 1949) and the Shannon-Wiener Index (Shannon and Weaver, 1949), both of which take into account species richness but also the relative abundance or "importance" of each species, such that rare species contribute less to the diversity index than common ones (Ricklefs and Miller, 2000). Overall, information on species richness (i.e., number) should be combined with surrogate or proxy information for the other characteristics mentioned earlier in this section (abundance, evenness, functional traits, etc.), to the extent possible, in order to more accurately portray biodiversity (Millenium Ecosystem Assessment, 2005).

Biodiversity can be affected by biofuel production in several ways. Direct impacts due to habitat loss and fragmentation occur when habitat that formerly supported native species is converted into agricultural land or biorefinery facilities. There can also be indirect effects on adjacent areas, even those that are not converted, because of noise, pollutant emissions, water withdrawals, or the introduction of invasive species. Indirect impacts can also be global in nature because of indirect land use (Searchinger et al., 2008). If biofuel crops are grown on land currently used for food crops, that could lead to increased cropland requirements, and farmers in other parts of the world might clear wild land to meet the displaced demand for food, resulting in increased GHG emissions and biodiversity loss (Fargione et al., 2008; Searchinger et al., 2008; United Nations Environment Programme (UNEP), 2009). These indirect effects could be greater than the direct impacts (Hellmann and Verburg, 2010). Both direct and indirect impacts could also result due to transportation, refining, and burning of fuels (Groom et al., 2008).

The most significant threat to biodiversity from biofuel production is habitat loss resulting from cropland expansion (Gasparatos et al., 2011). Clearing land of its native species so that it can be converted into agricultural land for crops or sites for biorefinery facilities decreases the amount of available suitable habitat for many species of wildlife and plants and reduces the level of ecosystem function (Groom et al., 2008). Land-use change can also result in the release of stored nutrients and the runoff of soil and pollutants onto adjacent lands or into nearby water bodies (Diaz-Chavez et al., 2011; Martinelli and Filoso, 2008). Emissions of nutrients and pollutants onto adjacent land may affect species composition and biodiversity in both terrestrial and aquatic systems (Groom et al., 2008; UNEP, 2009). Carbon dioxide released from the system (from burning, decomposing, and oxidizing) can offset some or all GHG benefits of biofuels, depending on the preexisting land-use type (Bailis and Baka, 2010; Fargione et al., 2008; Stratton et al., 2010; Tilman et al., 2009; UNEP, 2009), as described in the preceding sections on GHG life cycle analysis. A recent scenario analysis investigating cellulosic and first-generation biofuel production (among other energy options) projected that adverse impacts of biofuel production are likely to be greatest for temperate deciduous forests and temperate grasslands as compared with other ecosystems found in the United States (McDonald et al., 2009).

The introduction of novel biofuel crop species can further threaten biodiversity, as many of these species are selected for traits that also carry a high risk of making species invasive in natural areas (Buddenhagen et al., 2009; Invasive Species Advisory Committee, 2009; Lewis and Porter, 2014; Raghu et al., 2006; UNEP, 2009). Such invaders are not only costly to manage and control (Pimentel et al., 2000, 2005; Rejmanek and Pitcairn, 2002) but also tend to outcompete native species, reduce diversity, and disrupt local ecosystem processes (e.g., D'Antonio and Vitousek, 1992; Gordon, 1998; Mack and D'Antonio, 2003; Vaughn and Berhow, 1999). Genetically engineered bioenergy crops that are not themselves invasive may facilitate gene introgression to weedy relatives, resulting in range expansion and invasion by those species (for example, by transferring herbicide- or pest-resistance genes) (Binimelis et al., 2009; Cerdeira and Duke, 2006).

Biofuel production could potentially have positive impacts on biodiversity as well. For example, previous work has suggested that mixed native prairie grasses can be used to create a highly productive biofuel feedstock production system (Tilman et al., 2006) while providing habitat for birds and wildlife (Fargione et al., 2009). Careful management of even monocultures of perennial native grasses could provide habitat for birds and other wildlife provided that harvest is timed carefully to avoid the breeding season and rotated among fields (Harper and Keyser, 2008). Extraction of invasive species in areas where they are already established can assist with eradication and restoration efforts and enhance local native biodiversity; such uses are allowed under two Federal incentive programs for production of alternative fuels: the Environmental Protection Agency's Renewable Fuel Standard and the U.S. Department of Agriculture's Biomass Crop Assistance Program (Lewis and Porter, 2014). Finally, the goal of many alternative fuel programs is to reduce GHG emissions to minimize climate change, which would be expected to assist in maintaining biodiversity in a variety of ecosystems.

As described previously, biodiversity can be measured in many different ways at different levels of detail (genes to ecosystems), but for most purposes, it is generally measured by species richness and evenness. However, even this simplified process is time- and labor-intensive, and it can be difficult to get data over time. For this reason, current approaches to evaluating sustainability of biofuel production with regard to biodiversity for regulatory or incentive programs (e.g., the EPA's Renewable Fuel Standard or the European Renewable Energy Directive) tend to provide incentives to deter biodiverse land conversion in order to limit disturbance, destruction of native species and habitat, and GHG emissions. Voluntary sustainability programs such as that of the Roundtable for Sustainable Biofuels promote weed risk assessment and management techniques to limit potential introduction of invasive species and minimize impacts on soil, water, and air quality that might result in indirect impacts (Roundtable on Sustainable Biofuels, 2010).

To reduce impacts on biodiversity, growers should select biofuel crops that provide high yield for small area, require low agricultural inputs (e.g., fertilizer and pesticides) to limit runoff and impacts on aquatic systems and nontarget organisms (Groom *et al.*, 2008), and emphasize perennial and/or mixed species crops (Groom *et al.*, 2008; Tilman *et al.*, 2006, 2009). Careful species selection through the use of risk management tools such as weed risk assessment (International Union for Conservation of Nature (IUCN), 2009) and implementation of best management practices during cultivation can reduce the invasive species risks posed by nonnative or engineered biofuel crops (Barney and Ditomaso, 2008; Byrne and Stone, 2011; Davis *et al.*, 2010; IUCN, 2009; Lewis and Porter, 2014; Lonsdale and Fitzgibbon, 2011; National Invasive Species Council (NISC), 2008).The restoration and use of degraded lands that already offer minimal habitat can provide the necessary land for biofuel conversion without serious biodiversity impacts (Groom *et al.*, 2008).

## 13.5   CONCLUSIONS

One of the key goals of alternative jet fuel production (and alternative fuels in general) is the improvement of environmental performance, specifically in comparison with standard petroleum-based fuels. This chapter has explored only a few of the key environmental indicators that are potentially important for alternative jet fuel and feedstock production, and the sources of potential risk (for example, air quality impacts were not considered herein). In this chapter, we provide information on petroleum-based fuel greenhouse gas baselines as well as water-use information, but analyses for other environmental indicators are rarely performed for petroleum-based fuels on the same basis as is done for alternative fuels, making comparisons of overall environmental performance difficult. It is important that both petroleum-based fuels and alternative jet fuels be evaluated with the same criteria.

As a critical driver for alternative jet fuel production, environmental performance needs to be evaluated rigorously and by well-accepted methods (usually with outside verification) to ensure that environmental considerations are fully addressed. Environmental performance needs to be measured over the course of a life cycle basis, per unit energy for each fuel but local context and absolute performance are also crucial, particularly for impacts that are not global in nature (which is to say, everything except GHGs). Environmental performance is also not uniform across indicators, and a given fuel may exhibit excellent environmental performance in some areas (e.g., GHG emissions reductions) and poorer performance in other areas (e.g., water quality or use, or air quality), and these tradeoffs require societal prioritization among indicators at the appropriate level, as priorities at the international or national level might not reflect local priorities (e.g., water use in a desert or drought-stricken area).

Alternative jet fuels provide one possible approach to improving the environmental performance of the aviation sector as a whole. Careful evaluation of environmental performance during feedstock and fuel development and deployment will enable the aviation sector to take advantage of these fuels for achieving environmental, cost, and energy security goals.

## REFERENCES

Allen, D. T., C. Allport, K. Atkins, *et al.* 2009. Propulsion and power rapid response research and development (R&D) support. In *Delivery Order 0011: Advanced Propulsion Fuels Research and Development– Subtask: Framework and Guidance for Estimating Greenhouse Gas Footprints of Aviation Fuels,* AFRL–RZ–WP–TR–2009–2206.

Al-Riffai, P., B. Dimaranan, and D. Laborde. 2010. Global trade and environmental impact study of the EU biofuels mandate. ATLASS Consortium, Specific Contract No. S12.537.787 implementing Framework Contract No. TRADE/07/A2, p. 125.

Bailis, R. E., and J. E. Baka. 2010. Greenhouse gas emissions and land use change from *Jatropha Curcas-* based jet fuel in Brazil. *Environ. Sci. Technol.* 44:8684–8691.

Barney, J. N., and J. M. DiTomaso. 2008. Nonnative species and bioenergy: are we cultivating the next invader? *BioScience* 58:64–70.

Berndes, G. 2002. Bioenergy and water—the implications of large-scale bioenergy production for water use and supply. *Glob. Environ. Chang.* 12:253–271.

Binimelis, R., W. Pengue, and I. Monterroso. 2009. "Transgenic treadmill": responses to the emergence and spread of glyphosate-resistant Johnsongrass in Argentina. *Geoforum* 40:623–633.

Bond, J. Q., A. A. Upadhye, H. Olcay, *et al.* 2014. Production of renewable jet fuel range alkanes and commodity chemicals from integrated catalytic processing of biomass. *Energ. Environ. Sci.* 7: 1500–1523.

British Standards Institution. 2011. The guide to PAS 2050:2011—How to carbon footprint your products, identify hotspots and reduce emissions in your supply chain. *http://shop.bsigroup.com/upload/Shop/ Download/PAS/PAS2050Guide.pdf* (accessed December 1, 2016).

Buddenhagen, C. E., C. Chimera, and P. Clifford. 2009. Assessing biofuel crop invasiveness: a case study. *PLoS ONE* 4:e5261.

Buhmann, A., and J. Papenbrock. 2013. Biofiltering of aquaculture effluents by halophytic plants: basic principles, current uses and future perspectives. *Environ. Exp. Bot.* 92:122–133.

Byrne, M., and L. Stone. 2011. The need for 'duty of care' when introducing new crops for sustainable agriculture. *Curr. Opin. Environ. Sustain.* 3:50–54.

Caiazzo, F., R. Malina, M. D. Staples, *et al.* 2014. Quantifying the climate impacts of albedo changes due to biofuel production: a comparison with biogeochemical effects. *Environ. Res. Lett.* 9:024015.

Cardinale, B. J., J. E. Duffy, A. Gonzalez, *et al.* 2012. Biodiversity loss and its impact on humanity. *Nature* 486:59–67.

Carter, N. A. 2012. Environmental and economic assessment of microalgae-derived jet fuel. Cambridge, MA: Massachusetts Institute of Technology.

Carter, N. A., R. W. Stratton, M. K. Bredehoeft, *et al.* 2011. Energy and environmental viability of select alternative jet fuel pathways. AIAA 2011–5968.

Cerdeira, A. L., and S. O. Duke. 2006. The current status and environmental impacts of glyphosate-resistant crops: a review. *J. Environ. Qual.* 35:1633–1658.

Chapagain, A. K., A. Y. Hoekstra, and H. H. G. Savenije. 2006. Water saving through international trade of agricultural products. *Hydrol. Earth Syst. Sci.* 10:455–468.

Chiu, Y. W., and M. Wu. 2012. Assessing county-level water footprints of different cellulosic-biofuel feed-stock pathways. *Environ. Sci. Technol.* 46:9155–9162.

Commercial Aviation Alternative Fuels Initiative. 2014. Alternative jet fuel environmental sustainability overview. Version 4.1.

Commercial Aviation Alternative Fuels Initiative. 2016a. Environmental sustainability overview. *http:// www.caafi.org/information/fuelreadinesstools.html#EnvironmentalSustainability* (accessed September 7, 2016).

Commercial Aviation Alternative Fuels Initiative. 2016b. Environmental progression. *http://www.caafi.org/ information/fuelreadinesstools.html#EnvironmentalProgression* (accessed September 7, 2016).

Committee on Environment and Natural Resources. 2010. Scientific assessment of hypoxia in U.S. coastal waters. Washington, DC: Interagency Working Group on Harmful Algal Blooms, Hypoxia, and Human Health of the Joint Subcommittee on Ocean Science and Technology.

D'Antonio, C. M., and P. M. Vitousek. 1992. Biological invasions by exotic grasses, the grass/fire cycle, and global change. *Annu. Rev. Ecol. Evol. Syst.* 23:63–87.

Davis, A. S., R. D. Cousens, J. Hill, *et al.* 2010. Screening bioenergy feedstock crops to mitigate invasion risk. *Front. Ecol. Environ.* 8:533–539.

De La Torre Ugarte, D. G., L. He, K. L. Jensen, *et al.* 2010. Expanded ethanol production: implications for agriculture, water demand, and water quality. *Biomass Bioenergy* 34:1586–1596.

Diaz-Chavez, R., G. Berndes, D. Neary, *et al.* 2011. Water quality assessment of bioenergy production. *Biofuels Bioprod. Biorefin.* 5:445–463.

Doney, S. C., V. J. Fabry, R. A. Feely, *et al.* 2009. Ocean acidification: the other $CO_2$ problem. *Ann. Rev. Mar. Sci.* 1:169–192.

Dumortier, J., D. J. Hayes, M. Carriquiry, *et al.* 2011. Sensitivity of carbon emission estimates from indirect land-use change. Corrigendum, *Appl. Econ. Perspect. Pol.* 33:673.

El-Halwagi, M. M., K. R. Hall, and H. D. Spriggs. 2014. Integrated biofuel processing system. Patent US8802905 B2.

Fargione, J., J. Hill, D. Tilman, *et al.* 2008. Land clearing and the biofuel carbon debt. *Science* 319:1235–1238.

Fargione, J. E., T. R. Cooper, D. J. Flaspohler, *et al.* 2009. Bioenergy and wildlife: threats and opportunities for grassland conservation. *BioScience* 59:767–777.

Fingerman, K. R., M. S. Torn, M. H. O'Hare, *et al.* 2010. Accounting for the water impacts of ethanol production. *Environ. Res. Lett.* 5.

Fingerman, K. R., G. Berndes, S. Orr, *et al.* 2011. Impact assessment at the bioenergy-water nexus. *Biofuels Bioprod. Biorefin.* 5:375–386.

Food and Agriculture Organization/Bioenergy and Food Security. 2012. *Good environmental practices in bioenergy feedstock production: making bioenergy work for climate and food security*, ed. A. Rossi, Rome, Italy: Food and Agriculture Organization of the United Nations, Environment and Natural Resources Working Paper 49.

Futurepast. 2012. Alternative aviation jet fuel sustainability evaluation report. In *Task 2: Sustainability Criteria and Rating Systems for Use in the Aircraft Alternative Fuel Supply Chain*, Contract No. DTRT57–11–C–10038, Report No. DOT–VNTSC–FAA–13–04.

Garbisu, C., and I. Alkorta. 2001. Phytoextraction: a cost-effective plant-based technology for the removal of metals from the environment. *Bioresource Technol.* 77:229–236.

Gasparatos, A., P. Stromberg, and K. Takeuchi. 2011. Biofuels, ecosystem services and human wellbeing: putting biofuels in the ecosystem services narrative. *Agr. Ecosyst. Environ.* 142:111–128.

Gawel, E., and G. Ludwig. 2011. The iLUC dilemma: how to deal with indirect land use changes when governing energy crops? *Land Use Policy* 28:846–856.

Georgescu, M., D. B. Lobell, and C. B. Field. 2011. Direct climate effects of perennial bioenergy crops in the United States. *Proc. Natl. Acad. Sci.* 108:4307–4312.

Gerbens-Leenes, W., A. Y. Hoekstra, and T. H. van der Meer. 2009. The water footprint of bioenergy. *Proc. Natl. Acad. Sci.* 106:10219–10223.

Global Bioenergy Partnership. 2011. *The Global Bioenergy Partnership sustainability indicators for bioenergy.* Rome, Italy: Food and Agriculture Organization of the United Nations.

Gordon, D. 1998. Effects of invasive, non-indigenous plant species on ecosystem processes: lessons from Florida. *Ecol. Appl.* 8:975–989.

Gordon, M., N. Choe, J. Duffy, *et al.* 1998. Phytoremediation of trichloroethylene with hybrid poplars. *Environ. Health Perspect.* 106:1001–1004.

Groom, M. J., E. M. Gray, and P. A. Townsend. 2008. Biofuels and biodiversity: principles for creating better policies for biofuel production. *Conserv. Biol.* 22:602–609.

Hallgren, W., A. Schlosser, and E. Monier. 2012. Impacts of land use and biofuels policy on climate: temperature and localized impacts. MIT Joint Program on the Science and Policy of Global Change Report No 227.

Harper, C. A., and P. D. Keyser. 2008. SP704–A Potential impacts on wildlife of switchgrass grown for biofuels. University of Tennessee Biofuels Initiative Extension Program SP704A–5M–5/08 R12–4110–070–018–08 08–0167.

Harris, N., S. Grimland, and S. Brown. 2009. Global GHG emission factors for various land-use transitions. Technical Report submitted to U.S. EPA, April 2009. Little Rock, AR: Winrock International.

Harvey, B. G., and H. A. Meylemans. 2011. The role of butanol in the development of sustainable fuel technologies. *J. Chem. Technol. Biotechnol.* 86:2–9.

Havlik, P., U. A. Schneider, E. Schmid, *et al.* 2011. Global land-use implications of first and second generation biofuel targets. *Energ. Policy* 39:5690–5702.

Hellmann, F., and P. H. Verburg. 2010. Impact assessment of the European biofuel directive on land use and biodiversity. *J. Environ. Manage.* 91:1389–1396.

Hertel, T., A. Golub, A. Jones, *et al.* 2009. Global land use and greenhouse gas emissions impacts of U.S. maize ethanol: the role of market-mediated responses. GTAP Working Paper No. 55.

Hoekstra, A. Y., A. K. Chapagain, M. M. Aldaya, *et al.* 2011. The water footprint assessment manual: setting the global standard. Washington, DC: Earthscan.

Hoekstra, A. Y., M. M. Mekonnen, A. K. Chapagain, *et al.* 2012. Global monthly water scarcity: blue water footprints versus blue water availability. *PLoS ONE* 7:e32688.

International Organization for Standardization. 2006. Environmental labels and declarations—type III environmental declarations—principles and procedures. ISO 14025:2006.

International Organization for Standardization. 2014. Environmental management—water footprint—principles, requirements and guidelines. ISO 14046:2014.

International Union for Conservation of Nature. 2009. Guidelines on biofuels and invasive species. Gland, Switzerland: International Union for Conservation of Nature and Natural Resources. *https://cmsdata.iucn.org/downloads/iucn_guidelines_on_biofuels_and_invasive_species_.pdf* (accessed September 26, 2016).

Invasive Species Advisory Committee. 2009. Biofuels: cultivating energy, not invasive species. Washington, DC: National Invasive Species Council, Paper 11.

Isbell, F., V. Calcagno, A. Hector, *et al.* 2011. High plant diversity is needed to maintain ecosystem services. *Nature* 477:199–202.

Kenny, J. F., N. L. Barber, S. S. Hutson, *et al.* 2009. Estimated use of water in the United States in 2005. U.S. Geological Survey Circular 1344.

King, C. W., and M. E. Webber. 2008. Water intensity of transportation. *Environ. Sci. Technol.* 42: 7866–7872.

Klass, D. L. 1998. Biomass for renewable energy, fuels, and chemicals. Cambridge, MA: Academic Press.

Laporde, D. 2011. Assessing the land use change consequences of European biofuel policies. Final report of the ATLASS consortium for the European Commission. Specific Contract No. SI2. 580403 implementing Framework Contract No TRADE/07/A2.

Lewis, K. C., and R. D. Porter. 2014. Global approaches to addressing biofuel-related invasive species risks and incorporation into U.S. laws and policies. *Ecol. Monograph.* 84:171–201.

Lindstrom, M. J. 1986. Effects of residue harvesting on water runoff, soil erosion and nutrient loss. *Agr. Ecosyst. Environ.* 16:103–112.

Liu, D., X. Wu, J. Chang, *et al.* 2012. Constructed wetlands as biofuel production systems. *Nat. Clim. Chang.* 2:190–194.

Lonsdale, W. M., and F. Fitzgibbon. 2011. The known unknowns—managing the invasion risk from biofuels. *Curr. Opin. Environ. Sustain.* 3:31–35.

Mack, M. C., and C. M. D'Antonio. 2003. Exotic grasses alter controls over soil nitrogen dynamics in a Hawaiian woodland. *Ecol. Appl.* 13:154–166.

Maestre, F. T., J. L. Quero, N. J. Gotelli, *et al.* 2012. Plant species richness and ecosystem multifunctionality in global drylands. *Science* 335:214–218.

Martinelli, L. A., and S. Filoso. 2008. Expansion of sugarcane ethanol production in Brazil: environmental and social challenges. *Ecol. Appl.* 18:885–898.

McDonald, R. I., J. Fargione, J. Kiesecker, *et al.* 2009. Energy sprawl or energy efficiency: climate policy impacts on natural habitat for the United States of America. *PLoS ONE* 4:e6802.

Midgley, G. F. 2012. Biodiversity and ecosystem function. *Science* 335:174–175.

Millennium Ecosystem Assessment. 2005. Ecosystems and human well-being: biodiversity synthesis. Washington, DC: World Resources Institute.

Naeem, S., J. E. Duffy, and E. Zavaleta. 2012. The functions of biological diversity in an age of extinction. *Science* 336:1401–1406.

National Invasive Species Council. 2008. 2008–2012 National Invasive Species Management Plan. Washington, DC: NISC.

National Research Council of the National Academies. 2008. Water implications of biofuels production in the United States. Washington, DC: The National Academies Press.

Pearlson, M., C. Wollersheim, and J. Hileman. 2013. A techno-economic review of hydroprocessed renewable esters and fatty acids for jet fuel production. *Biofuels Bioprod. Biorefin.* 7:89–96.

Peters, M. W., and J. D. Taylor. 2015. Renewable jet fuel blendstock from isobutanol. U.S. Patent 8975461 B2.

Pimentel, D., L. Lach, R. Zuniga, *et al.* 2000. Environmental and economic costs of nonindigenous species in the United States. *BioScience* 50:53–65.

Pimentel, D., R. Zuniga, and D. Morrison. 2005. Update on the environmental and economic costs associated with alien-invasive species in the United States. *Ecol. Econ.* 52:273–288.

Plevin, R. J., M. O'Hare, A. D. Jones, *et al.* 2010. Greenhouse gas emissions from biofuels' indirect land use change are uncertain but may be much greater than previously estimated. *Environ. Sci. Technol.* 44:8015–8021.

Raghu, S., R. C. Anderson, C. C. Daehler, *et al.* 2006. Adding biofuels to the invasive species fire? *Science* 313:1742.

Rejmanek, M., and M. I. Pitcairn. 2002. When is eradication of exotic pests a realistic goal? In *Turning the tide: the eradication of invasive species*, eds. C. Veitch and M. Clout, Gland, Switzerland: International Union for Conservation Natural Resources, 249–253.

Ricklefs, R. E., and G. L. Miller. 2000. Ecology. Fourth ed., New York, NY: W.H. Freeman and Company.

Ridoutt, B. G., and S. Pfister. 2010. A revised approach to water footprinting to make transparent the impacts of consumption and production on global freshwater scarcity. *Glob. Environ. Chang.* 20:113–120.

Roundtable on Sustainable Biofuels. 2010. RSB principles & criteria for sustainable biofuel production, RSB-STD-01-001, Version 2.1, Lausanne, Switzerland: Round Table on Sustainable Biofuels. *http://rsb.org/pdfs/standards/11-03-08%20RSB%20PCs%20Version%202.1.pdf* (accessed September 26, 2016).

Scown, C. D., A. Horvath, and T. E. McKone. 2011. Water footprint of U.S. transportation fuels. *Environ. Sci. Technol.* 45:2541–2553.

Searchinger, T., R. Heimlich, R. A. Houghton, *et al.* 2008. Use of U.S. croplands for biofuels increases greenhouse gases through emissions from land-use change. *Science* 319:1238–1240.

Seber, G., R. Malina, M. N. Pearlson, *et al.* 2014. Environmental and economic assessment of producing hydroprocessed jet and diesel fuel from waste oils and tallow. *Biomass Bioenerg.* 67:108–118.

Serrano-Ruiz, J. C., and J. A. Dumesic. 2011. Catalytic routes for the conversion of biomass into liquid hydrocarbon transportation fuels. *Energ. Environ. Sci.* 4:83–99.

Shannon, C. E., and W. Weaver. 1949. *The mathematical theory of communication.* Urbana, IL: University of Illinois Press.

Sheehan, J. J. 2009. Biofuels and the conundrum of sustainability. *Curr. Opin. Biotech.* 20: 318–324.

Simpson, E. H. 1949. Measurement of diversity. *Nature* 163:688.

Solley, W. B., R. R. Pierce, and H. A. Perlman. 1998. Estimated use of water in the United States in 1995. In U.S. Geological Survey Circular 1200.

Staples, M. D., H. Olcay, R. Malina, *et al.* 2013. Water consumption footprint and land requirements of large-scale alternative diesel and jet fuel production. *Environ. Sci. Technol.* 47:12557–12565.

Staples, M. D., R. Malina, H. Olcay, *et al.* 2014. Lifecycle greenhouse gas footprint and minimum selling price of renewable diesel and jet fuel from fermentation and advanced fermentation production technologies. *Energ. Environ. Sci.* 7:1545–1554.

Stephanopoulos, G. 2007. Challenges in engineering microbes for biofuels production. *Science* 315:801–804.

Stratton, R. W., H. M. Wong, and J. I. Hileman, 2010. Life cycle greenhouse gas emissions from alternative jet fuel. Version 1.2, PARTNER Project 28 Report No. PARTNER–COE–2010–001.

Stratton, R. W., H. M. Wong, and J. I. Hileman. 2011. Quantifying variability in life cycle greenhouse gas inventories of alternative middle distillate transportation fuels. *Environ. Sci. Technol.* 45:4637–4644.

Subramaniam, R., S. Dufreche, M. Zappi, *et al.* 2010. Microbial lipids from renewable resources: production and characterization. *J. Ind. Microbiol. Biotechnol.* 37:1271–1287.

Tanner, C. C. 1996. Plants for constructed wetland treatment systems—a comparison of the growth and nutrient uptake of eight emergent species. *Ecol. Eng.* 7:59–83.

Tidwell, V., A. Cha-tein Sun, and L. Malczynski. 2011. Biofuel impacts on water. Sandia Report SAND2011–0168.

Tilman, D., J. Hill, and C. Lehman. 2006. Carbon-negative biofuels from low-input high-diversity grassland biomass. *Science* 314:1598–1600.

Tilman, D., R. Socolow, J. A. Foley, *et al.* 2009. Beneficial biofuels—the food, energy, and environment trilemma. *Science* 325:270–271.

United Nations Environment Programme. 2009. Towards sustainable production and use of resources—assessing biofuels. Nairobi, Kenya: UNEP.

U.S. Department of Energy. 2006. Energy demands on water: report to Congress on the interdependency of energy and water.

U.S. Environmental Protection Agency. 2010. Renewable Fuel Standard Program (RFS2) regulatory impact analysis. EPA–420–R–10–006.

U.S. Geological Survey. 1984. National water summary 1983: hydrologic events and issues. U.S. Geological Survey Water Supply Paper 2250.

Vasudevan, V., R. W. Stratton, M. N. Pearlson, *et al.* 2012. Environmental performance of algal biofuel technology options. *Environ. Sci. Technol.* 46:2451–2459.

Vaughn, S. F., and M. A. Berhow. 1999. Allelochemicals isolated from tissues of the invasive weed garlic mustard (*Alliaria petiolata*). *J. Chem. Ecol.*, 25:2495–2504.

Wang, M. Q., J. Han, Z. Haq, W. E. Tyner, *et al.* 2011. Energy and greenhouse gas emission effects of corn and cellulosic ethanol with technology improvements and land use changes. *Biomass Bioenerg.* 35:1885–1896.

Wilsey, B. J., and C. Potvin. 2000. Biodiversity and ecosystem functioning: importance of species evenness in an old field. *Ecology* 81:887–892.

World Resource Institute. 2011. Dataset of eutrophic and hypoxic coastal areas. *http://www.wri.org/resource/interactive-map-eutrophication-hypoxia* (accessed September 27, 2016).

Wu, M., M. Mintz, M. Wang, *et al.* 2009. Water consumption in the production of ethanol and petroleum gasoline. *Environ. Manage.* 44:981–997.

Yang, H., Y. Zhou, and J. Liu. 2009. Land and water requirements of biofuel and implications for food supply and the environment in China. *Energ. Policy* 37:1876–1885.

Yeh, S., G. Berndes, G. S. Mishra, *et al.* 2011. Evaluation of water use for bioenergy at different scales. *Biofuels Bioprod. Biorefin.* 5:361–374.

Zhang, X., R. C. Izaurralde, D. Manowitz, *et al.* 2010. An integrative modeling framework to evaluate the productivity and sustainability of biofuel crop production systems. *Glob. Change Biol. Bioenergy* 2:258–277.

Zhang, Y., H. Y. H. Chen, and P. B. Reich. 2012. Forest productivity increases with evenness, species richness and trait variation: a global meta-analysis. *J. Ecol.* 100:742–749.

# CHAPTER 14

## Perspectives on the future of green aviation

Jay E. Dryer

Drawing on all of the prior subject matter, this chapter provides an integrated forecast for the most likely development paths to sustainable green aviation. Required technology development for aircraft design/operation and alternative fuel production are highlighted, both for the short and long term. Land use and other feedstock issues are examined on a regional basis. The essential features of the move towards green aviation are delineated on the basis of projected public policy.

## 14.1  INTRODUCTION

Many people attempt to predict the future, but few really succeed, and when they do it may be more likely to be due to luck than skill. Nonetheless, an assessment of likely outcomes is still a useful exercise because it helps with planning and provides goals to guide research and technology development. In this chapter we analyze some of the key factors that are likely to influence the development of sustainable green aviation. Since the main focus of this book is to examine aircraft technologies and the development of alternative fuels, the future of each of these technologies is examined. While predicting the future based on observed trends is extremely difficult, predicting disruptive technologies is even harder, but we still try. We also summarize some plausible paths for sustainable green aviation against several possible future world scenarios.

## 14.2  KEY FACTORS AFFECTING THE FUTURE OF GREEN AVIATION

### 14.2.1  *Technology trends*

Since the majority of this text focuses on technology development for green aviation, we only examine some general trends. Overall, the speed of technology development continues to accelerate at a rapidly. Consider that in the late 1800s it took 35 years for 25% of the population to adopt the telephone. In the 1900s, it only took 18 years for 25% of the population to obtain color televisions. More recently, over 25% of the population was utilizing the Internet in only 7 years (McGrath, 2013). Of course, the pace of technology adoption is not the same for all areas. It is often 8 to 10 years between new airliners, whereas it is not uncommon to see a new cell phone released every year. Why is this pace slower for aircraft? One key reason is that this is an expensive business. The barrier of entry for a new company is high and each step of testing new technologies often requires complex and unique infrastructure. Another key factor is the high safety level of the entire aviation enterprise. Certification is often a long and costly endeavor, but it is critical for the stringent levels of safety that we expect.

Another trend is the increasing reliance on computational modeling in the development of aviation systems. Increased simulation capabilities have reduced, but not eliminated, the need for ground and flight testing. This increased capability is due to both the development of software with the improved capacity to capture physical phenomena and the rapid improvement in computational capability for the machines that run these codes. At the same time, the need to

Table 14.1.    NASA subsonic aircraft technology goals.

| Technology benefits[a] | Technology generations (TRLd, 4 to 6) | | |
|---|---|---|---|
| | N+1 (2015) | N+2 (2020b) | N+3 (2025) |
| Noise (cum margin relative to Stage 4), dB | −32 | −42 | −52 |
| Landing/takeoff nitrogen oxide ($NO_x$) emissions (relative to CAEP 6[c]), % | −60 | −75 | −80 |
| Cruise $NO_x$ emissions (relative to 2005 best in class), % | −55 | −70 | −80 |
| Aircraft fuel/energy consumption (relative to best in class), % | −33 | −50 | −60 |

[a]Projected benefits once technologies are matured and implemented by industry. Benefits vary by vehicle size and mission. N+1 and N+3 values are referenced to a 737–800 aircraft with CFM56–7B engines. N+2 values are referenced to a 777–200 aircraft with GE90 engines.
[b]$CO_2$ emission benefits dependent on life-cycle $CO_2$ emissions per megajoule for fuel and/or energy source used.
[c]Committee on Aviation Environmental Protection.
[d]Technology readiness level.

conduct ground-testing still exists. Because of the significant capital cost of building new facilities, it is unlikely that we will see an expansive growth in new wind tunnels, but improvements in both the existing facilities and measurement techniques will allow us to observe features and understand the behavior of systems much better than we can today. There has been a renewed interest in conducting flight testing to help prove new technologies. While it is true that there is not a good example of a recent large X-plane or technology demonstrator—especially for civil applications, there are still good examples of new technologies being tested in flight on smaller unmanned vehicles such as the X–48 and X–56. It is not unrealistic to expect that the use of such small vehicles will continue and may even grow; there is keen interest from both the National Aeronautics and Space Administration (NASA) and the U.S. Department of Defense to explore the development of larger X-planes. Such a vehicle may have a significant impact on the acceptance of new technologies and may also capture the attention of the general public.

When we consider aircraft technologies, it is easy to focus on how they will increase performance, improve efficiency, or reduce noise and emissions. Table 14.1 shows the NASA performance goals for future aircraft. However, cost can be a barrier to incorporating many of these advanced technologies. In general, new technologies tend to be more expensive. Specialized tooling is needed to build many structures—even for testing purposes. It is possible that trends in manufacturing technologies (e.g., additive manufacturing or improved composites machines) may facilitate the development of new technologies.

## 14.2.2    Economic trends

Over time, aviation growth has been consistent with trends in the gross domestic product (FAA, 2014), which is to be expected. As a whole, aviation has enjoyed steady growth over the years. Each year, the U.S. Federal Aviation Administration releases the Terminal Area Forecast, which is generally optimistic. As Figure 14.1 shows, actual growth has not quite kept up with forecasts (U.S. Government Accountability Office (GAO), 2016). General economic forecasts will have some bearing on aviation growth in general, but how does this apply to the growth in green aviation? As discussed in Chapter 9, the adoption of alternative fuels will be highly dependent on the price of oil. In 2008, the price of oil was over $120 per barrel, but it fell to less than $35 a barrel in 2016. The price of oil will undoubtedly influence the vigor of the aviation industry's drive to seek new, more efficient vehicles as well as the availability of alternative fuels. However, it is important to note that it is not just the current price of oil that has an impact, but also the stability of the market. As history has shown, the price of oil can fluctuate greatly over time. The impact of this fluctuation on aviation fuel price is shown in Figure 14.2.

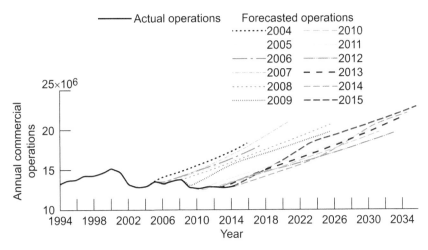

Figure 14.1.   Federal Aviation Administration's (FAA) Aerospace Forecasts, Commercial Operations (2004 to 2015). (From U.S. Government Accountability Office. 2016. Aviation forecasting: FAA should implement additional risk-management practices in forecasting aviation activity. GAO Report to Congress GAO–16–210. From GRA, Inc., based on analysis of FAA data. With permission.)

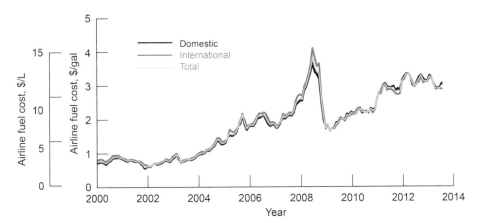

Figure 14.2.   Variation of aviation fuel prices 2000 to 2013. (From U.S. Department of Transportation. Bureau of Transportation Statistics. 2013. Form 41, Air Carrier Financial: Schedule P-12(a). Work of the U.S. Government.)

Another economic factor affecting green aviation is competition. For some time, the world has seen a relative duopoly between Boeing and Airbus in the large civil air transport market. Even this minimal competition has had an impact on what technologies are introduced and when new products are delivered. For example, Boeing's move to more composite materials in the aircraft structure in the Boeing 787 affected Airbus' design of the A350, and the timeline for delivering the new Boeing 737 MAX was influenced by the announcement of the A320NEO. This competitive field is expected to grow. China is attempting to break into the highly lucrative single-aisle commercial transport market with the Commercial Aircraft Corporation of China (COMAC) C919, and even companies that are more traditionally associated with regional jets, such as Bombardier and Embraer, have developed larger aircraft that are competing with some segments of the 737 and A320 class markets.

A positive feature of most green aviation technologies is that they not only decrease environmental impact, but they tend to reduce operations cost at the same time. More efficient vehicles

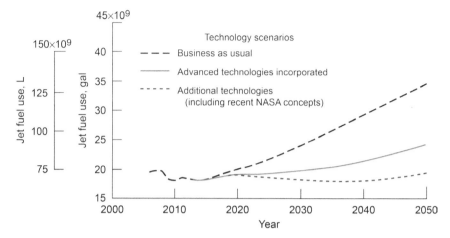

Figure 14.3   Various scenarios for the impact of technologies on fuel burn (NASA). 2011 terminal area forecast.

reduce emissions and fuel costs. However, the big questions are, "Will the market demand these new technologies, and will they be able to "buy their way" onto the next aircraft that is developed?" The current trend indicates that the market demands a modest improvement with each new aircraft type. However, this current improvement trend is not taking place at a rate that keeps pace with the overall growth in aviation. Figure 14.3 shows a NASA estimate of the impact of several technology solutions compared with the current trend of growth or "business as usual."

### 14.2.3  *Policy and regulatory trends*

There is growing global interest in addressing environmental issues, including climate change such as the 2015 Paris Climate Change Agreement. Although there have been numerous summits, meetings, and even agreements, there is not a clear impact to vehicles or operations. That being said, it is quite possible that policy decisions (e.g., carbon taxes) could have such an impact. However, this is still a controversial discussion, and there is not a clear consensus on the impacts and unintended consequences of such policy decisions.

Regulations are in place to not only ensure aircraft safety, but to also set limits for environmental performance in key areas such as noise and emissions. However, these regulations typically do not force radical changes in the adoption of new technologies in order to meet them. Instead, they typically follow technology trends. For example, a new nitrogen oxide ($NO_x$) emission standard will not be out of reach for the newest gas turbine engines that are currently flying. These regulations help to ensure that companies are motivated to continue improvement of their products. In addition, they have an impact on forcing the retirement or modification of older, less-efficient aircraft, thus helping to make the overall system more efficient. One area in which regulations could have an impact on green aviation is related to what is being regulated. In addition to several generations of noise and emissions (primarily $NO_x$ formation) standards, other constituents are being considered for regulation. Currently, the International Civil Aviation Organization is considering new standards on both $CO_2$ and particulate matter. The adoption of these standards could have an impact on the growth of green aviation. One of the key differences between the performance of alternative and standard aviation fuels is the significant reduction in particulate matter for alternative fuels, as demonstrated by NASA in recent ground and flight tests. New limitations on particulate emissions could tip the balance toward the adoption of alternative fuels.

14.2.4  *Social trends*

Social trends will also have an impact on green aviation, but it is hard to predict their level of impact in a concrete sense. Younger generations tend to be more environmentally conscious, but even among the population overall, there is a greater appreciation for reducing human impact on the environment. One reason is that there is significant publicity and debate on issues such as global warming. This trend is being realized in the adoption of new technologies in some markets, such as the rapid growth in hybrid and all-electric cars. However, it remains to be seen how this trend will manifest itself in aviation. There are often increased costs for being the early adopter of new aviation technologies. In a highly competitive environment, will a significant number of passengers choose to fly on an airline that is greener (e.g., from an extensive use of alternative fuels) at the expense of higher ticket prices?

Another social trend that is worth noting is that there is continued public attention on the development of new aviation technologies. Most people take air travel for granted, and there is nothing unique or overly exciting about taking a commercial flight these days. However, there is still a lot of public attention when a major new product is released, such as the Boeing 787. In addition, new aviation technologies are still catching the public eye. For example, in 2007, the X–48 Blended Wing Body experimental was named as a runner-up for invention of the year by Time magazine (losing to the iPhone® (Apple, Inc., Cupertino, California)) (Buechner *et al.*, 2007). This continued public attention will be important to both encourage taxpayer-supported research and motivate airlines to adopt new technologies.

14.3   REQUIRED TECHNOLOGY FOR AIRCRAFT DEVELOPMENT
AND DESIGN

There is no single technology to enable new generations of environmentally friendly aircraft. Instead, commitments will be needed to advance the state of the art in many key areas. In addition, there is not a single well-defined roadmap that will lead to new high levels of efficiency and environmental performance. Indeed, it is possible that in the future we may actually see a greater diversity of aircraft configurations and propulsion systems. In this section, we explore some of the potential advanced technologies that may have a significant impact, but this is by no means all-encompassing.

One area that has seen significant change in recent years is materials and manufacturing. The introduction of the Boeing 787 ushered in a new era in which the majority of an aircraft structure is composite instead of metallic. However, we have yet to take advantage of the full range of benefits afforded by composite construction. In many applications, the material is still used as "black aluminum," with structures that bear many striking similarities to metal-based versions. It will be important to show that we can fabricate advanced composite structures that clearly demonstrate benefits like excellent damage tolerance qualities, guaranteed lightning protection, and affordable manufacturing characteristics, to name a few.

Advances in structures and materials technologies will facilitate the adoption of new structural concepts. The casual observer probably does not understand the many improvements in aircraft design that have taken place since the introduction of the Boeing 707. It is hard for many to see the significant differences between these first "tube and wing" designs and the most modern aircraft. It is quite possible that in the near future we will have new configurations that not only look different, but have significant efficiency benefits. Throughout history, there have been numerous attempts to develop practical alternatives to more traditional designs. Recently, NASA initiated a series of advanced aircraft studies with the goal of enabling far-reaching improvements in efficiency and environmental performance. A thought-provoking result of these studies was that alternate configurations might be the key to making the next momentous leap in performance improvements. The purpose of these investigations was not to design a specific aircraft or develop a prototype *per se*, but rather to identify some of the essential enabling technologies

Figure 14.4.   Truss-braced wing concept in NASA Wind Tunnel (NASA).

Figure 14.5.   Hybrid wing body concept (NASA).

that are critical to opening the door to new designs (Greitzer *et al.*, 2010). Examples of these technologies include a new generation of high-aspect-ratio wings that may require the addition of ancillary structural support in the form of a truss brace wing such as shown in Figure 14.4. Other concepts are based on reshaping the fuselage to improve lift and achieve efficiency gains through ingestion of the boundary layer. Another configuration that has gained attention is the blended, or hybrid, wing body, which more fully integrates the wing and fuselage structure that in turn yields a more efficient lifting body as shown in Figure 14.5. In addition, this design has promise for reducing the acoustic signature of the aircraft through shielding made possible with engines mounted on top of the aft structure.

   Another key finding from recent advanced vehicle studies is the importance of the method of integrating the propulsion system with the airframe. Many new concepts no longer rely on the traditional design of engines mounted in nacelles below the wing. Instead, the engines may be mounted on top of either the wing or rear fuselage. One advantage of this placement is that it enables larger diameter engines with a higher bypass ratio. As mentioned earlier, some new configurations are taking advantage of ingesting and reenergizing the boundary layer over the fuselage area. From a configuration standpoint, this offers the potential for a notable improvement in efficiency, which has been demonstrated in wind-tunnel testing and computational analysis. However, in order to develop such a system, it is necessary to come up with new fan designs that are tolerant of much greater flow distortion without affecting engine performance. We may also see a move away from two or four large engines to new designs that utilize multiple smaller engines in a distributed fashion. There is a distinct possibility that the lines between the airframe and the propulsion system may continue to blur in the future.

In addition to advances in airframe or propulsion-airframe integration concepts, we expect to see continued improvements to the gas turbine engines themselves. Just as advanced materials are enabling the design of new structures, the same is true of improving engine performance. High-temperature ceramic matrix composites are just now beginning to be used in engine designs, and this trend is likely to continue. The trend toward enlarging fan size in order to increase the bypass ratio is continuing to grow. There is also an expectation that core sizes will become smaller. Such small-core engine concepts will complement designs for distributed propulsion. In addition to focusing on more efficient engine designs, there is an increasing emphasis on reducing $NO_x$ emissions. There are even greater challenges to designing low-$NO_x$ combustors in some of these small-core configurations, but recent research shows that this is indeed possible.

So what will it take to enable these new airframe and/or engine concepts? The path to the future will be paved with a coordinated approach of computational analysis, ground experiments, and flight testing. At least for the foreseeable future, the door to a new future cannot be unlocked without the complementary insights provided by all three techniques. Computational fluid and structural dynamics tools are becoming more accurate and more efficient, but there are still many challenges ahead before they can adequately capture the complex physics of flight that accompany advanced designs. For example, today's computational fluid dynamics (CFD) codes are fairly accurate when predicting cruise conditions, but they still struggle to accurately predict the highly separated flows that are typical of high-angle-of-attack flight. Advances in these tools are needed to help predict conditions that may be present in new configurations, such as the juncture flows that surround a truss brace wing. The NASA-sponsored CFD–2030 study focused on assessing the potential future needs for the next generation of fluid flow prediction tools and found that a greater integration of CFD and future design tools will be needed (Slotnick *et al.*, 2014). Most of the nation's wind tunnels have been in operation for many decades, but nonetheless they continue to be an integral facet of aviation research. Continued improvements in data acquisition and test techniques are allowing a more comprehensive understanding of how bodies behave in flight. These facilities are critical for both validating and benchmarking our computational tools as well as reducing risk by critically evaluating and optimizing design concepts before they are taken all the way to flight. Of course, both CFD and wind-tunnel testing are inherently simplified abstractions of the real world. The final arbiter of real-world performance can only come from actual flight testing, which offers unique perspectives on how new systems actually behave in the air. Even now, the United States is poised to develop a number of new X-planes to test some of these new concepts. This may be the most important step in breaking down the barriers that will allow us to move beyond more traditional "tube and wing" designs and to realize the significant improvements alluded to in this and previous chapters.

## 14.4   REQUIRED TECHNOLOGY FOR GREATER ALTERNATIVE FUEL UTILIZATION

The use of alternative fuels is not really a new concept. Even going back to World War II, the Germans devised Fischer-Tropsch processes to create aviation fuel. However, the idea of greater utilization of alternative fuels has gained renewed attention to reduce both the reliance on foreign petroleum and the net carbon footprint of aviation. The focus is on "drop-in" fuels that have properties and characteristics similar to those of petroleum-based jet fuel. Burning these fuels produces a level of $CO_2$ similar to that of traditional jet fuel, but the net carbon savings come from the utilization of feedstocks that are renewable and involve carbon capture of their own.

Perhaps the greatest barrier to the emergence of alternative fuel as a practical competitor to petroleum-based jet fuel is the availability of an adequate feedstock supply chain. Chapters 8 to 13 of this book show that there is not a single identifiable solution to this problem. Instead, it is more likely that we will see regional approaches to both feedstock choice and refinement techniques. An important consideration is choosing feedstock materials that do not incur additional environmental burden and do not compete directly with food sources. One example of potential future alternative fuel production could be of a biocrude from biomass or non-fossil feedstock along with co-processing of the biocrude and fossil-based crude oil in existing refineries. In addition, we may see the implementation of new refining processes such as the alcohol-to-jet process and the biochemical/catalytic conversion of sugars to hydrocarbons. Progress in this area will be highly dependent on industry as well as on the efforts of the U.S. Department of Agriculture and the U.S. Department of Energy. Greater use of alternative fuels has been a key area of international collaboration with agencies from the United States, Germany, and Canada and has even involved collaborating on flight testing of alternative fuels.

Another important step for greater alternative fuel utilization is being able to understand and reliably predict the performance of the fuel on the system and to ensure that there are no safety or performance drawbacks. In short, we need to be able to efficiently certify the use of these fuels. There are well-honed processes for conducting such certification (see Chapter 12), but further development of analytical capabilities will make this process faster and less costly. In addition, we must work to better understand the performance of these alternative fuels in flight in order to maximize their potential environmental benefits. Another potential benefit of alternative fuels is the reduction of particulate matter. It is quite possible that we may see new regulations pertaining to particulate matter in the near future, and greater use of alternative fuels may indeed help with meeting the new standards.

Finally, it will be important to continue to ensure that the propulsion system can reliably and efficiently handle a variety of fuels. A common term used is a "fuel-flexible combustor." Even though all of the fuels that are being considered are "drop-in," there are still differences in the formulations that affect combustor performance. Even fuel handling and storage systems will have to be evaluated to ensure that they can help capitalize on the potential benefits of these new fuels.

## 14.5    POSSIBLE DISRUPTIVE TECHNOLOGIES

Predicting the future is difficult, but predicting when disruptive technologies will come along and change the current path is more difficult. In this section we briefly explore several potential disruptive technologies that could have an impact on future green aviation.

To this point in the chapter, our analysis of propulsion systems has focused primarily on the gas turbine engine. The modern gas turbine engine is a remarkable machine that has proven to be quite efficient. However, it is also very difficult to produce. Few companies in the world have the technology and skill needed to build a machine with moving parts under heavy loads and with extreme tolerances. It is indeed possible that we will see propulsion cycles that are different in the near future. Just as hybrid- and all-electric systems are powering cars on the road today, we might someday see a similar revolution in flight. At this time, the most likely scenario is for a hybrid system that still relies heavily on the gas turbine for both thrust and energy generation—at least for transport-class aircraft. A logical step may be to extract energy from the engine to power one or more distributed fan systems, but additional technology development is needed to make this happen. For example, lightweight, reliable motor-generators will be critical. Further in the future, it may be possible to even better optimize the load sharing between the gas turbine and the stored electrical systems, depending on the phase of flight. However, significant advances in energy storage are needed to make this a reality. It is true that small (e.g., two-person) all-electric aircraft are flying today, but much more work is needed to scale these technologies.

Another potential change is that the way airlines operate could have an impact on the types of aircraft that operate within the system. The most numerous aircraft in the fleet today are the Boeing 737 and Airbus A–320. These aircraft offer a good compromise in performance for the range of missions that are flown most often. It is possible that, if there is a move away from the traditional "hub and spoke" operational model, there could be a desire to reconfigure aircraft based on the route being flown. Such a reconfigurable design could increase efficiency, and such a change in the operational model could generate a greater demand for smaller aircraft. These smaller aircraft may provide a good application for the hybrid-electric systems described in the previous paragraph or could take advantage of new configurations or propulsion-airframe integration approaches more easily because there is less risk in implementing a new design on a smaller aircraft.

Advances in autonomy could offer another means for improving the efficiency of aircraft. For example, autonomy applied to the air traffic control system might allow aircraft to fly more efficient routes and allow more dynamic routing so that aircraft could better take advantage of local conditions such as winds aloft. There is also the potential to apply autonomy to the control of an aircraft itself. For example, advanced sensing might allow a vehicle to better analyze its local environment and fly in a more efficient and perhaps even smoother manner. We could see more specific applications such as automatic gust load alleviation that would permit lighter structures to be developed that have the same safety margins that comparable structures have today. Ultimately we might even see the pilots replaced by computers that can process far more data that a human ever could in order to fly in the most efficient way possible.

## 14.6   FORECAST

Now that we have briefly analyzed some of the factors and key technologies that will have an impact on the future of aviation. What does this all mean, and where is it likely to lead? To better answer this question, we will crudely define three potential future states. Clearly the path that aviation takes will be heavily influenced by changes in the world and in society. It is hard to prognosticate about the future without providing some context. These future states can be generalized as pessimistic, steady-state, and optimistic scenarios for improving the environmental performance of aviation. In each of the following subsections, we start by quickly describing potential major factors in the world. Then we explain how these factors might affect aviation, with a focus on what they mean for the development of a "greener" aviation system. It should be obvious that the actual future will be something different. The goal is to not predict the future but rather to provide a case for how the system may change.

### 14.6.1   *The steady-state case (or business as usual)*

The first scenario assumes that things continue to progress along roughly the same path as they have over the last decade or two. This period has been marked with cycles in both the global economy and conflict around the world. In general, despite regional fighting, periods of economic downturn, and even significant terrorist events, both commerce and, as a result, the demand for air travel have steadily increased over time. In this scenario, environmental regulations continue to tighten but do so in a way that keeps up with technology advancement. Regulations themselves are not driving technology advancement. International competition remains strong, but it takes time for new entrants to achieve significant market share. Throughout this period, it is likely that the trend of growing urbanization will continue.

In this scenario the environmental performance of the aviation system is improving steadily. New major aircraft are entering service for the major companies about every 8 to 10 years. Economic factors tend to dictate market demand, and the fluctuating price of oil keeps pressure on aircraft manufacturers to improve efficiency and on energy producers to develop alternative fuel options. However, neither of these entities will take the large risks needed to make a significant leap in capability. It is possible that we could see the introduction of new configurations, but the timeline to do so will be unnecessarily long. It is also likely that, even though new alternative fuels will be produced, they will constitute a relatively low percentage of the market. The likely outcome is that technology advancements will at least keep pace with the growth in air travel and may even begin to show a net positive benefit because of new aircraft entering the fleet along with improvements in the air traffic management system.

### 14.6.2   *The pessimistic case*

There are a number of global events that could trigger a retreat from the current trend to steadily make the aviation system "greener." Examples of such factors are, war on a larger scale than the recent regional conflicts, prolonged terrorist activities—possibly involving weapons of mass destruction—or a prolonged global recession. To some extent, this scenario is similar to the global climate in the 1930s to mid-1940s. During times like these, people tend to focus on the main issue at hand (e.g., war), and ideas such as protecting the environment shift lower on the priority scale. There was a measurable drop in the demand for air travel following the horrors of September 11, 2001, so it is not inconceivable to imagine that wide-spread global terror could have an even more profound impact.

Dark times would be likely to involve overall reductions in air travel. Some operators could even go out of business, and the overall demand for new aircraft could decrease. This would have the effect of slowing the introduction of new models that have improved performance. In addition, as the large backorders currently experienced by most major companies are erased, the capital available for investment in new technology development will shrink, which also will have a slowing effect on becoming greener. Depending on the situation, there could also be a shift in resources to invest in the military. At the same time, the funding available for government labs may be re-prioritized to focus on defense or security-related issues. Although this would certainly not be an ideal situation, there could still be a silver lining when it comes to mitigating the environmental impact from aviation. In this situation, it is likely that the overall demand for air travel would decline. Even though the introduction of new environmentally beneficial technologies would also slow, the overall net effect would likely be a reduction of aviation-related emissions. Obviously, it would be preferable to see such a reduction occur because we have become more efficient and not from a general reduction in service. Another potential benefit is that an increase in investment in military aviation could ultimately yield benefits for the civilian sector. History is rife with examples where this has occurred. Many technological leaps (e.g., the introduction of the jet engine) were first realized in a military capacity. A period of time dominated by such negative factors would undoubtedly have an impact on the growth of green aviation, but it might be possible to rebound over time and get back on track to making a difference in civil aviation.

### 14.6.3   *The optimistic case*

Most people would agree that it is much more interesting and exciting to envision a positive future that is full of opportunities. Several situations or factors could contribute to an acceleration of global green aviation. Such a period is characterized by a favorable global economy with a few manageable periodic downturns. Global stability contributes to such a positive environment. With growing urbanization, there is a greater demand to connect the world's cities. Air travel can play a significant role in international commerce under such conditions. In addition, we assume that there is still strong, perhaps greater, attention on global environmental issues.

Several factors might prompt an increase in attention and action related to significantly improving the efficiency of aviation. An example of one such factor is increased international competition and demand for new aircraft. On one hand, an increase in demand for air travel will have an influence on the environmental footprint, but it would also mean that companies would be vying for improved products and would see a faster turnover of the fleet as new (higher performing) aircraft enter service sooner. Such an environment is also likely to motivate organizations (both government and industry) to take greater risks, which might encourage an earlier adoption of new configurations that open the door to more significant benefits. Another driving factor could be an intensified focus on regulations to expedite environmental improvements in aviation. Such an approach could have a negative impact increasing the cost of, and thus reducing the demand for, aviation as a whole. However, this would also have a positive influence because companies would increasingly compete based on the efficiency of their aircraft. A major contributor to the adoption of new technologies or ideas might be a major government initiative to reduce risk by flying new concepts or intensifying the drive to implement a more modern and efficient air traffic management system. A new era of X-planes could drive renewed growth in the sector just as it did when aviation grew significantly in the 1950s and early 1960s.

## 14.7   SUMMARY

There are many factors that affect the growth of aviation and impact how fast the world moves to implement more efficient and cleaner aircraft. In addition, there are a number of viable options on the horizon to realize significant gains in aircraft performance, modernize the air traffic control system, or even impact the fuel that we use to power aircraft. Aviation is inextricably linked to the global economy and is an important factor in our daily lives. It is paramount that we strive to ensure that this essential sector can grow in a safe and responsible matter and that we continue to reduce the environmental impact of aviation on a global scale.

## REFERENCES

Buechner, M. M., K. Dell, A. Dorfman, *et al.* 2007. Best inventions of 2007. *Time Magazine.*

Federal Aviation Administration. 2014. The economic impact of civil aviation on the U.S. economy.

Greitzer, E. M., P. A. Bonnefoy, E. De la Rosa Blanco, *et al.* 2010. N+3 aircraft concept designs and trade studies, final report. NASA/CR—2010-216794.

McGrath, R. 2013. The pace of technology adoption is speeding up. *Harvard Business Review.*

Slotnick, J., A. Khodadoust, J. Alonso, *et al.* 2014. CFD vision 2030 study: a path to revolutionary computational aerosciences. NASA/CR—2014-218178.

U.S. Department of Transportation. Bureau of Transportation Statistics. 2013. Form 41, Air Carrier Financial: Schedule P-12(a).

U.S. Government Accountability Office. 2016. Aviation forecasting: FAA should implement additional risk-management practices in forecasting aviation activity. GAO Report to Congress GAO–16–210.

# Acronym list

| | |
|---|---|
| 8×6 SWT | 8- by 6-Foot Supersonic Wind Tunnel |
| 9×15 LSWT | 9- by 15-Foot Low-Speed Wind Tunnel |
| AAFEX | Alternative Aviation Fuel Experiment |
| ABR | Aberdeen Regional Airport (SD) |
| AC | Alternating current |
| ACEE | Aircraft Energy Efficiency (Program) |
| ADS–B | Automatic Dependent Surveillance—Broadcast |
| ADS–C | Automatic Dependent Surveillance—Contract |
| AFM | Airplane Flight Manual |
| AFP | Airspace Flow Program |
| AIAA | American Institute of Aeronautics and Astronautics |
| ANOPP | NASA Aircraft Noise Prediction Program |
| ANSP | Air navigation service provider |
| APEX | Aircraft Particle Emissions eXperiment |
| ARI | Aerodyne Research, Inc. |
| ARP | Aerospace Recommended Practice |
| ARTCC | Air Route Traffic Control Center |
| ASDE–X | Airport Surface Detection Equipment—Model X |
| ASME | American Society of Mechanical Engineering |
| ASP | Aquatic Species Program |
| ASPIRE | Asia and Pacific Initiative to Reduce Emissions |
| ASSC | Airport Surface Surveillance Capability |
| AST | Advanced Subsonic Technology |
| ATAG | Air Transport Action Group |
| ATC | Air Traffic Control |
| ATCSCC | Air Traffic Control System Command Center |
| ATCT | Air Traffic Control Tower |
| ATJ | Alcohol to jet |
| ATM | Air traffic management |
| ATP | Advanced Turboprop (program) |
| avgas | Aviation gasoline |
| BAT | Biomass Assessment Tool |
| BCE | Bryce Canyon International Airport (UT) |
| BLI | Boundary layer ingestion |
| BOS | Boston Logan International Airport (MA) |
| BPF | Blade-Passing Frequency |
| BPR | Bypass Ratio |
| BSAS | Barium strontium aluminum silicate |
| CAAFI | Commercial Aviation Alternative Fuels Initiative |
| CAEE | Committee on Aviation Engine Emissions |
| CAS | Chemical Abstracts Service |
| CCN | Cloud condensation nuclei |
| CDG | Charles de Gaulle Airport (Paris) |
| CDM | Collaborative decisionmaking |

| | |
|---|---|
| CFC | Chlorinated fluorocarbon |
| CFD | Computational fluid dynamics |
| CFR | Code of Federal Regulations |
| CH | Catalytic hydrothermolysis |
| CLEEN | Continuous Lower Emissions, Energy, and Noise (program) |
| CMAS | Calcium-magnesium aluminosilicate |
| CMC | Ceramic matrix composite |
| CML | Continuous mold-line |
| CNT | Carbon nanotube |
| COMAC | Commercial Aircraft Corporation of China |
| CRC | Coordinating Research Council |
| CTFM | Collaborative Traffic Flow Management |
| DAC | Dual-Annular Combustor |
| DC | Direct current |
| DFW | Dallas/Fort Worth International Airport (TX) |
| DMC | Defense Manufacturing Conference |
| DNL | Day-Night average sound level |
| DOE | Department of Energy |
| E3 | Energy Efficient Engine (Project) |
| EBC | Environmental barrier coating |
| ECI | Engine Component Improvement (Project) |
| EDA | Efficient Descent Advisor |
| EI | Emissions Index |
| $EIC_2H_4$ | Emissions Index ethylene |
| EICHO | Emissions Index aldehyde |
| EICO | Emissions Index carbon monoxide |
| EIHC | Emissions Index for hydrocarbons |
| EIHCOOH | Emissions Index carboxyl group |
| EIHONO | Emissions Index nitrous acid |
| $EINO_x$ | Emissions Index $NO_x$ |
| EIS | Entry into service |
| EISA | Energy Independence and Security Act |
| EIUHC | Emissions Index unburned hydrocarbons |
| EO | Electrolytic oxide |
| EPA | Environmental Protection Agency |
| EPNL | Effective perceived noise level |
| ER | Equivalence Ratio |
| ERA | Environmentally Responsible Aviation |
| ERB | Engine Research Building |
| EU | European Union |
| EUETS | European Union Emissions Trading System |
| FAA | Federal Aviation Administration |
| FAAE | Fatty acid alkyl ester |
| FAME | Fatty acid methyl ester |
| FANS | Future Air Navigation System |
| FFP | "Fit-For-Purpose" |
| FICAN | Federal Interagency Committee on Aviation Noise |
| FICON | Federal Interagency Committee on Noise |
| FID | Flame ionization detector |
| FRL | Fuel Readiness Level |
| FSJF | Fully synthetic jet fuel |
| FSRL | Feedstock Readiness Level |
| FT | Fischer-Tropsch |

| GAO | Government Accountability Office |
| GD | Ground Delay (Program) |
| GDP | Gross Domestic Product |
| GE | General Electric |
| GHG | Greenhouse gas |
| GIS | Geographic information systems |
| GPS | Global Positioning System |
| GRC | Glenn Research Center |
| GREET | Greenhouse Gases, Regulated Emissions, and Energy Use in Transportation (software) |
| GS | Ground Stop |
| GTF | Geared turbofan |
| GTL | Gas to liquid |
| HA | Highly annoyed |
| HBR | High Bypass Ratio |
| HC | Hydrocarbon |
| HDCJ | Hydroprocessed Depolymerized Cellulosic Jet |
| HEFA | Hydroprocessed Esters and Fatty Acids |
| HITL | Human-in-the-loop (simulation) |
| HLFC | hybrid laminar flow control |
| HLN | Helena Regional Airport (MT) |
| HPC | High-Pressure Compressor |
| HPT | High-pressure turbine |
| HRJ | Hydrotreated Renewable Jet Fuel |
| HTL | Hydrothermal liquefaction |
| HWB | Hybrid wing-body |
| IAD | Washington Dulles International Airport (DC) |
| ICAO | International Civil Aviation Organization |
| ICCAIA | International Coordinating Council of Aerospace Industries Associations |
| IEA | International Energy Agency |
| IEP2 | (Second CAEP Noise Technology) Independent Expert Panel |
| IPK | Isoparaffinic Kerosene |
| ISO | International Organization for Standardization |
| JETS | Jet Emissions Testing for Specification |
| JFK | John F. Kennedy International Airport (New York City) |
| JFTOT | Jet fuel thermal oxidation tester |
| JPDO | Joint Planning and Development Office |
| LBO | Lean boilout |
| LCA | Life Cycle Assessment |
| LCI | Life Cycle Inventory |
| LDV | laser Doppler velocimetry |
| LE | Leading edge |
| LE | Lipid extraction |
| LEAP® | Leading Edge Aviation Propulsion |
| LFC | Laminar flow control |
| LGA | LaGuardia Airport (New York City) |
| LLNL | Lawrence Livermore National Laboratory |
| LPC | Low-pressure compressor |
| LPT | Low-pressure turbine |
| LTO | Landing and takeoff |
| MBM | Market-based measure |
| ME | Methyl ester |
| MINIT | Minutes-in-Trail |

| | |
|---|---|
| MIT | Massachussetts Institute of Technology |
| MIT | Miles-in-Trail |
| MPT | Multiple pure tone |
| MTOW | Maximum takeoff weight |
| MWW | Municipal waste water |
| NAS | National Airspace System |
| NEF | Noise exposure forecast |
| NextGen | Next Generation Air Transportation System |
| NOAA | National Oceanic and Atmospheric Administration |
| $NO_x$ | Nitrogen oxides |
| NRA | NASA Research Announcement |
| nvPM | Nonvolatile particulate matter |
| ODT | Optimal descent trajectory |
| OEM | Original equipment manufacturer |
| OPD | Optimized Profile Descent |
| OPEC | Organization of Petroleum Exporting Countries |
| OPR | Overall Pressure Ratio |
| ORPR | Open Rotor Propulsion Rig |
| ORY | Orly Airport (Paris) |
| OSI | Oxidative Stability Index |
| OTR | Over the rotor (acoustic treatment) |
| P&W | Pratt & Whitney |
| PAA | Propulsion airframe aeroacoustics |
| PAN | Peroxyacetyl nitrate |
| PAN | Polyacrylonitirle |
| PAR | Photosynthetically active radiation |
| PBN | Performance-Based Navigation |
| PBR | Photobioreactor |
| PM | Particulate matter |
| PMC | Polymer matrix composite |
| PNL | Perceived noise level |
| PNNL | Pacific Northwest National Laboratory |
| PRSEUS | Pultruded Rod Stitched Efficient Unitized Structure |
| PTP | Pointe-à-Pitre International Airport (Guadeloupe, France) |
| PTR-MS | Proton-transfer reaction mass spectrometry |
| PVdF | Polyvinylidene fluoride |
| PW | Pratt & Whitney |
| RAP | Rapid City Regional Airport (SD) |
| RF | Radiative Forcing |
| RFM | Rotorcraft Flight Manual |
| RFS | Renewable Fuel Standard |
| RNAV | Area Navigation |
| RNP | Required Navigation Performance |
| RQL | Rich-burn, quick-quench, lean-burn |
| RR | Rolls-Royce |
| SA | Separation assurance |
| SAC | Single-Annular Combustor |
| SAC | Sacramento International Airport (CA) |
| SAFN | Sustainable Aviation Fuels Northwest |
| SAK | Synthetic Aromatic Kerosene |
| SARS | Severe Acute Respiratory Syndrome |
| SESAR | Single European Sky ATM Research |

| | |
|---|---|
| SFC | Specific Fuel Consumption |
| SFW | Subsonic Fixed Wing (Project) |
| SIP | Synthesized Isoparaffin |
| SK | Synthetic kerosene |
| SKA | Synthetic paraffinic kerosene with aromatics |
| SMA | Shape memory alloy |
| SN | Smoke Number |
| SOA | State of the art |
| SPK | Synthetic Paraffinic Kerosene |
| SPL | Sound pressure level |
| SV | Soft vane |
| SWAFEA | Sustainable Way for Alternative Fuels and Energy in Aviation |
| TA | Terminal area |
| TAG | Triacylglycerol |
| TAPS | Twin-Annular Premixing Swirler |
| TBC | Thermal barrier coating |
| TE | Trailing edge |
| TF | Turbofan |
| TFM | Traffic Flow Management |
| TGO | Thermally grown oxide |
| TILDAS | Tunable Infrared Differential Absorption Spectroscopy |
| TMU | Traffic Management Unit |
| TRACON | Terminal Radar Approach Control |
| TRL | Technology Readiness Level |
| TSFC | Thrust-Specific Fuel Consumption |
| UDF | Unducted fan |
| UEET | Ultra Efficient Engine Technology |
| UHB | Ultrahigh-Bypass |
| UHBR | Ultrahigh Bypass Ratio |
| UHC | Unburned hydrocarbon |
| UMR | University of Missouri-Rolla |
| UN | United Nations |
| USA | United States of America |
| USDA | U.S. Department of Agriculture |
| UTRC | United Technologies Research Center |
| VAN | Variable-area nozzle |
| VDL | VHF Digital Link |
| VHF | Very high frequency |
| WFSP | World Fuel Sampling Program |
| ZAB | Albuquerque airspace center |
| ZAU | Chicago airspace center |
| ZBW | Boston airspace center |
| ZDC | Washington airspace center |
| ZDV | Denver airspace center |
| ZFW | Fort Worth airspace center |
| ZHU | Houston airspace center |
| ZID | Indianapolis airspace center |
| ZJX | Jacksonville airspace center |
| ZKC | Kansas City airspace center |
| ZLA | Los Angeles airspace center |
| ZLC | Salt Lake airspace center |
| ZMA | Miami airspace center |

| | |
|---|---|
| ZME | Memphis airspace center |
| ZMP | Minneapolis airspace center |
| ZNY | New York airspace center |
| ZOA | Oakland airspace center |
| ZOB | Cleveland airspace center |
| ZSE | Seattle airspace center |
| ZTL | Atlanta airspace center |

# Subject index

Advanced biofuels industry, 248
Aerodynamic noise, 3, 9
Aeropropulsion fuel efficiency, 49, 54–56
Aircraft emissions, 26, 33, 37, 42, 106,
    137, 166
Aircraft engine combustion products, 27
Aircraft noise, 3–4, 6, 7, 8–9, 10, 19, 22–23,
    83, 84, 91, 98, 101, 105
  airframe noise sources, 15
    flap (side-edge) noise, 17–18, 20
    landing gear noise, 16
    leading edge noise, 16–17
    slat noise, 16–17
    trailing edge noise, 15–16, 93–94
    wing spoilers, 18
  engine noise sources, 10
    core noise, 14
    fan noise, 13
    gear noise, 16, 17, 20, 94, 97, 100
    jet noise, 14–15
    propeller noise, 13–14
    reduction, see Engine noise reduction
    turbine noise, 14
  reduction goals, 101
    long-term goals, 101
    mid-term goals, 101
Airframe noise reduction, 100
  long-term technologies, 100–101
  mid-term technologies100
Air-fuel mixture, 14
Airspace operation, 166
  separation assurance, 167
  traffic flow management, 167
    national traffic flow management, 167–169
    regional traffic flow management, 169
Airspace systems technologies, 165–180
Airtraffic management, 165, 166, 171, 172,
    177, 184, 344, 345
Air transportation system technologies,
    176–180
Algae growth characteristics, 212–217,
    271–275
  biophysics, 215, 216, 271–273
  climate variability, 273–274

maximum theoretical productivity, 215
nitrogen deprivation, 216
productivity, 212–216, 274–275
sunlight, effect of, 215
Alternative fuel drivers
  economy, 240–242
  energy security, 242
  environment and human health, 239–240
Alternative jet fuels, see Aviation fuel
Aviation fuel
  blends, 197–207
  certification, 186–197, 295-303
  drop-in fuels, see Drop-in fuels
  FAAE fuel, 196–197, 207–209, 220
  FAME fuel, 196–197, 205, 219
  Fischer-Tropsch (FT) fuel, 40–45, 190,
      192, 193, 195–196, 202, 209, 242,
      249, 301, 303, 315, 321
  fossil fuel, 27, 174–175, 224, 278, 284,
      309, 310, 319, 321
  HEFA fuel, 40, 195, 196, 197, 201,
      243–244, 301, 312, 313, 316,
      318, 321
  HRJ fuel, 40, 41, 195
    properties, 187, 189–194, 196, 197–207
  SPK fuel 195, 201, 300–301
  Surrogate fuel, 202–203

Biofuel feedstocks, 209–216, 251, 256, 258,
    259, 260, 269, 326, 328
  algae, 212–216, 269–290
  hydrocarbon composition, 201, 210
  lipid content, 214–216
  processing, 217–219
  refining, 219–223, 256, 260, 263, 278
  seeds, 209–212, 252
Blade-passing frequencies, 12, 13, 14, 84, 86
Boundary-layer-ingesting engines, 69–71

Carbon nanotubes, 105, 109–112, 130, 132
Carbon-trading schemes, 241
Ceramic matrix composites, 125–128
CNT fibers, 110, 111, 130
Combustion noise, 14, 90, 99, 100

Combustion products, 26, 27, 155, 161, 224
Combustors
    GE Aviation's dual-annular combustor
        (DAC), 153, 155
    modern low-emission combustors, 90
    N+2 advanced low-$NO_x$ combustors, 155
    RQL combustor, 147, 148
Committee on Aviation Environmental
        Protection, 7, 25, 138, 139, 143, 144,
        146, 148, 156
Conversion pathways, *see* Biofuel refining
Crude oil extraction, 216–218, 321

Data communication, 178
Drop-in fuels, 40, 194, 296–297, 300

Ecological models, 247
    biodiversity-ecosystem functioning,
        258–259, 263
    food webs, 258, 259, 264
    Lotka-Volterra competition, 258
    patch-occupancy models, 257–258,
        263, 264
Edge noise, *see* Aircraft noise
Electrical components
    energy storage system, 106, 112, 129
    high-power-density electrical motor, 131
    high-power-density power electronics,
        130
    lightweight power transmission system,
        105, 130
    lightweight thermal management, 130
Emissions, *see* Aircraft noise; Gaseous
        emissions; Greenhouse gas emissions;
        Particle emissions
Emissions control, 142–145
Emission tradeoffs, 161
Energy crisis, 56–59
Engine-airframe integration, 10, 11
Engine core research, 71–73
Engine noise reduction, 98
    engine long-term technologies, 99–100
    engine mid-term technologies, 98–99
Engine $NO_x$ control strategies, 145–159

FAAE fuels, *see* Aviation fuels
Factors affecting the future of green aviation,
        335
    economic trends, 336–338
    policy and regulatory trends, 338
    social trends, 339
    technology trends, 335–336
FAME fuels, *see* Aviation fuel

Fischer-Tropsch (FT) fuels, *see* Aviation fuel
Flap side-edge noise, *see* Aircraft noise
Fossil fuels, 27, 175, 224, 284, 310, 321
Fuel properties
    composition and thermophysical
        properties, 198, 199, 204, 206
    emissions, 208–209

Gaseous emissions, 25–37
    carbon dioxide, 27–28
    carbon monoxide, 29
    methane, 37
    nitrogen oxides ($NO_x$), 26–27, 31, 37, 42,
        43–44, 74–77, 121, 138, 139–140,
        141–142, 145–146
    sulfur oxides ($SO_x$), 26–27, 28, 35–37, 41
    unburned hydrocarbons (UHCs), 26–27,
        29–34, 42
    water vapor, 33–35
Gear noise, *see* Aircraft noise
Green aviation, 105–132, 174, 183–227,
        335–345
Green crude, 188, 196, 197, 202, 212,
        216, 219
Greenhouse gas emissions, 39, 45, 129,
        179, 223, 226, 249, 250, 269, 296,
        309–322, 323, 326

HEFA fuels, *see* Aviation fuel
High-lift system, 96
High-temperature materials, 123–129
    ceramic matrix composites, 125–128
    high-temperature Ni-base superalloys,
        123–125
    turbine operating temperatures, 128–129
HRJ Fuel, *see* Aviation fuel
Hybrid wing body aircraft, 63–64, 155, 340
Hydroprocessing, 197, 219, 221–222,
        247, 249

Ingestion of airborne contaminants, 128, 132

Larger diameter ultra high bypass engines, 75
Life cycle analysis for aviation fuels,
        223–227, 278, 307–330
    conventional jet fuel, 313–315
    FT jet fuel, 315–316
    HEFA fuel, 316–317
    land use change, 211, 226, 252–253,
        279–280, 308, 311, 313, 319–320
    renewable fuel from sugars, 317–319
Low-emissions combustion, 40, 148, 152
Low-weight durable oxide nozzles, 75

Materials for electric aircraft, 106,
129–131
Microalgae feedstocks, *see* Biofuel
feedstocks
Modern open-rotor designs, 65, 68
MPT noise, 13, 87
Multifunctional material systems, 112

NASA
component test cases, 59–63
environmentally responsible aviation
project, 63, 66, 69, 73, 76,
154, 157
historical role in green aviation, 195
$NO_x$ emissions, 25, 28, 30, 43, 44, 71, 75, 76,
137, 140, 141, 142, 143, 145, 149,
150, 151, 154, 155, 156, 157, 158,
159, 160, 161, 162
$NO_x$ formation rate, 145, 146, 147, 148, 149,
160, 161
Noise emissions, *see* Aircraft noise
Noise mitigation strategies, 83–103

Particle emissions, 37–39, 42
Performance-based navigation, 176–177
Piezoceramic actuators, 120, 121
Piezoelectric materials, 116, 119–123

Resource modeling, 278

Sandwich structures, 105, 114–116, 132
Seed crops, *see* Biofuel feedstocks
SIP fuels, *see* Aviation fuel
SPK fuels, *see* Aviation fuel
Surface traffic operations, 171
Sustainability assessment, *see* Life cycle
analysis

Terminal area operations, 169–171
Trailing edge noise sources, *see* Aircraft noise
Transesterification, 197, 217, 219–221, 222
Turbine noise, *see* Aircraft noise
Turboelectric system, 129
Turbofan engines, 9, 11, 19, 50, 54, 56, 59,
86, 87, 91

UHC emissions, 30, 33, 35, 42, 137–138,
140, 141, 149
Ultra-high-bypass engine cycle research,
68–69
Ultra-lightweight core materials, 115
U.S. Environmental Protection Agency, 3, 37,
137, 240, 282, 312
U.S. Federal Aviation Administration, 165,
166, 168, 297, 302–303, 312, 336

# Sustainable Energy Developments

*Series Editor: Jochen Bundschuh*

ISSN: 2164-0645

Publisher: CRC Press/Balkema, Taylor & Francis Group

1. Global Cooling – Strategies for Climate Protection
   Hans-Josef Fell
   2012
   ISBN: 978-0-415-62077-2 (Hbk)
   ISBN: 978-0-415-62853-2 (Pb)

2. Renewable Energy Applications for Freshwater Production
   Editors: Jochen Bundschuh & Jan Hoinkis
   2012
   ISBN: 978-0-415-62089-5 (Hbk)

3. Biomass as Energy Source: Resources, Systems and Applications
   Editor: Erik Dahlquist
   2013
   ISBN: 978-0-415-62087-1 (Hbk)

4. Technologies for Converting Biomass to Useful Energy – Combustion, gasification, pyrolysis, torrefaction and fermentation
   Editor: Erik Dahlquist
   2013
   ISBN: 978-0-415-62088-8 (Hbk)

5. Green ICT & Energy – From smart to wise strategies
   Editors: Jaco Appelman, Anwar Osseyran & Martijn Warnier
   2013
   ISBN: 978-0-415-62096-3

6. Sustainable Energy Policies for Europe – Towards 100% Renewable Energy
   Rainer Hinrichs-Rahlwes
   2013
   ISBN: 978-0-415-62099-4 (Hbk)

7. Geothermal Systems and Energy Resources – Turkey and Greece
   Editors: Alper Baba, Jochen Bundschuh & D. Chandrasekaram
   2014
   ISBN: 978-1-138-00109-1 (Hbk)

8. Sustainable Energy Solutions in Agriculture
   Editors: Jochen Bundschuh & Guangnan Chen
   2014
   ISBN: 978-1-138-00118-3 (Hbk)

9.  Advanced Oxidation Technologies – Sustainable Solutions for Environmental Treatments
    Editors: Marta I. Litter, Roberto J. Candal & J. Martín Meichtry
    2014
    ISBN: 978-1-138-00127-5 (Hbk)

10. Computational Models for CO2 Geo-sequestration & Compressed Air Energy Storage
    Editors: Rafid Al-Khoury & Jochen Bundschuh
    2014
    ISBN: 978-1-138-01520-3 (Hbk)

11. Micro & Nano-Engineering of Fuel Cells
    Editors: Dennis Y.C. Leung & Jin Xuan
    2015
    ISBN: 978-0-415-64439-6 (Hbk)

12. Low Energy Low Carbon Architecture: Recent Advances & Future Direction
    Editor: Khaled A. Al-Sallal
    2016
    ISBN: 978-1-138-02748-0 (Hbk)

13. Geothermal, Wind and Solar Energy Applications in Agriculture and Aquaculture
    Editors: Jochen Bundschuh, Guangnan Chen, D. Chandrasekharam & Janusz Piechocki
    2017
    ISBN: 978-1-138-02970-5 (Hbk)

14. Green Aviation: Reduction of Environmental Impact Through Aircraft Technology and
    Alternative Fuels
    Editors: Emily S. Nelson & Dhanireddy R. Reddy
    2017
    ISBN: 978-0-415-62098-7 (Hbk)

T - #0170 - 111024 - C99 - 246/174/19 - PB - 9780367573041 - Gloss Lamination